Indoor Pollutants

Committee on Indoor Pollutants

Board on Toxicology and Environmental Health Hazards

Assembly of Life Sciences

National Research Council

NATIONAL ACADEMY PRESS
Washington, D.C. 1981

NOTICE: The project that is the subject of this report was approved by the Governing Board of the National Research Council, whose members are drawn from the councils of the National Academy of Sciences, the National Academy of Engineering, and the Institute of Medicine. The members of the committee responsible for the report were chosen for their special competences and with regard for appropriate balance.

This report has been reviewed by a group other than the authors according to procedures approved by a Report Review Committee consisting of members of the National Academy of Sciences, the National Academy of Engineering, and the Institute of Medicine.

The work on which this publication is based was performed pursuant to Contract 68-01-4655 with the Office of Research and Development of the Environmental Protection Agency.

Library of Congress Catalog Card Number 81-83813

International Standard Book Number 0-309-03188-5

Available from
National Academy Press
2101 Constitution Avenue, N.W.
Washington, D.C. 20418

Printed in the United States of America

COMMITTEE ON INDOOR POLLUTANTS

JOHN D. SPENGLER, Harvard School of Public Health, Boston, Massachusetts, Chairman

MICHAEL D. LEBOWITZ, University of Arizona Medical Center, Tucson, Arizona, Cochairman

RONALD W. HART, National Center for Toxicological Research, Jefferson, Arkansas

CRAIG D. HOLLOWELL, University of California, Berkeley, California

MORTON LIPPMANN, New York University Medical Center, New York, New York

DEMETRIOS J. MOSCHANDREAS, GEOMET Technologies, Inc., Gaithersburg, Maryland

JAN A. J. STOLWIJK, Yale University School of Medicine, New Haven, Connecticut

DAVID L. SWIFT, The Johns Hopkins University, Baltimore, Maryland

JAMES E. WOODS, JR., Iowa State University, Ames, Iowa

JAMES A. FRAZIER, National Research Council, Washington, D.C., Staff Officer

NORMAN GROSSBLATT, National Research Council, Washington, D.C., Editor

LESLYE B. GIESE, National Research Council, Washington, D.C., Research Assistant

JEAN E. PERRIN, National Research Council, Washington, D.C., Secretary

CONTRIBUTORS TO THE REPORT ON INDOOR POLLUTANTS

JAMES BERK, University of California, Berkeley, California

WILLIAM F. BRANDOM, University of Denver, Denver, Colorado

DAVID M. BURNS, University Hospital, San Diego, California

BENJAMIN BURROWS, Arizona Health Sciences Center, Tucson, Arizona

WILLIAM CAIN, Yale University School of Medicine, New Haven, Connecticut

ROY R. CRAWFORD, Iowa State University, Ames, Iowa

CHARLES W. DENNIGER, Stanley Consultants, Inc., Muscatine, Iowa

DOUGLAS DOCKERY, Harvard School of Public Health, Boston, Massachusetts

NURTAN ESMEN, University of Pittsburgh, Pittsburgh, Pennsylvania

HUGH EVANS, New York University Medical Center, New York, New York

ARTHUR FRANK, Mount Sinai School of Medicine, New York, New York

RALPH F. GOLDMAN, Institute of Environmental Research, U.S. Army, Natick, Massachusetts

JACK D. HACKNEY, Rancho Los Amigos Hospital, Downey, California

CHARLES M. HUNT, National Bureau of Standards, Washington, D.C.

GEORGE JAKAB, The Johns Hopkins University, Baltimore, Maryland

JOHN E. JANSSEN, Honeywell, Inc., St. Paul, Minnesota

EDUARDO A. B. MALDONADO, Iowa State University, Ames, Iowa

PRESTON E. McNALL, National Bureau of Standards, Washington, D.C.

BRIAN MOKLER, Lovelace Biomedical and Environmental Research Institute, Albuquerque, New Mexico

GERALDINE M. MONTAG, Iowa State University, Ames, Iowa

ANTHONY NERO, University of California, Berkeley, California

ANTHONY NEWMAN-TAYLOR, Brompton Hospital, London, England

WAYNE OTT, Stanford University, Stanford, California

GARY L. REYNOLDS, Iowa State University, Ames, Iowa

RICHARD RILEY, Petersham, Massachusetts

GEORGE ROYAL, American Institute of Architects, Washington, D.C.

ROBERT N. SAWYER, Yale University Health Services, New Haven, Connecticut

FREDRICK H. SHAIR, California Institute of Technology, Pasadena, California

DONALD SIBBERT, GEOMET Technologies, Inc., Pomona, California

SAMUEL SILBERSTEIN, National Bureau of Standards, Washington, D.C.

WILLIAM R. SOLOMON, University of Michigan Hospital, Ann Arbor, Michigan

JAMES STEBBINGS, JR., Los Alamos Science Laboratory, Los Alamos, New Mexico

THEODOR D. STERLING, Simon Fraser University, Burnaby, British Columbia, Canada

JAMES E. TROSKO, Michigan State University, East Lansing, Michigan

McDONALD E. WRENN, University of Utah, Salt Lake City, Utah

JOHN YOCOM, TRC Corporation of New England, Wethersfield, Connecticut

BOARD ON TOXICOLOGY AND ENVIRONMENTAL HEALTH HAZARDS

Ex Officio Members

EDWARD BRESNICK, University of Vermont, Burlington, Vermont

DAVID CLAYSON, Eppley Institute for Cancer Research, Omaha, Nebraska

JAMES F. CROW, University of Wisconsin, Madison, Wisconsin

JOHN DOULL, University of Kansas Medical Center, Kansas City, Kansas

ROGER O. McCLELLAN, Lovelace Biomedical and Environmental Research Institute, Albuquerque, New Mexico

ROBERT MENZER, University of Maryland, College Park, Maryland

ROBERT MILLER, National Cancer Institute, Bethesda, Maryland

SHELDON MURPHY, University of Texas, Houston, Texas

NORTON NELSON, New York University Medical Center, New York, New York

JOHN D. SPENGLER, Harvard School of Public Health, Boston, Massachusetts

JAMES L. WHITTENBERGER, Harvard School of Public Health, Boston, Massachusetts

ACKNOWLEDGMENTS

This document is a result of individual and coordinated efforts of the members of the Committee on Indoor Pollutants and the contributors selected to prepare various sections of the report.

Drs. John D. Spengler and Michael D. Lebowitz, Chairman and Cochairman of the Committee, prepared Chapters I, II, and III, on the basis of material submitted by the other members. Drs. Craig D. Hollowell and Demetrios J. Moschandreas coordinated the preparation of Chapters IV, V, and VI. Chapter VII was written under the direction of Drs. Lebowitz, Morton Lippmann and David L. Swift. Drs. Lebowitz and James E. Woods, Jr., collaborated in the preparation of Chapter VIII, and Dr. Woods prepared Chapter IX and Appendix B. Appendix A was compiled by Dr. Hollowell.

The whole manuscript was organized, reviewed, and approved by the full Committee.

A special acknowledgment should be paid to Dr. Ronald W. Hart, who chaired the Committee in its formative period and contributed thereafter as a member.

Particular thanks should be extended to Dr. Woods, who hosted a subcommittee at Iowa State University to coordinate the material in several chapters.

For providing resource material and other information, we note our gratitude to Dr. Joseph F. Cuba,* Director of Research at the American Society of Heating, Refrigerating and Air-conditioning Engineers, Inc.; Mr. Harry Thompson at the U.S. Department of Commerce; Mrs. Nancy Naismith at the Office of Technology Assessment; and Mr. James L. Repace and Dr. Robert J. M. Horton of the Environmental Protection Agency.

Assistance was given also by the staff of the Committee on Toxicology, National Research Council; the National Agricultural Library; the National Library of Medicine; and the George Washington University Library. Other persons assisted in many ways; our appreciation is extended to those not specifically mentioned.

---

*Deceased.

Dr. Alan P. Carlin was the project officer for the Environmental Protection Agency.

The staff officer for the Committee on Indoor Pollutants was Mr. James A. Frazier, who acknowledges the generous assistance of Mrs. Jean E. Perrin, secretary. The bibliographic references were verified and prepared for publication by Mrs. Leslye B. Giese. Typing was done by the Manuscript Processing Unit of the National Academy of Sciences, coordinated by Miss Estelle Miller. The entire report was edited by Mr. Norman Grossblatt.

CONTENTS

EXECUTIVE SUMMARY

Many people spend large amounts of each day indoors--in many cases, 80-90%--in a house, an automobile, a waiting room, an office or other workplace, or a confined space accessible to the general public, such as a store or a restaurant. It has been shown that indoor exposure to environmental pollutants can be substantial. Although there is little epidemiologic evidence on the health effects of indoor pollutants, indoor concentrations of some pollutants that already have primary ambient-air quality standards exceed those standards. Indoor exposure has been largely overlooked in research on the health effects of environmental pollutants, but it can constitute an important fraction of the total exposure to many pollutants.

Indoor pollution in residences, public building, and offices is created for the most part by the occupants' activities and their use of appliances, power equipment, and chemicals, by wear and tear and outgassing of some structural or decorative materials, by thermal factors, and by the intrusion of outdoor pollutants.

In some cases, the outdoor pollutants that penetrate to the indoors may represent the most important pollutant stress on human health and welfare, and such effects have been addressed at length in reports of previous National Research Council committees. This report is focused primarily on the indoor air contaminants that are liberated indoors. When they attain high concentrations, they may cause nuisances, irritation of sensitive tissues, illness, and death from acute as well as chronic exposures. Some pollutant sources--such as cigarette-smoking--have been recognized for a long time, but their importance has only recently been evaluated. Others arise from new products or from old products in new uses, such as building materials, pesticides, and insulation. A number of sources are of concern only in the indoor environment, e.g., cooking, use of consumer products, space-heating devices, and floor and wall coverings. The expanded use of wood and coal for residential space-heating, of home hobby and craft products, and of products that liberate organic substances is a potential contributor to the contamination of indoor environments. Infectious microbes and allergenic agents can grow indoors or be transmitted into indoor environments.

Ventilation alone (whether natural or forced) may not be sufficient to dilute indoor pollution to an acceptable point. Natural ventilation may be inappropriate because it is variable and generally not controllable in a precise way for many indoor settings. Exclusive dependence on forced ventilation is inappropriate because it is not universally available; moreover, introduction of untreated outdoor air may not always be desirable. Also, natural ventilation and forced ventilation may have substantial energy penalties, owing to heating and cooling losses. The adoption of energy-saving proposals to reduce ventilation rates could aggravate problems in indoor air quality, create new problems, and perhaps be generally detrimental to health and property, unless appropriate pollution-control measures are also taken.

Public-health laws are broad enough to permit evaluation and, when required, control of indoor environments. The public expects a safe and healthful outdoor environment, and the same expectation applies to public indoor environments. The regulatory authority to control contaminant sources, to set or recommend building codes, and to support or conduct research on or monitoring of indoor contaminants rests with diverse federal, state, and local government units, but no specific government unit has been directly charged with the responsibility for protecting the quality of indoor environments.

For all the reasons stated above, the present quality of the indoor environment and how this quality may change are matters of immediate and great concern.

SCOPE OF THE REPORT

This report was prepared, at the request of the Environmental Protection Agency (EPA), by the Committee on Indoor Pollutants, which was appointed by the National Research Council in the Board on Toxicology and Environmental Health Hazards, Assembly of Life Sciences. It is intended to characterize the quality of the indoor environment--primarily with respect to airborne pollutants, although others are discussed--and to determine the potential adverse health effects of indoor pollutants. The charge was to review, compile, and appraise the available knowledge. The Committee has also identified the research needed for abatement of indoor pollution. "Indoor" refers to the environments in homes, schools, public buildings, and similar spaces to which the public has access; industrial working environments, however, are excluded from consideration here.

It is beyond the scope of this report to list all the pollutants found indoors that are hazardous to human health. The examples given make it plain that humans are exposed to a variety of potentially hazardous indoor pollutants from diverse sources. It is hoped that this report will encourage researchers to broaden the list of hazardous indoor pollutants and to characterize the hazards, so that the general public and those responsible for pollution control and abatement can be informed.

Throughout this report, pollutants are mentioned without discussion of their health effects. This does not constitute an oversight on the part of the Committee, but rather reflects a decision that the discussion here be adequate to show that there are indoor pollutants that cause adverse health effects in humans. The reader's attention is directed to Chapter III, which offers some recommendations for further health research with respect to these pollutants, for further exposure studies, and for public education about effective ways of reducing exposure to many contaminants encountered indoors. The Committee on Indoor Pollutants and its contributors prepared concise reviews of such physical aspects as sources and concentrations and of such biologic aspects as the physiologic and toxicologic effects of a variety of contaminants encountered indoors. In addition, the effects of those contaminants that bear on human well-being in a more general way, such as soiling and corrosion, and the available means of controlling the presence of the contaminants are discussed in some detail. The report attempts to focus personal, corporate, and government attention on present and potential problems related to indoor contaminants.

The Committee notes that documentation of excessive indoor air pollution should not in itself be considered sufficient reason to relax standards for ambient air. The barriers between indoor air and outdoor air are not absolute, and ambient air contributes to indoor air. Furthermore, outdoor and indoor air pollutants may interact chemically and physiologically. The Committee recognizes the complexity of human exposures that have multiple sources. The development of effective and efficient strategies for mitigating hazardous contamination requires improved understanding of responses to exposure and of pollutant interactions.

The Committee has not attempted to set priorities for research on or regulation or control of indoor pollutants. Nor has it attempted to develop risk analyses for these pollutants. The order in which contaminants are presented in this report does not constitute a ranking of importance by the members of the Committee.

To set priorities for differentiating among indoor contaminants and to establish objectives for research and control programs, there must be a system for comparison. The dimensions of this system include the numbers of people exposed, the severity of exposures, and the consequences of the exposures. To be comprehensive, the system must also deal with ecologic and material damage, loss of productivity, degradation of artifacts, and other kinds of impact not related to health. Priorities could be derived from a ranking of these variables for pollutants of interest, but proper risk analysis would require measurement of exposures by population subgroups and weighting of exposure-response relationships by importance of outcome.

The Committee unanimously agreed that establishing firm priorities for research on indoor pollutants that ranks one contaminant as more important than another is premature. In most instances, we do not appreciate the extent of population exposures. Available reports on indoor air pollutants contain almost no data on the incidence of disease or even annoyance related to changes in pollutant concentrations. Nevertheless, the Committee has made judgments based on its interpretation of reported concentrations, of prevalence of sources, and of evidence of effects on health and welfare. The Committee offers guidance for research, development, and educational programs by specifying current gaps in information and scientific and engineering deficiencies. Priorities within these programs will have to depend on the extent of federal and private funding; the Committee believes that the determination of such priorities is urgent.

ORGANIZATION OF THE REPORT

The introduction, summary and conclusions, and recommendations of the Committee are in Chapters I, II, and III, respectively. This material is based on both the body of the report and Committee deliberations. Some conclusions and recommendations were derived from consideration of the documented information in specific sections of the report and the substantiating discussions therein. Others, based on Committee consensus, were formulated from a more comprehensive perspective on the subject of indoor pollutants.

After the introduction, summary and conclusions, and recommendations in Chapters I, II, and III, the treatment of indoor pollutants is presented in three primary chapters and three secondary chapters. The three primary chapters are Chapter IV, on the sources and characterization of indoor pollution; Chapter VII, on the health effects; and Chapter IX, on the control of indoor pollutants. They are the most voluminous chapters and respond most directly to the Committee's charge to review and appraise the available knowledge on indoor air pollution. The other chapters--Chapter V, on factors that influence indoor pollution; Chapter VI, on the measurement of indoor pollution and exposures; and Chapter VIII, on other non-health-related effects--offer important additional information. They elaborate on material presented in the primary chapters, and they introduce factual and conceptual material that the Committee feels essential for evaluating indoor pollution comprehensively.

Chapter IV, on the sources and characterization of indoor pollution, covers radon and its decay products, formaldehyde and other organic substances, asbestos and fibrous glass, combustion products, tobacco smoke, consumer products, odors, temperature and humidity, and other pollutants not specifically treated. The objective of the chapter, as of Chapter VII (on health effects), is not a global treatment of every possible hazard encountered indoors, but rather a selective treatment. The chief criterion for selection is direct or circumstantial evidence that a contaminant causes or is reasonably likely to cause human stress, illness, or material damage indoors.

Chapter V, on factors that influence indoor pollution, expands on the physical characterization of indoor pollutions in Chapter IV. Aspects of geography, building design, and human activity that lead to variations in ventilation rates and in the distribution of pollutant sources are presented.

Chapter VI, on the measurement and monitoring of indoor pollution and exposure, reviews the design and components of indoor and personal monitors, mathematical models for estimating indoor pollutant concentrations, and methods of estimating total personal exposure.

Chapter VII, on the health effects of indoor pollution, relates the current understanding of the toxicologic and physiologic effects of specific contaminants that are found at high enough concentrations and in a broad enough range of indoor environments to constitute an actual or reasonably likely challenge to the occupants of those environments. In some cases, as in Chapter IV, the discussion is structured by source, such as involuntary smoking or indoor combustion products; the health effects may be attributable to specific components of a mixture of gases and particles, or it may be attributable to the general, mixed exposure. The chapter also considers indoor airborne contagion and allergens.

Chapter VIII, on the effects of indoor pollution on human welfare, covers a number of items related to comfort, productivity, and material protection in indoor environments.

Chapter IX, on the control of indoor pollution, emphasizes the engineering aspects of air-conditioning and indoor air-cleaning. Ventilation codes and standards are reviewed, and mechanical systems for conditioning and cleaning air are described. The chapter discusses strategies for controlling contaminants to maintain acceptable indoor air quality in general. Some pollutants, because of their sources or their physical and chemical properties, cannot be treated with conventional control systems, and strategies for controlling these pollutants are described specifically.

Appendix A lists national primary ambient-air quality standards and occupational-health standards (for the industrial environment) established for the United States. In addition, it lists indoor-air pollution standards and guidelines of several foreign countries. Ventilation standards for dwellings are also listed. This appendix is not exhaustive with respect to relevant pollution or ventilation standards, but it does offer a point of reference for some of the more commonly used standards.

Appendix B presents an example of the interactions among energy conservation, comfort, and indoor air pollution in a residence. This simulation exercise illustrates the tradeoffs among energy-cost savings, retrofit costs, and thermal comfort under the constraints of maintaining various hypothetical conditions of indoor air quality in a particular kind of single-family residence.

PRINCIPAL FINDINGS ON SPECIFIC POLLUTANTS AND CLASSES OF POLLUTANTS

RADON

Radon and its alpha-emitting decay products contribute a major portion of the biologically significant dose associated with natural background radiation. Many natural substances contain radium, a precursor of radon gas. Soil, construction materials, and groundwater are the major sources of indoor radon. Indoor radon concentrations are often an order of magnitude greater than outdoor concentrations. The dose-response relationships for alpha-emitting radionuclides are not sufficiently well accepted to allow quantitation of the health risk associated with measured indoor concentrations, and surveys of radiation in homes are very limited. Expansion of the data base on the variation of indoor radon exposure with time and location is a necessary prerequisite for assignment of a health risk to indoor radon. In theory, techniques for controlling radon exposure are available; but they need development and evaluation before they can be applied economically on a large scale. The effectiveness of several contaminant control strategies, other than ventilation, has not been demonstrated in practice. Codes for new construction may be necessary to prevent the occurrence of high radon concentrations in modern or refurbished dwellings. In any event, prudent judgment concerning reduced ventilation in residences must be based on a better understanding of radiation exposures in present houses.

FORMALDEHYDE

The major indoor sources of formaldehyde have been identified. Aldehydes and other organic substances emanate from outgassing of urea-formaldehyde foam insulation, particleboard, plywood, fabrics, and, to a lesser extent, cigarettes and indoor combustion sources. The high surface-to-volume ratio of particleboard and plywood used as building materials in mobile homes, as well as lower air-exchange rates, causes the high measured formaldehyde concentrations. Formaldehyde can cause skin, eye, and throat irritation in occupants. Moreover, potential health problems associated with formaldehyde exposures are readily identified in acute cases when concentrations are high and tolerance is low. In addition to irritation, respiratory disorders and allergies have been associated with high formaldehyde concentrations. There is evidence of a decreased threshold of sensitivity with prolonged exposure. Recent studies have indicated that exposure of rats and mice to formaldehyde produces nasal cancer. Owing to the ubiquitous and increasing use of resins and solvents in building materials and furnishings, indoor formaldehyde concentrations have increased. Abatement technology is available, although at times expensive; new and less expensive techniques are being developed. The Committee is especially concerned with long-term and essentially continuous indoor exposures to low concentrations of formaldehyde.

ASBESTOS AND OTHER FIBERS

The health hazard posed by indoor exposure to asbetos has been perceived as a problem by virtue of the presence of asbestos fibers in insulating and decorative materials. Abrasion, mechanical vibration, or deliberate disruption of asbestos-containing surfaces can result in increased fiber concentrations in the indoor environment. There have been a small number of studies in which fiber counts have been documented in association with normal building use. Extrapolation from what we currently understand about the exposure-response relationships for asbestos fibers to the very low concentrations reported in indoor spaces, such as schools, suggests a small health risk under conditions of normal use. However, deliberate modification of surfaces to remove asbestos from buildings may create a risk of exposure of occupants and workers. Buildings in which asbestos exposure is likely to occur can be identified. The risk of exposure from dislodged fibers can be reduced by containment. The occurrence of mesothelioma (a specific form of cancer believed to result only from the inhalation of asbestos fibers) may provide a very sensitive indicator of the exposure of the general population. Home exposure to asbestos due to aging, cracking, or physical disruption of insulated pipes or asbestos-containing ceiling tiles and spackling compounds may be greater than public exposures in schools, which have received the most attention. Homes built before 1950 in northern climates are more likely to have pipes insulated with asbestos plaster. Given the very common use of asbestos in homes, schools,and other buildings, there is a need for further assessment to identify structures where actual asbestos exposure constitutes substantial risk to humans.

The extent of exposure of the general public to asbestos fibers has not been assessed; however, the occurrence of mesothelioma should be carefully monitored in the general population. Man-made fibers have produced skin irritation, but have not otherwise been demonstrated convincingly as hazardous to health. Epidemiologic and toxicologic investigation of synthetic fibers should continue. On the basis of present knowledge, synthetic fibers in the indoor environment should not cause undue concern.

TOBACCO SMOKE

Virtually every member of our society is exposed to tobacco smoke: 33% of the population smokes, and the rest are exposed to the smoke released by others. The constituents of tobacco smoke are well documented as hazardous, the prevalence of population exposures is very high, and there is an increased incidence of respiratory tract symptoms and functional decrements in children residing in homes with smokers, compared with those in homes without smokers. These considerations and recent evidence of increased lung-cancer rates among nonsmoking women living with smoking husbands have led us to conclude that indoor exposure to tobacco smoke has adverse effects. Coughing, headache, nausea, and irritation of eyes, nose, and throat are among the reported symptoms. Although many studies have measured various components of tobacco smoke indoors, total exposure has not been determined. Passive exposure to tobacco smoke may constitute an important exposure to respirable particles, such gaseous compounds as acrolein and formaldehyde, benzo[a]pyrene, and various trace metals. Reduced ventilation increases concentrations of tobacco smoke. As an energy-conserving compromise, smoking could be restricted to zones that are well ventilated. Public policy should clearly articulate that involuntary exposure to tobacco smoke has adverse health effects and ought to be minimized or avoided where possible. Under this framework, the prohibition or restriction of smoking in public buildings, offices, etc., is a control option to be considered with ventilation and air-cleaning.

INDOOR COMBUSTION

When fuel combustion occurs indoors--e.g., for heating, cooking, and power machinery, including automobiles--it gives rise to increased concentrations of gases and particles. Unvented gas cooking is probably responsible for a large portion of nitrogen dioxide exposures in our population. In many homes, chronic exposures to nitrogen dioxide indoors may exceed established national ambient-air quality standards. Shorter-term 1-h average concentrations indoors often exceed the highest hourly concentrations measured outdoors. The concentrations of nitrogen dioxide and carbon monoxide in residences have not been fully documented. However, some studies have shown an association between gas cooking and the impairment of lung function in children. Gas cooking appliances are also sources of carbon monoxide, carbon dioxide, formaldehyde, hydrogen cyanide, sulfate particles, organic particulate matter, and organic vapors. The problem of chronic or even peak exposures to combustion products indoors will be accentuated with decreased ventilation and the increased use of portable space-heaters, wood- and coal-burning stoves, and indoor venting of gas dryers. Carbon monoxide, nitrogen oxides, and particles from automobile exhaust can produce increased concentrations in office buildings and public areas. Concentrations exceeding 1-h carbon monoxide national ambient-air quality standards (NAAQS) by a factor of 2-4 have been reported in several ice-skating rinks that use gasoline-powered ice resurfacing machinery. Office buildings and apartment buildings with attached or underground garages can also have sustained high concentrations of carbon monoxide indoors. Because both carbon monoxide and nitrogen dioxide are odorless at typical concentrations, the presence of increased and possibly hazardous concentrations may go undetected.

Although confirmation is necessary, the available evidence suggests that important population exposures to nitrogen dioxide and carbon monoxide can occur indoors and may constitute a sufficient threat to the general public health to justify remedial action. Reducing exposure to those gases is relatively straightforward. Source removal or direct venting of combustion sources should be considered.

Efforts to conserve energy present other potential problems indoors. Effective energy-conservation measures can result in an overcapacity of existing heating equipment. Operation of such equipment at low load factors may decrease its overall combustion efficiency and increase emission of the products of combustion.

MICROORGANISMS AND ALLERGENS

Microorganisms are present in the indoor environment and are associated with human activity and the presence of domestic animals. The microorganisms include bacteria, viruses, and fungi. Many microorganisms--such as spores, molds, and fungi--multiply in the presence of increased humidity. It is possible that reduced ventilation and the increased use of untreated recirculating air could increase the concentrations of microorganisms. Many of these microorganisms can produce infection, disease, or allergic reactions. Respiratory viruses and bacteria can be transmitted from person to person in buildings and confined spaces. Certainly, respiratory infections are an important cause of morbidity that results in lost earnings and discomfort. It is reasonable to assume that some of the incidence of respiratory disease results from airborne transmission, but it is not at all clear what effect ventilation, air-conditioning, or air-cleaning will have on incidence. If the main transmission is between persons in contact with or close to each other, the mechanism and efficiency of disease transfer will be relatively insensitive to ventilation rates and other operating conditions of the air handling systems. However, to the extent that infectious and allergenic microorganisms remain viable and airborne, substantial reduction in ventilation rates will tend to increase concentrations and most likely the probability of infection and allergy.

MOISTURE

Water vapor in confined spaces is a product of metabolic and respiratory processes, as well as of indoor combustion and evaporation from clothes and dish-washing and bathroom functions. Condensation of water indoors has been shown to increase corrosive effects of absorbed gases. Decreases in ventilation tend to increase the indoor relative humidity during the heating seasons. Excess water vapor adsorbs or condenses on drier or colder surfaces, and that gives rise to increased deterioration or corrosion of building materials, furnishings, decorations, artwork, and other artifacts. Increased relative humidity may also promote the growth of molds, algae, and fungi. Thus, humidity control may become an important component of reduced-ventilation strategies. Some energy penalty may result that should be considered in relation to the energy savings that may be obtained through reduced ventilation.

RESPONSIBILITIES

The quality of the indoor environment is not the responsibility exclusively of any individual or government body. Even a single home built in isolation (away from outdoor sources of contaminants) has the potential to be contaminated by its occupants, soil, outgassing from piped-water use, building materials, cooking, space-heating, consumer products, pesticides, molds, and fungi. Responsibility for the indoor environment is on four levels: individuals; product manufacturers; building designers, contractors, and owners; and government.

Most sources of contamination are associated with human activity. Individuals also exercise some control over contamination sources and ventilation. Hence, the individual can directly affect his or her own exposure, as well as the exposure of those with whom the indoor space is shared. This is also true, to a lesser extent, in the nonindustrial workplace and in public buildings.

Building owners and operators are responsible for maintaining the indoor environment and ensuring that at least minimal ventilation standards are being met. It should be noted that there is little or no enforcement of ventilation standards, once building plans are approved. Architects, engineers, and contractors should treat indoor environmental quality as a design objective. There are many opportunities in building design to separate people from the sources of contamination or to remove the sources entirely.

Manufacturers have the responsibility to warn the consumer adequately of the potential hazards of products. As evidence on specific contaminants--such as formaldehyde and nitrogen dioxide from gas stoves--becomes available, less harmful substitutes can be considered, and results of research on the control of the sources can be made available.

Government shares the responsibility to ensure that the indoor environments to which the public has access are healthful. Clearly, in the assessment of indoor concentrations, in instrument development, and in the control of health-directed research on indoor pollutants, government can serve the public interest. Government can establish ordinances (regulations) to protect the general public from nuisances, contamination, or direct health damage; such ordinances may include performance standards related to building materials and ventilation codes for public and private buildings. In the same vein, government can require product certification in the case of known or potential hazards. Furthermore, government can establish concentration standards or source-control specifications. Radon, a naturally occurring substance, is a clear example of contaminants that would require government attention, as opposed to that of industry or the individual.

CONCLUSIONS

Definitive conclusions on the character of indoor air are prevented by the lack of systematic studies. The available data base has been generated by a series of pilot studies and does not fully characterize the variety of pollutants, indoor environments, and occupancy conditions. Furthermore, the implementation of energy-conservation measures and the introduction of new building materials have intensified the problem of indoor air contamination.

Studies explicitly addressing both long-term and episodic events have not been undertaken. Episodic release of contaminants in the indoor environment may be rare, but can lead to short-term high contaminant concentrations, which must be considered (in addition to long-term low concentrations) in assessing the overall health risk of indoor contaminants.

Measurement of indoor contaminants necessitates a sampling protocol that considers the spatial and temporal profile of several pollutants, as well as air diffusion and ventilation characteristics. In addition, measurement techniques for assessing indoor concentrations have to meet more rigorous requirements, particularly with regard to sensitivity and interferences. Unfortunately, many of the instruments required to characterize long-term and short-term indoor pollutant concentrations do not exist.

From a practical viewpoint, it would be desirable to determine the emission rate of an indoor pollutant by simple physical measurements and to infer the dose received by a human inhabiting the indoor space. But several intervening steps must be evaluated that involve degrees of uncertainty ranging from good estimates to total ignorance. The first process to be considered is the transport by diffusion and convection; transport is influenced initially by the fluid motion of the air near the source and throughout the indoor space. These modes of transport depend on a number of factors and are usually spatially and temporally variable. They lead to a concentration profile of the contaminant as a function of position and time. Measuring such profiles is virtually impossible, so the usual approach is to use mathematical models of dispersion. The human receptor is not stationary. Therefore, to obtain an exposure history, the spatial history of the receptor should be specified or estimated. Inexact knowledge of this function introduces a further degree of uncertainty. Dose to the receptor is related to exposure through deposition functions that express the fraction of the exposure that is available to reach specific receptor sites and produce effects. These deposition functions are themselves functions of several variables that are usually poorly specified or unknown. Thus, several layers of uncertainty are embedded between emission rate and receptor dose.

Indoor air pollutants generated or released indoors typically occur in concentrations and mixtures that are often episodic and generally vary over a wide range of time and from one space to another. As a result, human exposures are difficult to assess for individuals or groups. If, in addition, the adverse health effects are subtle, and especially if they are delayed, associations between indoor air pollutants and disease or premature mortality are unlikely to be discovered or demonstrated without a specific and substantial effort. Thus, efforts to improve indoor air quality most likely will have to be guided by information on the adverse health effects of pollutants demonstrated and studied in other settings, such as the occupational environment.

For a limited number of air contaminants that can be found in residential and public buildings, there is direct and circumstantial evidence that human exposures are large enough and common enough to account for substantial morbidity and premature mortality. These include radon progeny, sidestream tobacco smoke, formaldehyde, carbon monoxide, nitrogen dioxide, aeropathogens, and aeroallergens. However, our knowledge of the extent of the exposures and of the exposure-response relationships is inadequate to permit measurement or even estimation of the resulting mortality and morbidity. Furthermore, the kinds of health effects and the latent periods between exposures and effects are quite varied. For example, they extend from acute intoxication from carbon monoxide and formaldehyde to cancer that appears long after exposure to radon decay products and asbestos.

In evaluating health risks, it is reasonable to compare indoor exposure with ambient air-quality standards for pollutants for which there are such standards (carbon monoxide, sulfur dioxide, nitrogen dioxide, ozone, lead, and total suspended particles). Depending on how the standards were developed, however, they may or may not consider exposure of the most sensitive population groups (the ill or infirm or the very young). In addition, the composition of particulate matter indoors, the potential synergistic interaction of gaseous and particulate matter, and the time characteristics of exposures can differ widely between the indoors and the ambient environment. Hence, for some pollutants, the ambient air-quality standards may actually underestimate the health risks.

Current knowledge would permit the establishment of defensible indoor-air quality standards for only a few, if any, contaminants. In any case, the establishment of such standards would not necessarily lead to rational or enforceable controls beyond ventilation codes for dilution. Economical and reliable techniques for sampling and analyzing the airborne contaminants of interest at very low concentrations have not been developed; nor have methods been developed to relate spot samples from specific locations to integrated doses.

We conclude that the best approach to the reduction of health damage from exposure to indoor contaminants is to reduce the population exposure to those contaminants. Control strategies for some pollutants would target the high-exposure groups. For other pollutants, lowering the population-weighted mean exposure would, by best estimates, reduce the health hazards.

Because of the diversity of indoor pollutants and their sources and because of the unpredictability of the distribution of such pollutants over time and in different buildings, efforts to improve indoor air quality should concentrate on reducing the number and strength of the sources by substitution of other materials.

Control strategies based on the specifications of source control are preferred, whenever feasible, because they are generally the most dependable, with respect to their extent and reliability of exposure reduction. That suggests that reducing exposure through source control or removal or through material substitution must be tempered by the practical realities of existing sources, which might not be easy to eliminate, and by the unknown long-term toxicity of substitute materials. When those strategies cannot achieve the desired degree of control at reasonable cost, they can be supplemented or replaced by other engineering controls, such as dilution ventilation or air-cleaning. However, it should be recognized that the concentration reduction achieved through dilution may be less than that achievable with source control or air-cleaning, whereas the effectiveness of air-cleaning devices usually depends on frequent and effective application of maintenance procedures.

The specific source controls appropriate to each kind of contaminant can be expected to vary. For example, local exhaust control is most appropriate for nitrogen dioxide and carbon monoxide from gas ranges, the sealing of walls and floors for radon control, prohibition for asbestos-containing products, and specifications for care of furniture, drapery, and carpeting materials prepared with formaldehyde-containing substances.

Air-conditioning systems are generally designed to provide for all or most of the thermal environmental requirements (i.e., heating, cooling, and ventilating) of the occupied space. In the design of these systems, it is necessary to select components that will meet the particular requirements, such as heating coils or furnaces to meet winter design temperatures, evaporator coils and condensing units to meet summer design temperatures and humidities, and air-cleaners or ventilation air-flow rates to meet the air-quality requirements of the occupied space. The functional requirements of the space (i.e., residential, office, theater, etc.) also impose constraints on the type of system that might be selected. Because of the wide variety of functional requirements of indoor environments and the other constraints on design, a vast variety of control systems are used. For instance, lighting and acoustic requirements can influence the size and location of the air-conditioning system, the location of air supply and return devices, and air velocities in the air distribution system. Available information suggests that this trend will continue.

Although the requirements may be described discretely and some performance specifications are available for components of the system, the effectiveness of the system as a whole, including its impact on indoor air quality, must be evaluated. Unfortunately, very few data are available to indicate whether these systems, under actual loads, perform in accordance with their designs.

RECOMMENDATIONS

1. A staged assessment of the exposures of the general population to indoor pollutants and of the effects of such exposures on health and welfare should be conducted by the federal government in both residential and office buildings. Federal agencies with substantial interests in definition of the indoor exposures--i.e., the Environmental Protection Agency, the Department of Energy, the Consumer Product Safety Commission, the National Institute for Occupational Safety and Health, the Centers for Disease Control, the National Institute of Environmental Health Sciences, the Food and Drug Administration, the National Center for Toxicological Research, the

Department of Housing and Urban Development, etc.--should participate in the formulation of the study objectives and protocols and should provide technical assistance and financial support, as appropriate and feasible. Congress should enunciate clearly its interest in healthful and safe indoor environments.

2. Monitoring protocols and properly designed monitoring instruments must be developed to satisfy the special requirements involved in investigating the indoor environment.

3. The indoor pollutants that should receive the initial major focus in investigation of sources, concentrations, dispersion, and removal are radon and its progeny, tobacco smoke, formaldehyde, nitrogen dioxide, carbon monoxide, pesticides, water vapor, carbon dioxide, and airborne contagion, including allergens.

4. The health effects that should receive the initial major focus in investigation are respiratory-infection rates and respiratory mechanical function in relation to nitrogen dioxide, tobacco-smoke, and airborne-contagion exposures; respiratory-tract irritation and potential carcinogenic effects in relation to formaldehyde, tobacco smoke, asbestos, and alpha-emitting radon decay products; and acute intoxication, blood carboxyhemoglobin, and cardiovascular-disease aggravation in relation to carbon monoxide and nitrogen dioxide exposures.

5. The low-level acute and chronic complaints of malaise, headache, stuffiness, and eye and throat irritation that are reported with increasing frequency in large buildings deserve careful study.

6. The welfare and behavioral effects that should receive the initial major focus in investigation are material damage from mold formation in relation to relative humidity; corrosion and surface deterioration in relation to nitrogen dioxide, sulfur dioxide, and water vapor; fabric fading and deterioration in relation to solar radiation in combination with nitrogen dioxide, sulfur dioxide, house dust, and water vapor; soiling due to tobacco smoke; and the lowering of work productivity due to indoor air pollution and associated discomfort.

7. Responsibility for conducting a well-coordinated investigation of the influence of building design and operational factors on the concentrations of pollutants in both residential and commercial facilities should rest with the federal government, assisted by the appropriate professional and scientific organizations.

8. The building factors that should receive the initial major focus in investigation are as follows:

a. The effects of reducing infiltration rates in existing buildings on combustion efficiency of space-heating equipment and on increases in relative humidity and concentrations of indoor-generated air pollutants and airborne contagion.

b. The effects of materials of construction and furnishings on indoor-pollutant content--specifically, there should be systematic evaluations of outgassing and surface attrition of particleboard and plywood (for formaldehyde and other organic substances); of wall and floor coverings and fabrics (for organic substances); of masonry products (for radon and dust); of wallboard, plaster, and spackling compounds (for dust and fibers); and of the materials used for heat storage in dry solar systems (for radon, dust, surface molds, etc.).

c. The differences in air distribution, diffusion, mixing, etc., associated with the use of different climate-control systems, such as forced-air, baseboard, and radiant floor or ceiling systems.

d. The effectiveness of air-cleaning systems in capturing pollutants in recirculating air--specifically, this will require in-place testing of systems, rather than test-stand evaluation of components, and the effectiveness of a variety of commonly used systems should be evaluated for radon and radon progeny, formaldehyde and solvent vapors, and cigarette smoke.

9. The potential for consumer products to contaminate the indoor environment needs to be evaluated. Hazardous components of these products must be identified and tested. Adequate labeling, warning users of hazards associated with product use and misuse in enclosed spaces, should be required. Testing in homes is needed to assess the extent of contamination, allergic reactions, and other health effects of pesticides, residues, and consumer products.

I

INTRODUCTION

This report was prepared, at the request of the Environmental Protection Agency (EPA), by the Committee on Indoor Pollutants, which was appointed by the National Research Council in the Board on Toxicology and Environmental Health Hazards, Assembly of Life Sciences. It is intended to characterize the quality of the indoor environment, primarily with respect to airborne pollutants, and to determine the potential adverse health effects of indoor pollutants. The charge was to review, compile, and appraise the available knowledge. The Committee has also identified the research needed for abatement of indoor pollution. "Indoor" refers to the environments inside homes, schools, public buildings, and similar spaces to which the public has access; industrial working environments, however, are excluded from consideration here.

It is beyond the scope of this report to list all the pollutants found indoors that are hazardous to human health. The examples given make it plain that humans are exposed to a variety of potentially hazardous indoor pollutants from diverse sources. It is hoped that this report will encourage researchers to broaden the list of hazardous indoor pollutants and to characterize the hazards, so that the general public and those responsible for pollution control and abatement can be informed.

Throughout this report, pollutants are mentioned without discussion of their health effects. This does not constitute an oversight on the part of the Committee, but rather reflects a decision that the discussion here be adequate to show that there are indoor pollutants that cause adverse health effects in humans. The reader's attention is directed to Chapter III, which offers some recommendations for further health research with respect to these pollutants, for further exposure studies, and for public education about effective ways of reducing exposure to many contaminants encountered indoors.

Attention has recently been drawn to the problems of specific pollutants that originate indoors, e.g., formaldehyde released from urea-formaldehyde foam insulation and from urea-formaldehyde resins used to bind laminated-wood products, asbestos in building materials, and radon and its progeny. Efforts to reduce infiltration of outdoor air for energy conservation have heightened the interest in indoor pollution. It is hoped that this report will be useful to the Administrator of EPA and other agencies and individuals in considering indoor environments as a source of exposure of members of the general public to hazardous pollutants. Some of the same pollutants, of course, are now regulated as pollutants in the outdoor atmosphere and in the workplace (see the lists of air-quality standards in Appendix A).

The Committee's report outlines the scope of the problems regarding indoor pollutants and discusses their sources, their effects on human health and welfare (human comfort, productivity, soiling, and corrosion), the technologies available for their control or abatement, and concerns about the effects of energy-conservation strategies on the indoor concentrations of pollutants. It approaches the subject of indoor pollution from three viewpoints:

- Physical factors, such as indoor-pollutant sources and concentrations and population exposures to those pollutants.
- Biomedical evidence on the effects of several pollutants found in the indoor environment.
- Engineering, air-handling and -cleaning systems, and other control options for reducing indoor exposures to pollutants.

The report reviews current understanding of these subjects, assesses the quantity and quality of available information, and offers recommendations for additional studies where appropriate.

Because of the multidisciplinary and complex nature of the indoor pollution question, this document could not possibly treat all pertinent subjects. One important exclusion from the assessment is the indoor industrial environment. It is recognized that many of the pollutants found in areas to which the public has access are also common to industrial settings, often in higher concentrations. The Committee chose to consider only indoor environments to which the general population has access; these include residences, public facilities, recreational facilities, vehicles and transportation-related buildings, educational facilities, and many work settings. Examples of workplaces to which the public has access and in which the public may be compromised by indoor pollution include service stations, automobile showrooms with attached maintenance areas, banks, offices, and buildings with multiple uses.

This document reviews the information on the health and welfare effects of selected indoor pollutants, with emphasis on air pollutants. It includes a critical appraisal of reported measurement and exposure studies, but it does not attempt a quantitative assessment of exposure to the hazardous pollutants in the indoor environment, because in most cases current methods are inadequate for that.

There is no discussion here of the legal, social, or economic implications of regulating the indoor environment in public buildings or homes. Sociopolitical controls of pollution are quite complex and beyond the scope of this document.

Throughout the Committee's deliberations, and reflected in its conclusions and recommendations, were the following questions:

• Do indoor pollution exposures adversely affect the health, welfare, productivity, or sense of well-being of the population or any portion of the population?

• Does the indoor environment constitute an important component of exposure to pollutants?

• Are some groups or individuals at risk by virtue of high indoor concentrations of air pollutants or by virtue of susceptibility?

• What is known about the relative magnitudes of indoor and outdoor pollutant concentrations? Are the sources, ventilation rates, and reaction and removal factors that influence the indoor-outdoor relationships sufficiently well known to predict indoor concentrations and prescribe controls?

• What control strategies are effective for reducing population exposures to specific indoor pollutants?

• Will future changes in housing materials, products, ventilation codes, and activity patterns adversely affect health and welfare through changes in indoor exposures to air pollution?

It is very important that health and welfare problems related to indoor pollution be clearly differentiated from perceived problems or pseudoproblems. This requires measurements that are both accurate and sufficiently representative to identify or estimate the population at risk. And it requires that health research provide reasonable assurance that current or projected exposures can cause unacceptable effects in a portion of the population. Only when these two components are present can prudent judgments on recommended concentrations and control strategies be made.

Efforts to improve the public health and protect the public from hazardous airborne pollutants have been directed primarily toward improving the ambient and industrial environments. Improvements in outdoor (ambient) air have been achieved fundamentally through source control or removal; dilution by tall stacks and source relocation are not considered control strategies. In the indoor industrial environment, however, ventilation or dilution with outdoor air has usually proved to be the most cost-effective way of reducing worker exposure.

The indoor concentrations of airborne contaminants depend on five factors: the generation rate (for indoor-generated pollutants) or the ambient concentration (for outdoor-generated pollutants), the volume of the indoor environment, the air-exchange rate, the mixing efficiency of the indoor space, and the decay (removal) rates of the pollutants.

Until recently, the air in most buildings has been controlled for comfort and odor considerations, not for contaminants. Depending on heating, cooling, and humidity requirements for the indoor environment, the natural or forced infiltration of outdoor air to displace "conditioned" indoor air may entail a considerable energy penalty. Diluting the indoor air with outdoor air reduces concentrations of airborne pollutants generated by indoor sources, such as building materials, appliances, and tobacco-smoking. But outdoor air can also introduce pollutants of outdoor origin, and these pollutants may react with surfaces or other indoor contaminants. Conversely, reducing air exchange without compensating with air-cleaning will result in increased concentrations of contaminants that are generated indoors.

The current growing interest in the quality of the indoor environment is in part a result of efforts to reduce ventilation for energy conservation. In the United States, an estimated 20-50% of energy consumed is for space-heating and -cooling. In many buildings, the energy used to move and condition ventilating air can be as much as 90% of the total energy demand. Buildings lose energy by conduction and radiation through windows, walls, and ceilings and by exchange of indoor conditioned air with unconditioned outdoor air. Reducing ventilation in residential and commercial buildings can be a cost-effective way to achieve energy conservation. However, it is reasonable to expect concentrations of contaminants generated by building occupants, equipment, appliances, and materials to increase when ventilation is reduced. Predicting the results of ventilation changes is limited in some cases by lack of knowledge of the nature and behavior of contaminant sources, of the existing concentrations of pollutants, and of their chemical and physical reactions and removal rates. Even for current situations, the significance of the potential indoor-pollution problem is undefined for many pollutants, because the populations at possible risk have not been identified and the physiologic, behavioral, or welfare effects of various degrees of exposure have not been determined.

Although there is considerable mass-media coverage of the quality of indoor environments, the concern for indoor pollution in the nonindustrial setting is not new. Some countries have tried to regulate pollutants in nonindustrial environments (see Appendix A). Asbestos-fiber contamination in homes and schools has been monitored and in some cases contained or removed. Ozone generation by office copying machines has been regulated. Minimal acceptable ventilation rates for smoke and odor control are incorporated into municipal and state building codes. Several countries have set standards for residential or public facilities to limit exposures to formaldehyde, carbon dioxide, and radon. There have been surveys, but not systematic evaluation, of indoor pollutant concentrations in a variety of locations. The indoor environment has become an issue for the public, government, scientific groups, and corporations because of three phenomena: energy-conservation efforts, which may exacerbate an indoor-pollution problem; the realization that little is known on the hazards of many compounds that are commonly found indoors and outdoors, including consumer products and fuels in common use (note, for example, the resurgence in residential wood- and coal-burning); and the evaluation of pollutant hazards by federal agencies, which have begun to recognize the need to understand the total exposure.

For a large proportion of the population, normal activity occurs disproportionately indoors. In a consideration of integrated pollutant exposures, indoor concentrations are relevant. Between 80% and 90% of an average person's day or year is spent in enclosed areas. On the average, people spend approximately 16 hours/day in their homes. And a rather consistent 1-2 hour/day is spent in transit. Thus, for at least some pollutants, the indoor concentrations are the most important, with respect to potential health effects or material damage.

The time-integrated exposure is perhaps important in determining chronic effects, such as corrosion. But the short-term peak or transient pollutant exposures may be more important, causing or contributing to both acute and chronic effects. Using the average amount of time a person is outdoors or indoors or the time-averaged concentrations may be misleading, if the concern is for peak-exposure effects. Peak exposures may occur indoors or outdoors. They may be encountered only during specific activities or in locations occupied only infrequently. In fact, short-term peak concentrations may contribute only a small proportion of a person's total time-integrated exposure. Both time-integrated concentrations and short-term, transient high concentrations must be considered, whether they occur indoors or outdoors.

Although the indoor and outdoor environments have not been sufficiently assessed to characterize all pollutant constituents comprehensively, it is useful to categorize indoor pollutants into three groups. Table I-1 groups pollutants by source. Those in the first group are principally of outdoor origin; thus, their concentrations are generally higher outdoors. This group includes sulfur dioxide; ozone; many elemental, inorganic, and organic species of particles; pollen; and some organic vapors. They are encountered indoors primarily because they are carried in with infiltrating air. Some may be carried indoors on surfaces. Once inside, particles can be resuspended, or organic substances may volatilize because temperatures and partial pressures are different. The higher indoor surface-to-volume ratios increase the removal rates of many of these pollutants.

Pollutants in the second group have both indoor and outdoor sources. Generally considered as belonging to this class are pollutants produced during combustion, such as carbon dioxide, carbon monoxide, nitrogen oxides, and some components of suspended particulate matter (primarily fine particles--diameter less than 3.0 μm). Because of the limited indoor mixing volume and longer residence times, concentrations of these and other combustion products often exceed outdoor concentrations. This group also includes organic vapors from solvents that can be used outdoors, as well as indoors. Biologic materials, such as fungal spores, have both indoor and outdoor sources. Fibers, including asbestos fibers, have indoor and outdoor sources. Serpentine rock, brake linings, and industrial facilities contribute asbestos fibers to the outdoor air. Insulation, fireproofing, and decorative materials used indoors may contain asbestos. Similarly, water vapor, sound, and nonionizing radiation can be considered to belong to this group.

TABLE I-1

Typical Sources of Some Pollutants Grouped by Origin

Pollutants	Sources
Group 1--Sources predominantly outdoor:	
Sulfur oxides (gases, particles)	Fuel combustion, smelters
Ozone	Photochemical reactions
Pollens	Trees, grass, weeds, plants
Lead, manganese	Automobiles
Calcium, chlorine, silicon, cadmium	Suspension of soils or industrial emission
Organic substances	Petrochemical solvents, natural sources, vaporization of unburned fuels
Group II--Sources both indoor and outdoor:	
Nitric oxide, nitrogen dioxide	Fuel-burning
Carbon monoxide	Fuel-burning
Carbon dioxide	Metabolic activity, combustion
Particles	Resuspension, condensation of vapors and combustion products
Water vapor	Biologic activity, combustion, evaporation
Organic substances	Volatilization, combustion, paint, metabolic action, pesticides, insecticides, fungicides
Spores	Fungi, molds
Group III--Sources predominantly indoor:	
Radon	Building construction materials (concrete, stone), water
Formaldehyde	Particleboard, insulation, furnishings, tobacco smoke
Asbestos, mineral, and synthetic fibers	Fire-retardant, acoustic, thermal, or electric insulation
Organic substances	Adhesives, solvents, cooking, cosmetics, solvents
Ammonia	Metabolic activity, cleaning products
Polycyclic hydrocarbons, arsenic, nicotine, acrolein, etc.	Tobacco smoke
Mercury	Fungicides, in paints, spills in dental-care facilities or laboratories, thermometer breakage
Aerosols	Consumer products
Viable organisms	Infections
Allergens	House dust, animal dander

The third group of pollutants contains those whose sources are predominantly indoor. To this third group we may add pollutants whose concentrations are high enough only indoors to warrant concern for their effects. These pollutants are either generated by the occupants or associated with building materials, appliances, machines, consumer products, or art and craft materials. They include radon, formaldehyde, other organic substances from a variety of materials, asbestos and other fibers, odors, molds, and the numerous compounds identified in tobacco smoke.

Greater attention recently has been drawn to the third group of indoor pollutants. There have been reports of complaints about formaldehyde indoors after application of urea-formaldehyde foam insulation and particleboard and the installation of furnishings. Higher formaldehyde concentrations in European homes were reported in the early seventies. Radon and its progeny have been found in high concentrations in homes built on land reclaimed from phosphate mining and in other areas. Building materials, concrete, granite, and groundwater enriched in uranium are the apparent sources of radon. A plaster-resin material containing 10-30% asbestos has been used for fireproofing, acoustics, and, in some cases, decorative purposes. Asbestos concentrations above U.S. occupational concentrations occasionally have been found indoors.

The three general groups of contaminants found indoors are listed in Table I-1. Those in groups II and III are the prime focus of this report. Chapter IV discusses their sources and concentrations, and Chapter V, factors that affect indoor concentrations and personal exposures. The current understanding is reported with an illustrative but not exhaustive review of pertinent related work. Those two chapters discuss the relationships among sources, personal activity patterns, building factors, and ventilation that influence indoor concentrations and individual pollutant exposure. Temperature, light, and especially relative humidity also help to determine concentrations, chemical activity, and effects. Measurement of these effects to the point of predicting the ramifications of altering ventilation or introducing new products is not possible for all pollutants of interest, and in many cases the measurements have not been made. For other pollutants, the data will not be available until instruments are developed. For still others, the sources may be known, but their prevalence and distribution in buildings are not known.

Table I-2 summarizes some typical pollutant concentrations found in the indoor environment and compares them with outdoor concentrations. An indoor-to-outdoor ratio greater than 1 does not imply that hazardous concentrations occur indoors. This table shows that high concentrations of some pollutants have been reported in a variety of buildings that are commonly used during normal daily activities.

Water vapor is not reported in Table I-2 as a contaminant, but it is very important in the indoor environment. At low relative humidities, odors, particles, and such vapors as acrolein may be more irritating. Higher relative humidities favor mold and mite growth,

TABLE I-2

Sources, Possible Concentrations, and Indoor-to-Outdoor Concentration Ratios of Some Indoor Pollutants

Pollutant	Sources of Indoor Pollution	Possible Indoor Concentration[a]	I/O Concentration Ratio	Location
Carbon monoxide	Combustion equipment, engines, faulty heating system	100 ppm	>>1	Skating rinks, offices, homes, cars, shops
Respirable particles	Stoves, fireplaces, cigarettes, condensation of volatiles, aerosol sprays, resuspension, cooking	100-500 $\mu g/m^3$	>>1	Homes, offices, cars, public facilities, bars, restaurants
Organic vapors	Combustion, solvents, resin products, pesticides, aerosol sprays	NA	>1	Homes, restaurants, public facilities, offices, hospitals
Nitrogen dioxide	Combustion, gas stoves, water heaters, dryers, cigarettes, engines	200-1,000 $\mu g/m^3$	>>1	Homes, skating rinks
Sulfur dioxide	Heating system	20 $\mu g/m^3$	<1	Removal inside
Total suspended particles without smoking	Combustion, resuspension, heating system	100 $\mu g/m^3$	1	Homes, offices, transportation, restaurants
Sulfate	Matches, gas stoves	5 $\mu g/m^3$	<1	Removal inside
Formaldehyde	Insulation, product binders, particleboard	0.05-1.0 ppm	>1	Homes, offices
Radon and progeny	Building materials, groundwater, soil	0.1-30 nCi/m^3	>>1	Homes, buildings

Table I-2 (contd)

Pollutant	Sources of Indoor Pollution	Possible Indoor Concentration[a]	I/O Concentration Ratio	Location
Asbestos	Fireproofing	<1 fiber/cc	1	Homes, schools, offices
Mineral and synthetic fibers	Products, cloth, rugs, wallboard	NA	--	Homes, schools, offices
Carbon dioxide	Combustion, humans, pets	3,000 ppm	>>1	Homes, schools, offices
Viable organisms	Humans, pets, rodents, insects, plants, fungi, humidifiers, air conditioners	NA	>1	Homes, hospitals, schools, offices, public facilities
Ozone	Electric arcing, UV light sources	20 ppb 200 ppb	<1 >1	Airplanes Offices

[a]Concentrations listed are only illustrative of those reported indoors. Both higher and lower concentrations have been measured. No averaging times are given. NA, not appropriate to list a concentration.

greater formaldehyde release from particleboard, and deterioration of many materials.

Exposure of some members of the population to many of these pollutants may be determined by the frequency and duration of activities that place them in particular buildings. Hence, knowing use and activity patterns of the population and how these patterns change with age, sex, socioeconomic status, race, and geographic region is important, if we are to assess the population exposure to pollutants.

Seven classes of environmental factors with indoor sources have been identified as having substantial known or reasonably likely effects on human health: sidestream cigarette smoke, radon and radon progeny, mineral and vitreous fibers, formaldehyde, products of indoor combustion, agents of contagion and allergy, and extremes of temperature and humidity.

Chapter VII presents the evidence on health effects of the seven classes of indoor environmental factors. The seven classes were identified as particularly relevant to indoor exposures of the general population. For other contaminants that may represent special concerns for particular indoor locations--such as exposures to organic compounds found in pesticides, cleaning products, varnishes, or craft and hobby products--the reader is referred to the literature developed by the National Institute of Occupational Safety and Health, the Environmental Protection Agency (Office of Toxic Substances), the Consumer Product Safety Commission, the Food and Drug Administration, and the National Center for Toxicological Research. In the case of other compounds that may be in products found in homes or institutional buildings, not enough is known about their concentrations or their effects to evaluate their health effects.

Although this report does not recommend specific standards for the indoor environment, it discusses standards that have already been established for the outdoor, indoor working, and indoor public environments. It is clear that there is a divergence of opinion in the national and international health and regulatory communities as to what constitutes a safe exposure to contaminants and which contaminants are hazardous. Comparison with reported indoor concentrations makes it evident that--by some established ambient, occupational, or indoor standards--current exposures to some contaminants indoors could constitute a health risk to occupants.

A full risk assessment of these pollutants that would identify the population exposed and assign a health-damage function aimed at determining current and projected health consequences of indoor pollution has not been attempted. In many instances, the review of health-effects literature on specific pollutants produces conclusions that are similar. At higher concentrations, these pollutants have known carcinogenic, allergenic, respiratory, or other physiologic effects. However, except for some contaminants that cause irritation, the evidence of direct or important health damage at reported concentrations is not well established. The evidence in some cases--as in passive smoking and the use of gas appliances--is a statistical association between a health response and the source. For other indoor contaminants, such as radon and asbestos, the direct health effects have been demonstrated in experimental animals and in occupational studies. And for still others, such as formaldehyde, information is from experimental conditions and anecdotal reports of complaints. That these noncriteria pollutants do or will cause harm through current or projected exposures of the general public has not been demonstrated by epidemiologic studies. Indeed, direct evidence from epidemiologic studies may not be forthcoming; epidemiologic studies would be compromised from the outset by uncertainties in indoor pollution concentrations and personal exposure. Nevertheless, if there is consistency of toxicologic and occupational evidence of the harmful nature of specific pollutants at the reported indoor concentrations, then there is reason for serious concern.

In the absence of a confirmed dose-response relationship, careful judgment is required. We should cautiously consider secondary consequences of conservation strategies to the indoor environment. Some pollutants may exert effects only at concentrations above a threshold; others may have no threshold. There may be synergism between pollutants or between pollutants and temperature, humidity, or disease organisms. Some pollutants may manifest effects subtly in behavioral changes. Others may have long latent periods between exposure and effects. In view of the uncertainty in the myriad potential outcomes, one fundamental relationship is clear: if, either deliberately or inadvertently, we systematically modify indoor environments by reducing ventilation or by increasing sources of indoor contaminants without ameliorating efforts, we will be increasing the population exposure to pollutants of indoor origin.

A review of indoor pollutant concentrations and possible health significance would not be complete without a discussion of the implications of these exposures for epidemiologic studies of ambient-air pollution. Several substances generated indoors are present in both indoor and outdoor air, including carbon monoxide, nitric oxide, nitrogen dioxide, and particulate matter. Recent investigations have confirmed that personal exposures to nitrogen dioxide and respirable particles are not well represented by ambient measurements if there are substantial indoor sources. For pollutants of outdoor origin, the evidence indicates that personal and indoor exposures are less severe than outdoor exposures.

These observations have implications for epidemiologic studies attempting to establish a relationship between ambient concentrations and health responses. Air-pollution epidemiology attempts to establish a statistical relationship between the dependent health variable and the independent variable of pollution exposure, correcting for other influential variables, such as age, sex, smoking, occupation, and socioeconomic factors. The air-pollution exposure most often chosen is derived from ambient monitoring appropriate to the study population. Leaving aside the question of spatial representation, consider the potential misclassification of exposure that may result from indoor pollution. Depending on study design and pollutant investigated, the results could overestimate, underestimate, or simply incorrectly estimate the relationship between air-pollution exposure and health. Table I-3 illustrates by examples the potential bias imposed by indoor air pollution. The effects on the study can reflect a systematic bias or a random bias in exposure. Regardless of whether the pollutant is primarily outdoor or both indoor and outdoor in origin, the effects of a random misclassification of population exposure are the same. It tends to reduce the statistical power of the association.

The imprecision in air-pollution health-effects data may be due in part to indoor air-pollution concentrations. Indoor air-pollution exposures may sufficiently complicate epidemiologic investigations of the effects of outdoor pollutants so that assessments of indoor exposures, and thus larger study populations, will be needed to discern effects.

In Chapter VIII, the objective is to discuss the welfare effects of contaminants in existing enclosures of all general types and the impact of energy-conservation measures on indoor environmental quality. The effects on human comfort and productivity are presented in separate sections, and the effects of contaminants causing soiling in another section. Chapter IX discusses some of the relevant ventilation codes and standards. (Appendix A lists national primary ambient-air quality standards and occupational-safety and -health standards.) The effects of air-cleaning equipment and air diffusion control are also in separate sections, followed by a general discussion on the strategies used for control of indoor pollutants. In an effort to exemplify this complex interaction of choices, Appendix B presents some hypothetical assumptions for a residence.

TABLE I-3

Possible Consequences of Indoor Air Pollution in Epidemiology

Type of Pollution	Source of Bias	Result of Bias
Outdoor pollutant--indoor concentration lower than outdoor; air-conditioning causes further reduction (ozone, sulfur dioxide)	Systematic bias--air-conditioned homes not uniformly distributed	Overstated relationship: if studied (polluted) population has no air-conditioning, this overstates relationship between outdoor pollutant and health effect
		Understated relationship: if studied (polluted) population has more air-conditioning than other population, relationship between pollutant and health effects will be understated
	Random bias--distribution of air-conditioned homes not known	Underestimated relationship: effect of pollutant on health will be understated because unknown number of people have reduced exposure
		Increased error of estimate
Indoor pollutant--indoor concentration higher than outdoor (nitrogen dioxide, respirable particles)	Systematic bias--gas-cooking homes not randomly distributed	Overstated relationship: if gas-cooking homes are found mostly in studied (polluted) area, this will overstate effect of outdoor pollutant

	Incorrect relationship: if different pollutants are studied and outcome health variable is influenced by indoor pollutant concentrations, effect may be understated or not detected and attributed to wrong pollutant
	Increased power: if distribution of indoor air pollution sources or ventilation factors are known, analysis is stratified by exposure and statistical power to measure effects may increase
	Understated relationship: if gas-cooking homes are found mostly in clean area, effect of outdoor concentration will be understated
Random bias: gas-cooking homes and those with cigarette-smokers randomly distributed among populations studied	Understated relationship: effect of outdoor pollutant may be understated if indoor concentration dominates exposure and exerts an effect
	Increased statistical power: if distribution of smokers or gas stoves is known and study is large enough, statistical power to discern effects may increase

II

SUMMARY AND CONCLUSIONS

CHARACTERIZATION OF INDOOR AIR POLLUTION

The air quality of the indoor environment has been characterized in a limited number of pilot studies. Because of the large variety of distinct indoor environments--single and multifamily residences, offices, hospitals, restaurants, schools, recreational facilities, transportation facilities, etc.--there is a major difficulty in characterizing "the indoor air environment." Moreover, even within one indoor environment differences in structure, in the operation and strength of emission sources, and in human activities add to the complexity of characterizing air quality. The available data, mostly from the residential environment, amply demonstrate the diversity of characteristics of indoor air and help in identifying subjects that warrant further research.

RADIOACTIVITY (pp. 58-82)

The data base on sources and source strengths of indoor radon is just beginning to be established. Initial attention focused on building materials and groundwater. Recent evidence from regional studies in the United States points to ground soils (under buildings) as perhaps the major source of radon. Only a small number of buildings in the United States have been measured for radon and radon progeny. Indoor concentrations are affected by various factors, including ventilation rate, deposition of radon progeny on indoor surfaces, and interactions of radon progeny with fine particles from various sources (e.g., tobacco smoke and house dust).

Data from several studies indicate that indoor radon-222 concentrations vary by at least two orders of magnitude, with average values of about 1 nCi/m^3. Such a large range is not surprising, inasmuch as the studies included various types of buildings, building materials, underlying materials, and ventilation rates and used many different measurement techniques. Radon progeny concentrations are often given as potential alpha-energy concentrations (PAEC), expressed in working levels (WL). Limited measurements indicate that typical average radon progeny concentrations in residential buildings range from 0.004 to 0.02 WL in some houses. Concentrations are much higher indoors than outdoors.

ALDEHYDES (pp. 82-100)

Aldehyde concentrations are almost always higher indoors than outdoors. Formaldehyde is the most important aldehyde. Sources of aldehydes include building materials (particleboard and plywood), urea-formaldehyde (UF) insulation, and, to a lesser extent, combustion appliances, tobacco smoke, and other consumer products. Variations in indoor aldehyde concentrations are not well understood, and emanation rates from the various sources are not well quantified. Owing to the time spent inside residences (including mobile homes), offices, and other indoor environments, human exposures to indoor formaldehyde are markedly higher than exposures to outdoor formaldehyde.

Typical indoor formaldehyde concentrations in buildings with products containing urea-formaldehyde resins range between 0.05 and 0.3 ppm, although in some unusual instances concentrations of a few parts per million have been measured in houses with UF foam insulation. In residences with sources of high rates of emission of formaldehyde-containing products, the concentrations typically range from 0.01 to 1 ppm.

CONSUMER PRODUCTS (pp. 100-111)

Many consumer products may emit gaseous and particulate contaminants into the indoor environment during their use or even during storage. Most of the chemicals in these gases and particles may be known or can be identified, but the chemical products resulting from mixtures and interactions of them are not known. Likely exposures and durations are poorly understood, even for cases in which the products are used as directed. Willful abuse of aerosols or careless use of solvents in enclosed spaces have resulted in acute and delayed disorders and in death. The carcinogenicity of some compounds, such as benzene and vinyl chloride, has led to voluntary removal from consumer products, but many chemicals with potentially toxic effects are still in wide use. The use of insecticides, pesticides, and herbicides is widespread. Even when applied outdoors, some compounds have been measured indoors and have persisted over a considerable time.

ASBESTOS AND OTHER FIBERS (pp. 111-128)

Asbestos is a widespread component of the structural environment in schools, homes, and private and other public buildings. Its release in the indoor environment depends on the cohesiveness of the asbestos-containing material and the intensity of the disturbing force. Most contamination is episodic, activity-related, and local. Fiber counts and mass concentrations of fibers have been measured and shown to exceed those outdoors, and on occasion they may approximate the occupational limit of 2 fibers per cubic centimeter. Fortunately, during normal use, buildings containing asbestos have not shown indoor fiber counts higher than outdoor counts. Current data are very limited and apply mostly to schools and a few office buildings, but it appears that the general public exposure to asbestos fibers is exceedingly low in public buildings. A systematic and comprehensive survey of indoor asbestos fiber contamination is needed and will require reliable, portable, and continuous monitors. Asbestos control technologies have been applied in various indoor environments. Asbestos removal requires a complex protocol to be carefully applied, because the very activity of removal may cause severe asbestos contamination.

INDOOR COMBUSTION (pp. 134-149)

Unvented combustion appliances, especially gas stoves, are major sources of indoor air pollution. Although emission rates from a small number of gas stoves have been determined for several pollutants, the data base is very limited. Indoor concentrations of carbon monoxide and nitrogen dioxide associated with incomplete combustion have been observed to exceed current ambient-air quality standards. Carbon dioxide emission from unvented combustion appliances may build up to concentrations in the range of occupational air quality standards. Local exhaust ventilation appears to be the most effective control strategy for reducing pollutants from combustion. Improved combustion efficiency and source elimination (i.e., adsorbers or a change to the use of electric ranges) are two additional control approaches. Residential wood and coal stoves are also potential sources of indoor contamination. Attached and underground garages can contribute to indoor carbon monoxide, nitrogen dioxide, and particle concentrations.

Carbon Monoxide

Indoor carbon monoxide concentrations are often higher than corresponding outdoor concentrations. High indoor concentrations may be attributed to emission from such sources as gas cooking facilities, attached garages, faulty furnaces, and cigarette-smoking. Typical average indoor carbon monoxide concentrations in residences vary between 0.5 and 5 ppm; observed peak values reach 25 ppm. In public buildings, the indoor concentrations are usually lower than observed residential concentrations, except under conditions of exceptionally heavy smoking, as in bars, or in office buildings with underground garages and improperly designed or malfunctioning HVAC systems.

Carbon Dioxide

The indoor-to-outdoor ratio is greater than 1 for at least 90% of the total monitored hours. Hourly indoor carbon dioxide concentrations often exceed 2,000 ppm. Observed typical outdoor carbon dioxide concentrations are approximately 400 ppm. The principal sources of indoor carbon dioxide are the metabolic activity of occupants and unvented combustion appliances.

Nitrogen Oxides

Emission from cooking appliances and emission from unvented heaters are the principal contributors of oxides of nitrogen in the indoor environment. The range of observed hourly indoor (residential) nitric oxide concentrations is 30-300 ppb, with a maximum of about 500 ppb. Indoor hourly concentrations of nitrogen dioxide vary between 50 and 500 ppb; indoor peaks of 700 ppb have been measured. Typical weekly indoor concentrations of nitrogen dioxide range from 20 to 100 ppb. The upper values in all the ranges just noted are associated with unvented gas appliances.

SMOKING (pp. 149-168)

Passive exposure of many nonsmokers to the contaminants in tobacco smoke occurs repeatedly. The indoor concentrations of tobacco-smoke compounds that have other sources exceed the concentrations found outdoors. For many people, the main or sole exposure to numerous gaseous and particulate compounds results from passive exposure to tobacco smoke. Children of smoking parents are among the largest identifiable groups in this category. For the most part, however, the specific contribution of tobacco combustion products to personal exposures has not been documented. Most nonchamber measurements have been of the survey type; many have measured a single component of smoke without reference to outdoor concentrations, ventilation, or air dispersion.

Smoking is the major source of indoor particles, but other human activities (e.g., cooking and vacuum cleaning) also contribute indoor particles. Particulate matter has variable composition, and the data base indicates that there are no constant ratios of indoor to outdoor concentrations. The ratio of observed daily indoor concentrations of total suspended particles (TSP) to corresponding outdoor TSP concentrations varies from 0.3 to 4. Residences occupied by families with pre-school-age children and smokers often have higher indoor than outdoor concentrations. The TSP 24-h ambient-air quality standard, which must not be exceeded more than once a year, is 260 $\mu g/m^3$. The typical range of observed indoor residential 24-h TSP concentrations is 30-100 $\mu g/m^3$, with an observed maximum of 600 $\mu g/m^3$.

Concentrations of fine particles (diameter, less than 2.5 μm) range from 10 to more than 260 $\mu g/m^3$ for a 24-h sample. The higher concentrations are almost always associated with smoking. Concentrations in bars, offices, and cars with smoking can be higher than 500 $\mu g/m^3$.

ODORS (pp. 168-202)

Odors arising from occupants and their activities figure in indoor-air quality issues predominantly on the basis of comfort, rather than health. Such routine indoor activities as cooking, smoking, bathroom use, and maintenance give rise to odors that are often disagreeable and in some cases offensive. To a varying degree, almost all building materials and furnishings are sources of odor. The determination of odor attributes--such as intensity, character (pleasantness/unpleasantness), duration, and perceptual threshold--is complex, but can be effectively accomplished with a combination of instrumentation and the use of panels of human observers. Odor controls increase in complexity from good housekeeping to ventilation to masking and, finally, to air-cleaning.

OTHER CHEMICAL POLLUTANTS (pp. 82-111, 134-149)

Nonmethane Hydrocarbon

The ratio of indoor to outdoor total nonmethane hydrocarbon (NMHC) concentrations is greater than 1 for about 90% of the total monitored hours; that is, the NMHC concentrations observed in the residential environment are almost always higher than the outdoor concentrations. Fluctuations in the indoor concentrations may be associated with cooking, cleaning, and other activities. Typical concentrations in residential buildings vary between 0 and 8.0 ppm, whereas typical outdoor concentrations are between 0 and 3.5 ppm. Measured NMHC concentrations in new office buildings often exceed 10 ppm and reach as high as 50 ppm; this may be attributed to the extensive use of synthetic organic building materials and furnishings in new office buildings, as well as cleaning solvents and maintenance materials.

Ozone

Indoor ozone concentrations are generally lower than outdoor. Unless there is an indoor generation source of ozone from electric arcing or ultraviolet radiation (such as an electrostatic precipitator or a document copier), the ratio of corresponding hourly indoor to outdoor concentrations is almost always less than 1. Ozone is primarily a product of outdoor photochemical reactions. Precursor pollutants leading to the formation of ozone are primarily of automotive origin, but other sources include the combustion of fuels for heat and electricity, the burning of refuse, the evaporation of petroleum products, and the handling and use of organic solvents. Ozone is highly reactive and decays rapidly by absorption on indoor surfaces. Indoor ozone has been measured at up to 120 ppb; typical concentrations are between 0 and 20 ppb.

Sulfur Dioxide

Indoor sulfur dioxide concentrations are usually lower than corresponding outdoor concentrations. Sulfur dioxide emission indoors is usually small, and, because it is a relatively reactive contaminant, it is absorbed by indoor surfaces. Indoor hourly sulfur dioxide concentrations are typically below 20 ppb.

Particulate Chemical Composition

There is very limited information on the chemical composition of indoor particles. Measured lead concentrations in residences are commonly low--often below 0.5 $\mu g/m^3$. Lead concentrations as high as 2 $\mu g/m^3$ have been measured in residences with wall paints that contain lead compounds or in residences that are near major roads.

Indoor residential concentrations of nitrates are quite low and are driven mainly by the outdoor concentrations. Observed daily indoor concentrations of nitrates do not vary widely--between 1.0 and 5 $\mu g/m^3$, with typical values at the lower end of the range.

The data base on sulfates shows that the indoor 24-h sulfate concentration is usually lower than the corresponding outdoor concentration. The type of fuel used for cooking and heating is important in determining the indoor-outdoor relationship; houses with gas appliances have a slightly higher indoor/outdoor ratio than houses without gas appliances. Sulfur-containing compounds are added to residential gas for detecting leaks of the otherwise odorless fuel. Indoor daily sulfate concentrations range between 2.0 and 15.0 $\mu g/m^3$, with typical values at the lower end of the range.

AIRBORNE MICROORGANISMS AND ALLERGENS (pp. 394-417)

For indoor biogenic pollutants, the sparseness of satisfactory measurement methods and the resulting lack of an adequate quantitative data base constitute serious problems. In contrast with other indoor pollutants, biogenic pollutants bear complex and varied organic structures that defy automatic chemical assay. Biogenic agents exhibit limited direct toxicity, more often provoking infection or allergic responses. Bacterial and viral agents can produce infections in humans; however, the indoor transmission of these agents is not fully understood. A broad array of fungi, algae, actinomycetes, arthropod fragments, and dusts have been confirmed as airborne antigen sources that evoke human allergic responses. Indoor biologic pollutants--most notably bacteria and fungi--also play important roles in the deterioration of surfaces and spoilage of stored materials.

MONITORING AND MODELING OF INDOOR POLLUTION

Indoor air quality monitoring, in addition to pollutant sampling, must involve ventilation-rate measurements and daily activity logs of occupants. In addition, meteorologic data and outdoor pollution measurements may also be needed for the monitoring and assessment of indoor pollution.

Most indoor monitoring studies have relied on instrumentation developed for monitoring workplace or ambient air. The use of conventional monitoring instrumentation is frequently awkward, expensive, and suitable only for a limited number of comprehensive indoor air quality studies. Owing to the special requirements, instruments and sampling strategies are being developed specifically for indoor residential and office environments. The advent of personal monitors has permitted, in a few cases, the startup of monitoring and exposure studies for specific pollutants--nitrogen dioxide and radon. Personal and portable monitors are being developed for carbon monoxide, formaldehyde, and particulate matter. Monitoring the indoor environment, either with fixed-location sampling devices or with personal monitors, requires special protocols addressing pollutant sampling, instrument calibration, source operations, and occupant activity. When indoor monitoring takes place under normal occupancy conditions, the protocol must ensure that the act of monitoring itself avoids influencing those occupancy conditions.

Indoor-air pollution simulation models provide a theoretical framework for relating outdoor pollutant concentrations, meteorologic factors, building factors, ventilation rates, and indoor source and sink dimensions with indoor pollutant concentrations. Most importantly, a validated simulation model must accurately predict a desired concentration for conditions other than those tested experimentally. Depending on ventilation conditions and the geometry of the structure, a single room, a floor, or a whole building may be adequately approximated as a single air-quality compartment (entity). However, if sources and sinks are not uniformly distributed and if the indoor environment is large, pollutant stratification occurs within a building and a multicompartment numerical model is required to simulate the indoor-air pollution concentrations. Almost all numerical models are mass-balance equations that simulate the dynamic relationships among indoor pollutant concentrations, outdoor concentrations, indoor sources, and sinks (including ventilation).

FACTORS THAT AFFECT EXPOSURE TO INDOOR POLLUTION

Exposure is a dynamic concept that is defined as the joint occurrence of two perhaps independent events: the presence of a person in a specific environment and the presence of a pollutant at a specific concentration in the same environment. Because both human activities and air pollutant concentrations vary spatially and temporally, pollutant concentrations obtained from outdoor monitoring networks are inadequate for determining human exposure. Human activities are among the factors that must be addressed in assessing exposure to air pollutants. They have been studied by many researchers, mostly sociologists, to determine population mobility patterns and time budgets. The results indicate that, on the average, employed Americans spend 90% of the day indoors, whereas homemakers and retired people spend up to 95% of their time indoors. General sociologic studies may be used in air-pollution research, but do not address specific topics of interest for the assessment of human exposure to air pollutants. The exact indoor location (or environmental mode) is of paramount importance in exposure studies. Of all environmental types, the in-transit mode has been studied more extensively than any other microenvironment.

Indoor air quality, and therefore exposure to pollutants, varies geographically as a function of outdoor regional air quality and as a function of the rural, urban, or suburban character of the location of the indoor environment in question. In many residences, the indoor air quality does not vary substantially. In larger buildings with many ventilation zones, indoor air quality may vary in accordance with the function (utility) of each zone. Building factors that influence exposure include the site conditions, such as microclimate and proximity to major outdoor pollution sources, building design (age, size, ventilation systems), occupancy, and building operations. The exact nature of the cause-and-effect relationships between these factors and indoor air quality has not been established.

HEALTH EFFECTS OF INDOOR POLLUTION

Several classes of pollutants with major indoor sources were identified as having important known or reasonably likely effects on human health: sidestream cigarette smoke, radon and radon progeny, mineral and vitreous fibers, formaldehyde, indoor combustion products, agents of indoor contagion and allergens, and, to a lesser extent, temperature and humidity extremes, noise, and odors.

Other classes of indoor pollutants may have impacts on human health, such as consumer-product aerosols and pollutants from hobby, interior-decorating, and maintenance activities (e.g., solvents and pigments). Because the evidence of their effects on health is meager, the Committee could not determine whether specific effects were attributable to them and concluded that effective review at an appropriate depth was not feasible. Many airborne solvents, pigments, mineral dusts, and other products used in hobbies and interior decoration are present in the indoor air. The best data base on the effects of exposure to those substances is that drawn from studies of the industrial workplace, and the reader is therefore referred to the occupational-health literature.

INVOLUNTARY SMOKING (pp. 364-382)

Tobacco smoke is a major source of both gaseous and particulate pollution in the indoor environment, and the nonsmoker absorbs measurable amounts of carbon monoxide and nicotine, as well as small amounts of other smoke constituents, owing to involuntary smoking. The carbon monoxide absorbed varies from negligible in well-ventilated office buildings to amounts that raise the carboxyhemoglobin (COHb) concentration by 2-3% in an exposure of 1-2 h.

The COHb produced by the most severe involuntary-smoking exposure likely to occur in everyday living is capable of reducing the maximal exercise capacity of normal healthy adults, but does not measurably affect submaximal exercise capacity. Carbon monoxide has been shown in one study to reduce the amount of exercise that patients with hypoxic chronic obstructive lung disease can perform before the onset of dyspnea.

Patients with angina pectoris have a reduced exercise tolerance after involuntary smoking that may be a combination of psychologic stress and a carbon monoxide-induced decrease in oxygen delivery to the myocardium. Carbon monoxide clearly reduces the amount of exercise possible before the onset of angina in patients with angina pectoris.

Small changes in visual and auditory vigilance have been demonstrated at COHb concentrations that can be produced by involuntary smoking, but no change in tests of complex function has been demonstrated. Involuntary smoking has not been shown to produce acute changes in lung volumes or in a number of small-airway resistance measurements in normal healthy adults. Long-term exposure to cigarette smoke has been related to small-airway dysfunction in healthy nonsmoking adults.

Children whose parents smoke have been shown in some studies to be more likely to have respiratory symptoms, bronchitis, and pneumonia as infants. This relationship has been found in some studies to be independent of parental symptoms, socioeconomic class, and the smoking habits of other children in the household. It shows, in those studies, a dose-response relationship with the number of cigarettes smoked per day by the parents. To the extent that these associations may be due to cigarette smoke, it is reasonable to assume that the particle mass or a specific compound contained therein, rather than nitrogen dioxide or carbon monoxide, is responsible.

A twofold risk of cancer mortality in nonsmoking women has been associated (in a Japanese study) with having husbands who smoke. Apparently, the risk is proportional to the amount of passive smoking.

RADON AND RADON PROGENY (pp. 307-322)

The radon gas that diffuses out of radium-bearing building materials, subsurface soil beneath buildings, and well water into the indoor air undergoes radioactive decay. As a result, the indoor air contains both radon gas and alpha-emitting decay nuclides in particulate form, herein referred to as "radon progeny."

The health effects of radon and radon progeny are well established from studies of workers. Exposure to radon and its progeny at high concentrations has resulted in several hundred excess cases of lung cancer among uranium miners in the western United States. The health effects of much smaller amounts of radon progeny from indoor exposures can be estimated on the basis of a linear, no-threshold dose model, which yields upper-limit estimates of excess cancer in populations exposed to various indoor concentrations of radon and radon progeny. Lifetime cumulative exposures to radon progeny that result from current indoor exposures are lower by approximately a factor of 100-10,000 than those received by the U.S. uranium miners who have been studied.

The reliability with which the uranium-miner lung-cancer experience can be extrapolated to the effects of indoor exposures to radon progeny among the general population is limited by several important differences between the populations and by uncertainty about the extent of the effect of cigarette-smoking on the incidence and latent period for lung cancer related to radon progeny. The population differences include: (1) an adult, male, healthy working population versus a general population that includes the very old, the very young, and the chronically ill; (2) coexposures to relatively high concentrations of silica dust and diesel exhaust among the miners versus coexposures to relatively low concentrations of household pollutants and consumer products among the general population; and (3) differences in the ethnic and social backgrounds and smoking histories among the different populations.

ASBESTOS AND OTHER FIBERS (pp. 339-350)

The inhalation of asbestos fibers can lead, many years later, to pulmonary fibrosis, lung cancer, and mesothelioma of the pleura and peritoneum. All these diseases have been seen in humans who had chronic occupational exposures to airborne asbestos fibers, and they have all been reproduced in animals. Lung cancer and mesothelioma have also been seen in humans who had no occupational exposures, but who lived either in the same households as asbestos-workers or in neighborhoods where the ambient air had increased asbestos-fiber concentrations resulting from proximity to an asbestos-related industry or a geologic anomaly that acted as a source of airborne fiber.

Asbestos and asbestos-containing products--such as ceiling tiles, floor tiles, pipe insulation, and spackling compounds--were widely used in homes and public buildings because of their excellent thermal and acoustic insulation and structural properties. When these materials and products are displaced or disturbed by abrasion of deteriorating surfaces during housekeeping and maintenance operations, renovations, redecorating, or, especially in public buildings, malicious mischief, asbestos fibers can be released into the air. Concern about the inhalation of fibers that can result has led to extensive and expensive programs to remove asbestos, under controlled conditions, from accessible regions of public buildings, such as schools and libraries.

Fibrous materials used as substitutes for asbestos include glass fiber, rock wool, and slag wool. They have been shown, in animal injection and implantation studies, to be capable of producing lung fibrosis and mesothelioma. However, they are much less important in this regard than asbestos, and there is no corresponding human-health evidence associated with the forms in which they are used in industrial and consumer products. Thus, their substitution for asbestos appears to be beneficial, inasmuch as such substitution reduces the risk associated with asbestos exposure.

FORMALDEHYDE (pp. 322-338)

Formaldehyde has been the subject of numerous complaints regarding irritation of the eyes and respiratory tract, nausea, headache, rash, tiredness, and thirst. These symptoms have been reported mainly by residents of mobile and conventional homes in which formaldehyde-yielding products have been identified. Documented cases of bronchial asthma due specifically to formaldehyde are few; more commonly, asthma is aggravated by the irritating properties of formaldehyde.

Aqueous solutions of formaldehyde damage the eye and irritate the skin on direct contact. Repeated exposure to dilute solutions may lead to allergic contact dermatitis. Poisoning from ingestion is uncommon, because the irritancy of formaldehyde makes ingestion unlikely.

A preliminary report from the Chemical Industry Institute of Toxicology has shown that formaldehyde induces nasal cancer in laboratory rats and in some of the laboratory mice similarly exposed at the high dose. Nasal cancer has developed in the group of rats exposed at 15 ppm and 6 ppm, and dose-related histologic changes of the nasal mucosa in rats exposed at 2 and 6 ppm. Although the human mutagenic and teratogenic potential of formaldehyde is not known, it has exhibited mutagenic activity in a wide variety of organisms.

Data on the health effects of other environmental factors and their interactions--such as cigarette-smoking history, variability of health status, age, and genetic predisposition (which may modify responses to formaldehyde)--have not been adequately evaluated. That makes it difficult to assess accurately the health risks attributable to exposure to formaldehyde. However, the complaints of residents of homes with formaldehyde-containing products are similar to complaints made by persons studied in the laboratory at similar formaldehyde concentrations; hence, these health complaints may be related to formaldehyde exposure in the home. Accordingly, a substantial proportion of the U.S. population may be likely to develop symptoms as a result of exposure to formaldehyde at low concentrations. It has been estimated, on the basis of laboratory tests and various kinds of population surveys, that perhaps 10-20% of the general population may be susceptible to the irritant properties of formaldehyde at extremely low concentrations. For example, some persons report mild eye, nose, and throat irritation and other symptoms at concentrations less than 0.5 ppm, and some note symptoms at concentrations as low as 0.25 ppm.

These concentrations could also cause bronchoconstriction and asthmatic symptoms in some susceptible persons, and chronic exposure to low concentrations can result in sensitization. There appears to be a wide range of individual susceptibility to formaldehyde exposure. We cannot determine the exact numbers of susceptible people residing in indoor environments where exposure to formaldehyde could produce adverse responses. On the basis of estimates of the number of susceptible persons among the general population and the estimate that about 11 million persons in the United States now reside in mobile homes of varied age, construction, and quality, it may be concluded that a substantial number of persons are at risk of developing adverse health effects associated with formaldehyde.

INDOOR COMBUSTION (pp. 350-364)

The combustion of fossil fuels in air results in the generation of effluent streams containing carbon monoxide, nitric oxide, nitrogen dioxide, formaldehyde, carbonaceous particles, and other products of incomplete combustion, as well as the products of complete combustion--carbon dioxide, water, and sulfur dioxide. When the effluents are not vented to the outside, as in the case of most gas ranges and some space-heaters, the effluents are mixed into the indoor air.

The percentage increase in the indoor concentration of the combustion effluents resulting from such indoor sources is generally greatest for nitric oxide and nitrogen dioxide. For homes with gas ranges, indoor nitrogen dioxide concentrations are frequently twice as high as outdoor concentrations. The long-term integrated concentrations can exceed the national annual ambient-air quality standard (NAAQS) of 100 $\mu g/m^3$ (0.05 ppm) in some houses. Although chronic animal inhalation studies and community air-pollution epidemiology studies using central monitoring-station data have not established that exposures at or near the NAAQS for nitrogen dioxide produce measurable health effects, several recent studies of the health status of children living in homes with gas ranges have shown that they had more respiratory illness and poorer respiratory function than children living in comparable homes with electric ranges.

Increases in carbon monoxide sufficient to cause measurable health effects are usually associated with improperly operated flames or especially prolonged use of unvented space-heaters. Both carbon monoxide and nitric oxide bind with hemoglobin and reduce tissue oxygenation. Carbon monoxide from indoor combustion sources and sidestream cigarette smoke can be shown to cause measurable increases in COHb of exposed persons, but the health implications of such increases remain speculative. The importance of increased carbon monoxide and formaldehyde concentrations in indoor air was discussed above.

INDOOR AGENTS OF CONTAGION AND ALLERGY (pp. 382-417)

There is considerable evidence that a number of contagious-disease organisms--including those associated with influenza, Legionnaires' disease, tuberculosis, measles, mumps, and chicken pox--are capable of airborne transmission in the indoor environment. Other respiratory diseases, such as the common cold and pulmonary infections, involve airborne transmission. Because of the important role of respiratory diseases in overall acute morbidity, airborne transmission of contagious agents is important in the indoor environment.

The droplet-nucleus theory--whereby liquid particles emitted from the human respiratory tract evaporate to a particle size that can remain airborne for a period sufficient to be carried by natural air currents or convective ventilation flows and later deposited in the human airways--is generally accepted and used as a basis for transmission models. The effect of reduced ventilation in residences and offices on the incidence of infections is unknown.

Only a few airborne allergens are found in enclosed spaces. Their health effects are difficult to estimate, although their impact is sometimes appreciable.

EFFECTS OF INDOOR POLLUTION ON HUMAN WELFARE

Effects on human welfare are taken to include loss of productivity, human discomfort, and effects on materials, primarily soiling and corrosion of exposed surfaces.

SOCIOECONOMIC STATUS (pp. 419-421)

Members of low income classes are more likely to live in poorly insulated housing with higher air-exchange rates. Several reports have indicated that gas stoves or unvented gas or kerosene heaters are used for supplemental space-heating in northern cities. The percentage of homes with smokers appears to be inversely related to parental educational level. Lead intoxication in children occurs disproportionately in lower-income urban populations; higher ambient airborne-lead concentrations may contribute. However, some potential sources of indoor pollution may occur more frequently in the middle and upper income brackets. Many consumer products, as well as coal and wood stoves, exemplify such sources. Although the distributions of these and other factors may be functions of socioeconomic status that cause some segments of society to be more or less disadvantaged with respect to a hazardous indoor environment, the available data allow little more than speculation.

PRODUCTIVITY (pp. 431-437)

There is a growing recognition of the difficulties in clearly demonstrating the linkage between environmental quality and productivity. Perhaps as a result of these difficulties, there appears to be a slackening of research in this subject. Anecdotal or observational evidence can be found to support the conclusion that improving air quality should improve productivity, but objective documentation does not appear to exist or to be readily available. The most promising avenues for research appear to be those which demonstrate direct health effects of the various pollutants. Nevertheless, under the modern, broad definition of "productivity," a reduction in productivity is an almost certain consequence of pollution itself.

SOILING AND CORROSION (pp. 437-445)

Reduced indoor environmental quality can result in degradation and deterioration of indoor materials, furnishings, and artifacts. As efforts required for maintenance and housekeeping increase, the costs of owning or operating a building increase. To protect property or reduce costs of operation, more stringent control of indoor environmental quality may be required than may be indicated for protection of the health of occupants.

DISCOMFORT (pp. 421-431)

Control of indoor environments in nonindustrial facilities is designed to provide a degree of comfort acceptable to the occupants. When a stimulus (whether odor, temperature, humidity, noise, or air pollution) is changed beyond an acceptance threshold, there can be adverse effects. This kind of environmental control may be difficult to provide. However, the comfort-discomfort relationship may be one of the more important aspects to consider in evaluating the performance of indoor environmental control systems.

CONTROL OF INDOOR POLLUTION

CONTROL STRATEGIES (pp. 488-498)

Three basic strategies have been identified: source removal, dilution, and air-cleaning. In each of these classifications, methods can be selected that will reduce exposure. The appropriate strategies to be used, either separately or in combination, must be selected with respect to other and interacting factors--thermal, acoustic, energy-conservation, and economic.

CODES AND STANDARDS (pp. 451-465; Appendix A)

Minimal requirements of acceptability are often stated in building codes and standards in terms of air-exchange rules, temperature limits, and so forth. These documents tend to cause minimal requirements to be established for such direct effects as temperature, humidity, and odors, but may not be sufficient to provide for other effects, such as air pollution or noise. Nor do these codes consider the interactions that can occur among these factors and other system features such as lighting, thermal load, and spatial requirements.

AIR DIFFUSION CONTROL (pp. 465-471)

Indoor air quality is most commonly controlled by forced-air systems. However, if diffusion control is designed without considering possible stratification of air within a room or a building, there may be local violations of thermal, humidity, or air-quality criteria for acceptability, and occupants may be exposed to conditions other than expected from the design.

INDOOR ENVIRONMENTAL CONTROL SYSTEMS (pp. 465-471)

Control methods for indoor environments require specification of environmental criteria and definition of the control variables. The environmental criteria that are identified in this document are health, comfort, welfare, energy consumption, and costs. The control variables identified are spatial requirements, lighting factors, thermal factors, air quality, and acoustic factors. Although environmental criteria and control variables can be identified and described, the capability of sensing the appropriate variables and controlling the system to meet the specified criteria is severely limited. Moreover, most indoor environmental control systems must attempt to respond activity or passively to all five of the control variables simultaneously.

Residential air-conditioning systems are conventionally designed to respond to spatial, thermal, and air-quality variables and, to a limited extent, acoustic variables. For larger facilities, such as offices and schools, air-conditioning systems must also respond to variations in occupancy and lighting loads, in addition to spatial, thermal, air-quality, and acoustic factors. For other functional spaces (e.g., concert auditoriums, art galleries, museums, and hospitals), some or all of the variables must be controlled with additional precision.

For many years, air-conditioning systems were designed to meet the required environmental criteria (primarily thermal) at minimal first cost. Operating costs were not considered important as first costs, because energy was relatively inexpensive, compared with labor and material. However, as the costs of energy increased rapidly during the last decade, operating costs became a major factor in environmental control. Energy-conservation measures were implemented in many buildings to reduce operating costs. Some of these measures (e.g., improved system efficiency through better maintenance) had no impact on environmental control, but others (e.g., reduced ventilation, heat, and lighting) had potentially adverse effects. One reason for the occurrence of adverse effects was lack of understanding, by building operators and owners, of the interrelationships among ventilation rates, lighting, and health responses. As an example, changes in lighting can affect thermal loads, which affect ventilation rates. Conversely, results of some energy-conservation measures have indicated that indoor environmental quality need not be degraded and, in fact, may be enhanced by these changes (e.g., reduced stratification within occupied spaces). Thus, two general conclusions can be drawn: control methods may not be capable of adequately responding to environmental changes as energy-conservation measures and cost contraints are applied; and the quality of the indoor environment need not be degraded, but can be enhanced if care is exercised in the selection and implementation of the energy and cost constraints.

AIR-CLEANING EQUIPMENT (pp. 471-488)

Air-cleaning equipment for residential and commercial applications is generally limited to particle filtration. Some gas and vapor removal equipment is available, primarily for commercial applications. Methods of rating or evaluating the performance of the gas and vapor removal systems are not yet available. Methods are available for rating and evaluating particle removal equipment, but they are simplistic and outdated. Moreover, in-place methods of system evaluation are available only for special cases, such as hospitals and laboratories.

COST EFFECTIVENESS (Appendix B)

Several economic models are available that can be used to evaluate the costs associated with various control strategies. Cost-effectiveness models that incorporate life-cycle costing are needed for decision-making.

An approach to estimating the impact of residential energy-conservation measures on air quality is discussed in Appendix B. The approach has not been validated or put into practical use, but is presented for illustration and discussion.

III

RECOMMENDATIONS

The Committee on Indoor Pollutants recognizes that decisions affecting the quality of the indoor environment are being made by manufacturers, government agencies, builders, building operators, professional organizations, and private individuals. The decisions encompass a broad range of activities in our society, with important and long-term consequences. Federal agencies are planning energy-conservation programs in buildings, contemplating the banning of some products, and estimating the health risk associated with indoor pollutant exposures. State and local government units are considering revisions of building codes, ordinances to prohibit smoking in public buildings, and requirements that asbestos insulation be removed from schools. It can be presumed that there are similar examples of decision-making at various levels in the private sector that affect indoor environmental quality. In view of the possible impact of these decisions, the Committee is concerned that policy, research, and economic decisions be formulated with proper understanding of their implications for the quality of the indoor environment. For specific indoor contaminants, two basic inadequacies in the available data must be resolved rapidly: poor definition of population exposures and lack of understanding of the health and welfare consequences of exposure to contaminants in the indoor environment. This chapter presents the Committee's general and specific recommendations for remedying these inadequacies.

We have observed that the existing data base is, for the most part, derived from pilot studies or anecdotal reports. The results of the pilot studies reveal the complexity and diversity of the information that must be looked at in evaluating the quality of indoor environments. In some cases, the potential health significance of exposure to indoor contaminants has been alluded to; but the full extent of a potential problem with respect to types of contaminants, concentrations, and numbers of people exposed has not been determined.

We believe that the research problem is large and requires national coordination. A national coordinated research program would have the following advantages:

- It would provide reasonable allocation of research efforts among various federal agencies, national laboratories, and academic institutions.
- It would provide standards for study design and quality assurance that are sufficient for various decision-making purposes.
- It would provide integration of research activities and dissemination of the information derived.
- It would provide maximal effectivenss of available funding.

The objectives of such a research program should be these:

- To determine the various sources and distribution of selected pollutants, to measure their ranges of concentrations, and to identify populations at risk.
- To understand how the contaminants move inside buildings, how they mix and react, and the rate at which they are removed or dissipated under various conditions.
- To characterize indoor pollutant emission source strengths under actual conditions.
- To develop and test the effectiveness of control technologies.
- To determine the effects of energy-conservation measures on indoor air quality.

To meet some of these objectives, there must be research aimed at developing improved instruments and at strategies for using them in the study of indoor pollutants. Instruments used to characterize indoor air quality reflect the early stages of this scientific endeavor. The sampling devices used in indoor studies were originally designed for sampling outdoor or industrial air. Many commercially available ambient-air monitors are bulky, noisy, and expensive and have not been tested for interferences that may be encountered indoors. These devices have mostly been judged to be inadequate for investigating indoor air quality, because their sensitivity, accuracy, and precision are not sufficient for the measurement of pollutants in the small volumes of air in indoor spaces. The Committee recognizes the importance of recent developments in the field of personal monitors; but it also recongizes the need for a simultaneous effort devoted to the development of air sampling devices designed specifically for indoor environments. The new instruments must be designed to record short-term peaks in concentrations.

Instrumentation alone is not sufficient for characterizing the indoor environment. Numerous variables indoors (perhaps more than outdoors) must be considered: some sources of pollutants are peculiar to some indoor environments; there are differences in structure type, ventilation, and other characteristics that affect pollutant concentrations; and activity patterns and time budgets of occupants vary. Human activity patterns and time budgets, which are essential in determining total exposures to air pollutants, are now derived from population surveys. These surveys do not require (and therefore do not obtain) information relevant specifically to indoor air quality. The Committee sees the need for analysis of these variables and their effects and recommends the development of model strategies (for sampling, etc.) or protocols as guidance for future research. (The formulation of model sampling strategies, for example, is not meant to stifle innovation in sampling designs, but rather to establish the model to be used for comparing data obtained by researchers whose objectives differ.) The Committee recommends the formulation of strategies and protocols for measuring the strength of various indoor pollutant-emitting sources and for assessing the effectiveness of control devices and procedures in abating pollution.

The Committee recommends the formulation of a standard format for reporting data and the development of protocols for standardized statistical approaches that will require only minimal analysis to be used in validating numerical models. These would help to reduce difficulties in comparing existing data and facilitate the development of valid conclusions; conclusions now are often based on exceedingly small samples.

Even comprehensive information about the quality of the indoor environment would not permit determination of total pollutant exposure. It must be recognized that people are exposed to many of the same pollutants outdoors, in transit, in the occupational-industrial environment, and elsewhere. The relative importance of each kind of environment can be established and priorities can be set if and only if pollution exposures in all distinct environments are characterized. Lack of a complete assessment may lead to inefficient allocation of scientific effort and control funds in each kind of environment. The Committee believes that the research efforts to characterize indoor air pollution and human exposures indoors must continue and intensify, if we are to determine total human exposure to pollutants and understand environmental contamination and its effects on health and the quality of life.

The remainder of this chapter presents specific recommendations for research, grouped by class of indoor contaminant discussed in the body of the report, and discusses the need for increased understanding of indoor pollutants in general and the need for consumer protection.

RADON

Nationally coordinated investigations on radon and its progeny should take place on two levels. A well-funded and coordinated national survey of radon concentrations in a representative sample of residential buildings is necessary to estimate the exposure of the total population to radon and radon progeny. Monitors that use the track-etch plastic chip may be adequate for integrated measurements for such national surveys, because they are inexpensive and are specific to radon. However, the performance of these and other passive devices needs to be carefully evaluated. Inexpensive instruments for measuring radon concentrations on a short-term basis need to be developed. These instruments should be available to local health agencies and others for spot surveys. On another level, research on the transport and transformation of radon inside buildings deserves special attention. Rates of emanation of radon from various sources, building materials, soil, and groundwater should be evaluated in a variety of on-site and controlled conditions. The effectiveness of strategies for abating or eliminating indoor radon (including the use of material sealants and fine-aerosol collection devices) must be evaluated.

There is an urgent need to study the health effects of radon and radon progeny. On the basis of known effects in miners exposed to radon and radon progeny at relatively high concentrations, a plausible case can be made that a substantial fraction of the lung-cancer incidence in nonsmokers is due to the alpha-radiation dose to the respiratory tract epithelium from inhaled and deposited radon progeny particles. It is urgent that this observation be examined quantitatively by studies of appropriate human populations already known to be exposed at below 100 WLM and preferably in the range of 20-50 WLM. It is known that in some geographic areas large populations are being exposed to radon gas and radon progeny particles in their residences at concentrations that, although much lower than those in uranium mines, are substantially higher than those in most residences. Epidemiologic studies of populations exposed to radon and radon progeny are reasonable and can provide the information necessary for the establishment of realistic and needed exposure-response relationships. These research studies should include examinations to determine early pathologic changes other than tumors (e.g., changes in sputum cytology and chromosomal aberrations), with special attention to the possibility of a relationship of those changes to the eventual development of lung cancer. Such studies should be performed as soon as possible.

FORMALDEHYDE

Simple and reliable passive monitors that would easily satisfy the requirements of large surveys of buildings for formaldehyde emission do not exist. Monitoring formaldehyde is extremely difficult, because of the influences of temperature, humidity, and some analytic problems that affect its detection. Both continuous and passive monitors are needed with sensitivities in the range of 10-30 parts per billion. A national survey for indoor formaldehyde exposures is <u>not</u> needed, but a systematic study of formaldehyde concentrations in a variety of indoor locations is needed to estimate potential exposure of humans. This study would also identify sources of indoor formaldehyde by type of building and decorative materials and would evaluate the effects of ventilation rates and other variables on the concentration. Regular measurements over specified periods would help to identify formaldehyde emission rates of insulating building materials and furnishings as functions of temperature, humidity, ventilation rate, and material age.

Formaldehyde emitted from buildings and consumer products has resulted in complaints of adverse health effects by people in some mobile homes and in some conventional residences and other buildings.

Some of this emission occurs over long periods. Thus, long-term effects in humans continuously exposed to formaldehyde at low concentrations need to be studied. There is a particular need to assess the carcinogenic potential in the concentration ranges of human exposures, inasmuch as one study in one strain each of rats and mice has shown that long-term exposure (lasting 24 mo) caused nasal cancer. Humans have been and are now being exposed to formaldehyde in several types of occupations and in a variety of structures. Epidemiologic investigations are needed to assess the human health effects of formaldehyde, the magnitude and duration of exposure, and the influence of cigarette-smoking habits and the presence of other contaminants. The mutagenic, embryotoxic, and teratogenic effects must be included in the epidemiologic and animal studies. In humans exposed to formaldehyde, the mechanisms of airway and target cell responses must be evaluated and characterized as to sensitization and adverse effects in susceptible population groups, such as asthmatics and persons with chronic obstructive lung disease. Exposure-effect relations and the mechanisms involved in the biologic effects require further animal toxicologic research. Formaldehyde should be restricted to the extent that household consumer products and building products in normal use will not release potentially hazardous or irritating amounts of formaldehyde into indoor air.

TOBACCO SMOKE

Tobacco smoke has shown some evidence of being a major contaminant in many indoor environments. Involuntary exposure to tobacco smoke should be assessed to identify locations and populations with high exposure and to determine the factors that contribute to high exposures indoors. Physical and biologic evaluation of tobacco-smoke constituents should be continued. Tobacco-smoke constituents should be tested for their toxic effects, their ability to act as mutagens or promoters of carcinogenesis, and their effects in combination with other indoor pollutants. In addition, such properties of tobacco smoke as mass and age, chemical composition, irritation factors, and odor components should be examined to learn how they are affected by ventilation rate, occupancy, extent of smoking, air-cleaning, and other control strategies.

The extent to which passive exposure to sidestream tobacco smoke produces respiratory tract symptoms and functional decrements in nonsmokers, especially children, needs further documentation and measurement. Prospective studies of children in homes with smokers would be especially desirable to determine rates of lung maturation and illness frequency during childhood and adolescence.

Information on the potential health effects of exposure of nonsmokers to tobacco smoke should be widely disseminated. The "energy-cost penalty" of providing adequate ventilation in indoor environments that permit smoking should be analyzed in a variety of public buildings. Increased cigarette taxation as a mechanism of reimbursement for the cost of the additional air-conditioning needed to remove tobacco smoke should be explored by governments at all levels.

ASBESTOS AND ASBESTIFORM FIBERS

A systematic survey is needed for the evaluation of the distribution, integrity, and concentrations of asbestos in buildings that contain or are thought to contain asbestos material. However, before this survey can be conducted, there is a need to develop new instruments to record fiber counts continuously, with size determination and possibly asbestiform-fiber identification, because current sampling and analytic techniques are inadequate.

Synergistic and interactive toxic effects of asbestos fibers in combination with other air pollutants, particularly organic vapors, should be examined in animal toxicologic and mutagenicity studies. Although some asbestiform fibers themselves do not appear to constitute an immediate health concern, their role as initiators or promoters in various disease processes should be studied. The incidence of mesothelioma in humans should be monitored via a registry and appropriate surveillance methods, to detect cases associated with substantial nonindustrial exposure to asbestiform fibers.

Guidelines should be developed for the control of exposure to airborne asbestos fibers during maintenance, renovation, and reconstruction in buildings that contain asbestos and asbestos-bearing shingles, tiles, plaster, etc.

COMBUSTION

Indoor combustion produces a number of contaminants. Among the contaminants that deserve special attention are nitrogen dioxide, carbon monoxide, respirable particles, nitrosamines, and polynuclear aromatic hydrocarbons. The rates of their emission from sources of indoor combustion have not been adequately evaluated. The Committee recommends that controlled chamber experiments be conducted to determine the products and their rates of emission from various types of combustion under various conditions. These experiments should focus principally on gas and electric cooking appliances and supplemental heating systems, such as natural-gas, propane, and kerosene heaters and coal- and wood-burning stoves. Air-venting and air-cleaning systems should be studied as means of reducing indoor concentrations of contaminants.

Indoor concentrations of combustion products have only recently been surveyed. Combustion products are present in many indoor locations, such as restaurants, cafeterias, homes, hotels, buildings with attached garages, and recreational facilities that use gasoline-powered equipment. More comprehensive and systematic surveys are needed to identify the range of combustion-product concentrations encountered indoors and the numbers of people exposed to them. These studies should determine the population exposure to nitrogen dioxide over both the short and the long term. In addition, the applicability of ambient-air fixed-location monitors for recording nitrogen dioxide and carbon monoxide concentrations indoors and for assessing individual exposures should be studied.

Chemical reactions and rates of removal of emitted gases should be determined. Nitrogen dioxide formation and removal should be examined in detail. This will involve the simultaneous measurement of other gaseous compounds. The effects of reduced air-exchange rates, zone ventilation, and source modification on pollutant reactions should be assessed. As with other indoor contaminants, there is a general need to improve instrumentation. Both nitrogen dioxide and carbon monoxide monitors are available for passive integrating sampling and for continuous monitoring. However, for indoor use, they need to be evaluated with respect to interferences. To evaluate short-term personal exposures, lightweight continuous monitors for oxides of nitrogen need to be developed. Evaluation of personal exposures to respirable particles is currently limited to integrated samples. Lightweight portable samplers or direct-reading monitors that can measure mass concentration over shorter periods are needed.

Nitrosamines can be formed during cooking and smoking. However, very few detailed investigations of the concentrations, mechanisms of formation, and potential control methods have been done.

The polynuclear aromatic hydrocarbons can be formed during high-temperature combustion of organic matter. Some of them have been found indoors as a result of emission from self-cleaning ovens and fireplaces. Pilot studies should be initiated to evaluate the extent of emission of polynuclear aromatic hydrocarbons and their indoor concentrations.

The magnitude and prevalence of decreases in pulmonary function and increases in respiratory tract infection rates among children living in homes with gas ranges and homes with electric ranges need to be determined more accurately, and there are several related issues that require clarification:

- Whether the effects are due entirely to the increased nitrogen dioxide concentration in the gas-stove homes or are influenced by the presence of other combustion effluents from the stoves, such as carbon monoxide, formaldehyde, and particles.
- Whether the effects can be related more closely to peak concentrations or to long-term average exposures.
- Whether the effects of exposures to sidestream cigarette smoke and to nitrogen dioxide are additive or synergistic.
- Whether exposure-response relationships can be developed and, if so, whether they indicate an effective threshold concentration for peak or average exposures.

The influence of reductions in air infiltration rates in existing buildings on indoor concentrations of combustion products needs to be determined. Among the potentially serious health consequences of reductions in infiltration are:

- Creation of greater pressure differentials between the indoors and the outdoors, which could reduce the effectiveness of the venting of combustion sources, e.g., furnaces and water-heaters. Problems in proper venting could cause substantial exposures to carbon monoxide (which could lead to severe intoxication and fatalities), as well as greater exposures to carbon dioxide, nitrogen dioxide, formaldehyde, and particles.
- Increase in chronic exposures to carbon monoxide, nitrogen dioxide, formaldehyde, and particles at low concentrations from unvented combustion sources; this could lead to increases in respiratory infections in children and decreases in their lung function.

CONSUMER PRODUCTS

The use of some consumer products can lead to the release of aerosols and gases indoors. The gaseous compounds of concern are mostly organic vapors. Among the compounds of principal concern are aldehydes and polynuclear aromatic hydrocarbons that evolve from plasticizers; nitrosamines and hydrocines from rubber products, combustion products, and cleaning agents; polychlorinated biphenyls (PCBs) from burnt-out ballasts in fluorescent lights; and a variety of other middle- and higher-molecular-weight organic substances from pesticides. A few studies have noted the presence of many of these compounds indoors. However, no systematic survey has been done. With regard to the evolution of organic molecules from pesticides, organochlorinated pesticides should be examined first, including aldrin, dieldrin, endrin, benzene hexachloride, pentachlorophenol, kepone, chlordane, and DDT. The emphasis should be on the determination of body burdens of these compounds and specifically the relative contributions of inhalation, ingestion, and absorption to body burdens.

The contents of consumer products should be investigated, and the chemical constituents should be tested for their toxic, mutagenic, carcinogenic, and teratogenic properties. In particular, there should be toxicologic studies of solvents, vapors, aerosols, and particulate compounds present in these products. Their synergistic and interactive effects with other indoor pollutants should be tested.

Consumer products whose use involves the release of gaseous and particulate materials to the indoor atmosphere should be so labeled, with their components. Warning labels for consumer products that can seriously pollute the indoors should state that they are to be used only in areas with adequate ventilation and should stipulate the possible consequences of their use when there is inadequate ventilation.

AEROPATHOGENS AND ALLERGENS

Little is known about the sources, concentrations, and survival rates of many aeropathogens in homes and other buildings. Relationships among the incidence of respiratory infections, concentrations of aeropathogens, and air-exchange rates in buildings must be examined. The urgency arises from the recent modifications in building ventilation codes that recommend reduced ventilation rates in residential and commercial structures. A sample of commercial, institutional, and residential buildings should be evaluated for the types and concentrations of aeropathogens under a variety of conditions of occupancy, human activity, ventilation, humidity, temperature, and contaminant control. Special attention should be given to the newer energy-efficient buildings and buildings with drastically reduced ventilation. The potential for infectious contamination from air-cleaning filters, air heat-exchangers, air humidifying systems, and air-conditioning systems deserves special attention.

Other agents in the indoor environment known to produce allergic responses include pollens, household mites, molds, animal dander and excreta, and bacterial spores. Further work is needed to characterize the size distribution of allergen aerosols, their sources, and the conditions that are conducive to their generation. The airborne concentration of allergens in the indoor environment has been determined in only a few instances, and the relationship between indoor concentration and response is poorly understood. Case-control epidemiologic and immunologic studies are needed to clarify exposure-effect relationships. Such studies will require improved instrumental and analytic techniques to facilitate characterization of concentrations of allergens and of the variety of microorganisms in the indoor environment. Synergism of biologic and nonbiologic agents should be explored in animal toxicology studies.

Some acute allergic responses, such as "humidifier fever," are of unknown etiology. The pathologic agents in immunologic and case-control epidemiologic studies need to be identified. The pathogenic process by which repeated small exposures to some allergens often lead to irreversible fibrotic lesions, as in bird-fancier's disease, should be elucidated, and the potential of other, more common indoor pollutants to produce such disease states should be evaluated.

VENTILATION STANDARDS AND CONTROL STRATEGIES

Knowledge of ventilation rates is of primary importance in studies of indoor contaminant concentrations. Given the variety of residential living units and other public and private facilities, it is not surprising that very little information exists to characterize air-exchange rates. Studies should begin to characterize air-exchange rates in existing buildings by building type, geographic location, occupant life styles, building operation, and observed average pollutant concentrations during the different seasons. Smaller-scale studies over a considerable period are recommended to characterize air exchange and its effects on occupant behavior in a representative number of buildings. This information would help to show the relationships among air-exchange rates, pollutant concentrations, pollutant generation, occupant behavior, and physiologic effects, as well as the effects of energy-conservation programs aimed at reducing air infiltration rates.

With the objective of maintaining indoor air quality, while not adversely affecting heating and cooling costs, there is a need for engineering studies to determine alternative strategies for air dilution of pollutants. Case studies of specific buildings may be useful in determining the efficiency of selected filtration-ventilation schemes. For example, it may be economically preferable to filter (or scrub) recirculated air or a mixture of recirculated and makeup air to maintain indoor air quality. Studies should be done on the effectiveness and energy-conservation implications of pollutant sensors used to activate air-dilution or air-cleaning systems. Engineering studies on both air handling and air treatment systems are encouraged. Although specific buildings may be studied in detail, the application of the findings should be generic; i.e., the lessons learned should be applicable to other buildings.

Standard methods should be developed and applied to evaluate the performance of in-place environmental control systems and components. Improved methods of providing acceptable and efficient air diffusion for thermal and contaminant control should be developed. Life-cycle cost evaluations should be made on the basis of current and projected energy costs to characterize these costs for future use in alternative designs of residences and large buildings. Air-conditioning, heating, and ventilation systems especially should be evaluated, in connection with other building characteristics.

EXPOSURE STUDIES

Proper assessment of indoor contamination needs a perspective that only total-exposure studies can provide. The relative contributions to individual and population exposures of the contaminants encountered indoors, with both indoor and outdoor sources, must be evaluated. For contaminants with multiple entry routes, the contribution of inhalation of indoor pollutants must be compared with all other contributions to body burdens. Only with this evidence can research and control efforts be prudently allocated. Under current conditions, studies of total exposure to many contaminants are limited by the available instrumentation and, to a lesser extent, by analytic methods. Such efforts are expensive. The extrapolation of their results is constrained by the smallness and often the unrepresentativeness of the samples of participants, by the inadequacy of the information on activity patterns in the population at large, and by lack of understanding of the distribution of indoor and outdoor pollutant sources and the pathways that contribute to body burdens. However, studies of total exposure and an understanding of activity-related concentration data will eventually advance our knowledge of pollutant exposure. This knowledge is a prerequisite to rational allocation of resources for warranted reductions in population exposures.

The Committee urges investigation in the behavioral aspects of indoor environments, specifically the relationships among performance, sense of well-being, contaminant concentrations, and stress. Temperatures, odors, and noise outside preferred ranges can reduce productivity, especially in self-pacing tasks. The relationship between productivity and the quality of the indoor environment needs to be determined. It is recognized that relationships between the behavioral variables and pollutant concentrations may be difficult to establish. Simultaneous measurements of trace organic vapors, water-vapor content, conductivity, noise, light, temperature, and air exchange rates should be pursued.

EDUCATION

Public education offers an effective way of reducing exposure of the population to many contaminants encountered indoors. People informed about the potential for exposure to pollutants from consumer products, tobacco smoke, combustion products, etc., will exercise some control to reduce the pollutant concentrations in their environments. For the most part, their options for controlling these pollutants are limited to source maintenance, ventilation control, and, to some extent, air-cleaning. Information about maintaining a clean indoor environment and assessing indoor spaces for potential contamination before purchasing or renting a structure and a variety of suggestions could be disseminated through health-maintenance organizations, regional health-planning agencies, public-affairs offices, the Environmental Protection Agency, the Department of Housing and Urban Development, the Department of Energy, the Consumer Product Safety Commission, and a variety of other federal and state agencies. The General Services Administration, the armed forces, and the Department of Housing and Urban Development are responsible for many residences and other buildings. Through certification of minimal acceptable occupancy standards, these federal organizations could develop strategies to ensure that indoor spaces under their jurisdiction are free from hazardous concentrations of contaminants.

Professional and trade associations could be instrumental in developing and disseminating information. These associations are encouraged to establish standards for acceptable practice, with respect to manufacturing, designing, building, and using products, equipment, and structures that influence the quality of the indoor air.

IV

SOURCES AND CHARACTERIZATION OF INDOOR POLLUTION

This chapter addresses several chemical pollutants with respect to their sources, concentrations, and indoor-outdoor relationships. In addition, with the aim of characterizing the general quality of the indoor environment, it considers temperature, humidity, unwanted sound, and electromagnetic radiation, such as the radiofrequency, infrared, visible, ultraviolet, and x-ray portions of the spectrum.

In the case of some pollutants, information on health effects is scanty, at best. To the extent possible, the health effects of such pollutants are discussed here. Detailed discussion of the health effects of other pollutants, on which more information is available, is to be found in Chapter VII.

Radioactivity and formaldehyde emitted indoors from building products are discussed in the first two sections of this chapter. Consumer products, a generic source of indoor pollutants of many types, are discussed next. The chapter proceeds with sections on asbestos and fibrous glass (which occur in different forms in many indoor environments), combustion processes (especially of unvented cooking and heating appliances), and tobacco smoke (a highly complex and ubiquitous mixture of pollutants). Several indoor air pollutants can be recognized by their odors. Such odors are often the first indications of deterioration in air quality and may themselves affect people's well-being adversely; hence, they are treated as a distinct category of pollutant in this chapter. Air temperature, radiant temperature, and air velocity and humidity affect the quality of the indoor environment through physiologic and sensory responses, so the thermal environment is also discussed in a separate section. Other physical factors of the indoor environment, such as noise and electromagnetic radiation, are discussed briefly in a final section.

The diversity of subjects discussed in this chapter is evident. Some of the pollutants considered here may be associated with voluntary behavioral patterns, such as tobacco-smoking, whereas others may be related to involuntary and unavoidable exposure, such as exposure to substances emitted from building materials. The reader should not infer any order of priority among the pollutants discussed here. An effort to attach priorities would require judgments on exposures and effects, and the order of discussion is not intended to indicate the application of such judgment.

RADIOACTIVITY

INTRODUCTION

Radioactivity and ionizing radiation occur naturally throughout the biosphere, both because of the presence of primordial radioactive elements and their decay products in the earth and because of natural processes (primarily cosmic radiation) that produce radionuclides or direct radiation fields. These natural sources expose humans to radiation both outdoors and in buildings. The magnitudes of various contributions to total radiation dose vary from place to place and between outdoors and indoors, and the type of radiation dose depends on the radiation source. At one extreme, the cosmic-radiation field delivers a dose to the entire body; this dose is not affected greatly by the presence of a building and may be characterized primarily on the basis of altitude. At the other extreme, airborne alpha-emitting radionuclides may deliver doses specifically to the lungs, and their concentrations indoors may be strongly affected by the nature of building materials and other sources, such as soil and water, and by building operations, such as ventilation. As an intermediate case, the gamma-radiation field arising from radionuclides that are fixed in place typically exposes the whole body and is affected by radionuclide concentration, proximity, and shielding.

In the discussion that follows, we refer to radioactivity concentrations and radiation fields and, by inference, to radiation doses from sources that are inside and outside the body. Radioactivity is given in curies; 1 Ci = 3.7×10^{10} becquerels, so 1 pCi = 0.037 Bq. Radiation fields can be specified in terms of energy flux; but it is more conventional in the present context to use units of dose rate, in which case the type of radiation has to be indicated. We use the rad as the unit of (absorbed) dose when specifying gamma-radiation fields (1 rad = 0.01 J/kg, so 1 mrad = 1×10^{-5} J/kg). For gamma doses, the dose in rads is numerically equal to the dose equivalent (DE) in rems. A distinction must be drawn between the "tissue dose," that actually received by tissue and therefore including self-shielding by the body, and the "air dose," that deposited in air in the space under consideration.

It is useful to summarize the dose-rate contribution in the United States from radiation arising outside buildings. Three recent summaries are those of the National Council on Radiation Protection and Measurements[34] and the U.N. Scientific Committee on the Effects of Atomic Radiation,[49] which depended heavily on Oakley[38] for U.S. data, and the 1980 BEIR report of the National Research Council.[35] External radiation, that arising from sources outside the body, may be divided into two categories, cosmic and terrestrial. The average tissue dose rate outdoors from cosmic radiation is approximately 28 mrads/yr; the dose rate indoors is slightly reduced by overhead shielding (the NCRP report assumed a 10% reduction in average exposures). This contribution has a substantial altitude dependence, increasing from about 26 mrads/yr at sea level to about 50 mrads/yr at 1,600 m, the altitude of Denver. The average outdoor population-weighted tissue dose rate from terrestrial radionuclides--due principally to gamma rays from potassium-40, the thorium-232 series, and the uranium-238 series--is approximately 35 mrads/yr. This dose rate varies substantially because of geographic variations in the distribution of these radionuclides. For estimating average terrestrial dose rates, the NCRP assumed that indoor dose rates were 20% lower than outdoor rates. (It also assumed that the tissue dose was 20% less than the air dose.) Internal radionuclides contribute important beta and gamma doses (about 15 mrads/yr to most of the body, primarily from potassium-40) and an important alpha dose (even if that to the lungs from radon and its progeny is excluded). The alpha dose arises primarily from internally deposited uranium-238 and -234, radium-226 and -228, and polonium-210 and varies greatly with body organ. One of the larger contributions, about 3 mrads/yr, is the polonium-210 alpha dose to the cells lining the bone surfaces. However, alpha particles have a greater biologic effectiveness than gamma rays, so the absorbed alpha dose contributes a DE some 10 times greater than that of the same (absorbed) dose of gamma radiation. Table IV-1 shows estimates of various contributions to DE rates, in millirems per year, which are numerically equal to tissue dose rates (in millirads per year) for gamma and beta radiation. For alpha radiation, a quality factor of 10 was assumed (based on relative biologic effectiveness), although 20 is now recommended.[19] The value given for lung dose from inhaled radionuclides assumed a radon-222 concentration in air of 0.15 nCi/m^3 (and slightly less than equilibrium amounts of its radioactive decay products, or progeny). The resulting DE has the largest value in the table. Nonetheless, this value appears more appropriate for outdoor than for indoor air, in which higher radon concentrations are found.

All indoor dose rates from natural radiation sources are affected by buildings, and those from inhaled radionuclides are affected most strongly. The only natural airborne radionuclides of importance are radon and its progeny, principally the series beginning with radon-222, the alpha-decay product of radium-226 (a member of the uranium-238 series). Radon is a noble gas that can move from the site of its formation, giving it a substantial opportunity to reach air that is inhaled by humans. The short-lived decay products of radon--polonium, lead, and bismuth--are chemically active and thus can be collected in the lungs, either directly or through particles to which they attach. The most important dose arises from alpha decay of the polonium isotopes. The decay sequence beginning with radium-226 is shown in Figure IV-1, and, from the biomedical point of view, effectively ends with lead-210, because of its half-life of about 20 yr. Because the alpha energy associated with decays of the short-lived products to lead-210 poses the main risk, progeny concentrations are often expressed as the associated "potential alpha-energy concentration" (PAEC) in air. The unit conventionally used for PAEC is the working

TABLE IV-1

Summary of Average Dose Equivalent Rates from Various Sources of Natural Background Radiation in the United States[a]

Radiation Source	Gonads	Lung	Bone Surfaces	Bone Marrow	GI Tract
Cosmic[b]	28	28	28	28	28
Cosmogenic radionuclides	0.7	0.7	0.8	0.7	0.7
External terrestrial[c]	26	26	26	26	26
Inhaled radionuclides[d]	--	100[e]	--	--	--
Radionuclides in the body[f]	27	24	60	24	24[g]
Totals (rounded)	80	180	120	80	80

[a]Reprinted with permission from NCRP.[34]

[b]With 10% reduction for structural shielding.

[c]With 20% reduction for shielding by housing and 20% reduction for shielding by the body.

[d]Lung only; doses to other organs included in "Radionuclides in the body."

[e]Local DE rate to segmental bronchioles = 450 mrems/yr.

[f]Excluding cosmogenic contributions.

[g]Excluding contribution from radionuclides in gut contents.

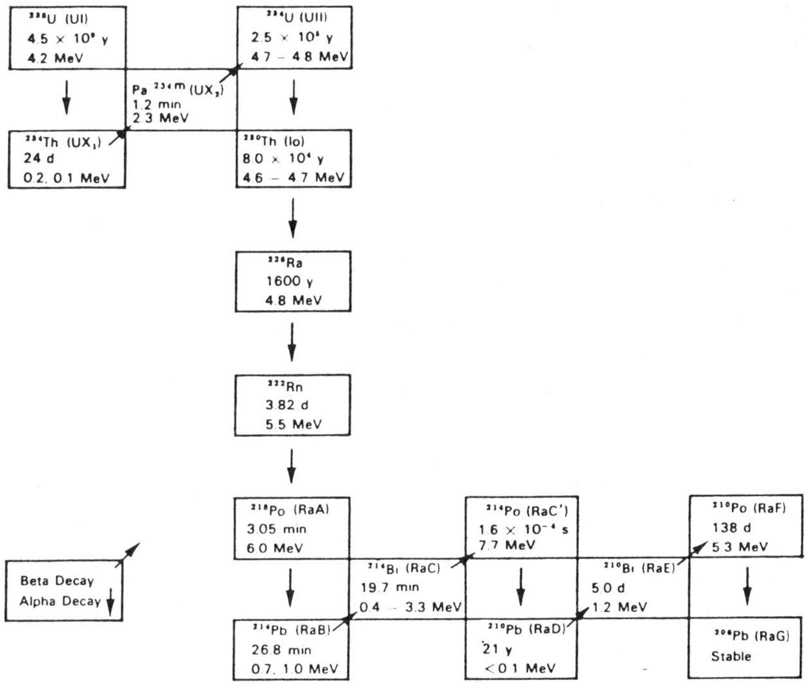

FIGURE IV-1 Principal decay scheme of uranium-238 to radon-222 to lead-206, showing alpha and beta decay; decay energies in millions of electron volts. Reprinted with permission from National Council on Radiation Protection and Measurements.[34(p. 4)]

level (WL), defined as 1.3×10^5 MeV/L, the PAEC if radon-222 at 100 nCi/m^3 is present with equilibrium amounts of its progeny. Dose (and DE) rates may be inferred from the PAEC on the basis of relatively complicated modeling, provided that the progeny particle size distribution and other factors are prescribed.

The character of a building may affect occupant radiation exposure in three principal ways: the building serves as a container for indoor-generated radon and its associated progeny, whether from building materials, underlying soil, or water and gas; the building materials contain natural gamma-emitters (potassium-40, the thorium-232 series, and the uranium-238 series); and the building shields occupants from cosmic or external terrestrial radiation. The last two effects tend to cancel one another. The building structure may, in unusual circumstances, also protect occupants from outdoor radon-progeny concentrations. However, the indoor concentration is ordinarily larger than the outdoor, and outdoor-generated radon usually contributes a small additive term to indoor concentrations. If this term is ignored, the steady-state indoor radon concentration for a fixed indoor radon source strength is inversely proportional to the air-exchange rate, the rate at which the indoor air is exchanged for outdoor air. The air-exchange rate for most U.S. buildings is around 1/h, with 0.5/h to 1.5/h typical for residences (windows closed). The air-exchange rate and other removal mechanisms also affect the ratios of radon-progeny concentration to radon concentration. Lack of removal implies activity ratios of 1, but substantially lower values have been observed. An equilibrium factor (F) is often defined as the ratio of the actual PAEC to the PAEC that would be associated with a specific radon concentration if the progeny were in equilibrium with this concentration.

This section characterizes indoor airborne radionuclides and radiation, summarizes measurements of actual concentrations or radiation fields, briefly indicates control measures, and suggests subjects for further research. The major emphasis is on radon and its progeny. The radionuclides in this decay chain, even at typical outdoor concentrations, cause larger radiation doses to internal organs than all other airborne radionuclides. Furthermore, the radon and progeny concentrations may be substantially higher indoors, particularly in buildings with low air-exchange rates. In addition, building occupants receive external whole-body radiation from radionuclides fixed in building materials and soil, and these doses are also given substantial treatment. This radiation arises principally from several primordial radionuclides--potassium-40 and members of the thorium-232 and uranium-238 decay series--with concentrations of around 0.1 pCi/g or greater in rocks, soil, and derivative building materials. These are also the decay chains in which radon-220, radon-222, and their progeny occur.

SOURCES OF RADIONUCLIDES AND RADIATION

Building Materials

Radionuclide Content. Few measurements and no wide-scale surveys of the radionuclide content of U.S. building materials have been made. Surveys of materials in Europe are summarized in UNSCEAR 1977,[49(p. 50)] which gives activity concentrations of potassium-40, radium-226, and thorium-232. As examples, average values for the concrete sample groups examined range from 0.9 to 2.0 pCi/g for radium-226, 0.8 to 2.3 pCi/g for thorium-232, and 9 to 19 pCi/g for potassium-40. By comparison, the ranges for brick are about 50% higher; those for cement are similar, except for potassium-40 (which is 50% less); and those for natural plaster are lower by about a factor of 5.

Available U.S. data (Table IV-2) show concentrations in the same range, assuming that the series radionuclides are sufficiently close to equilibrium to permit comparison. In a number of cases, U.S. workers have examined the radionuclide contents of concrete in the course of selecting materials for low-background facilities for use in radiation-counting;[29] the values obtained are consistent with the European data, although somewhat lower. The observed concentrations are also within the range of values typical for major rock types and soils. Concentrations for building materials not derived from crustal components, such as wood, are much lower.

Measurement programs have recently been initiated to characterize the radionuclide contents of building materials as a basis for understanding the resulting effect on the indoor radiation environment. Kahn et al.[25] have reported measurements of concentrations in various building materials in the Atlanta area; potassium-40, radium-226 progeny, and thorium-232 progeny concentrations for samples of concrete, brick, and tile are given in Table IV-2. Lawrence Berkeley Laboratory has begun to survey concretes and other materials as part of a program on indoor air quality; radionuclide contents for concrete and rock-bed samples from a number of areas are given in the table.[18]

Considerably greater radionuclide concentrations may be found in building materials that contain residues from industrial processes. The principal example of such materials in the United States is concrete blocks incorporating phosphate slag (essentially calcium silicate), a byproduct of phosphate production. As discussed by Roessler et al.,[42] this slag contains most of the radium-226 and uranium-238 found in the phosphate ore. For the electric furnace process used in Florida, concentrations in the ore are about 60 pCi/g, and the slag has similar concentrations. A plant in Alabama (using Florida and Tennessee phosphate ores) sold slag to companies in Alabama, Mississippi, Tennessee, Georgia, and Kentucky. The concrete produced by these companies has radium-226 concentrations estimated, and in some cases measured, to be about 20 pCi/g.[25] Phosphogypsum (essentially calcium sulfate produced by treatment of phosphate ores with sulfuric acid) may also be used for building materials, particularly wallboard. In this treatment, radium-226 follows the

TABLE IV-2

Average Radionuclide Content of U.S. Building Materials

Material	Concentration,[a] pCi/g			Comments
	Uranium-238 Series[b]	Thorium-232 Series[b]	Potassium-40	
Concrete[29]	0.29-1.32	0.28-1.58	6.6-9.8	Summarized measurements to select counting-room materials
Concrete[25]	1.4	1.5	21	Atlanta area
Brick[25]	1.8	1.8	17	Atlanta area
Tile[25]	1.9	1.1	8	Atlanta area
Concrete[18]	0.2-1.0	0.2-1.0	5-12	Nine metropolitan areas (preliminary values)
Solar rock bed[18]	1.5	1.4	25	New Mexico (preliminary values)
Concrete[49]	0.9-2.0	0.8-2.3	9-19	European concretes[c]

[a]Except where noted, each entry is average for sample group; range is given if several sample groups were examined.

[b]Because various members of decay series were detected, results in each column are directly comparable only if series equilibrium may be assumed.

[c]For comparison.

calcium, leading to tens of picocuries per gram in the gypsum; but such gypsum has not been used on a large scale in U.S. wallboard. In contrast, concrete that incorporates phosphate slag may have been used in approximately 100,000 homes.[25] Finally, some fly ash from coal-fired power plants has been used in cement production, and this use may continue. Heretofore, it has not been thought to contribute substantially to the radionuclide content of the resulting building material.[25] Emanation measurements on fly-ash concretes are now being performed at Lawrence Berkeley Laboratory.

Radon Emanation. The effective radon-222 generation rate in building materials depends on the radium-226 content, which varies widely, and on the percentage of radon formed that does not remain lodged in the matrix of the material. Radon that is not fixed in place may move through the matrix by diffusion or, if the material contains large air spaces, by convection. Diffusive movement depends on the diffusion length of the material in question and on its thickness. The extent to which these processes occur depends not only on the material's characteristics, but also on environmental conditions--pressure, temperature, and moisture content. A rule of thumb sometimes cited (e.g., UNSCEAR[49]) is that 1% of the radon-222 generated from materials in walls and ceilings escapes into the adjacent air space. However, recent measurements at Lawrence Berkeley Laboratory and elsewhere have indicated that a considerably higher fraction can escape, e.g., from concrete. Ingersoll et al.[18] cited escape-to-production ratios of 0.08-0.25 for radon-222 from concrete. (Radionuclide contents for the sample groups examined are indicated in Table IV-2.)

Of most direct interest for indoor air quality is the actual emanation rate, often given as picocuries per square meter per second and sometimes as picocuries per gram per second. Measurements for various materials give emanation rates over a wide range. For example, European gypsum board and bricks yield radon-222 at about 0.3×10^{-4} pCi/m^2-s, whereas rates for European concretes range from 0.001 to 0.2 pCi/m^2-s.[24,32] Preliminary measurements of radon-222 emanation rate per unit mass for sample groups of concrete from U.S. metropolitan areas (Table IV-2) give averages that range from 0.4 to 1.2 pCi/kg-h (0.8 pCi/kg-h yields approximately 0.03 pCi/m^2-s for 0.1-m-thick concrete). Several rock samples from solar-heat storage beds averaged 0.5 pCi/kg-h, although radium-226 contents were considerably higher than those for the concrete samples.[18] The resulting indoor radon-222 concentrations depend on the amount of such material in the structure, the interior volume, and the air-exchange rate. For an air-exchange rate of 1/h and a ratio of indoor emanating surface to indoor volume of 0.5 m^2/m^3, an emanation rate of 0.03 pCi/m^2-s corresponds to a radon-222 concentration of about 0.04 nCi/m^3. If the equilibrium factor is 0.5, this would yield a PAEC of about 0.0002 WL. Direct measurement of emanation rates of materials made with industrial byproducts (such as phosphate-slag concrete) is underway, but results are not available. Because these materials may contain 20 times as much radium-226 as a typical concrete, radon-222 contributions of up to several nanocuries per cubic meter of radon-222 and a corresponding increase in the PAEC could be expected if the same emanation ratio pertains.

Measurements of emanation rate vary by more than an order of magnitude,[49] so it is difficult to use radium content to predict the contribution of a particular material to indoor radon content. For this reason, more comprehensive information on diffusible fraction, diffusion length, etc., and their dependence on material or environmental factors is required before we can characterize building materials on the basis of radionuclide content. If this information becomes available, radionuclide contents may then be helpful in characterizing indoor concentrations on a broad scale, e.g., by geographic area. However, the dependence of diffusion and emanation rates on environmental factors, such as pressure and temperature, and on the moisture content of the material may limit the possibility for such characterization.

In some cases, radon-220 ("thoron") and its progeny, ordinarily present at much lower concentrations than radon-222 and its progeny, may assume importance, particularly when mechanisms exist for transporting emanating radon-220 rapidly into the air space of interest. In comparison with the half-life of radon-222, the much shorter half-life of thoron, 55 s, causes the measured radioactivity in curies to be a characteristic of secondary interest. However, the PAEC still gives a relatively direct indication of possible dose to the lung. One WL of radon-222 progeny has the same PAEC as that associated with progeny in equilibrium with thoron at 7 nCi/m^3. To the extent that uranium-238 and thorium-232, which have similar half-lives, have similar activities in source materials, the PAEC from their progeny, radon-220 and -222, can reach similar values if rapid transport mechanisms exist. This may occur, for example, in solar buildings that sweep air through rock or concrete thermal-storage beds. A few efforts have begun to measure thoron emanation rates, but results are not yet available.

Gamma Radiation. The energies and intensities of photons from decay of natural radionuclides have been well characterized. The external dose from radionuclides in building materials is due to the gamma rays emitted and depends on the geometry of the structure and attenuation by the materials, as well as the gamma-ray energies. A simple expression may be derived for the gamma-ray air dose in a hole in an infinite uniform medium:[25]

$$\dot{X}_\infty = (2.43\ \mu\text{rad/h})(E_U C_U + E_{Th} C_{Th} + E_K C_K),$$

where C_U, C_{Th}, and C_K are the concentrations (in picocuries per gram) of uranium-238 and its progeny, thorium-232 and its progeny, and potassium-40, respectively, and E_U, E_{Th}, and E_K are the average gamma-ray energies per disintegration of the same radionuclides (including disintegration of the progeny for the uranium and thorium series). Using E_U = 1.72 MeV, E_{Th} = 2.36 MeV, and E_K = 0.156 MeV,[25] $\dot{X}_\infty = 4.2C_U + 5.7C_{Th} + 0.38C_K$, in microrads per hour. The stated dose contributions from the uranium and thorium series are slightly less than those cited elsewhere, e.g., by Krisiuk et al.,[27] who may have used older information on decay schemes. For the radionuclide contents cited in Table IV-2, the three terms in the expression for $\dot{X}_\infty$ contribute comparable amounts. (An analogous expression for the dose from a flat plane is cited in the section on soil.)

For an actual structure, the geometry is complex and varied; in addition, the building materials may attenuate the external radiation dose from other sources. Moreover, radon-222 and its progeny may be present in the material at less than equilibrium values, thereby decreasing the corresponding gamma-ray dose. The radon-222 escape-to-production ratio is most often in the range of low to 0.25, causing a small reduction in the value of $\dot{X}$. The effects of geometry and attenuation cannot be so simply characterized. Dose-rate expressions from various workers, pertaining to a variety of structures, have been summarized.[37] Some of these expressions account for reduction of the dose rate from outdoor sources. Moeller et al.[33] described a computer program suitable for analysis of varied geometries.

The infinite-geometry case yields air dose rates of about 8 μrads/h for a potassium-40 concentration of 8 pCi/g and uranium-238 and thorium-232 series concentrations of 0.5 pCi/g. An infinitely thick slab of such material would contribute about half this dose rate at its surface. As discussed earlier, a typical outdoor tissue dose rate from terrestrial radionuclides is 35 mrads/yr or 4 μrads/h. (Owing to shielding by the body, the tissue dose rate is about 20% less than the air dose rate.)

Soil and Groundwater

Radionuclide Content. Radionuclide concentrations of major rock types and soil have been summarized.[34] U.S. soil values of 0.6, 1.0, and 12 pCi/g have been stated for uranium-238, thorium-232, and potassium-40, respectively, on the basis of 200 measurements of gamma-ray dose rate cited by Lowder et al.[30] These values vary by a factor of around 3 from place to place. Values for crustal rocks[34] typically lie within this same range, but are considerably higher for some formations. For example, the phosphate rocks of Florida contain the uranium-238 series at tens of picocuries per gram, but normal amounts of thorium-232; commercial uranium ore bodies in the United States have uranium-238 concentrations of hundreds of picocuries per gram and higher.

Radon Emanation and Transport. The uranium-238 series, typically present in soils and rocks at concentrations of about 1 pCi/g, includes radium-226, the source of radon-222. The actual radon-222 emanation rate from the ground depends, as for building materials, on the percentage of diffusible radon, diffusion length, and other transport mechanisms (including groundwater) in the soil. A review of available measurements of radon-222 indicates a mean emanation rate from the soil of 0.42 pCi/m^2-s.[50] Given this value for the ground under a one-story house, and assuming that the emanated radon finds its way into the indoor air, the soil could account for indoor radon-222 at about 1 nCi/m^3 at a typical air-exchange rate of 1/h. Because emanation rates vary by at least a factor of 10 from place to place, this potential contribution can also be expected to vary substantially among U.S. buildings.

The soil as a source of radon-222 can be characterized directly by emanation measurements or, if disequilibrium and transport mechanisms (including groundwater) are known, indirectly by measurements of members of the uranium-238 series. Because of the relative ease of measuring gamma rays, the indirect methods may be more appropriate for large-scale surveys intended to characterize the contribution of soil radon by geographic area. Gamma-ray source measurements may also be less sensitive to changes in pressure, temperature, and moisture content than emanation-rate measurements (see UNSCEAR[49] and NCRP[34]). Moreover, variations in emanation rate may correlate with factors that affect air-exchange rates and may thus complicate assessment of the importance of soil as a source of indoor radon.

The mechanisms by which radon may be transported into buildings have been studied little. Soil-gas measurements, which have yielded results of 100-2,000 pCi/L (Kraner;[26] Scott;[44] and unpublished measurements by Lawrence Berkeley Laboratory), may be relevant to this question, because they may help in characterizing the radon content of air trapped beneath buildings. Emanation rates themselves are useful only for placing an upper limit on the potential of soil as an indoor source. However, a more detailed understanding of radon transport in soil could provide a basis for using emanation data to estimate the amount of radon that may accumulate beneath houses and be transported indoors. Such collection and transport mechanisms may be greatly affected by changes in barometric pressure, soil moisture content, temperature gradients, and wind.

The actual pathway by which radon enters a building from the soil appears to vary substantially with building design and construction practice. In houses with concrete basements that are closed to the outdoors, radon may enter by diffusion through the basement floor, by convection within basement walls, and by movement through cracks, designed openings, or penetrations in either of these components. Even in communities where numerous measurements have been performed, it has not been possible to determine the relative importance of these mechanisms.[12] In some mining communities, sealing of cracks has proved relatively successful in reducing indoor radon content, but the effectiveness of this method in general has not been evaluated. The movement of radon from the point of entry to other parts of the building depends on internal construction and building use. Even in buildings with ventilated crawlspaces, the radon concentration in the crawlspace air may be considerably higher than outdoors, and a substantial amount of the radon emanating from the soil may reach the interior space by transport from the crawlspace.

More comprehensive information on how radon is transported is needed for the development of techniques to prevent radon from entering buildings and for establishment of a correlation between the radium-226 content in soil and the indoor radon content attributable to this source.

Gamma Radiation. The gamma dose from radionuclides in soil may be expressed in a fashion analagous to that for building materials; the air dose rate (μrad/h) at 1 m above the ground due to natural emitters uniformly distributed in the soil has been given as $\dot{X}_{plane} = 1.82C_U + 2.82C_{Th} + 0.179C_K$ for C_U, C_{Th}, C_K in picocuries per gram.[4] More current data on decay schemes may alter this slightly. As noted above, concentrations of natural radionuclides in soil and rock vary from place to place, causing comparable variations in dose rates. The air dose rate is estimated at 2.6 μrads/h on the coastal plain (the Atlantic and Gulf coastal areas), 10.2 μrads/h on the Colorado Plateau, and 5.2 μrads/h on the rest of the contiguous United States (NCRP,[34] based on nuclear-plant site surveys). The materials in a building can provide significant shielding of occupants from gamma rays from local radionuclide concentrations, but the radionuclide content of the materials may more than compensate for this shielding.

Radon from Utilities

Water. Measured concentrations of radon-222 in well water in Maine and New Hampshire average 53,000 and 101,000 pCi/L, respectively.[31] More recent measurements have been performed in Maine.[12,14] Lawrence Berkeley Laboratory has found concentrations of 100-7,500 pCi/L in tapwater from wells or underground reservoirs associated with houses and has correlated use of such water with increases in indoor radon content. Radon-222 in water can quickly transfer to air, with efficiencies of 30-90%, depending on water use;[11] a concentration of 10,000 pCi/L can raise average indoor radon-222 content by about 1 nCi/m^3. It is not known how widespread such water concentrations are, nor how closely they correlate with high radium content in surface soils and rocks.

Natural Gas. Concentrations of radon-222 in natural gas in the Houston area have been found to average approximately 50 pCi/L at STP.[10] Concentrations in distribution lines at various points in the United States were found to average about 20 pCi/L.[22] The resulting concentrations in U.S. residences due to natural-gas combustion have been estimated to be less than 0.1 nCi/m^3, even with unvented burners.

INDOOR CONCENTRATIONS AND RADIATION FLUXES

Airborne Radionuclides

Radon Concentrations. Data from several sources tabulated by UNSCEAR[49] indicated that indoor radon-222 concentrations vary by two orders of magnitude, with average values of about 1 nCi/m^3. Such a large range is not surprising, considering that the studies included various types of buildings, building materials, underlying materials, and ventilation conditions and used many measurement techniques. More recent measurements have confirmed this wide variation.

A wide variation is expected even for conventional housing, because the air-exchange rate typically ranges from about 0.5/h to 1.5/h in such buildings, and further variation in air-exchange rates occurs because of window or door openings and mechanical ventilation systems. The soil under a house can be expected to be the principal contributor to the indoor radon concentration in most cases. As noted earlier, a typical soil emanation rate, if the radon goes into the interior of a house with an air exchange rate of 1/h, would contribute radon-222 at about 1 nCi/m^3. Inasmuch as soil emanation rates and effective capture by the house vary by an order of magnitude and air-exchange rates vary widely, a large range of indoor concentrations would result.

As indicated in Table IV-3, homes monitored in New York and New Jersey were found to have an annual average radon-222 concentration of 0.3-3.1 nCi/m^3 in the living space, with a geometric mean of about 0.8 nCi/m^3.[9] Similar measurements in Austria yielded a geometric mean of 0.42 nCi/m^3.[46] In these studies the mean indoor concentrations were 3-4 times as great as local outdoor concentrations. Spot measurements of homes in the San Francisco area, made during the summer with windows closed and with an average air-exchange rate of 0.4/h, showed concentrations averaging 0.3 nCi/m^3.[15] Spot measurements in Illinois showed a substantial incidence of concentrations greater than 5 nCi/m^3; six of 22 houses had concentrations of 10 nCi/m^3 or more.[43]

High radon-222 concentrations have been found in uranium-mining areas and in buildings that use materials high in radium. In houses monitored in Bancroft, Ontario, 50% of the sample had concentrations greater than 3 nCi/m^3, over 25% had concentrations greater than 7 nCi/m^3, and about 6% had concentrations greater than 15 nCi/m^3.[21] High concentrations have also been found in homes in mining areas in the United States; at Grand Junction, Colorado, PAECs corresponding to radon-222 at up to hundreds of nanocuries per cubic meter have been measured. In a survey of several Swedish houses built with alum-shale concrete, the average radon-222 concentration was 7 nCi/m^3;[48] more recent data showed average concentrations of 15 nCi/m^3 or more for residences built entirely of alum shale.[47]

Radon-222 concentrations of 0.6-22 nCi/m^3 have been found during spot measurements of energy-efficient homes, many of which had low air-exchange rates; these measurements were taken with windows closed, and the air-exchange rates were measured simultaneously.[16] Concentrations and air-exchange rates have also been measured in

TABLE IV-3

Selected Radon and Radon-Progeny Measurements in U.S. Residences[a]

Location	Radon-222 Concentration,[b] nCi/m^3	Progeny PAEC Concentration,[b] WL	No. Residences	Type of Measurement	Comments
Ordinary Areas:					
Tennessee	--	0.008(0.0008-0.03)	15	Grab	Shale area; mostly concrete construction
Boston	0.07(0.005-0.2)	(up to 0.002)	7	Grab and ventilation	Single family; air change rate, 1-6/h
New York-New Jersey	0.8[c](0.3-3.1)	0.004[c](0.002-0.013)	21	Several integrated measurements over year	17 single family, 3 multiple family, 1 apartment building
Illinois	(0.3-33)	--	22	Grab	Wood-frame construction, unpaved crawl spaces (windows closed)
San Francisco area	(0.4-0.8)	--	26	Grab and ventilation	Air change rate, 0.02-1.0/h (windows closed)
U.S.-Canada	(0.6-22)	--	17	Grab and ventilation	Energy-efficient houses; air change rate, 0.04-1.0/h (windows closed)
Special Areas:					
Grand Junction, Colorado	--	0.006[c]	29	Integrated year round	Controls for remedial-action program (which has included houses in range 0.02-1 WL)

Table IV-3 (contd)

Location	Radon-222 Concentration,[b] nCi/m^3	Progeny PAEC Concentration,[b] WL	No. Residences	Type of Measurement	Comments
Florida	--	0.004	28	Integrated year round	Controls on unmineralized soils
	--	0.004(0.0007-0.014)	26	Integrated year round	Controls on unmineralized soils
	--	0.014	133	Integrated year round	Houses on reclaimed phosphate lands
Butte, Montana	--	0.02	56	Integrated year round	Intensive mining area
Anaconda, Montana	--	0.013	16	Integrated year round	Intensive mining area

[a]Data from W. J. Barnes, personal communication, cited in George and Breslin.[9] Single-family residences except where noted.

[b]Averages; values in parentheses are ranges. All measurements are in living space; values in basements are typically higher.

[c]Geometric mean.

conventional houses in England[5] and in houses at Elliot Lake, Ontario;[45] the measured radon concentrations were consistent with those observed for conventional houses elsewhere.

Radon-Progeny Concentrations. Radon-progeny concentrations are often measured as potential alpha-energy concentrations (PAEC), given in working levels. Indoor concentrations of radon-222 progeny were measured at the Environmental Measurements Laboratory (EML) in New York City.[8] The concentrations were 0.02 WL in the EML basement and about 0.01 WL in the building's fifth floor, both with progeny activity ratios of polonium-218, lead-214, and bismuth-214 of about 1:0.5:0.3. For 21 New York and New Jersey houses, the mean annual-average PAEC for progeny of radon-222 was about 0.004 WL in the living space, with a range of values from one house to another of 0.002-0.013 WL; equilibrium factors averaged slightly above 0.6 in the living space.[8]

Measurements in Florida houses built on reclaimed phosphate land yielded average radon-progeny concentrations of about 0.01 WL, but the range extended to above 0.05 WL.[13] Houses in Grand Junction, Colorado, in which remedial action has been recommended, had PAECs ranging from 0.02 to 1 WL. Sets of control houses monitored in Florida and Colorado had an average PAEC similar to that in New York and New Jersey (see Table IV-3). Measurements have also been performed in homes in the vicinity of uranium-mining operations.[1,2]

A few measurements of individual radon-progeny concentrations have been made, often to correlate such concentrations with possible removal processes.[17,36] These processes constitute potential control techniques.

Some work has been done on characterizing the particle size distribution of indoor radon progeny, as well as the dependence of concentrations and distributions on various characteristics, including location, particulate mass concentration, air-exchange rate, and air-mixing rate. The fraction of radon progeny that is unattached to particles, as well as the size distribution of attached progeny, was measured at the EML building and in homes.[8,9] Such measurements have also been performed in uranium mines. The diffusion coefficients of radon progeny have been measured,[41] and their interactions with particles have been examined theoretically.[40]

Lawrence Berkeley Laboratory has performed a few measurements of radon progeny in solar homes in New Mexico and found PAECs of about 0.005 WL (J.G. Ingersoll, personal communication).

The simplest models of indoor radon and radon-progeny concentrations use a set of simple equations connecting the indoor radon source strength, outdoor concentrations, and rate of air exchange, assumed to be the only removal mechanism (other than radioactive decay). An example is a computer program of Kusuda,[28] which permits step variations in air-exchange rate. Models may also simulate diffusion of radon into a house,[33] but transport has not been modeled in any comprehensive way. Models have been made of radon-progeny diffusion and attachment processes[40,51] and of the effect of such processes on progeny concentrations and unattached fractions.[17,20,39] However, no realistic models of radon and progeny behavior in buildings, by which actual concentrations (or the effect of control measures) might be simulated, has been attempted. More experimental information will evidently be required to develop and validate such models.

Gamma-Radiation Fluxes and Shielding Effects from Building Materials

As discussed above, gamma radiation from terrestrial radionuclides may arise from both building materials and nearby soil and rock, although the radionuclide content of these two sources may vary significantly. Moreover, the structural materials shield occupants both from gamma rays from soil and rock and, to a lesser degree, from cosmic rays. As a result, the building may affect external dose rates of occupants in various ways and degrees. Given information on a particular building, the net effect may be calculated in a way similar to that used by Moeller et al.,[33] based on the gamma-ray dose-rate expressions given above and on estimation of shielding effects.

In some cases, the structure may have little effect on terrestrial or cosmic dose rates. Exclusive use of materials that do not contain substantial radioactivity, such as wood, has the effect of shielding the terrestrial gamma flux (tissue dose, about 35 mrems/yr) by about 20 or 30% and has little effect on the cosmic-ray dose (about 28 mrems/yr). A concrete foundation (slab floor or basement) would have no effect on the cosmic-ray dose and, if its radionuclide content were similar to that of surrounding soil or rock, little effect on the terrestrial dose. That is, although concrete substantially attenuates gamma radiation from the soil or rock, it contributes a gamma-ray flux that compensates for this reduction.

However, if a building also uses concrete in the walls and ceilings and has a radionuclide content similar to that of local soil and rock, an approximate doubling of the terrestrial dose rate would occur. As some compensation, concrete walls and ceilings would tend to shield occupants from cosmic rays in many cases by only about 20%, but by larger fractions for large buildings.

Ordinarily, then, building materials with crustal components whose radionuclide contents are similar to those of local soil and rock may increase external dose rates for occupants by up to tens of millirems per year or may decrease cosmic-ray rates by a somewhat smaller amount. For building materials and surrounding soil or rock that contain higher radionuclide contents, the dose-rate differences between outdoors and indoors would be correspondingly larger.

CONTROL TECHNIQUES

From the few available indoor measurements of radon-222 progeny, it appears that variations of 0.01 WL from one building to another, depending on air-exchange rates and on building or ground materials, are not unusual. The full range of values for conventional houses is considerably larger than this, and measures that reduce the air-exchange rate can be expected to change it further. A progeny concentration of 0.01 WL, if experienced two-thirds of the time, corresponds to an exposure of about 0.3 WLM/yr--less by about a factor of 10 than the occupational limit of 4 WLM/yr. (Exposure of a person to 1 WL for 170 h, a working month, yields one working-level month, or 1 WLM.) But variations in external dose rate due to ordinary building materials are around 10 mrems/yr, less than one-hundredth of the whole-body occupational dose limit of 5 rems/yr. If these occupational limits correspond to similarly valued risks, it appears that the effect of the structure on radon-progeny exposures (given in working-level months per year) is far more important than the effect on external whole-body dose rates (given in rems per year). Health effects are discussed elsewhere, but this simple comparison indicates one basis for emphasizing methods for controlling radon-progeny exposures. Of these methods, only material substitution may be used for control of gamma-ray dose rates, particularly where materials have unusually high radionuclide contents.

Techniques for controlling indoor concentrations of radon-222 or its progeny include measures that decrease radon sources, reduce transport from sources, remove radon or its progeny from indoor air, or exchange indoor air for outdoor air. The easiest technique to implement in many cases is to increase the air-exchange rate--for example, by opening windows or installing fans. For reasons of comfort or energy efficiency, other methods, sometimes equally straightforward, may often be preferable. In general, not enough is known about the cost, effectiveness, and applicability of various measures for a judgment of their importance in the general building stock.

Material Selection or Site Preparation

In construction of a building, the use of materials whose radon-222 emanation rates are low affects the source strength directly. However, in situations where the surrounding soil and rock contribute most of the radon, opportunities for controlling the source strength are limited, especially because the diffusion length of radon-222 is relatively large and radon source strength is not often a criterion for site selection. Attention to building materials or site materials (underlying and surrounding soil) in new construction has a substantial effect in cases where the emanation rate from either of these may be unusually high. Replacing such materials (on a remedial basis) is often difficult or expensive, so other measures may be favored.

Reducing Transport

The principal means of reducing the transport of radon to building interiors are the sealing of materials that have high emanation rates and, for the case of transport from surrounding soil, the plugging of cracks and holes through which air with a high radon-222 content (e.g., soil gas) moves. Materials may be sealed by epoxy resins or other coatings with up to 90% effectiveness.[3 6] Sealing surfaces, filling holes with impervious materials, and stopping transport by installing plastic or other barriers have proved effective in some cases that required remedial action (see, for example, Atomic Energy Control Board[1]), but they all require integrity of the barrier for long-term reduction of transport. The general applicability or effectiveness of these measures as long-term passive controls is not known. It should be noted that confinement of radon by diffusion or convection barriers also permits buildup of radon and its progeny behind the barrier, causing an increase in gamma radiation from building materials. Nevertheless, this increase appears less important than the associated decrease in airborne radon-222 and its progeny.[7] Transport may also be reduced by ventilating crawlspaces or basements or (in new construction) by designing transport routes that bypass slab floors or basements.

Removal of Progeny from Indoor Air

Methods for removing radon-222 progeny from indoor air include filtration with fiber, electrostatic, or charcoal filters; mixing of indoor air to cause deposition within the structure or ventilation system; and space-charging to remove progeny ions. Filtration systems are effective in reducing airborne particulate mass concentrations. However, depending on the system, they may thereby raise the concentration of unattached progeny ions, especially polonium-218;[13] for some particle size distributions, this would raise the ratio of lung dose to PAEC. Nazaroff[31] observed a substantial decrease in PAEC from operation of the furnace fan (which thereby activated the system's filter), but the unattached fraction was not measured. Holub et al.[17] and Jonassen[23] have performed related experiments on air-mixing, ventilation, and filtration. Finally, in many measurement techniques, charged radon-222 progeny are collected by voltage differentials, but it does not appear that this principle can easily be applied as a control measure.

Exchange of Indoor and Outdoor Air

Use of air-to-air heat exchangers to remove indoor air while conserving potentially lost energy is being investigated by Lawrence Berkeley Laboratory. Preliminary results[36] indicate that this method is effective, in at least one configuration, in reducing radon-222 and progeny concentrations. This method is particularly attractive because it can be applied in both new and existing buildings and because it is effective in reducing concentrations of other indoor contaminants.

RESEARCH NEEDS

Substantial research efforts are needed in three subjects: the characterization of radon sources and of the indoor concentrations and behavior of radon and its progeny, the development and testing of control techniques, and the modeling of radon and its progeny in structures. These efforts need to be supported by development of measurement instrumentation, followed by an evaluation of indoor concentrations, control measures, and building energy-conservation measures, among other factors. In addition, evaluative efforts will require further work on the health effects of radon, which have not been discussed here.

Programs to characterize building materials by radon emanation rate or radionuclide content should be more widespread and complete. It is even more important to survey soil and groundwater with respect to radionuclide content, radon emanation, and radon transport. A rapid effort should be undertaken to determine the feasibility of geologic or geographic characterization of soil. As part of efforts to characterize materials, attention should be given to the effects of moisture, pressure, and temperature. Community water supplies should also be surveyed.

Studies of indoor radon and progeny concentrations should be undertaken with two major purposes: to learn the range and distribution of radon and its progeny in the building stock, and to understand the behavior of radon and its progeny in buildings. The first purpose requires surveys of many buildings of a variety of types and in various geographic areas. These surveys may be implemented by associating them with other large-scale efforts, such as those for energy-conservation retrofits or for insurance purposes. They may measure either radon concentrations or potential alpha-energy concentrations (PAECs), but the former may be measured more easily and may in fact be preferable, in that an adequate understanding of progeny behavior could be used to infer PAECs in a way that lends itself to generalization.

This interpretative basis must be developed through intensive measurements to characterize indoor radon and progeny behavior. Such intensive work at only a few sites would serve as a basis not only for improving measurement techniques but also for developing control techniques. Particular attention must be given to progeny-particle interactions and removal processes. Results of intensive investigations would be validated by less-detailed field measurements at a larger number of sites. Ultimately, the results would serve as a basis for estimates of health effects.

Many measurement programs will have to be supported by instrumentation development. More convenient portable instruments for field source measurements based on alpha-scintillation techniques or on sodium iodide gamma-ray detectors could be developed. Further work on integrating devices for large-scale surveys of indoor concentrations is warranted, as is development of simple and quick progeny monitors, presumably based on semiconductor detectors. For intensive investigation of progeny behavior at a few sites, more versatile special-purpose systems must be designed to measure infiltration rate, radon, individual radon progeny, particle concentrations, and environmental conditions automatically.

Substantial efforts to develop and study control techniques are required. The effects of techniques to clean the air (rather than control the source) would have to be studied in the manner indicated above for detailed investigations of progeny behavior.

These measurement programs must be accompanied by corresponding modeling efforts. Models that characterize sources (on a geologic and geographic basis) and transport (by site and building type) are needed. Although models for physical processes involving radon progeny have begun to be developed, much more work is needed, especially for understanding progeny-particle interactions and control techniques. Models of indoor-air quality that use the source and progeny models appropriately could then be developed. Finally, the models of indoor-air quality could be combined with models of the building stock to represent current radon and progeny concentrations and the effects of changes in building design and of potential control measures. Models of indoor-air quality and the building stock will be necessary for any indoor air pollutant and for evaluation of potential strategies for controlling indoor-air quality.

REFERENCES

1. Atomic Energy Control Board [Canada]. Workshop on Radon and Radon Daughters in Urban Communities Associated with Uranium Mining and Processing, Elliot Lake, Ontario, March 7, 1978. Ottawa, Ont., Canada: Atomic Energy Control Board, 1979.
2. Atomic Energy Control Board [Canada]. Second Workshop on Radon and Radon Daughters in Urban Communities Associated with Uranium Mining and Processing, Bancroft, Ontario, March 12-14, 1979. Ottawa, Ont., Canada: Atomic Energy Control Board, 1980.
3. Auxier, J. A., W. H. Shinpaugh, G. D. Kerr, and D. J. Christian. Preliminary studies of the effects of sealants on radon emanation from concrete. Health Phys. 27:390-392, 1974.
4. Beck, H. L., J. A. DeCampo, and C. V. Gogolak. In Situ Ge(Li) and NaI(Tl) Gamma-Ray Spectrometry. U.S. Department of Energy, Health and Safety Laboratory Report HASL-258. Washington, D.C.: U.S. Department of Energy, 1972. Available from National Technical Information Service, Springfield, Va., as HASL-258.
5. Cliff, K. D. Assessment of airborne radon daughter concentrations in dwellings in Great Britain. Phys. Med. Biol. 23:696-711, 1978.
6. Culot, M. V. J., K. J. Schiager, and H. G. Olson. Radon Progeny Control in Buildings. Final Report. U.S. Atomic Energy Commission Report COO-22734. Fort Collins, Col.: Colorado State University, 1973. 277 pp.
7. Culot, M. V. J., K. J. Schiager, and H. G. Olson. Prediction of increased gamma fields after application of a radon barrier on concrete surfaces. Health Phys. 30:471-478, 1976.
8. George, A. C. Indoor and outdoor measurements of natural radon and radon daughter decay products in New York City air, pp. 741-750. In J. A. S. Adams, W. M. Lowder, and T. F. Gesell, Eds. The Natural

Radiation Environment II. Proceedings of the Second International Symposium on the Natural Radiation Environment, August 7-11, 1972, Houston, Texas, U.S.A. Available from National Technical Information Service, Springfield, Va., as CONF-720805-P-2.

9. George, A. C., and A. J. Breslin. The distribution of ambient radon and radon daughters in residential buildings in the New Jersey-New York area, pp. 1272-1292 (includes discussion). In T. F. Gesell and W. M. Lowder, Eds. Natural Radiation Environment III. Vol. 2. Proceedings of a Symposium Held at Houston, Texas, April 23-28, 1978. Oak Ridge, Tenn.: U.S. Department of Energy, Technical Information Center, 1980.

10. Gesell, T. F. Some radiological health aspects of radon-222 in liquified petroleum gas, pp. 612-629. In R. E. Stanley and A. A. Moghissi, Eds. Noble Gases. U.S. Energy Research and Development Administration Report CONF-730915. Washington, D.C.: U.S. Government Printing Office, 1975.

11. Gesell, T. F., and H. M. Prichard. The contribution of radon in tap water to indoor radon concentrations, pp. 1347-1363. In Natural Radiation Environment III. Vol. 2. Proceedings of a Symposium Held at Houston, Texas, April 23-28, 1978. Oak Ridge, Tenn.: U.S. Department of Energy, Technical Information Center, 1980.

12. Gesell, T. F., H. M. Prichard, and C. T. Hess. Epidemiologic Implications of Radon in Public Water Supplies. Paper presented at Specialist Meeting on the Assessment of Radon and Daughter Exposure and Related Biological Effects, Rome, Italy, March 3-7, 1980.

13. Guimond, R. J., Jr., W. H. Ellett, J. E. Fitzgerald, Jr., S. T. Windham, and P. A. Cuny. Indoor Radiation Exposure due to Radium-226 in Florida Phosphate Lands. Washington, D.C.: U.S. Environmental Protection Agency Report No. EPA 520/4-78-013. Revised printing. Washington, D.C.: U.S. Government Printing Office, 1979. [211] pp.

14. Hess, C. T., R. E. Casparius, S. A. Norton, and W. F. Brutsaert. Investigations of natural levels of radon-222 in groundwater in Maine for assessment of related health effects, pp. 529-546. In Natural Radiation Environment III. Vol. 1. Proceedings of a Symposium Held at Houston, Texas, April 23-28, 1978. Oak Ridge, Tenn.: U.S. Department of Energy, Technical Information Center, 1980.

15. Hollowell, C. D., et al. Building Ventilation and Indoor Air Quality. Annual Report. Lawrence Berkeley Laboratory Report No. LBL 10390. Berkeley, Cal.: Lawrence Berkeley Laboratory, 1979.

16. Hollowell, C. D., J. V. Berk, M. L. Boegel, P. A. Hillis, J. G. Ingersoll, D. L. Krinkel, and W. W. Nazaroff. Radon in Energy Efficient Residences. Lawrence Berkeley Laboratory Report LBL-9560. Berkeley, Cal.: Lawrence Berkeley Laboratory, 1980.

17. Holub, R. F., R. F. Droullard, W.-L. Ho, P. K. Hopke, R. Parsley, and J. J. Stukel. The reduction of airborne radon daughter concentration by plateout on an air-mixing fan. Health Phys. 36:497-504, 1979.

18. Ingersoll, J. G., B. D. Stitt, and G. H. Zapalac. A Survey of Radionuclide Contents and Radon Emanation Rates in U.S. Building Materials. Lawrence Berkeley Laboratory Report LBL-11771. Berkeley: Lawrence Berkeley Laboratory, University of California, 1981.

19. International Commission on Radiological Protection. Recommendations of the International Commission on Radiological Protection. New York: Pergamon Press, 1977.

20. Jacobi, W. Activity and potential α-energy of ^{222}radon and ^{222}radon-daughters in different air atmospheres. Health Phys. 22:441-450, 1972.

21. James F. MacLaren Limited. Investigation and Implementation of Remedial Measures for the Reduction of Radioactivity Found in Bancroft, Ontario, and Its Environs. Report to Atomic Energy Control Board [Canada]. Willowdale, Ont., Canada: James F. MacLaren Limited, 1979. 104 pp. (unpublished)

22. Johnson, R. H., Jr., D. E. Bernhardt, N. S. Nelson, and H. W. Galley. Radiological health significance of radon in natural gas, pp. 532-539. In R. E. Stanley, and A. A. Moghissi, Eds. Noble Gases. U.S. Energy Research and Development Administration Report CONF-730915. Washington, D.C.: U.S. Government Printing Office, 1975.

23. Jonassen, N. Measurement of Radon and Radon Daughters. Paper presented at Specialist Meeting on the Assessment of Radon and Radon Daughter Exposure and Related Biological Effects, Rome, Italy, March 3-7, 1980.

24. Jonassen, N., and J. P. McLaughlin. Exhalation of radon-222 from building materials and walls, pp. 1211-1224. In T. F. Gesell and W. M. Lowder, Eds. Natural Radiation Environment III. Vol. 2. Proceedings of a Symposium Held at Houston, Texas, April 23-28, 1978. Oak Ridge, Tenn.: U.S. Department of Energy, Technical Information Center, 1980.

25. Kahn, B., G. G. Eichholz, and F. J. Clarke. Assessment of the Critical Populations at Risk due to Radiation Exposure in Structures. Atlanta: Georgia Institute of Technology, School of Nuclear Engineering, 1979.

26. Kraner, H. W., G. L. Schroeder, and R. D. Evans. Measurements of the effects of atmospheric variables on radon-222 flux and soil-gas concentrations, pp. 191-215. In J. A. S. Adams, and W. M. Lowder, Eds. The Natural Radiation Environment. Chicago: University of Chicago Press, 1964.

27. Krisiuk, E. M., E. P. Lisachenko, S. I. Tarasov, V. P. Shamov, and N. I. Shalak. A Study on Radioactivity in Building Materials. Leningrad: Ministry of Public Health of the U.S.S.R., Leningrad Research Institute for Radiation Hygiene, 1971.

28. Kusuda, T., S. Silberstein, and P. E. McNall. Modeling of radon and its daughter concentrations in ventilated spaces. J. Air Pollut. Control Assoc. 30:1201-1207, 1980.

29. Lloyd, R. D. Gamma-ray emitters in concrete. Health Phys. 31:71-73, 1976.

30. Lowder, W. M., W. J. Condon, and H. L. Beck. Field spectrometric investigations of environmental radiation in the U.S.A. In J. A. S.

Adams, and W. M. Lowder, Eds. The Natural Radiation Environment. Chicago: University of Chicago Press, 1964.

31. Lucas, H. F. A fast and accurate survey technique for both radon-222 and radium-226. In J. A. S. Adams, and W. M. Lowder, Eds. The Natural Radiation Environment. Chicago: University of Chicago Press, 1964.
32. McLaughin, J. P., and N. Jonassen. The effect of pressure drops on radon exhalation from walls, pp. 1225-1236. In T. F. Gesell and W. M. Lowder, Eds. Natural Radiation Environment III. Proceedings of a Symposium Held at Houston, Texas, April 23-28, 1978. Oak Ridge, Tenn.: U.S. Department of Energy, Technical Information Center, 1980.
33. Moeller, D. W., D. W. Underhill, and G. V. Gulezian. Population dose equivalent from naturally occurring radionuclides in building materials, pp. 1424-1443. In T. F. Gesell and W. M. Lowder, Eds. Natural Radiation Environment III. Vol. 2. Proceedings of a Symposium Held at Houston, Texas, April 23-28, 1978. Oak Ridge, Tenn.: U.S. Department of Energy, Technical Information Center, 1980.
34. National Council on Radiation Protection and Measurements. Natural Background Radiation in the United States. NCRPM Report No. 45. Washington, D.C.: National Council on Radiation Protection and Measurements, 1975. 163 pp.
35. National Research Council, Committee on the Biological Effects of Ionizing Radiations. The Effects on Populations of Exposure to Low Levels of Ionizing Radiation: 1980. Washington, D.C.: National Academy Press, 1980. 524 pp.
36. Nazaroff, W. W., M. L. Boegel, C. D. Hollowell, and A. D. Roseme. The Use of Mechanical Ventilation with Heat Recovery for Controlling Radon and Radon-Daughter Concentrations. Lawrence Berkeley Laboratory Report LBL-10222. Paper presented at Third Workshop on Radon and Radon Daughters in Urban Communities Associated with Uranium Mining and Processing, Port Hope, Ontario, Canada, March 12-14, 1980.
37. Nuclear Energy Agency. Exposure to Radiation from the Natural Radioactivity in Building Materials. Paris: Organization for Economic Cooperation and Development, Nuclear Energy Agency, 1979.
38. Oakley, D. T. Natural Radiation Exposure in the United States. U.S. Environmental Protection Agency Report ORP/SID 72-1. Washington, D.C.: U.S. Environmental Protection Agency, Office of Radiation Programs, Surveillance and Inspection Division, 1972. 77 pp. Available from National Technical Information Service, Springfield, Va., as PB-235 795.
39. Porstendörfer, J., A. Wicke, and A. Schraub. The influence of exhalation, ventilation, and deposition processes upon the concentrations of radon (Rn-222), thoron (Th-222), and their decay products in room air. Health Phys. 34:465-473, 1978.
40. Raabe, O. G. Concerning the interactions that occur between radon decay products and aerosols. Health Phys. 17:177-185, 1969.
41. Raghunath, B., and P. Kotrappa. Diffusion coefficients of decay products of radon and thoron. J. Aerosol Sci. 10:133-138, 1979.

42. Roessler, C. E., Z. A. Smith, W. E. Bolch, and R. J. Prince. Uranium and radium-226 in Florida phosphate materials. Health Phys. 37:269-277, 1979.
43. Rundo, J., F. Markun, and N. J. Plondke. Observation of high concentrations of radon in certain houses. Health Phys. 36:729-739, 1979.
44. Scott, A. G. The source of radon in Elliot Lake. In Workshop on Radon and Radon Daughters in Urban Communities Associated with Uranium Mining and Processing, Elliot Lake, Ontario, March 7, 1978. Ottawa, Ont., Canada: Atomic Energy Control Board, 1979.
45. Smith, D. Ventilation rates and their influence on equilibrium fraction. In Second Workshop on Radon and Radon Daughters in Urban Communities Associated with Uranium Mining and Processing, Bancroft, Ontario, March 12-14, 1979. Ottawa, Ont., Canada: Atomic Energy Control Board, 1979.
46. Steinhäusler, F., W. Hofmann, E. Pohl, and J. Pohl-Rüling. Local and temporal distribution pattern of radon and daughters in an urban environment and determination of organ dose frequency distributions with demoscopical methods, pp. 1145-1162 (includes discussion). In T. F. Gesell and W. M. Lowder, Eds. Natural Radiation Environment III. Vol. 2. Proceedings of a Symposium Held at Houston, Texas, April 23-28, 1978. Oak Ridge, Tenn.: U.S. Department of Energy, Technical Information Center, 1980.
47. Swedjemark, G. A. Radioactivity in Houses Built of Aerated Concrete Based on Alum Shale. Paper presented at Specialist Meeting on the Assessment of Radon and Radon Daughter Exposure and Related Biological Effects, Rome, Italy, March 3-7, 1980.
47. Swedjemark, G. A. Radon in dwellings in Sweden, pp. 1237-1259 (includes discussion). In T. F. Gesell and W. M. Lowder, Eds. Natural Radiation Environment III. Vol. 2. Proceedings of a Symposium Held at Houston, Texas, April 23-28, 1978. Oak Ridge, Tenn.: U.S. Department of Energy, Technical Information Center, 1980.
49. United Nations Scientific Committee on the Effects of Atomic Radiation. Sources and Effects of Ionizing Radiation. Report to the General Assembly, with Annexes. New York: United Nations, 1977. 725 pp.
50. Wilkening, M. H., W. E. Clements, and D. Stanley. Radon 222 flux measurements in widely separated regions, pp. 717-730. In J. A. S. Adams, W. M. Lowder, and T. F. Gesell, Eds. The Natural Radiation Environment II. Proceedings of the Second International Symposium on the Natural Radiation Environment, August 7-11, 1972, Houston, Texas, U.S.A.
51. Wrenn, M. E., M. Eisenbud, C. Costa-Ribeiro, A. J. Hazle, and R. D. Siek. Reduction of radon daughter concentrations in mines by rapid mixing without makeup air. Health Phys. 17:405-414, 1969.

FORMALDEHYDE AND OTHER ORGANIC SUBSTANCES

The infiltration of outdoor air is one source of formaldehyde and other organic substances in the indoor environment, but the primary sources are in the indoor environment itself--building materials, combustion appliances, tobacco smoke, and a large variety of consumer products. A buildup of formaldehyde may be exacerbated in buildings that have been subjected to energy-efficiency measures intended to reduce infiltration and, thus, energy consumption. Emission rates for formaldehyde and other organic pollutants emitted in the indoor environment are generally unknown. Analytical techniques have been applied mostly in measuring concentrations in indoor air. Very little work has been done in measuring specific source strengths or ranking emission sources, except on a broad relative basis.

FORMALDEHYDE

In general, indoor formaldehyde concentrations exceed those outdoors. The contribution of formaldehyde in outdoor air to indoor air appears to be minor.

A recent National Research Council report deals in great detail with the sources and effects of formaldehyde and other aldehydes.[15] The reader is referred to that report for a more comprehensive treatment than is feasible here.

Sources and Emission Rates

Insulation. Urea-formaldehyde (UF) foam is used as thermal insulation in the side walls of existing buildings, especially single-family residential buildings. UF foam is a convenient substance for retrofitting existing buildings, because it is injected directly into wall cavities through small holes that are then sealed up.

UF foams were developed in 1933 and first used as an insulating material in the 1960s. UF foam has been used for thermal insulation in Europe for many years, but is relatively new in the United States. In the early 1970s, interest in and use of this material increased as the cost of energy mounted and resulted in increased demand for residential insulation. A production peak occurred in 1977, when the demand for insulating products created shortages of other insulating materials, such as cellulose and fiberglass.[19] Approximately 170,000 houses were insulated in 1977. The need for efficient thermal insulation in housing has increased dramatically in the last few years. About 150,000 houses a year are now being insulated with UF foam. Industry representatives reportedly believe that more than 200,000 homes will be insulated with UF foam during 1980.[19]

Installation involves mixing partially polymerized UF resin with a surfactant (foaming agent) and an acid catalyst under pressure that forces air into the mixture to create a foam. The foam hardens within minutes and cures and dries completely within a few days. Building codes in the United States, concerned with the fire-safety aspects of UF-foam insulation, rate it as a combustible material. The codes require that UF foam, when used on the inside of buildings, must be protected by a thermal barrier of fire-resistant material. In England

and Holland, UF insulation materials are certified for use only in masonry cavities of buildings.

If the ingredients of UF-foam insulation are improperly formulated or mixed, formaldehyde may be released into the building. Long et al.[12] enumerated some of the factors that affect the release of formaldehyde from UF foam:

- Excessive formaldehyde in the resin-concentrate solution.
- Excessive acid catalyst in the foaming agent.
- Excess foaming agent (surfactant).
- Foaming during periods of high humidity and high temperature.
- Foaming with cold chemicals (optimal temperature, 50-80°F).
- Improper use of vapor barriers.
- Improper use of foams (in ceilings, etc.).

Owing to the diversity of time factors and the complexity of their interrelationships, the quantity and rate of formaldehyde relesase from a house insulated with UF foam is difficult to predict.

Particleboard and Plywood. The superior bonding properties and low cost of formaldehyde polymers make them the resins of choice for the production of building materials, especially plywood and particleboard. Among the various formaldehyde resins used in building materials--urea-formaldehyde, phenol-formaldehyde, and melamine-formaldehyde--urea-formaldehyde resin is the most common adhesive used in indoor plywood and particleboard. Plywood is composed of several thin sheets of wood glued together with UF resin. Particleboard is made by saturating small wood shavings with UF resin and pressing the resulting mixture, usually at high temperature, into the final form. Particleboard can emit formaldehyde continuously for a long time (several months, or even years). In buildings in which these wood products are used for partition walls or furniture, formaldehyde may reach concentrations sufficient to cause eye and upper respiratory irritation. In cases of extensive use of these products where air-exchange rates are low, the concentration can reach 1 ppm or more. This is due in part to the high surface-to-volume ratio of particleboard and plywood used as building materials. The emission rate depends on a number of factors--the original manufacturing process, quality control of fabrication, porosity, ambient temperature, humidity, cutting of the board for final use, etc.

UF resins contain some free formaldehyde; in addition, the resin may hydrolyze and release free formaldehyde at high temperature and humidity. The phenol-formaldehyde resins used for wood products that require greater moisture resistance (i.e., exterior plywood) do not release formaldehyde as readily as products bound with UF resin. Phenol-formaldehyde resins are not generally used for indoor wood products, because of their higher cost.

Combustion Appliances. Recent studies have reported on combustion-generated indoor air pollutants, specifically contaminants from gas stoves and heating systems in residential buildings.

Laboratory studies have shown that gas stoves emit substantial quantities of aldehydes and that formaldehyde is the major component of the aldehydes measured (Schmidt and Götz;[17] G. W. Traynor, Lawrence Berkeley Laboratory, personal communication). Reported formaldehyde emission from a single gas stove under test conditions has been measured as approximately 15,000 and 25,000 μg/h for each top burner and the oven, respectively (Traynor, personal communication).

Tobacco Smoke. Tobacco smoke is a source of several chemicals, including several aldehydes other than formaldehyde. It may contribute formaldehyde to the indoor environment. The smoker's exposure to these chemicals results principally from smoke inhaled directly into the lungs (mainstream smoke). The smoke that is not directly inhaled into the lungs enters the space surrounding the smoke (sidestream smoke). It is the sidestream smoke that is the major contributor to indoor pollution.

Analysis by Osborne *et al.*[16] indicated that acrolein was an important component of tobacco smoke; this finding was confirmed by Jermini *et al.*,[9] whose studies were conducted on a smoking machine in an environmental chamber. Formaldehyde and acetaldehyde have also been identified in cigarette smoke.

Harke *et al.*[7,8] measured concentrations of nicotine, carbon monoxide, acrolein, and other aldehydes (expressed as acetaldehyde) in the air of an unventilated room in which a series of experiments with a smoking machine had been performed. Substantial concentrations of all four compounds were observed with this extremely low ventilation.

Other Sources. Several products that are potential sources of formaldehyde emission are mentioned below. Because there is no information on the rates or quantities of such emission, it is not known which are important sources.

UF resins are used in the paper industry to increase the wet strength of various grades of paper. Paper products typically treated with UF resins are grocery bags, wax paper, facial tissues, napkins, paper towels, and disposable sanitary products.

Formaldehyde polymers are used extensively in the manufacture of floor coverings and as carpet backing. UF resins are used in the textile industry as binders for pigments, fire retardants, or other materials to cloth and to impart stiffness, wrinkle resistance, and water repellency to fabrics.

Fertilizers and pesticides used for commercially grown plants also may use aldehydes and, theoretically, could contribute to the aldehyde content of ambient air locally.

Urea-formaldehyde fertilizers are used not only to obtain a more uniform release rate than is possible with soluble sources of nitrogen but also to minimize the hazards of water pollution by nitrates leached out of the soil. These compounds have been used with field crops, turfgrass, pine seedlings, and geranium. The extent of their indoor use and the amounts and rates of formaldehyde release are unknown.

Formaldehyde is used in numerous places, such as biologic laboratories, hospitals, and hobby and craft areas.

Concentrations

Indoor monitoring data for U.S. homes are few; there are limited monitoring data for European homes, particularly in the northern European countries, and they show higher indoor formaldehyde concentrations than in the United States. Table IV-4 summarizes recent monitoring data. Most of the measurements of organic substances in the indoor environment have been made on aldehydes--specifically on formaldehyde. Studies of the indoor-outdoor relationships of formaldehyde show that indoor concentrations generally exceed outdoor.

There have recently been several studies measuring indoor formaldehyde concentrations in which emission was from particleboard and plywood furnishings and UF-foam insulation in houses. Measurements in Denmark,[2] Sweden (J. Sundell, personal communication; T. Lindvall, personal communication), West Germany (B. Seifert, personal communication; M. Deimel, personal communication), and the United States (P.A. Breysse, personal communication) have shown that indoor concentrations often exceed 0.1 ppm and may even exceed the then-established 8-h time-weighted average safe exposure limit (3 ppm) for workroom air.[23] In the 23 Danish houses, the average formaldehyde concentration (the predominant source was particleboard) was 0.62 mg/m^3 (about 0.5 ppm), and the range was 0.08-2.24 mg/m^3 (about 0.07-1.9 ppm).[2]

As a result of occupants' complaints, formaldehyde was measured in more than 200 mobile homes in the United States; the concentrations reported ranged from 0.03 to 2.4 ppm (about 0.04-2.9 mg/m^3) (Breysse, University of Washington, personal communication). A study of formaldehyde concentrations in new office trailers with air-exchange rates as low as 0.16/h found formaldehyde concentrations of 0.15-0.20 ppm,[5] in contrast with outdoor concentrations of less than 0.01 ppm.

Formaldehyde concentrations build up in mobile homes, not only because of emission from some building materials used in their construction, but also because mobile homes are often more tightly constructed than conventional homes, thus decreasing ventilation.

Aldehydes (measured by the MBTH method) were monitored in a study of 19 homes across the United States.[14] Indoor concentrations of aldehydes were always higher than outdoor concentrations, typically by a factor of 6 and quite often by a factor of about 10. Figure IV-2 shows the data collected in this study from a gas-cooking home with one smoker. Although the source strengths were not determined in this study, the highest concentrations were observed in the two mobile homes; in general, the plywood and particleboard appeared to be the primary sources.

In a more recent study, formaldehyde and total aliphatic aldehydes (formaldehyde plus other aliphatic aldehydes) were measured at several energy-efficient research houses at various geographic locations in the United States.[11] Results showed that, at low infiltration rates (<0.3 air change per hour, or ach), indoor formaldehyde concentrations often exceeded 0.1 ppm, whereas outdoor concentrations typically remained at 0.016 ppm (20 μg/m^3) or less. Typical air-exchange rates for single-family residential buildings are between

TABLE IV-4

Summary of Aldehyde Measurements in Nonoccupational Indoor Environments[a]

Sampling Site	Concentration,[b] ppm Range	Mean	Method of Analysis
Danish residences	1.8 (peak)	--	Unspecified
Netherlands residences built without formaldehyde releasing materials	0.08 (peak)	0.03	Unspecified
Residences in Denmark, Netherlands, and Federal Republic of Germany	2.3 (peak)	0.4	Unspecified
Two mobile homes in Pittsburgh, Pa.	0.1-0.8[b]	0.36	BTH bubblers
Sample residence in Pittsburgh, Pa.	0.5 (peak)[b]	0.15	MBTH bubblers
Mobile homes registering complaints in state of Washington	0-1.77	0.1-0.44	Chromotropic acid (single impinger)
Mobile homes registering complaints in Minnesota	0-3.0	0.4	Chromotropic acid (30-min sample)
Mobile homes registering complaints in Wisconsin	0.02-4.2	0.88	Chromotropic acid
Public buildings and energy-efficient homes (occupied and unoccupied)	0-0.21	--	Pararosaniline and chromotropic acid
	0-0.23[b]	--	MBTH bubblers

[a]Reprinted from National Research Council.[15(p. 5-13)]

[b]Formaldehyde, unless otherwise indicated.

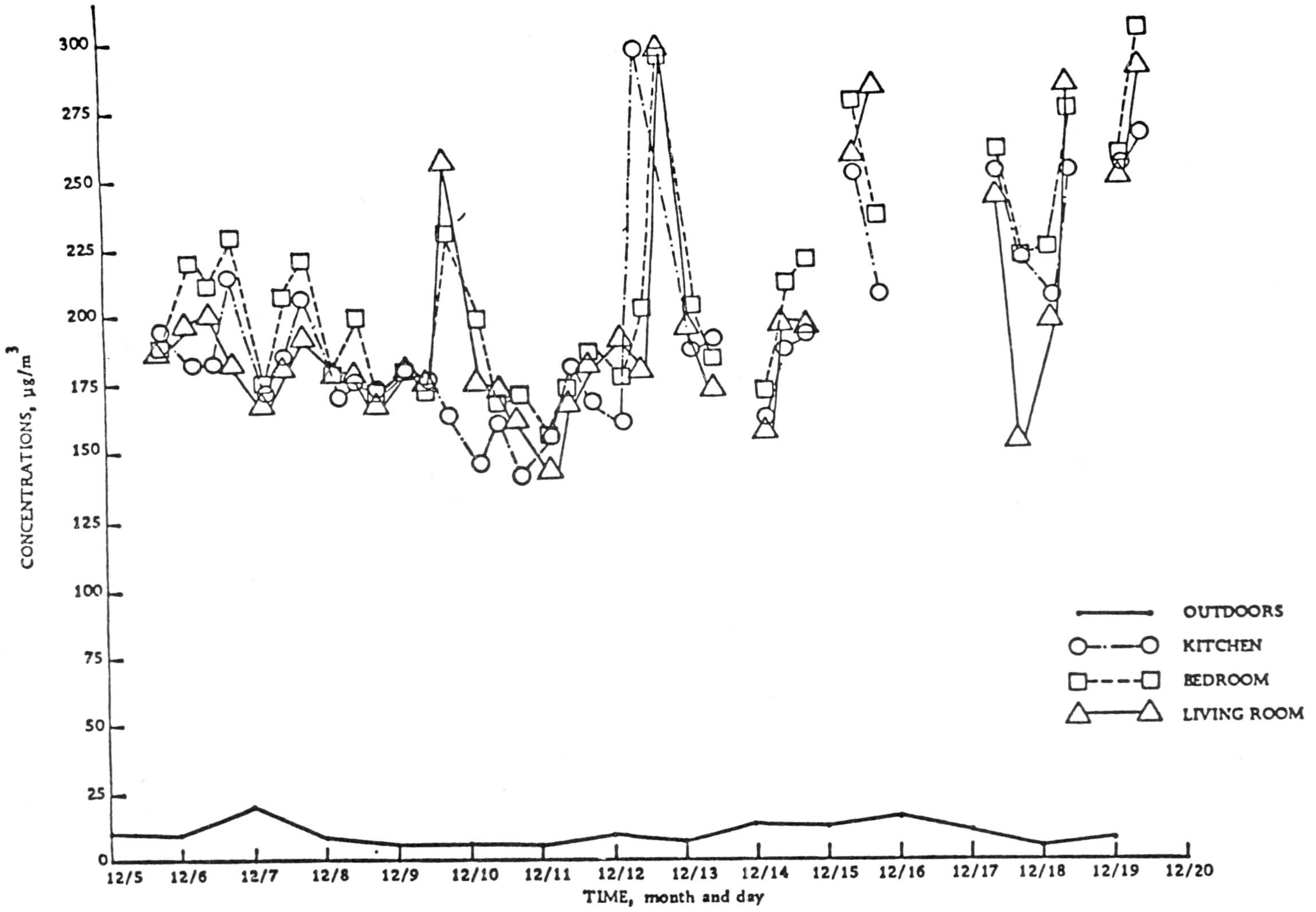

FIGURE IV-2 Diurnal variation in indoor and outdoor aldehyde concentrations at single-family house in Chicago. Reprinted from National Research Council.[15(p. 5-14)]

0.6 and 1 ach. Figure IV-3 is a histogram showing the frequency of occurrence of formaldehyde and total aliphatic aldehyde concentrations measured at an energy-efficient house with an average air-exchange rate of 0.2 ach. Data taken at an energy-efficient house in Mission Viejo, California, are shown in Table IV-5. As shown, when the house did not contain furniture, formaldehyde content was 80 $\mu g/m^3$; when furniture was added, formaldehyde rose almost threefold. A further increase was noted when the house was occupied, very likely because of such activities as gas cooking. When occupants opened windows to increase ventilation, the formaldehyde content dropped substantially. Although high, aldehyde contents observed in the majority of the energy-efficient dwellings monitored were generally below 200 $\mu g/m^3$.

Indoor and outdoor formaldehyde and aldehyde concentrations were found to be about the same at a public school in Columbus, Ohio, and a large medical center in Long Beach, California, and were well below 0.1 ppm (120 $\mu g/m^3$). Both buildings have high ventilation rates; this probably explains the low indoor concentrations, essentially equivalent to outdoor concentrations.

Because many of the data reported from these field-monitoring studies involved houses whose occupants had complained of indoor air quality, these findings may not be representative of all homes. However, when data from a random sample in Wisconsin are compared with those from the Washington mobile-home sample, most of the differences in aldehyde concentrations can be explained by differences in the age of the homes. The mobile homes in the complaint sample are much newer than those in the random sample. Tabershaw Associates[21] analyzed the complaint data for mobile homes in Washington, and found that there was no statistically valid relationship between the severity of symptoms reported by occupants and the concentration of formaldehyde and that, regardless of the actual exposure concentration, all persons in the mobile homes reacted in substantially the same manner.

Foreign (particularly Danish and Swedish) houses monitored for formaldehyde appear to have much higher concentrations than U.S. houses. These findings probably represent differences in construction and, hence, cannot be considered as representative of U.S. houses.

Andersen et al.[2] formulated a mathematical model that estimates the indoor air concentration of formaldehyde. Although use of Danish studies may not be appropriate for U.S. houses, the treatment of Andersen et al. illustrates the many variables that must be considered. By conducting climate-chamber experiments, Andersen et al. found the equilibrium concentration of formaldehyde from particleboard to be related to temperature, water-vapor concentration in the air, ventilation, and the amount of particleboard present. From this work, a mathematical model was established for room-air concentration of formaldehyde.

When this mathematical formulation was applied to the room-sampling results, a correlation coefficient of 0.33 was found between the observed and predicted concentrations--not a particularly good predictive ability. The authors then modified the adjustable constants by calculating them for each room on the basis of monitoring results. The modified values led to a correlation coefficient of 0.88--a considerable improvement in predictive ability.

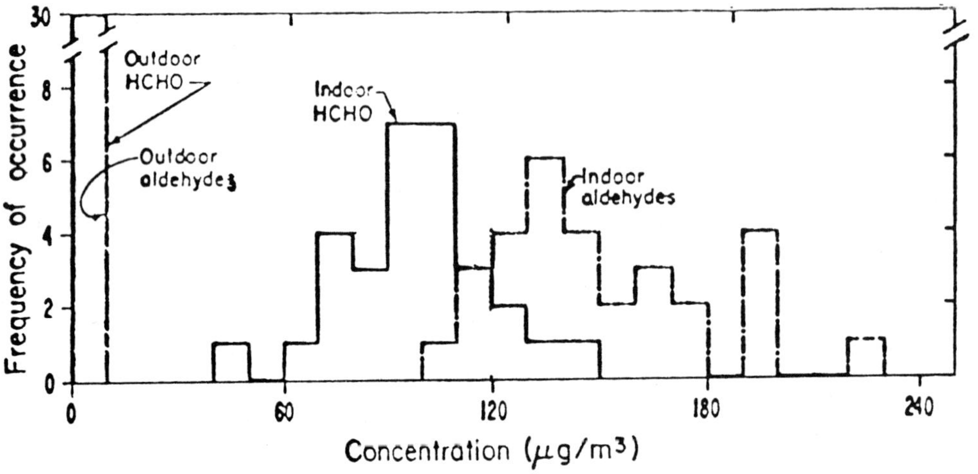

FIGURE IV-3 Indoor and outdoor formaldehyde and other aldehyde concentrations at single-family house in Maryland. Histogram showing the frequency of occurrence of formaldehyde and total aliphatic aldehydes at an energy-efficient house with 0.2 ach. Reprinted from National Research Council.[15(p. 5-17)]

TABLE IV-5

Indoor and Outdoor Formaldehyde and Aliphatic Aldehyde Concentrations Measured at a Single-Family House (California)[a]

Condition	Number of Measurements	Sampling Time	Formaldehyde (μg/m³)[b]	Aliphatic Aldehydes (μg/m³)[c]
Unoccupied, without furniture	3	12	80 ± 9%	90 ± 16%
Unoccupied, with furniture	3	24	223 ± 7%	294 ± 4%
Occupied, day[d]	9	12	261 ± 10%	277 ± 15%
Occupied, night[e]	9	12	140 ± 31%	178 ± 29%

[a]Reprinted from National Research Council.[15(p. 5-18)]

[b]Determined using pararosaniline method (120 μg/m³ ≈ 100 ppb). All outside concentrations <10 μg/m³.

[c]Determined using MBTH method, expressed as equivalents of formaldehyde. All outside concentrations <20 μg/m³.

[d]Air exchange rate ≃ 0.4 ach.

[e]Windows open part of time; air exchange rate significantly greater than 0.4 ach and variable.

Formaldehyde release from interior particleboard occurs at a decreasing rate with an increase in product age. Eventually the rate of formaldehyde evolution decreases to an imperceptible point. The time necessary for this phenomenon to occur (perhaps several years) depends on the atmospheric conditions to which the board has been subjected, as well as the degree of cure of the resin. The more unstable groups degrade first, followed by the more stable free methylol groups.

A 1977 field study in which field tests and a mathematical model were used to determine the half-life of formaldehyde in particleboard typically used in Scandinavian home construction reported it to be about 2 yr when the ventilation rate in the home is 0.3 ach (C.D. Hollowell, personal communication). Suta has analyzed the effect of home age on formaldehyde concentrations in Danish houses. These data give the following relationship of formaldehyde concentrations as a function of house age when no corrections are made for other pertinent factors, such as the amount of particleboard in the home, temperature, humidity, and ventilation: $C = 0.50e^{-0.012A}$, where C = formaldehyde concentration (ppm) and A = home age (mo). On the basis of this formula, the half-life of formaldehyde concentration is 58 mo. The difference between half-life values derived from test data and those from house-monitoring may result partly from the fact that particleboard is often added to older homes for repair and improvement.

Monitoring data for the 65 randomly selected mobile homes in Wisconsin show a similar decrease in formaldehyde concentration with increasing home age (M. Woodbury, personal communication). In these studies, the reported formaldehyde half-life was 69 mo, which is quite similar to that found in the Denmark study. Monitoring data on 45 complaint mobile homes in Wisconsin also showed a decrease in formaldehyde concentration with increase in home age; the indicated half-life in this sample was 28 mo. When these data are combined, the formaldehyde half-life becomes 53 mo, or approximately 4.4 yr.

Residences are not all expected to have the same formaldehyde concentration. As suggested earlier, there is variation even among homes of the same age, depending on the amount and type of particleboard, plywood, and UF-foam insulation used in the construction, as well as on temperature, humidity, and ventilation. Therefore, monitored concentrations from a sample of similar homes will be characterized by a frequency distribution that can be approximated by a known statistical distribution, which, in turn, can be used to estimate the range of human exposures to formaldehyde in the residential environment.

The average atmospheric formaldehyde concentration appears to be approximately 0.2 ppm in both mobile homes and homes insulated with UF foam. Little information is available on conventional houses that do not contain UF-foam insulation or that were not designed to be energy-efficient. The average formaldehyde concentration in conventional houses appears to be 0.01-0.1 ppm and may be only slightly higher than outdoor concentrations. Houses containing more particleboard would fall on the high side of this concentration range, and houses with no particleboard would fall on the low side.

Outdoor atmospheric formaldehyde concentrations are generally much lower than 0.1 ppm in U.S. cities. Examples of annual average concentrations are 0.05 ppm for Los Angeles,[1 20 22] 0.004-0.007 ppm for four New Jersey cities,[4] 0.04 ppm for Wisconsin cities (Hanrahan, personal communication), and less than 0.03 ppm in Raleigh, North Carolina, and Pasadena, California.[6] Formaldehyde concentrations at four Swiss locations ranged from 0.007 to 0.014 ppm; these concentrations are about one-fifth the corresponding indoor Swiss concentrations.[25] A mean value of 0.004 ppm has been reported for mainland Europe.[3]

Thus, formaldehyde concentrations in mobile homes and for homes insulated with UF-foam resin are considerably higher than those of the corresponding outdoor atmosphere.

Control Techniques

Several measures have been used in attempts to correct problems associated with formaldehyde release from building materials, including:

- Ventilation (opening doors and windows).
- Mechanical ventilation coupled with the use of heat exchangers.
- The use of impregnated charcoal in furnace or air-conditioning filters.
- The evaporation of household ammonia in closed and overheated rooms to neutralize formaldehyde, followed by ventilation.
- Injection of ammonia into insulation through holes in walls to neutralize formaldehyde.
- Spraying of air filters or floors with a specified odor absorbent provided by the manufacturer.
- Use of a "masking agent" available from the manufacturer.
- Application of vinyl wallpaper or a low-permeability paint to interior walls.
- Removal of all or some of the insulation from the home.

Although many manufacturers claimed that they had successfully used these remedial measures, no studies have shown that the measures will reliably control the release of formaldehyde gas.

OTHER ORGANIC SUBSTANCES

In addition to formaldehyde, many other organic contaminants can be present in indoor environments. Very little work has been done to identify or measure organic contaminants that may be harmful. Nevertheless, these compounds may provide a partial explanation for complaints registered by people in indoor environments where it is determined that formaldehyde and other indoor-pollutant concentrations are low or undetectable.[13]

The experimental tasks associated with characterizing organic contaminants in indoor environments are formidable. The contaminants are usually present as complex mixtures of many compounds at low concentrations. Enough work has been done to outline broadly the nature of the problem and the available information.

Sources and Emission Rates

Four major sources of organic contaminants can be identified. People emit one category of organic contaminants termed "bioeffluents." Building materials can also emit organic contaminants. The other two categories, personal consumer products (including insecticides, pesticides, and herbicides) and tobacco-smoking, are discussed later in this chapter.

Bioeffluents. Humans, through normal biologic processes, emit a category of organic substances known as "bioeffluents." Wang[24] studied bioeffluents in a school auditorium seating over 400 people. Many organic contaminants were observed, but the major ones associated with people were methanol, ethanol, acetone, and butyric acid. As an example, the absolute concentrations and the emission rate per person during a class lecture are shown in Table IV-6. Emission rates of bioeffluents increased sharply during a class examination, considered by Wang to be a period of stress. The findings of Wang were corroborated in part by Johansson,[10] who studied two schoolrooms in Sweden. He observed that acetone and ethanol were associated with the presence of schoolchildren. No emission rates were reported.

Building Materials. Such products as adhesives, paints, and sealants contain solvents and other agents that can be released during and immediately after application. The health hazards associated with these short-term releases of organic contaminants are acknowledged in the warning labels regarding the use of adequate ventilation, which are commonly applied to these products.

Less well understood is the potential for long-term emission of organic contaminants from building materials. Slow release of residual solvents and other agents (e.g., catalysts, surfactants, and plastic monomers) is one possibility, as is the gradual production of contaminants by degradation (e.g., air oxidation, photoinitiated reactions, and retropolymerization reactions).

Preliminary data indicate that concentrations of organic compounds in new buildings are generally higher than outdoor concentrations.[13,18] Figure IV-4 shows comparative gas chromatograms of an indoor air sample and an outdoor air sample taken simultaneously at the same building. Classes of compounds consistently observed in indoor air include hydrocarbons, alkylated aromatic compounds, and chlorinated hydrocarbon solvents. Table IV-7 lists organic compounds identified in several office buildings at concentrations at least 5 times greater than outdoor concentrations.[18]

TABLE IV-6

Average Concentrations and Emission Rates of Organic Bioeffluents in a Lecture Class (389 People at 9:30 a.m.)

Bioeffluent	Concentration, ppb	Emission Rate, mg/day per person
Acetone	20.6 ± 2.8	50.7 ± 27.3
Acetaldehyde	4.2 ± 2.1	6.2 ± 4.5
Acetic acid	9.9 ± 1.1	3.6 ± 3.6
Allyl alcohol	1.7 ± 1.7	19.9 ± 2.3
Amyl alcohol	7.6 ± 7.2	21.9 ± 20.8
Butyric acid	15.1 ± 7.3	44.6 ± 21.5
Diethylketone	5.7 ± 5.0	20.8 ± 11.4
Ethyl acetate	8.6 ± 2.6	25.4 ± 4.8
Ethyl alcohol	22.8 ± 10.0	44.7 ± 21.5
Methyl alcohol	54.8 ± 29.3	74.4 ± 5.0
Phenol	4.6 ± 1.9	9.5 ± 1.5
Toluene	1.8 ± 1.7	7.4 ± 4.9

[a]Data from Wang.[24]

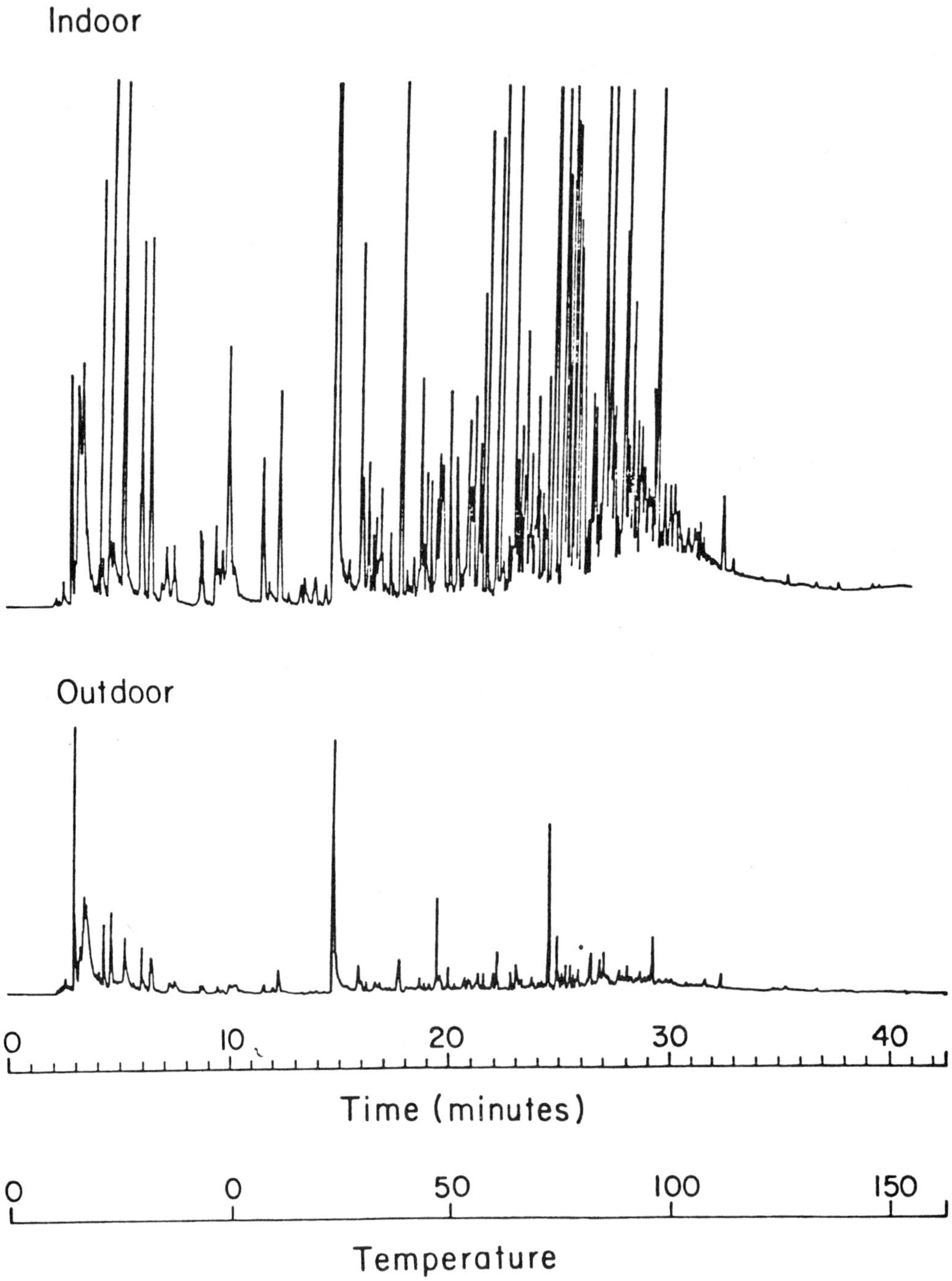

FIGURE IV-4 Comparison of chromatograms of samples taken inside and outside an office for trace organic substances. From C. D. Hollowell, personal communication.

TABLE IV-7

Organic Substances Detected in Offices[a]

Substance	OSHA Permissible Exposure Limit, ppm
Hydrocarbons:	
n-Hexane	500
n-Heptane	500
n-Octane	500
n-Nonane	--
n-Undecane	--
2-Methylpentane	--
3-Methylpentane	--
2,5-Dimethylheptane	--
Methylcyclopentane	--
Ethylcyclohexane	--
Methylcyclohexane	500
Pentamethylheptane	--
Aromatics:	
Benzene	1
Xylenes	100
Toluene	200
Halogenated hydrocarbons:	
Trichloroethane	350
Trichloroethylene	100
Tetrachloroethylene	100
Miscellaneous:	
Hexanal	--
Methylethylketone	200

[a]Data from Schmidt *et al.*[18]

In general, concentratons of specific organic compounds in nonindustrial indoor environments are well below occupational exposure limits established by OSHA. The health hazard from the effects of organic compounds at concentrations observed in indoor environments cannot now be assessed. It is important to note that OSHA criteria were established for the industrial environment, where high exposures to single compounds are encountered. The possibility cannot be overlooked that cumulative exposure to several compounds at low concentrations, or synergistic effects, may explain the complaints of building occupants.

REFERENCES

1. Altshuller, A. P., T. A. Bellar, and S. P. McPherson. Hydrocarbons and Aldehydes in the Los Angeles Atmosphere. Presented at Air Pollution Control Association Annual Meeting, May 2, 1962, Chicago, Illinois. Cincinnati: U.S. Department of Health, Education, and Welfare, Division of Air Pollution, Public Health Service, 1962.
2. Andersen, I., G. R. Lundqvist, and L. Molhave. Indoor air pollution due to chipboard used as a construction material. Atmos. Environ. 9:1121-1127, 1975.
3. Cauler, H. Some problems of atmospheric chemistry. In Compendium of Meteorology. Baltimore: Waverly Press, Inc., 1951.
4. Cleveland, W. S., T. E. Graedel, and B. Kleiner. Urban formaldehyde: Observed correlation with source emissions and photochemistry. Atmos. Environ. 11:357-360, 1977.
5. Fanning, L. Z. Formaldehyde in Office Trailers. Lawrence Berkeley Laboratory Report No. LBID-084. Berkeley, Cal.: Lawrence Berkeley Laboratory, Energy and Environment Division, 1979. 7 pp.
6. Hanst, P. L., W. E. Wilson, R. K. Patterson, B. W. Gay, Jr., L. W. Chaney, and C. S. Burton. A spectroscopic study of California smog, pp. 17-70. In Proceedings of the 6th Annual Symposium. Trace Analysis and Detection in the Environment, 29 April-1 May, 1975. Edgewood Arsenal Special Report EO-76001.
7. Harke, H.-P. The problem of "passive smoking." Münch. Med. Wochenschr. 112:2328-2334, 1970 (in German; English summary)
8. Harke, H.-P., A. Baars, B. Frahm, H. Peters, and C. Schultz. Passive smoking. Concentration of smoke constituents in the air of large and small rooms as a function of number of cigarettes smoked and time. Int. Arch. Arbeitsmed. 29:323-339, 1972. (in German)
9. Jermini, C., A. Weber, and E. Grandjean. Quantitative determination of various gas-phase components of the side-stream smoke of cigarettes in the room air as a contribution to the problem of passive-smoking. Int. Arch. Occup. Environ. Health 36:169-181, 1976. (in German; English summary)
10. Johansson, I. Determination of organic compounds in indoor air with potential reference to air quality. Atmos. Environ. 12:1371-1377, 1978.

11. Lin, C.-I., R. N. Anaclerio, D. W. Anthon, L. Z. Fanning, and C. D. Hollowell. Indoor/Outdoor Measurements of Formaldehyde and Total Aldehydes. Presented at the 178th National Meeting of the American Chemical Society, Washington, D.C., September 9-14, 1979. Abstract No. 112 in Abstracts of Papers. 178th American Chemical Society National Meeting. Vol. I. Washington, D.C.: American Chemical Society, 1979.
12. Long, K. R., D. A. Pierson, S. T. Brennan, C. W. Frank, and R. A. Hahne. Problems Associated with the Use of Urea-Formaldehyde Foam for Residential Insulation. Part I: The Effects of Temperature and Humidity on Formaldehyde Release from Urea-Formaldehyde Foam Insulation. Oak Ridge National Laboratory Report No. ORNL/SUB-7559/I. Washington, D.C.: U.S. Government Printing Office, 1979. 89 pp.
13. Mølhave, L. Indoor air pollution due to building materials, pp. 89-110. In P. O. Fanger, and O. Valbjørn, Eds. Indoor Climate. Effects on Human Comfort, Performance, and Health in Residential, Commercial, and Light-Industry Buildings. Proceedings of the First International Indoor Climate Symposium, Copenhagen, August 30 - September 1, 1978. Copenhagen: Danish Building Research Institute, 1979.
14. Moschandreas, D. J., J. W. C. Stark, J. E. McFadden, and S. S. Morse. Indoor Air Pollution in the Residential Environment. Vol. 1. Data Collection, Analysis, and Interpretation. U.S. Environmental Protection Agency Report No. EPA 600/7-78-229a. Research Triangle Park, N.C.: U.S. Environmental Protection Agency, Environmental Monitoring and Support Laboratory, 1978. 201 pp.
15. National Research Council, Committee on Aldehydes. Formaldehyde and Other Aldehydes. Washington, D.C.: National Academy Press, 1981. [354] pp.
16. Osborne, J. S., S. Adamek, and M. E. Hobbs. Some components of gas phase of cigarette smoke. Anal. Chem. 28:211-215, 1956.
17. Schmidt, A., and H. Götz. Die Entstehung von Formaldehyd bei der Verbrennung von Erdgas in Haushaltsgeräten. GWF--Gas/Erdgas 118:112-115, 1977. (in German)
18. Schmidt, H. E., C. D. Hollowell, R. R. Miksch, and A. S. Newton. Trace Organics in Offices. Lawrence Berkeley Laboratory Report 11378. Berkeley, Cal.: Lawrence Berkeley Laboratory, 1980.
19. Sheldrick, J. E., and T. R. Steadman. Product/Industry Profile and Related Analysis on Formaldehyde and Formaldehyde-Containing Consumer Products. Part II. Products/Industry Profile on Urea Formaldehyde. Report to U.S. Consumer Product Safety Commission. Columbus, Ohio: Battelle Columbus Division, 1979. [24] pp.
20. Stahl, Q. R. Preliminary Air Pollution Survey of Aldehydes. A Literature Review. National Air Pollution Control Administration Publication No. APTD 69-24. Raleigh, N.C.: U.S. Department of Health, Education, and Welfare, National Air Pollution Control Administration, 1969.

21. Tabershaw, I. R., H. N. Doyle, L. Gaudette, S. H. Lamm, and O. Wong. A Review of the Formaldehyde Problems in Mobile Homes. Report to National Particleboard Association. Rockville, Md.: Tabershaw Occupational Medicine Associates, P.A., 1979. 19 pp.
22. U.S. Department of Health, Education, and Welfare, National Air Pollution Control Administration. Air Quality Criteria for Hydrocarbons. DHEW Publications No. (HSM) 72-7516. Washington, D.C.: U.S. Government Printing Office, 1970. 118 pp.
23. U.S. Department of Labor, Occupational Health and Safety Administration. Occupational Safety and Health Standards. Subpart Z--Toxic and hazardous substances. Code of Federal Regulations, Title 29, Part 1901.1, July 1, 1980.
24. Wang, T. C. A study of bioeffluents in a college classroom. ASHAE Trans. 81 (Pt. 1):32-44, 1975.
25. Wanner, H. U., A. Deuber, J. Satish, M. Meier, and H. Sommer. Air pollution in the vicinity of streets, pp. 99-107. In M. W. Benarie, Ed. Proceedings of the 12th International Colloquium on Atmospheric Pollution, 1976. Amsterdam: Elsevier, 1976.

CONSUMER PRODUCTS

A wide array of activities and products that affect air quality can be found in contemporary residential spaces. The full range is often wider than is found in most occupational or public spaces. We are concerned here with the air-pollution aspects of such activities and consumer products.

Pollutants emitted into the indoor environment by consumer products are usually dissipated by dilution and surface deposition. For a given amount of such a pollutant, the rate of discharge, the volume of the space, the ventilation rate or infiltration rate, and the presence of occupants determine the severity and duration of human exposure. The specifics of control strategies are discussed in Chapter IX, but it will be clear that even an identical activity could produce radically different exposures in different spaces, depending on the application and efficacy of control strategies.

Consumer products introduce pollutants into the indoor air in a number of forms and ways. For example, spray paint and sprayed-on oven cleaners introduce an aerosol of the chemical products whenever they are used. There are many spray products; it has been estimated that the average U.S. residence at any given time contains 45 aerosol sprays.[12] Aerosols can also be produced indirectly, as in grinding, sanding, cleaning, and some hobby activities. Other consumer products and the activities associated with them introduce pollutants by evaporation or sublimation of solvents or active ingredients, such as paint solvents, cleaners, bleaches, and disinfectants. Still other products--e.g., plastics, paints, and textiles with artificial fibers and conditioners--release, or "outgas," small amounts of volatile chemicals over long periods. The association of emission with these types of products or activities is shown in Table IV-8. The table represents a compilation of kinds of emission encountered in a search

TABLE IV-8

Types of Emission of Indoor Air Pollutants Associated with Various Activities and Consumer Products

Activity or Product	Intentional Aerosol Production	Unintentional Aerosol Production	Evaporation or Sublimation	Unintentional Outgassing
Cleaning	X	X	X	X
Painting	X		X	X
Polishing	X		X	X
Stripping		X	X	
Refinishing	X		X	X
Hobbies and crafts	X	X	X	X
Deodorizer	X		X	
Insecticide	X		X	
Disinfectant	X		X	
Personal grooming product	X		X	X
Plastic				X

of the literature and in lists in consumer catalogs; it is clearly incomplete.

AEROSOL-PRODUCING PRODUCTS

Pressurized aerosol cans have found widespread use in a great variety of applications. Aerosol cans typically contain a propellant gas under a relatively low pressure--about 40 psi (3 kg/cm^2)--with a vapor pressure at normal room temperature that allows some of the propellant to be in equilibrium in the liquid phase. Until recently, dichlorodifluoromethane and trichlorofluoromethane (fluorocarbon-12 and fluorocarbon-11) were used widely as propellants, but their effect on the atmospheric ozone layer has led to a prohibition of this application in the United States. Propane, butane, nitrous oxide, methylene chloride, and others are currently used as propellants. Most propellant gases are biologically inactive or active only at high concentrations. Some are extremely flammable and could reach explosive concentrations in enclosed spaces.

Active ingredients vary from product to product, and a complete list of propellants and active ingredients rarely appears on a can. Cleaning agents include sodium or potassium hydroxide in oven-cleaners, ammonium hydroxide in window-cleaners, and tetrachloroethylene and petroleum-derived solvents in spot-cleaners. Spray paints often contain toluene, xylene, methylchloride, and other volatile organic substances, as well as pigments and a vehicle. Dravnieks[26] measured the organic chemicals present in indoor air of a high-rise apartment in which aerosol products had been used. The use of a scented oven-cleaner released at least 13 organic chemicals into the residential space. An unscented aerosol deodorant and a scented aerosol furniture polish released similarly large numbers of chemical species throughout the residence.

Cote et al.[7] surveyed the composition of a range of aerosol products for propellants and active ingredients. Mokler et al.[23] reported that under "worst reasonable conditions" some aerosol products introduce respirable particles smaller than 6 μm in aerodynamic diameter at a concentration of over 50 mg/m^3--10 times the threshold limit value for daily average exposure in an industrial environment for biologically inactive nuisance dust.[3] The same investigators found that the conditions of discharge did not have a great effect on the size characteristics.[24] There was no evidence of animal toxicologic effects after a series of studies of cosmetic aerosols.[19]

Marier and co-workers[21] exposed 20 human subjects to a number of aerosol products daily for four consecutive weeks. These products included deodorants, hair spray, frying-pan spray, room-freshener, insect repellent, window-cleaner, insecticide, furniture polish, bathroom-cleaner, and a depilatory. All products were used according to the manufacturer's instructions; after 3 wk of exposure, all subjects were evaluated for cardiac function, respiratory function, and hematologic and clinical biochemical characteristics. None of the tests showed evidence of toxic effects, and no fluorocarbon was found at any time in any blood sample. All the spray products used fluorocarbon as propellant.

In an epidemiologic study of 3,800 people in Tucson, Arizona, Lebowitz reported in 1976 that he found respiratory effects associated with frequent use of aerosols.[17] In a followup study 5 yr after the first one, he found that use of aerosol products was lower by a factor of 10, and the association between aerosol use and respiratory effects was no longer seen.[18]

Stewart _et al._[32] evaluated physiologic responses to different propellant gases (isobutane, propane, fluorocarbon-12, and fluorocarbon-11). They were inhaled at concentrations of 250, 500, and 1,000 ppm for periods of 1 min to 8 h. None of the subjects showed a decrease in pulmonary function or a change in cardiac rhythm as a result of these exposures.

Deliberate inhalation of some aerosol sprays often leads to serious consequences, or even death. Lipid sprays[10] have been shown under such circumstances to lead to acute functional and anatomic disintegration of alveolar surfactant. The resulting alveolar collapse can cause fatal hypoxemia. In one preliminary study,[29] the statistical association found between the indoor use of spray adhesives and congenital malformations was sufficient to warrant further study.

PARTICLES PRODUCED AS A BYPRODUCT

The major production of particles in the indoor environment is undoubtedly due to tobacco-smoking and food-preparation (which are considered elsewhere, as is malfunctioning heating equipment). In the course of cleaning of floors and furniture, there is likely to be a resuspension of dust particles that had previously settled. Maintenance and cleaning activities have been identified as a source of asbestos fibers in spaces that contain accessible asbestos. Resuspension of settled dust in the absence of ventilation or air-cleaning effectively increases the exposure of occupants.[20]

Even minimal exposures to asbestos fibers in hobby materials[6] have been found to result in identifiable asbestos bodies in the lung, and the asbestos fibers found in mesothelioma have been ascribed to minimal exposures in the pursuit of hobbies.[5]

Hobby and home craft activities can cause substantial production of particles.[22] The use of lead glazes by potters[14] can lead to high lead intakes, and an unknown fraction of this intake will be from inhalation. The use of solder and flux in stained-glass fabrication and jewelry-making results in the aerosolization of lead, cadmium, and flux.[15] The effects of inhalation of lead and cadmium are known better from occupational exposures, but some residential exposure must be expected.

Some phases of woodworking result in the production of airborne wood dust from sawing, grinding, and sanding. Acheson and co-workers studied the risk of nasal cancer in woodworkers in the furniture industry[1] and concluded that exposure to wood dust, rather than varnish or polish, caused an increased risk of nasal cancer. For occupational exposures, the risk of nasal cancer was as much as 500 times the normal incidence.

PRODUCTS AND ACTIVITIES ASSOCIATED WITH EVAPORATION OR SUBLIMATION

Solvents are used in the indoor environment for a variety of purposes--in clothing- and furniture-cleaners; as carriers for polishes, paints, and varnishes; and as chemical strippers in the refinishing of furniture--and they are inherent in many adhesives, personal grooming products, disinfectants, and deodorizers. Most of the chemical species involved are used in the occupational environment, and threshold limit values have been adopted for that environment.[2 3] Table IV-9 contains a nonexhaustive list of substances that evaporate or sublimate from solvents and household products and their occupational exposure limits.[3]

The containers of many of the products containing substances listed in Table IV-9 do not list their constituents or concentrations, although hazard warnings are generally provided. The indoor air concentrations are rarely measured; even when they are, there is disagreement concerning the typical conditions. An example of this is provided by two reports on methylene chloride. The first deals with exposures to methylene chloride in paint-removers in home workshops.[31] One person developed acute myocardial infarction in three separate episodes after the use of methylene chloride for furniture-stripping in a basement workshop. The last exposure was fatal. In the case of a healthy younger man, it was accidentally discovered that he had a high carboxyhemoglobin (COHb) concentration (6-8%) on each of several mornings after 2-h exposures to paint-removers at home on the previous evening. Controlled exposures of volunteers to methylene chloride at 50, 250, and 986 ppm for 3 h produced COHb at concentrations of 2%, 4.5%, and 15%, respectively, measured 1 h after the end of exposure. In the second study,[30] volunteers were exposed to methylene chloride at 450 ppm for 26 min; the authors considered that a typical exposure for the use of spray paint with methylene chloride as propellant. They found "a clinically insignificant increase" in COHb 7 h after exposure. These two reports represent a range of reactions to methylene chloride--from nearly undetectable to accidentally detected to fatal--that depends on individual vulnerability, conditions of use, and criteria used for detection. Measurements of actual concentrations of chemicals associated with household products are scarce; where they are reported, they can be very high, as in a report of benzene at 130 ppm found in a double garage during furniture-stripping.[39]

The use of mercury compounds as fungicides in latex paints gives rise to long-term emission of mercury into the indoor spaces in which such paints have been applied. Taylor[34] used a radioactive tracer and found that 20-25% of the mercury was lost in the first 90 d after application. Foote[11] measured background ambient atmospheric mercury concentrations in a number of homes at slightly over 2 ng/m^3 and in rooms recently painted with latex paint at up to 3,000 ng/m^3. In

TABLE IV-9

Threshold Limit Values of Various Substances[a]

Chemical	TLV-TWA,[b] mg/m^3	TLV-STEL,[c] mg/m^3	Source
Acetone	2,400	3,000	Lacquer solvent
Ammonia	18	27	Cleaner
Benzene	30	--	Adhesive, spot cleaner, paint remover
Carbon tetrachloride	65	130	Spot cleaner, dry cleaner
Chlorine	3	9	Cleaner
Methanol	260	310	Paint, spot cleaner
Trichloroethane	1,900	2,380	Cleaning fluid
Methylene chloride	700	870	Paint remover
Ozone	0.2	0.6	Copying machine, electrostatic air cleaner
Trichloroethylene	535	800	Dry-cleaning agent
Turpentine	560	840	Paint, finish
Xylene	435	655	Solvent, paint carrier, shoe dye
Toluene	375	560	Solvent, paint carrier, dry cleaning

[a]Data from American Conference of Governmental Industrial Hygienists.[3]

[b]Threshold limit value--time-weighted average.

[c]Threshold limit value--short-term exposure limit.

rooms painted 3 yr before, he found concentrations of 68 ng/m^3. The threshold limit value adopted by the American Conference of Governmental Industrial Hygienists[3] for 8-h workdays is 50,000 ng/m^3. Sibbett et al.[28] found room-air concentrations of around 1,000-2,000 ng/m^3 in rooms with fresh latex paint--also well below the TLV and not likely to produce clinical symptoms. In comparison, spilt mercury from broken clinical thermometers[11] produced mercury-vapor concentrations in room air of about 5 μg/m^3, continuous exposure to which might produce clinical symptoms.

Residential or commercial use of insecticides, pesticides, and herbicides both inside and outside has the potential for contaminating the indoor environment. Some organochlorinated or organophosphated pesticides have specific agricultural applications, but others are used widely in urban areas by both private property-owners and municipalities. The organic-based compounds can find their way indoors by various routes. Some are applied directly indoors for rodent and insect control. The long-term effectiveness of many of these compounds is achieved by prolonged sublimation. Various studies have found many of these compounds indoors.[8,9,13,16] In spite of the widespread use of residential and commercial insecticides, pesticides, and herbicides, no systematic survey has been done to identify indoor concentrations for the numerous compounds likely to be present.

SOME MECHANISMS OF BIOMEDICAL EFFECTS

The wide variety of the chemicals in consumer products makes it difficult to anticipate all the possible adverse health effects. Some classes of chemicals have common characteristics that cause them to attack particular organ systems.

Various environmental chemicals have structural similarities that suggest that they may have similar effects on the myocardium; these chemicals have a lung-tissue half-life that could represent a long-term hazard. Some examples are polyhalogenated hydrocarbons used as insecticides and industrial chemicals, such as polychlorinated and polybrominated biphenyls (PCB and PBB). These may produce sudden death. The polyhalogenated hydrocarbons bind to estrogen receptors and are estrogenic in animal systems. This may increase cholesterol and triglyceride concentrations and so increase the risk of coronary heart disease (CHD), of mortality from CHD, or of CHD-related death.[35] Myocardial fibers may also be damaged by toxic agents, such as ozone.[27] Coronary-arterial-disease mortality has also been shown to be related to concentrations of suspended particulate matter in the external environment.[27] Thus, there may be many sources of cardiotoxic agents in the indoor environment.[4]

Organophosphorus pesticides, such as parathion, lead to clinical symptoms resembling strong cholinergic stimulation due to a nonreversible blockade of cholinesterase and leading to an accumulation of endogenous acetylcholine.[37] This accumulation disrupts the transmission of impulses from nerve fibers to muscles.[36] (In the case of parathion, skin may be the route of entry.) Chronic effects on cerebrospinal fluid may also occur. Other disturbances of the nervous system may occur through exposure to such chemicals as PCB, which may be ingested and may be stored in fatty tissue. PCB inhibits growth in cultured cells and interferes with the activity of a variety of enzymes.[36] Pesticides occur in indoor environments through the spraying of pesticides or herbicides and through the contamination of items brought into the home, including foods and flowers.[25]

Solvents, especially chlorinated hydrocarbons, may damage the kidneys and liver.[38] Although the skin acts as an effective barrier, serious toxicologic effects may result from exposure of the skin to such substances as some organic phosphates, lead compounds, and acid. Dermatitis--especially in the form of dry, scaly or fissured reactions--can be caused by recurrent contact with solvents, emulsifying substances, dehydrators, or detergents. Acidic or alkaline gases and aerosols are readily dissolved in the aqueous protective film of the eye, on the mucous membranes of the nose and mouth, and on skin that is moist with sweat. Such exposure may also erode teeth and change hair structure.[33]

These mechanisms have been demonstrated; because the exposures described can occur in nonoccupational indoor spaces, their potential impact should be considered. However, the importance of such exposures is still difficult to evaluate.

SUMMARY AND CONCLUSIONS

Among the sources of pollutants in the indoor air that are due to consumer products or hobby or craft activities, many can harm exposed occupants. Such products usually bear labels with hazard warnings and instructions for use that, if followed carefully, will reduce pollutant exposures to a point that is presumably acceptable for healthy users. Willful abuse, as in the case of direct inhalation of aerosol products or careless use of solvents in enclosed spaces, can result in acute or delayed disorders or even death. The prohibition of fluorocarbons as propellants in aerosol spray products has resulted in substitution of other propellants that may be found to be toxicologically less desirable. The recognition of the carcinogenicity of vinyl chloride and benzene has resulted in the banning of these chemicals from consumer products, but a number of chemicals with serious toxic potential continue to be used.

Many consumer products are used only intermittently by a given person, and those in different households are likely to use different products for a given purpose. For safety, most products are formulated to avoid acute discomfort or irritation and because such acute effects will reduce marketability. Risks of long-term or delayed adverse health effects are not as likely to become apparent to the consumer and are not as likely to be incorporated in hazard warnings on products.

If the constituents of the consumer products are known, it may be possible to use a combination of occupational threshold limit values and likely exposure concentrations and exposure durations in making assessments of the impact of consumer products on indoor air quality.

Disclosure of product composition, assessment of acute and chronic consequences, and labeling with composition, directions for use, and hazard warnings specific for a particular formulation amount to one of several strategies that should be evaluated.

Monitoring or surveillance techniques now in use by such responsible agencies as the Consumer Product Safety Commission, the Environmental Protection Agency, and the Centers for Disease Control are more likely to discover acute consequences than delayed adverse health effects. The assessment of the human exposure and adverse health consequences due to the storage and use of consumer products is made difficult by the irregular, sporadic, and highly variable exposures, scarcity of measurements, and limited knowledge about composition of many of the products. Except for studies of accidental poisoning, epidemiologic assessments are almost completely impossible, owing to the episodic and irregular nature of exposures. It will be necessary to rely on knowledge of and experience with the use of the constituents of consumer products in the workplace. Assessment of the impact of consumer products on nonoccupational indoor air quality must, then, be based on constructed risks and potential exposures.

Labeling of consumer products with lists of constituents, instructions for safe use, and hazard warnings is often inadequate, and in any case it may be disregarded by the user and is ineffective when the products are handled by children.

REFERENCES

1. Acheson, E. D., R. H. Cowdell, E. Hadfield, and R. G. Macbeth. Nasal cancer in woodworkers in the furniture industry. Br. Med. J. 2:587-596, 1968.
2. American Conference of Governmental Industrial Hygienists. Documentation of the Threshold Limit Values for Substances in Workroom Air. Cincinnati: American Conference of Governmental Industrial Hygienists, 1977.
3. American Conference of Governmental Industrial Hygienists. TLVs. Threshold Limit Values for Chemical Substances in Workroom Air Adopted by ACGIH for 1980. Cincinnati: American Conference of Governmental Industrial Hygienists, 1980. 93 pp.
4. Burch, G. E. Toxic agents, cardiovascular disease, and the polluted home. Am. Heart J. 87:679-680, 1974.
5. Chen, W., and N. K. Mottet. Malignant mesothelioma with minimal asbestos exposure. Hum. Pathol. 9:253-258, 1978.
6. Churg, A., and M. L. Warnock. Analysis of the cores of asbestos bodies from members of the general population: Patients with probable low-degree exposure to asbestos. Am. Rev. Respir. Dis. 120:781-786, 1979.
7. Cote, W. A., W. A. Wade, III, and J. E. Yocom. A Study of Indoor Air Quality. Final Report. U.S. Environmental Protection Agency Report No. EPA-650/4-74-042. Washington, D.C.: U.S. Environmental Protection Agency, 1974. 282 pp.

8. Davies, J. E., W. F. Edmundson, and A. Raffonelli. The role of house dust in human DDT pollution. Am. J. Public Health 65(1):53-57, 1975.
9. Davis, J. H., J. E. Davies, A. Raffonelli, and G. Reich. The investigation of fatal acrylonitrile intoxications, pp. 547-555. In W. B. Deichmann, Ed. Pesticides and the Environment: A Continuing Controversy. Chronic Toxicology, Ecological Effects, Carcinogenesis, Mutagenesis, Teratogenesis, Drug Interactions. Vol. II. New York: Intercontinental Medical Book Corporation, 1973.
10. Fagan, D. G., J. B. Forrest, G. Enhörning, M. Lamprey, and J. Guy. Acute pulmonary toxicity of a commercial fluorocarbon-lipid aerosol. Histopathology 1:209-223, 1977.
11. Foote, R. S. Mercury vapor concentrations inside buildings. Science 177:513-514, 1974.
12. Fritsch, A. J., Ed. The Household Pollutants Guide. Garden City, N.Y.: Anchor Press/Doubleday, 1978. 309 pp.
13. Holleman, J. W., M. G. Ryon, and A. S. Hammons. Chemical Contaminants in Nonoccupationally Exposed U.S. Residents. U.S. Environmental Protection Agency Report No. EPA-600/1-80-001. Research Triangle Park, N.C.: U.S. Environmental Protection Agency, Health Effects Research Laboratory, 1980. 150 pp.
14. Koplan, J. P., A. V. Wells, H. J. P. Diggory, E. L. Baker, and J. Liddle. Lead absorption in a community of potters in Barbados. Int. J. Epidemiol. 6:225-229, 1977.
15. Kronoveter, K. J., and C. R. Meyer. Industrial hygiene study in a stained glass workshop, pp. 28-35. In M. McCann and G. Barazani, Eds. Proceedings of the SOEH Conference on Health Hazards in the Arts and Crafts. Washington, D.C.: Society for Occupational and Environmental Health, 1980.
16. Leary, J. S., W. T. Keane, C. Fontenot, E. F. Feichtmeir, D. Schultz, B. A. Koos, L. Hirsch, E. M. Lavor, C. C. Roan, and C. H. Hine. Safety evaluation in the home of polyvinyl chloride resin strip containing dichlorvos (DDVP). Arch. Environ. Health 29:308-314, 1974.
17. Lebowitz, M. D. Aerosol usage and respiratory symptomatology. Arch. Environ. Health 31:83-86, 1976.
18. Lebowitz, M. D. The Effects of Cosmetic Aerosols on Respiratory Physiology. Final Contract Report. Washington, D.C.: U.S. Food and Drug Administration, 1980.
19. Lovelace Foundation. Inhalation toxicological studies of aerosolized products. Final Contract Report. Washington, D.C.: U.S. Food and Drug Administration, 1979. Available from National Technical Information Service as PB 89 108509.
20. Lum, R. M., and T. E. Graedel. Measurements and models of indoor aerosol size spectra. Atmos. Environ. 7:827-842, 1973.
21. Marier, G., H. MacFarland, G. S. Wiberg, H. Buchwald, and P. Dussault. Blood fluorocarbon levels following exposure to household aerosols. Can. Med. Assoc. J. 111:39-42, 1974.
22. McCann, M., and G. Barazani, Eds. Proceedings of the SOEH Conference on Health Hazards in the Arts and Crafts. Washington, D.C.: Society for Occupational and Environmental Health, 1980. 232 pp.

23. Mokler, B. V., B. A. Wong, and M. J. Snow. Respirable particulates generated by pressurized consumer products. I. Experimental method and general characteristics. Am. Ind. Hyg. Assoc. J. 40:330-338, 1979.
24. Mokler, B. V., B. A. Wong, and M. J. Snow. Respirable particulates generated by pressurized consumer products. II. Influence of experimental conditions. Am. Ind. Hyg. Assoc. J. 40:339-347, 1979.
25. Morse, D. L., E. L. Baker, and P. J. Landrigan. Cut flowers: A potential pesticide hazard. Am. J. Public Health 69:53-56, 1979.
26. Moschandreas, D. J., Ed. Indoor Air Pollution in the Residential Environment. Vol. II. Field Monitoring Protocol, Indoor Episodic Pollutant Release Experiments and Numerical Analyses, pp. 198-220. U.S. Environmental Protection Agency Report No. EPA-600/7-78-229b. Research Triangle Park, N.C.: U.S. Environmental Protection Agency, Environmental Monitoring and Support Laboratory, 1978.
27. Shy, C., J. Goldsmith, J. Hackney, M. D. Lebowitz, and D. Menzel. Statement on the health effects of air pollution. ATS Newsletter, 4:22-62, 1978.
28. Sibbett, D. J., R. H. Moyer, and G. H. Milly. Emission of mercury from latex paints. Presented Division of Water, Air and Waste Chemistry, Americal Chemical Society Boston, Mass., April 1972.
29. Silberg, S. L., D. R. Ransom, J. A. Lyon, and P. S. Anderson, Jr. Relationship between spray adhesives and congenital malformations. South. Med. J. 72:1170-1173, 1979.
30. Stevenson, M. F., G. L. Cooper, and M. B. Chenoweth. Effect on carboxyhemoglobin of exposure to aerosol spray paints with methylene chloride. Clin. Toxicol. 12:551-561, 1978.
31. Stewart, R. D., and C. L. Hake. Paint-remover hazard. J. Am. Med. Assoc. 235:398-401, 1976.
32. Stewart, R. D., P. E. Newton, E. D. Baretta, A. A. Herrmann, H. V. Forster, and R. J. Soto. Physiological response to aerosol propellants. Environ. Health Perspect. 26:275-285, 1978.
33. Stokinger, H. E. Mode of action of toxic substances, pp. 13-26. In W. M. Gafafer, Ed. Occupational Diseases. A Guide to Their Recognition. U.S. Department of Health, Education and Welfare, Public Health Service Publication No. 1097. Washington, D.C.: U.S. Government Printing Office, 1964.
34. Taylor, C. G. The loss of mercury from fungicidal paints. J. Appl. Chem. 15:232-236, 1965.
35. U.S. Department of Health Education, and Welfare, National Heart, Lung and Blood Institute (NHLBI). Working Group on Heart Disease Epidemiology. DHEW (NIH) Publication No. 79-1667. Washington, D.C.: U.S. Department of Health, Education, and Welfare, 1979.
36. Waldbott, G. L. Health Effects of Environmental Pollutants. Saint Louis: The C.V. Mosby Company, 1973. 316 pp.
37. Westerman, E. Accumulation of environmental agents and their effects in the body, pp. 16-27. In D. H. K. Lee and D. Minard, Eds. Physiology, Environment and Man. New York: Academic Press, Inc., 1970.
38. World Health Organization. Health Hazards of the Human Environment. Geneva: World Health Organization, 1972. 387 pp.

39. Young, R. J., R. A. Rinsky, P. F. Infante, and J. K. Wagoner. Benzene in consumer products. Science 199:248, 1978.

ASBESTOS

"Asbestos" is a collective term for the fibrous or asbestiform types of various minerals. Characteristics of flexibility, strength, and durability have brought these mineral fibers into numerous and varied applications involving potential exposure of large populations. Both widespread use and increasing investigations of the health effects of asbestos exposure have created intense interest in asbestos as an environmental contaminant. The health effects and toxicologic impact of mineral fibers are covered in detail in Chapter VII.

Because the asbestiform minerals have been used in numerous construction materials, consumer products, and appliances, the nonoccupational environment has become an area for investigation of asbestos contamination and human exposure. A potential for contamination from some types of these materials in structures during construction, renovation, demolition, and even normal use has been demonstrated. Repair and maintenance of household appliances, furnaces, stoves, and asbestos-cement pipes can also result in release of fibers into the air.

DEFINITION OF ASBESTOS

"Asbestos" is applied to chemically varied, naturally occurring mineral silicates of the serpentine and amphibole classification that are separable into fibers that are flexible and incombustible and usually have large length-to-diameter ratios. These asbestos, or asbestiform, mineral fibers have high tensile strength and desirable thermal and electric insulating properties and resist chemical degradation. Asbestos minerals with commercial and exposure importance are serpentine chrysotile and the amphibole group of asbestos minerals: amosite, crocidolite, anthophyllite, and actinolite-tremolite.[6,52]

IMPORTANT CHARACTERISTICS OF ASBESTIFORM MINERAL FIBERS

The characteristics of durability, airborne lifetime, and fiber dimension are especially important in determining the potential of exposure and biologic effect.

Durability

Asbestiform fibers retain physical integrity in nearly all uses and applications and within human tissue.[14,52]

Potential to Remain Airborne

This potential strongly affects exposure probability. Settling velocity depends heavily on fiber diameter and to a lesser extent on fiber length.[4 36] Settling in still air in a 3-m-high room, a fiber 5 μm long and 1 μm in diameter will remain airborne for approximately 4 h. A fiber of the same length with a 0.1-μm diameter will remain airborne for up to 20 h.[36] Such settling times could be prolonged in turbulent air, and, like other suspended particles, fibers can be transported by air currents. Disruptive mechanical forces cause predominantly longitudinal cleavage of fibers or fiber bundles into a larger population of particles with smaller diameters and increased persistence in the air.[3] However, the extent of fiber cleavage by natural forces in the environment is unknown.

Fiber Dimensions

The deposition and retention of fibers in the respiratory tract depend on fiber dimensions, breathing conditions, and airborne fiber concentration. Most fibers retained in human lungs are shorter than 5 μm, and have diameters less than approximately 2.5 μm,[12 14 26 40] but some thin fibers up to 200 μm have been found in lung samples.[40]

ASBESTOS PRODUCTION AND APPLICATION

Uses

The characteristics of durability, flexibility, strength, and resistance to wear bring the asbestiform minerals into thousands of applications. They are used in roofing and flooring products, textiles, papers and felts, friction materials, filters and gaskets, cement, panels, pipes, sheets, coating materials, and thermal and acoustic insulation.[38] Asbestos production began late in the nineteenth century, when it was used as thermal insulation for steam engines. Worldwide production is now nearing 5 million tons/yr, with chrysotile the principal fiber type.[42]

Production

Approximate consumption of asbestos in the United States was 600,000 tons in 1979[42] and is expected to be 400,000-900,000 tons in 2000. Over 90% of the asbestos used in the United States is imported, and over 90% of the imported asbestos is Canadian chrysotile. Over 70% of the asbestos is used in the construction industry.[2]

Spray Application

Of all the uses in the construction industry, the spray application of asbestos onto structural surfaces is the most important in the development of potential contamination situations. Such sprayed material is usually friable or susceptible to damage and disintegration by hand pressure. Sprayed material has been applied extensively to steel work to retard deformation during fire and to other structural surfaces for thermal and acoustic insulation, decoration, or condensation control.

Spray application of asbestos fibers began in the 1930s and allowed the rapid covering of irregular surfaces without the use of mechanical support or extensive preparation. Early spray applications in the United States were mainly for decoration and acoustic insulation. In 1950, the Underwriters Laboratories approved the use of sprayed asbestos where concrete had been required for prevention of deformation of steel from fire in multistory buildings. This approval brought about an intense use of sprayed asbestos material in new construction.[28,29]

However, evidence of the health hazards of asbestos exposure was accumulating.[22] In 1972, the New York City Council banned asbestos spray application because of the health hazard to spray operators, other construction workers, and the general public.[21] After failure of attempts at on-site asbestos-contamination control,[50] the EPA, in 1973, banned sprayed asbestos application for structural insulating or fireproofing.[48] Decorative materials and some heavy mix materials were not included in the ban. In July 1978, the EPA banned spray-on application of materials, except those in which the asbestos fibers are encapsulated with a bituminous or resinous binder during spraying and that are not friable after drying.[47] A rough estimate of the total amount of asbestos-containing materials sprayed over the 28-yr period is 500,000 tons.[36] Although the spraying of asbestos-containing materials in construction ceased, such friable material in existing structures remains a widespread asbestos-fiber source with potential for indoor contamination (see Lumley;[17] Sawyer;[33] and H. V. Brown, UCLA, personal communication).

ASBESTOS CONTAMINATION OF THE ENVIRONMENT

Fiber release depends on both material cohesiveness and the disruptive energy applied. The majority (85% or more) of asbestos in current use is immobilized in strong binding materials, such as cement or tiles; however, any asbestos-containing materials will release fibers when sufficiently disrupted, and the hard materials will liberate fibers if ground, sanded, or cut. The remaining asbestos--including that in insulation, troweled asbestos plaster, and pipe lagging--releases fibers upon minor disturbance.[36] These friable materials are the most important source of asbestos contamination in structures. Friable materials can be found on open and visible ceilings, walls, and structural members and on hidden surfaces accessible to maintenance, renovation, or ventilatory air flow.

The proportion of asbestos in such material is generally 10-30% by weight, but may vary from trace amounts to nearly 100%.[36] Other fibrous components include fibrous glass, mineral wools, and cellulose. Friable materials can also contain vermiculite, talc, perlite, diatomaceous earth, organic fibers, clays, quartz, gypsum, and various adhesives.

Environmental contamination from asbestos-containing materials can occur in three general ways: fallout, contact disruption, and reentrainment of previously released but settled fibers.[35] Fallout is, except for very friable material, negligible. Contact disruption and reentrainment are activity-dependent and can result in substantial contamination and exposure.

Fibers enter occupied spaces at a relatively low rate, depending on material friability and exposed surface area. Variations in the fallout rate are due to structural vibration, air movement, and changes in cohesiveness. Fallout can result in the accumulation of surface deposits of fibers over long periods; such accumulations are then available for later disturbance and reentrainment.

Any asbestos-containing material will release fibers if the energy applied to it is adequate. Contact may be intentional during demolition, renovation, or vandalism, unavoidable during maintenance, or accidental during routine activity. Fiber release depends on probability and energy of contact. Contamination is episodic and local and can be intense.[36]

The disturbance of released and accumulated fibers can cause repeated cycles of settling and resuspension. Such reentrainment contamination may occur after any disturbance, but can be important in custodial activities.[36]

ENVIRONMENTAL SAMPLING FOR ASBESTOS

Analysis of Bulk Materials

Identification of asbestos is relatively simple with mineralogic specimens that are generally uniform in type and composition. With samples of construction materials, identification is more difficult. The amount of asbestos may be small, and construction materials may contain other fibrous components with a variable collection of nonfibrous components.

The primary method for asbestos identification in bulk materials is polarized-light microscopy (PLM). X-ray diffraction (XRD) is used for quantitative analysis of fiber type.[18,31] Transmission electron microscopy has only limited application.[36]

The petrographic microscope is a transmitted-polarized-light instrument widely used for identification and characterization of substances based on their optical and crystallographic properties. The techniques are established, and the equipment is inexpensive. However, a high degree of skill and experience is required of the microscopist.[18,31,36]

X rays diffracted by crystalline material produce a characteristic pattern. The technique usually yields information with a high degree of diagnostic reliability and a printed record. It is usually used to confirm results of petrographic microscopy. X-ray diffraction requires a large investment in equipment, references, mineral standards, and technical expertise. X-ray diffraction of bulk construction material cannot define particle shape and may fail to detect concentrations of asbestos much below 5%. Moreover, other silicates or crystalline phases can interfere with asbestos identification.[18 31]

Specific and accurate fiber identification can be achieved by examination of the structure of individual particles, especially in conjunction with electron diffraction or energy-dispersive x-ray analysis. The extrapolation of precise electron-microscope data, however, to bulk sample information is inefficient and costly. Its use in identification is usually confined to resolving ambiguities raised by petrographic microscopy and x-ray diffraction. The most important use of such analytic techniques is the identification and analysis of inorganic particles in tissue.[14 36]

Errors in asbestiform-mineral analysis have potentially serious consequences and are not uncommon. False-negative results will lead to a continuation of unnecessary environmental contamination, and false-positive results can precipitate unnecessary action. Errors arise from analyst inexperience and from the use of phase-contrast microscopy, rather than the appropriate polarized-light instrument.[37]

Measurement of Airborne Asbestos

A pump is used to draw a measured volume of air through a membrane filter. The pump and filter are either stationary or carried on a person, with the sampling orifice in the respiratory zone. Common sampling rates are 2.0 L/min for personnel monitoring and 10 L/min for general environmental sampling. Sampling times vary from minutes to many hours, depending on anticipated fiber concentrations. Filter segments can be examined by various methods and observers for comparison or verification and can be stored indefinitely.[15]

Estimation of the amount of asbestos collected on the sampling filters is performed by one of two methods: counting fibers by optical microscopy with a phase-contrast light microscope and counting fibers by electron microscopy.

The standard technique for fiber enumeration is specified by the National Institute for Occupational Safety and Health (NIOSH) for determination of airborne asbestos in occupational settings.[15 19] Air is pumped through a membrane filter (effective pore size, 0.8 μm). A filter segment is examined with a microscope that has phase-contrast illumination at a magnification of 400-450. Particles longer than 5 μm and with a length-to-width ratio (aspect ratio) greater than 3 are counted. Results are presented as the numbers of fibers per milliliter of air.[15] This method enumerates only particles of defined aspect ratio and length. It is not capable of accurate identification of asbestos. Both the resolution limits of the optical microscope and the 5-μm length limit preclude enumeration of some smaller fibers, which may be present and in numbers greater by an order of magnitude or more.[9 30] However, the short fibers that are not counted may be less hazardous than the longer ones that are.[39 51]

Electron microscopy is the definitive method for both identifying and counting small asbestos particles. When it is combined with selected-area electron diffraction (SAED) and energy absorption, accurate identification of a particle is possible.[13 27] Laboratories vary in techniques of sample preparation, magnification, and mass estimation, and comparison of results has been discouraging.[16 36] Furthermore, there are no standards for interpretation of exposure data derived this way. Investigation of asbestos contamination by electron microscopy is both expensive and extremely time-consuming. Results are usually given in nanograms per cubic meter for mass estimations, but can also appear as fibers per milliliter where only enumeration is performed.[13 27 36]

Other Contamination Assessment Methods

The most relevant estimate of asbestos hazard is based on the concentration of airborne, respirable asbestos fibers in the immediate environment of building occupants. Under the usual conditions and with the standard NIOSH optical-microscopy technique, this task is difficult. The technique was originally intended for use in areas of recognized contamination, such as asbestos production facilities. The airborne contaminant was known, and relatively high concentrations were readily measured. In this setting, the optical technique is effective and appropriate. However, in the assessment of exposure situations in other, nonoccupational structures, it has become apparent that it is not entirely satisfactory.[37] The optical-microscopy method is truncated in its limit of resolution both physically and by regulation. The 5-μm lower limit of counted particles will preclude the enumeration of many small fibers in the sample environment.[30] Fiber emission is nearly always local, sporadic, and activity-related. Routine air-sampling commonly fails to describe this contamination situation.[36 37]

An alternative approach to assessment by air sampling involves a subjective observational ranking system to provide guidance in potential contamination situations.[11 49] This process evaluates relevant contamination factors that can contribute to total contamination potential. Rating systems are an approach to a complex process involving fiber aerodynamics, material characteristics, structure effects, and human activity. However, they are not exact, are subject to variations in factor estimation, and cannot be easily evaluated with existing methods. They do not provide a precise benchmark for selecting appropriate corrective actions, but can provide consistent guidance in assessment and evaluation.

An adequate system should reflect factors that influence material fallout, direct material disturbance, and reentrainment of fibers. These modes of contamination are functions of material characteristics, structure configuration, and user activities. One example is a system that considers eight factors that influence exposure.[11] It is predictive to the extent of estimating the probability of contamination and exposure potential. The factors are:

- Total asbestos content of material.
- Friability of material.
- Condition of material.
- Extent of damage of material by water.
- Accessibility of material to activity and contact.
- Surface area of material exposed.
- Activity or movement in environment.
- Plenum or other airstream effects.

The factors are individually weighted and used in a formula to generate a single score. Contamination evaluation by this method will provide guidance in two ways: (1) It indicates corrective action; scores exceeding a given value indicate that hazard potential is substantial. (2) It establishes relative priority, the most useful feature of the scaling system; the higher the score, the greater the need for corrective action, and scores can be used to establish a logical sequence of corrective actions within indoor spaces or buildings.

ASBESTOS AIR DATA

Studies have been performed in structures containing various types of asbestos material (classified as either friable or bonded) and under a series of activity conditions. Tables IV-10 and IV-11 list airborne-fiber concentration data obtained by optical microscopy (with the NIOSH standardized technique[15]) and electron microscopy. Nearly all the data were collected in nonoccupational settings in apartment buildings and private homes and in offices and schools. Both surveillance and reenactment studies are included.

The airborne-fiber concentrations determined by optical microscopy range from zero in background, quiet, and some routine activity to over 100 fibers/ml in stripping dry-spray-applied ceiling material. Fiber contamination has an expected relationship to activity and proximity.

The data in Table IV-11 obtained by transmission electron microscopy are expressed as nanograms per cubic meter. Thus, they do not distinguish between fibrous and nonfibrous asbestos, nor between long and short fibers. Estimation of the health significance of mass-concentration data from Table IV-11 is impossible. There is a lack of methodologic standardization, exposure standards, and applicable epidemiologic information that relates fiber number concentration to mass concentration. The electron-microscopy results vary from zero to nearly 2,000 ng/m^3. There is a progression in contamination from background through routine and custodial activity.

Exposure probabilities can be estimated to some degree by consideration of modes of contamination and activity in a building.

TABLE IV-10

Airborne Asbestos in Structures: Optical Data

Main Mode of Contamination	Activity	Mean Count, fibers/ml	n	Range or SD	References
Friable asbestos-containing material:					
Mixed fallout	Background, city	0.0	42	0.0	32,33
	Quiet conditions	0.0	65	0.0	32,33
Reentrainment	Routine activity:				
	Dormitory	0.1	10	0.0-0.8	a
	Schools, general	0.0	79	0.1	32,33
	Offices	0.0	60	0.0-0.6	32,33
	Dry sweeping, floor	1.6	5	0.7	33
	Dry dusting near face	4.0	6	1.3	33
	Bystander to cleaning	0.3	3	0.3	33
	Heavy dusting	2.8	8	1.6	36
	Laundry (contaminated clothing)	0.4	12	0.0-1.2	33
Contact with material	Maintenance:				
	Relamping	1.4	2	0.1	33
	Plumbing	1.2	6	0.1-2.4	32
	Cable movement	0.4	24	0.2-6.4	32

Mixed: contact reentrainment	Renovation:				
	Ceiling repair	17.7	3	8.2	32
	Track light	7.7	6	2.9	33
	Hanging light	0.3	12	0.8	33
	Partition construction	3.1	4	1.1	33
	Pipe lagging	4.1	8	1.8-5.8	36
Contact	Ceiling damage by vandalism	12.8	5	8.0	32
	Stripping dry ceiling material	82.2	11	22.4-117.0	33
	Stripping wet (amended water) ceiling material	1.2	96	5.2	32,33[a]
	Drilling, machining	3.4	7	1.0-5.8	32
	Abatement by encapsulation	0.0	28	0.7	32
Bonded asbestos-containing material:					
Contact	Stripping cementitious by wet method (amended water)	0.1	26	1.0	32
	Machining:				
	Sanding tiles	1.2	2	1.2-1.3	20
	Sanding concrete	7.2	6	2.1	32
	Cutting concrete	6.3	14	2.3	32
	Grinding concrete	0.3	6	0.2	32
	Sanding taping	5.3	11	1.3-16.9	30

[a]H. V. Brown, personal communication.

TABLE IV-11

Airborne Asbestos: Electron-Microscopy Data

Location and Activity	Mean, ng/m^3	n	Range, ng/m^3	Reference
Urban outdoors:				
48 U.S. cities	<10	187	--	23
N.Y. City	17	22	2-65	25
N.J. schools	14	3	3-30	25
Indoors:				
Friable-asbestos, structural surface:				
Office building	79	3	40-110	36
N.Y. City schools	99	5	9-135	24
Mass. schools	151	5	38-260	24
N.J. schools	217	27	9-1,950	25
Office buildings	2.5-200	116	0-800	24
Custodial activity:				
N.J. apartment	296	1	--	36
Conn. school	643	2	186-1,100	36
N.J. school	1,950	1	--	25

Quiet or background conditions may represent the usual extent of activity in areas with low probability of either contact or reentrainment of asbestos. Under these conditions, contamination is low. Routine activities in a structure containing sprayed-asbestos surfaces will not usually result in detectable fiber concentrations. Routine activity can result in intense contamination in some situations. A school population's routine activity in a building with accessible asbestos surfaces may cause environmental contamination. Increased fallout, occasional contact, and reentrainment may all contribute to the highly variable fiber concentrations found under these conditions.

Custodial work can cause disturbance and reentrainment of accumulations of asbestos fibers. Reentrainment can be high during custodial activity, depending on cleaning methods. Maintenance work may involve direct contact with asbestos surfaces. Such activities may result in marked fiber dissemination.[33,36]

Uncontrolled removal of sprayed-asbestos surfaces during renovation not only causes high fiber concentrations for the duration of the work, but also increases the released-fiber burden in the structure. In such cases, exposures involve the renovation worker and the routine building-user as well.[33] Before a decision on building renovation to remove asbestos, the potential contamination during and after renovation must be evaluated. Both contact and reentrainment release mechanisms are involved, and very high concentrations occur during actual contact.

STANDARDS

Estimation of the hazard associated with the airborne fiber concentrations of Table IV-10 can be only approximate. No exposure standards to evaluate hazard have been developed for the general indoor environment. The only existing standard is that of the Occupational Safety and Health Administration (OSHA). Other recommendations for occupational exposure limits are those of NIOSH and the American Conference of Governmental Industrial Hygienists (ACGIH). Although these apply to and were intended for only occupational exposures, their consideration in the general indoor environment may have some merit:

- The standard optical-microscopy method is used, and the air data of Table IV-10 are comparable with the standard limits.
- The occupational exposure limits represent a distillate or summary of both exposure and epidemiologic information. Recently proposed changes reflect additional relevant epidemiologic evidence.

The use of the occupational exposure limits is considered by some investigators to be acceptable for approximating the exposure hazard.[34]

Table IV-12 outlines the occupational exposure limits from 1972 to the present. Comparison of the data in Table IV-10 with the exposure limits of Table IV-12 demonstrates that some activities can exceed

TABLE IV-12

Occupational Asbestos Exposure Limits

Limit	Time-Weighted Average (8 h/d), fibers/ml	Ceiling, fibers/ml
OSHA original, 1972[46]	5.0	10.0
OSHA present, 1976[45,46]	2.0	10.0
OSHA proposed, 1975[43]	0.5	5.0
NIOSH revised, 1977[41]	0.1	0.5
ACGIH adopted, 1980:[1]		
Amosite	0.5	--
Chrysotile	2.0	--
Crocidolite	0.2	--
Other forms	2.0	--

present time-weighted average (TWA) OSHA limits. Occasional events--such as removal, renovation, and vandalism (contact-mode categories)--exceed the 15-min excursion limit of 10 fibers/ml.

The present TWA and ceiling limits were set by the 1972 regulations to become effective in 1976.[46] The more recent limits, the 1975 proposed and the 1977 revised recommended, reflect the increasing awareness in asbestos-disease epidemiology and are more stringent. With each successive regulation, the range of activities that could be considered hazardous becomes more inclusive. The activities remaining outside the 1977 limits are only in categories of quiet and nonspecific routine. Occupational standards can potentially be exceeded during renovation, maintenance, and custodial activities that disturb applied material or accumulated fibers.

The optical-microscopy data indicate that contamination can sometimes exceed concentrations considered hazardous. Exposures occur in existing structures, and the population involved is large and varied in age, occupation, and behavior.

Children attending schools that contain friable asbestos material constitute a population of special concern. The schoolchildren population differs from other nonoccupational populations in age, population density, and behavior. Any exposure would occur early in their life, leaving a long period for development of asbestos-related diseases. A large number of students can be exposed at one time to asbestos that is released from asbestos-containing materials in the school building. The school population is also very active. Friable asbestos-containing materials can be damaged during routine activities and as a result of capricious behavior. Many cases of badly damaged asbestos-containing materials have been found in schools.[37]

REGULATIONS

Most asbestos handling, control, and disposal in the structural environment are subject to regulation by the EPA and the OSHA. Other federal agencies that regulate asbestos in various settings are the Mine Safety and Health Administration (MSHA), Consumer Product Safety Commission (CPSC), Food and Drug Administration (FDA), and Department of Transportation (DOT).

In accordance with section 112 of the Clean Air Act ("National Emission Standards for Hazardous and Air Pollutants"), the EPA promulgated regulations on asbestos in 1973 (40 CFR 61, Subpart B, "National Emission Standard for Asbestos").[48]

These regulations apply to the renovation or demolition of friable asbestos materials and to the spraying of asbestos. They specify procedures for removal and stripping of friable sprayed-asbestos fireproofing and insulation materials. The required work practices include EPA notification, material-wetting, containment, container labeling, and disposal of the removed material in an approved landfill. Fiber concentrations are not specified, but the regulations require that there be no visible emission outside the structure.

Actions taken by the CPSC have banned the use of asbestos textiles in general-use garments, asbestos in artificial-fireplace materials, and asbestos-containing spackling and taping compounds.[7, 8] Voluntary actions by manufacturers controlled the use of asbestos in hand-held hair-dryers.

CONTROL OF CONTAMINATION POTENTIAL

Where a potentially hazardous situation has been identified in a structure, asbestos-contamination control or elimination methods are indicated.[36, 49] There are four approaches to corrective action: preventive management (a specific management system initiated to prevent disturbance of asbestos-containing material, with no direct action taken on the material itself), removal and disposal by burial, encapsulation (asbestos-containing material is coated with a sealant), and enclosure (asbestos-containing material is separated from the building environment by barriers, such as sealed suspended ceilings).

The corrective methods can be used separately or in combination, and each has its own advantages. Removal eliminates the source of exposure to asbestos. Both enclosure and encapsulation are containment methods; because the asbestos material remains in the building, enclosure and encapsulation should be considered as temporary control methods, for use until the building is renovated or demolished.

The surface of asbestos-containing material can be damaged--causing the friable fragments to be released--by inadvertent or uninformed maintenance, repair, or renovation. A management system should be implemented to control any activity by either structure personnel or contractors. Any necessary work would be performed under controlled conditions to protect involved personnel and other building users and to prevent contamination of the building environment. Renovation or demolition should include elimination of friable asbestos-containing materials under safe conditions and protection for the worker, the building users, and the community.

Corrective action for surfaces considered very hazardous should be planned with priority appropriate to their contamination potential. Removal is the ultimate solution and will end contamination potential. Encapsulation with either penetrating or bridging sealants will enhance the cohesiveness of the material, eliminate fallout, and protect from minor damage. The technology of encapsulating asbestos has recently been the subject of intense interest, and effective sealants are now available. Management and control should be initiated for all areas that contain friable asbestos material, no matter what other action is planned. An asbestos management system can be implemented immediately, has low cost, and is highly effective in exposure control.[37, 49]

SUMMARY

Asbestos is a widespread component of the structural environment. Release of asbestiform mineral fibers from structural components depends on the cohesiveness of the asbestos-containing material and the intensity of the disturbing force. Asbestos-containing material that is friable is most readily released in structures. High-energy or machine disruption is necessary for the release of fibers from hard or bound asbestos-containing materials. Durability and aerodynamic capability have combined to produce a persistent and important contaminant for human exposure. Contamination can be due to slight fiber release in fallout, relatively great release by contact or direct material disruption, or reentrainment of fallen and accumulated fibers. Most contamination is episodic, activity-related, and local. Documented concentrations have been compared with existing and proposed occupational standards. Extensive disruption of asbestos-containing material results in substantial hazard potential. Removal, renovation, and demolition of friable material and machining of hard asbestos-containing material result in a high degree of environmental contamination in the vicinity of the disturbance. Maintenance work or custodial care involving either friable material or accumulated fibers can cause airborne contamination that should be considered hazardous.

REFERENCES

1. American Conference of Governmental Industrial Hygienists. TLVs. Threshold Limit Values for Chemical Substances in Workroom Air Adopted by ACGIH for 1980. Cincinnati: American Conference of Governmental Industrial Hygienists, 1980. 93 pp.
2. Asbestos Information Association of North America. Asbestos--General Information. Washington, D.C., 1975.
3. Assuncao, J., and M. Corn. The effects of milling on diameters and lengths of fibrous glass and chrysotile asbestos fibers. Am. Ind. Hyg. Assoc. J. 36:811-819, 1975.
4. Bragg, G. M., L. van Zuiden, and C. E. Hermance. The free fall of cylinders at intermediate Reynold's numbers. Atmos. Environ. 8:755-764, 1974.
5. British Occupational Hygiene Society, Committee on Hygiene Standards. Hygiene standard for chrysotile asbestos dust. Ann. Occup. Hyg. 11:47-49, 1968.
6. Campbell, W. J., R. L. Blake, L. L. Brown, E. E. Cather, and J. J. Sjoberg. Selected Silicate Minerals and Their Asbestiform Varieties. Mineralogical Definitions and Identification-Characterization. Bureau of Mines Information Circular 8751. College Park, Md.: U.S. Department of Interior, Bureau of Mines, College Park Metallurgy Research Center, 1977. 64 pp.
7. Consumer Product Safety Commission. Consumer patching compounds and artificial emberizing materials (embers and ash) containing respirable free-form asbestos. Fed. Reg. 42:63354-63365, December 15, 1977.
8. Consumer Product Safety Commission. General use garments containing asbestos are banned hazardous substances. Code of Federal Regulations, Title 16, Part 1500.17 (a7), 1970.

9. Dement, J. M., R. D. Zumwalde, and K. M. Wallingford. Discussion paper: Asbestos fiber exposures in a hard gold mine. Ann. N.Y. Acad. Sci. 271:345-352, 1976.
10. Dreessen, W. C., J. M. Dallavalle, T. I. Edwards, J. W. Miller, and R. R. Sayers. A study of Asbestosis in the Asbestos Textile Industry. Public Health Bulletin No. 241. Washington, D.C.: U.S. Treasury Department, Public Health Service, 1938. 126 pp.
11. Ferris, G., Jr. Asbestos Numerical Rating System. Boston: Commonwealth of Massachusetts Special Legislature Commission on Asbestos, September 18, 1978.
12. Fondimare, A., et al. Quantitative study of the deposition of asbestos in the lung and pleura of subjects with diverse exposures. In Proceedings of the Symposium on the Pathology of Asbestos, Rouen, France, October 28, 1975.
13. Langer, A. M. Approaches and constraints to identification and quantitation of asbestos fibers. Environ. Health Perspect. 9:133-136, 1974.
14. Langer, A. M., R. Ashley, V. Baden, C. Berkley, E. C. Hammond, A. D. Mackler, C. J. Maggiore, W. J. Nicholson, A. N. Rohl, I. B. Rubin, A. Sastre, and I. J. Selikoff. Identification of asbestos in human tissues. J. Occup. Med. 15:287-295, 1973.
15. Leidel, N. A., S. G. Bayer, R. D. Zumwalde, and K. A. Busch. USPHS/NIOSH Membrane Filter Method for Evaluating Airborne Asbestos Fibers. DHEW(NIOSH) Publication No. 79-127. Cincinnati: U.S. Department of Health, Education, and Welfare, National Institute for Occupational Safety and Health, 1979. 89 pp.
16. Levine, R. J., Ed. Asbestos: An Information Resource, p. C-9. DHEW Publication No. (NIH)79-1681. Washington, D.C.: U.S. Government Printing Office, 1978. [190] pp.
17. Lumley, K. P. S., P. G. Harries, and F. J. O'Kelly. Buildings insulated with sprayed asbestos: A potential hazard. Ann. Occup. Hyg. 14:255-257, 1971.
18. McCrone, W. C. Evaluation of asbestos in insulation. Am. Lab. 11(12):19-31, 1979.
19. Millipore Corp. Monitoring Airborne Membrane Filter. Application Procedure. Bedford, Mass.: Millipore Corp., 1972.
20. Murphy, R. L., B. W. Levine, F. J. Al Bazzaz, J. J. Lynch, and W. A. Burgess. Floor tile installation as a source of asbestos exposure. Am. Rev. Respir. Dis. 104:576-580, 1971.
21. New York City Council. Air Pollution Control Code. Local Law, Section 1403.2-9.11(B). New York: New York City Council, 1971.
22. Nicholson, W. J., A. N. Rohl, and E. F. Ferrand. Asbestos air pollution in New York City, pp. 136-139. In H. M. Englund and W. T. Beery, Eds. Proceeedings of the Second International Clean Air Congress. New York: Academic Press, Inc., 1971.
23. Nicholson, W. J., A. N. Rohl, R. N. Sawyer, E. Swoszowski, and J. D. Rodaro. Measurement of Asbestos in Ambient Air. Final Report. Contract CPA 70-92. National Air Pollution Control Administration, 1971.
24. Nicholson, W. J., A. N. Rohl, and I. Weisman. Asbestos contamination of building air supply systems. Paper No. 29-6 in Proceedings. International Conference on Environmental Sensing and

Assessment. Vol. II. Institute of Electrical and Electronics Engineers Ann. No. 75CH1044-I-29-6. Piscataway, N. J.: Institute of Electrical and Electronics Engineers, Inc., 1976.

25. Nicholson, W. J., E. J. Swoszowski, Jr., A. N. Rohl, J. D. Todaro, and A. Adams. Asbestos contamination in United States schools from use of asbestos surfacing materials. Ann. N.Y. Acad. Sci. 330:587-596, 1979.
26. Pooley, F. D. Electron microscope characteristics of inhaled chrysotile asbestos fibre. Br. J. Ind. Med. 29:146-153, 1972.
27. Pooley, F. D. The identification of asbestos dust with an electron microscope microprobe analyser. Ann. Occup. Hyg. 18:181-186, 1975.
28. Prust, R. S. Future problems to be anticipated: Demolition, repair and disposal. Ann. N.Y. Acad. Sci. 330:545-547, 1979.
29. Reitze, W. B., W. J. Nicholson, D. A. Holaday, and I. J. Selikoff. Application of sprayed inorganic fiber containing asbestos: Occupational health hazards. Am. Ind. Hyg. Assoc. J. 33:178-191, March 1972.
30. Rohl, A. N., A. M. Langer, I. J. Selikoff, and W. J. Nicholson. Exposure to asbestos in the use of consumer spackling, patching and taping compounds. Science 189:551-553, 1975.
31. Rohl, A. N., A. M. Langer, and A. G. Wylie. Mineral characterization of asbestos-containing spray finishes, pp. 59-64. In U.S. Environmental Protection Agency. Asbestos Containing Material in School Buildings: A Guidance Document. Part 1. U.S. Environmental Protection Agency, Office of Toxic Substances Publication No. C00090. Washington, D.C.: U.S. Environmental Protection Agency, 1979.
32. Rohl, A. N., and R. N. Sawyer. Airborne Fiber Levels in Asbestos Abatement Projects. To be presented at International Symposium on Indoor Air Pollution, Health and Energy Conservation, Amherst, Mass., October 14, 1981, sponsored by Harvard University School of Public Health, Energy and Environmental Policy Center.
33. Sawyer, R. N. Asbestos exposure in a Yale building. Analysis and resolution. Environ. Res. 13:146-169, 1977.
34. Sawyer, R. N. Indoor asbestos pollution: Application of hazard criteria. Ann. N.Y. Acad. Sci. 330:579-586, 1979.
35. Sawyer, R. N. Yale art and architecture building: Asbestos Contamination: Past, Present, and Future. Institute of Electrical and Electronics Engineers, Inc., Ann. No. 75CH1044-1-20-5. Piscataway, N.J.: Institute of Electrical and Electronics Engineers, Inc., 1976.
36. Sawyer, R. N., and C. M. Spooner. Sprayed Asbestos-Containing Material in Buildings. A Guidance Document. Part 2. U.S. Environmental Protection Agency Report No. EPA-450/2-78-014. Research Triangle Park, N.C.: U.S. Environmental Protection Agency, 1978.
37. Sawyer, R. N., and E. J. Swoszowski, Jr. Asbestos abatement in schools: Observations and experiences. Ann. N.Y. Acad. Sci. 330:765-776, 1979.
38. Speil, S., and J. P. Leineweber. Asbestos minerals in modern technology. Environ. Res. 2:166-208, 1969.
39. Stanton, M. F., and C. Wrench. Mechanisms of mesothelioma induction with asbestos and fibrous glass. J. Nat. Cancer Inst. 48:797-821, 1972.

40. Timbrell, V. Inhalation and biological effects of asbestos, pp. 429-445. In T. T. Mercer, P. E. Morrow, and W. Stober, Eds. Assessment of Airborne Particles. Fundamentals, Applications, and Implications to Inhalation Toxicity. Proceedings of the Third Rochester International Conference on Environmental Toxicity. Springfield, Ill.: Charles C Thomas, Publisher, 1972.
41. U.S. Department of Health, Education, and Welfare, National Institute for Occupational Safety and Health. Revised Recommended Asbestos Standard. DHEW (NIOSH) Publication No. 77-169, 1977. Washington, D.C.: U.S. Government Printing Office, 1977. 96 pp.
42. U.S. Department of Interior, Bureau of Mines. Mineral Industry Surveys. Asbestos. Washington, D.C.: U.S. Department of Interior, Bureau of Mines, 1979.
43. U.S. Department of Labor, Occupational Safety and Health Administration. Occupational exposure to asbestos. Notice of proposed rulemaking. Fed. Reg. 40:47651-47665, October 9, 1975.
44. U.S. Department of Labor, Occupational Safety and Health Administration. Occupational safety and health standards. Emergency standard for exposure to asbestos dust. Fed. Reg. 36:23207-23208, December 7, 1971.
45. U.S. Department of Labor, Occupational Safety and Health Administration. Occupational safety and health standards. Recodification of air contaminant standards. Fed. Reg. 40:23072-23073, May 28, 1975 (29 CFR 1910.1001).
46. U.S. Department of Labor, Occupational Safety and Health Administration. Occupational safety and health standards. Standard for exposure to asbestos dust. Fed. Reg. 37:11318-11322, June 7, 1972.
47. U.S. Environmental Protection Agency. National emission standards for hazardous air pollutants. Amendments to asbestos standard. Fed. Reg. 43:26372-26374, June 19, 1978.
48. U.S. Environmental Protection Agency. National emission standards for hazardous air pollutants. Asbestos. Fed. Reg. 38:8820-8823, 8829-8830, April 6, 1973.
49. U.S. Environmental Protection Agency, Office of Toxic Substances. Asbestos Containing Material in School Buildings: A Guidance Document, Part 1. U.S. Environmental Protection Agency, Office of Toxic Substances Publication No. C00090. Washington, D.C.: U.S. Environmental Protection Agency, 1979.
50. Villecco, M. Spray fireproofing faces controls or ban as research links asbestos to cancer. Archit. Forum 133(5):50-52, 1970.
51. Wagner, J. C., G. Berry, J. W. Skidmore, and V. Timbrell. The effects of the inhalation of asbestos in rats. Br. J. Cancer 29:252-269, 1974.
52. Zoltai, T. Asbestiform and acicular mineral fragments. Ann. N.Y. Acad. Sci. 330:621-643, 1979.

FIBROUS GLASS

Fibrous glass is a man-made inorganic fiber with widespread application and distribution in the fabrication, textile, and construction industries. It is used in thermal insulation (for structures, appliances, and pipe), acoustic insulation, textiles, plastic-material reinforcement, tire cord, yarns, matting, and filters. It is a common component of the structural environment. Although the production of fibrous glass is a half-century old, interest in potentially serious adverse health effects has occurred fairly recently.

DEFINITION

The term "fibrous glass" generally includes particles composed of glassy material with a length-to-width (or length-to-diameter) ratio that exceeds 3. The composite elements of the glassy material form an amorphous structure and are not well ordered or crystalline, as in the asbestiform minerals. Contemporary conventional fibrous-glass production blends silica sand, limestone, and soda ash as raw materials in a continuous process. Before 1950, glassy fibrous materials, commonly known as "mineral wools," were commonly produced by melting the slag of ore-smelting processes (slag wool) or naturally occurring rocks (rock wool). Mineral-wool production began serious development in this country after 1920 and reached a peak in the 1950s, before modern continuous glass processes reduced the use of slag and rock as primary materials.[13, 16]

There are two major categories of fibrous-glass products: continuous-filament glass and glass wool.

Continuous-filament glass is used in textiles and fabrics; as reinforcement in plastics, rubber, and paper; and in numerous other applications. It is produced by extruding molten glass through dimensioned orifices to yield fibers with fairly well-defined diameters. Continuous glass fibers can be selectively sized to provide the strength, hardness, or thermal properties desired for the intended application. Most continuous-filament operations produce fibers roughly 6 μm in diameter, with some glass reinforcing fibers having diameters of over 10 μm (see Pundsack;[13] Smith;[16] and W. Rietze, Johns Manville Corporation, personal communication).

The most important example of glass wool is the fibrous-glass thermal-insulation material used extensively in construction and equipment fabrication. Contemporary glass-wool production is a continuous processing of raw materials through melting, fiberization, and packaging. Fiberization devices use rotors or high-velocity gas jets either alone or in combination to produce particles of a desired dimension range. The diameter is important in thermal-insulation production, because the effectiveness of the material varies inversely with the fiber diameter. The fiber population produced in this way does not have the well-defined diameter of the continuous-filament process. A glass-wool fiberization system produces a range of sizes following a frequency distribution characteristic of the process.[19] Most commercial fibrous-glass insulation has mean fiber diameters in the range of 4.0-9.0 μm. For special applications, a small percentage of glass wool is produced at mean diameters of 1.0 μm or less.[13, 16]

CONCERN OVER POTENTIAL ADVERSE HEALTH EFFECTS

The increasing concern over the potentially serious health effects of fibrous-glass exposure is a consequence of a number of factors:

• Interest in the carcinogenicity of asbestiform mineral fibers has raised questions concerning a possible similar effect of other fibrous materials, including fibrous glass.[17]

• The fibrogenicity and carcinogenicity of fibrous glass had been demonstrated in animal inoculation studies in which mesotheliomas were produced by intrapleural and intraperitoneal implantation of fibrous glass.[19,22]

• Some studies of the mechanism of carcinogenicity associated with fibers have indicated that fiber dimensions are more important than physical or chemical properties.[18] The demonstration of the apparent influence of particle size and shape indicated the possibility of common mechanisms and effects among fibrous materials, specifically asbestos and fibrous glass.

• Glass fibers 1.0 μm or less in diameter are termed "microfibers." In animal studies, fibers of pathologic importance are in the microfiber range, with diameters less than approximately 0.5 μm.[19] Fibers considered respirable have diameters of approximately 3.5 μm or less.[18,20] Furthermore, the aerodynamic capability and potential for respiration increase mainly as a function of decreasing fiber diameter.[20,21] This implicates microfibers as the most suspect in pathologic importance, with respect to cellular effect, respirability, and aerodynamic capability.[4] Microfibers are being intentionally produced for special applications, and there is incidental microfiber production in some glass-wool processes.

• The use of fibrous glass is widespread and increasing. There is heavy consumption of fibrous glass by the construction industry, including friable materials that readily release fibers in the structural environment. The substitution of fibrous glass for asbestos and the demand for fibrous-glass insulation products for energy conservation will increase the use of these friable materials. The forms, uses, and distribution of fibrous-glass materials implies a substantial impact if the material has marked adverse health effects.

There is concern that human exposure to fibrous glass may cause disease. However, studies of mortality and morbidity and radiographic examinations have failed to demonstrate discernible hazard in occupational populations with exposure to fibrous-glass particles.[8,9,12,23]

Glass microfibers could be expected to have more airborne persistence, respirability, and cellular effect. However, no human epidemiologic data support the concept of microfiber pathogenicity. The industrial use of microfibers is relatively small, and there have been few accumulated years of human exposure.[4] However, in the older mineral-wool production, with a wide distribution of fiber dimensions, there has been a substantial population of particles that meet the microfiber definition. In the mineral-wool industry, the accumulated exposure experience is large in numbers and years.[13] The health significance of microfiber exposure in human populations is not known.

IMPORTANCE OF CHARACTERISTICS OF FIBROUS GLASS

The potential for environmental contamination and exposure is influenced by the dimensions of fibrous-glass particles. Both airborne persistence and respirability of the environmental contaminant are less than those of asbestos, because of the relatively large diameters of the glass particles.

Emission rates of fibrous glass depend on the proximity of the source material, general characteristics of cohesiveness and friability, and the intensity of the force causing the disruption. Studies of the general environment, including space ventilation systems, have demonstrated extremely low concentrations, less than 3 fibers/L.[3] Glass-fiber concentrations in occupational and production environments vary widely with the nature of the production process. Most studies have shown normal production-facility concentrations well below 1 fiber/ml as measured by the OSHA-NIOSH standard method for airborne asbestos.[5-7,10] As would be anticipated from aerodynamic considerations, studies in microfiber production facilities have documented airborne concentrations orders of magnitude higher than those in conventional fibrous-glass processes.[4,10]

The more friable forms, such as thermal insulation, are potential sources for environmental contamination. Studies of concentrations of airborne fibrous particles during removal of friable insulation material have shown high concentrations of airborne fibrous glass in the vicinity of worker activity. During the removal of friable spray-applied material (20% chrysotile asbestos and 70% fibrous glass), fiber counts in excess of 100 fibers/ml were encountered.[15]

ANALYSIS

The polarized-light microscope can be an effective instrument for identification of fibrous glass in construction materials. Characteristic shape, transparency, and lack of birefringence distinguish fibrous glass from asbestos mineral fibers. Modern fibrous glass usually appears as isotropic particles of fairly uniform diameter, rod-like appearance, and high length-to-diameter ratio. Some mineral-wool products may have highly variable dimensions, teardrop shapes, and spherical glass "shot" of relatively large diameters.[11,14]

STANDARDS

A TLV/TWA of 10 mg/m^3 for fibrous glass or dust has been listed by the ACGIH.[24] This had been listed as a nuisance particle, with the occupational exposure limit for fibrous glass given as 30×10^6 particles/ft^3, or 10 mg/m^3 of air.[1] In a recent industry-wide survey, concentrations of airborne particulate matter were generally less than 2.5 mg/m^3.[5]

CONTROL

In consideration of the uncertainties of carcinogenicity, relevance of the experimental tumor-production studies in animals, and the characteristics of glass-fiber exposures of people, it appears prudent to reduce microfiber exposure to the lowest possible point permitted by available technology.

REFERENCES

1. American Conference of Governmental Industrial Hygienists. Industrial Ventilation. A Manual of Recommended Practice, p. 13-13. 14th ed. Lansing, Mich.: American Conference of Governmental Industrial Hygienists, 1977.
2. American Conference of Governmental Industrial Hygienists. TLVs. Threshold Limit Values for Chemical Substances in Workroom Air Adopted by ACGIH for 1980, p. 19. Cincinnati: American Conference of Governmental Industrial Hygienists, 1980.
3. Balzer, J. L. Environmental data; airborne concentrations found in various operations, pp. 83-89. In U.S. Department of Health, Education, and Welfare, National Institute for Occupational Safety and Health. Occupational Exposure to Fibrous Glass. Proceedings of a Symposium. HEW Publication No. (NIOSH) 76-151. Washington, D.C.: U.S. Government Printing Office, 1976.
4. Dement, J. M. Environmental aspects of fibrous glass production and utilization, pp. 97-109. In U.S. Department of Health, Education, and Welfare, National Institute for Occupational Safety and Health. Occupational Exposure to Fibrous Glass. Proceedings of a Symposium. HEW Publication No. (NIOSH) 76-151. Washington, D.C.: U.S. Government Printing Office, 1976.
5. Esmen, N., M. Corn, Y. Hammad, D. Whittier, and N. Kotsko. Summary of measurements of employee exposure to airborne dust and fiber in sixteen facilities producing man-made mineral fibers. Am. Ind. Hyg. Assoc. J. 40:108-117, 1979.
6. Esmen, N. A., Y. Y. Hammad, M. Corn, D. Whittier, N. Kotsko, M. Haller, and R. A. Kahn. Exposure of employees to man-made mineral fibers: Mineral wool production. Environ. Res. 15:262-277, 1978.
7. Fowler, D. P., J. L. Balzer, and W. C. Cooper. Exposure of insulation workers to airborne fibrous glass. Am. Ind. Hyg. Assoc. J. 32:86-91, 1971.
8. Gross, P., J. Tuma, and R. T. P. deTreville. Lungs of workers exposed to fiber glass. A study of their pathologic changes and their dust content. Arch. Environ. Health 23:67-76, 1971.
9. Hill, J. W., W. S. Whitehead, J. D. Cameron, and G. A. Hedgecock. Glass fibres: Absence of pulmonary hazard in production workers. Br. J. Ind. Med. 30:174-179, 1973.

10. Konzen, J. L. Results of environmental air-sampling studies conducted in Owens-Corning fiberglas manufacturing plants, pp. 115-120. In U.S. Department of Health, Education, and Welfare, National Institute for Occupational Safety and Health. Occupational Exposure to Fibrous Glass. Proceedings of a Symposium. HEW Publication No. (NIOSH) 76-151. Washington, D.C.: U.S. Government Printing Office, 1976.

11. McCrone, W. C. Evaluation of asbestos in insulation. Am. Lab. 11(12):19-31, 1979.

12. Nasr, A. N. M., T. Ditchek, and P. A. Scholtens. The prevalence of radiographic abnormalities in the chests of fiber glass workers. J. Occup. Med. 13:371-376, 1971.

13. Pundsack, F. L. Fibrous glass--manufacture, use, and physical properties, pp. 11-18. In U.S. Department of Health, Education, and Welfare, National Institute for Occupational Safety and Health. Occupational Exposure to Fibrous Glass. Proceedings of a Symposium. HEW Publication No. (NIOSH) 76-151. Washington, D.C.: U.S. Government Printing Office, 1976.

14. Rohl, A. N., A. M. Langer, and A. G. Wylie. Mineral characterization of asbestos-containing spray finishes, pp. 59-64. In U.S. Environmental Protection Agency, Office of Toxic Substances. Asbestos Containing Material in School Buildings: A Guidance Document. Part 1. Washington, D.C.: U.S. Government Printing Office, 1979.

15. Sawyer, R. N. Asbestos exposure in a Yale building. Analysis and resolution. Environ. Res. 13:146-169, 1977.

16. Smith, H. V. History, processes, and operations in the manufacturing and uses of fibrous glass--one company's experience, pp. 19-26. In U.S. Department of Health, Education, and Welfare, National Institute for Occupational Safety and Health. Occupational Exposure to Fibrous Glass. Proceedings of a Symposium. HEW Publication No. (NIOSH) 76-151. Washington, D.C.: U.S. Government Printing Office, 1976.

17. Stanton, M. F. Fiber carcinogenesis: Is asbestos the only hazard? J. Nat. Cancer Inst. 52:633-634, 1974.

18. Stanton, M. F. Some etiological considerations of fibre carcinogenesis, pp. 289-294. In P. Bogovski, G. Gilson, V. Timbrell, and J. C. Wagner, Eds. Biological Effects of Asbestos. Scientific Publications No. 8. Lyon, France: International Agency for Research on Cancer, 1973.

19. Stanton, M. F., and C. Wrench. Mechanisms of mesothelioma induction with asbestos and fibrous glass. J. Nat. Cancer Inst. 48:797-821, 1972.

20. Timbrell, V. Aerodynamic considerations and other aspects of glass fiber, pp. 33-50. In U.S. Department of Health, Education, and Welfare, National Institute for Occupational Safety and Health. Occupational Exposure to Fibrous Glass. Proceedings of a Symposium. HEW Publication No. (NIOSH) 76-151. Washington, D.C.: U.S. Government Printing Office, 1976.

21. Sawyer, R. N., and C. M. Spooner. Sprayed Asbestos-Containing Material in Buildings. A Guidance Document. Part 2. U.S. Environmental Protection Agency Report No. EPA-450/2-78-014. Research Triangle Park, N.C.: U.S. Environmental Protection Agency, 1978.
22. Wagner, J. C., G. Berry, and V. Timbrell. Mesotheliomata in rats after inoculation with asbestos and other materials. Br. J. Cancer 28:173-185, 1973.
23. Wright, G. W. Airborne fibrous glass particles. Chest roentgenograms of persons with prolonged exposure. Arch. Environ. Health 16:175-181, 1968.

COMBUSTION SOURCES

Whenever unvented combustion takes place indoors or venting systems attached to stoves, boilers, or heaters are malfunctioning, a wide range of combustion products can be discharged directly into the indoor atmosphere. This section summarizes essential information pertaining to indoor sources of combustion products and their effects on indoor air quality. The emphasis is on residential buildings, where research efforts on indoor air quality have been concentrated, but results of limited studies of combustion-generated pollution in other types of buildings are also presented.

In general, the data presented here represent isolated situations whose characteristics are highly specific to the site and the combinations of activities that produced the effects; therefore, transfer of these findings to other situations must be done only with extreme care.

Smoking (which is discussed elsewhere), is the most widely encountered source of combustion products indoors. Besides smoking, the primary sources of combustion byproducts in residential buildings are usually space heaters, gas stoves, and gas water heaters. Exhaust from automobiles in attached garages can also be a source of combustion byproducts in buildings, as can wood fires, oil and kerosene lamps, and candles.

The major pollutants associated with indoor combustion are carbon monoxide, nitric oxide, nitrogen dioxide, aldehydes and other organic compounds, and fine particles. These combustion products usually occur in low concentrations, compared with the major combustion products--carbon dioxide and water vapor. Inefficient combustion from unvented or poorly vented space heaters, fireplaces, and lamps can also emit carcinogenic hydrocarbon particles. Carbon dioxide and water vapor are also produced as a result of normal metabolic processes of building occupants and add to the burden associated with gas appliances. Humans produce 30-60 g of carbon dioxide and a similar amount of water vapor per hour. An unvented space heater rated at 10,000 Btu/h, or about 2,500 kcal/h, produces around 750 g of carbon dioxide per hour. Accordingly, depending on occupant density, space limitations, and the extent of ventilation and infiltration, carbon dioxide concentration in the indoor atmosphere can rise substantially above the normal range of 0.03-0.06%. Respiration is affected when the concentration of carbon dioxide in the air rises above 1.5%, and concentrations above 3% can cause headache, dizziness, and nausea. Above about 6-8%, carbon dioxide causes stupor to the degree that exposed persons are unable to take steps for self-preservation.[4] The Threshold Limit Value in the occupational environment is 0.5%[1]--a value which has also been applied to submarine crews, whose incidence of illness increased after long-term exposure to concentrations of 0.5-1%.[19 22 25] In the residential environment under occupied conditions, carbon dioxide concentrations are typically 0.07-0.20%.

RESIDENTIAL BUILDINGS

Space Heating

There are many documented cases of health problems and even deaths resulting from excessive carbon monoxide released by unvented or improperly vented heating systems;[2 16 18 27] however, few systematic studies have provided detailed measurements of indoor air-quality problems in such houses. Most space heating in U.S. houses is by externally vented heating systems (central furnaces or space heaters). When the heating system is properly designed, maintained, and functioning, combustion products that could directly affect indoor air quality do not enter the indoor environment. However, if a negative pressure develops in the interior space or if there is a faulty exhaust system (e.g., a cracked heat exchanger or blocked flue), there can be direct and serious degradation of indoor air quality.

Some space heating of homes is generally provided by unvented gas and kerosene heaters. This type of heating is more commonly used in rural areas and warm climates, such as the southern United States, and is especially dangerous, because it emits its combustion products directly into the living space.

In their study of indoor air quality in several homes in Rotterdam, Biersteker and associates determined that sulfur dioxide concentrations indoors are not normally affected by heating systems that are kept in good condition.[3] However, in one older home with a faulty heater, the indoor concentration was 3.8 times the outdoor concentration.

In a 1969-1970 study of indoor-outdoor air quality in the United States, Yocom and co-workers selected four homes, two public buildings, and two office buildings for analysis.[28 29] One of the two homes with coal heating had an antiquated central heating system with a leaky flue. Sulfur dioxide in this home approached 1 ppm and carbon monoxide exceeded 50 ppm over periods of 1 h and longer, coinciding with periods when coal was added to the fire and the fire was stoked.

As part of a pilot study to assess indoor air quality in buildings, Hollowell and co-workers showed that a gas-fired heating system in one home, although vented to the outside, produced higher nitric oxide and nitrogen dioxide concentrations indoors than outdoors.[9] In this work and in the earlier work of Yocom et al.[28 29] and Biersteker et al.,[3] no attempt was made to measure the emission rates of the combustion source or the rates at which pollutant gases entered the indoor environment.

In a 1973-1974 study of indoor sources of air pollutants, Cote and co-workers determined the emission rates of carbon monoxide, nitric oxide, and nitrogen dioxide from an unvented gas-fired space heater.[6] Table IV-13 shows the emission data obtained.

Yamanaka and co-workers measured nitrogen oxides from various unvented and vented space heaters commonly used in Japan.[26] Results on both radiant and convection types of unvented kerosene heaters, as well as on various water heaters and gas stoves, were reported. Nitrogen dioxide emission from the radiant kerosene-fired space heater averaged 46 μg/kcal (0.011 g/MJ); that from the convection type averaged 251 μg/kcal (0.060 g/MJ).

In all the studies noted, the purpose was to measure emission of "typical" units. The units tested were not necessarily representative of the entire class of devices from which they were selected, nor was there any attempt to conduct an exhaustive study of their combustion characteristics.

Homeowners in many parts of the country are returning to the use of wood as a heating fuel, because of the increasing cost of oil and natural gas. This trend is especially strong in the northeastern states, which have depended largely on oil as a heating fuel. (The price of No. 2 fuel oil increased by a factor of 6 or more between 1965 and 1980.)

Wood stoves and fireplaces are vented to the outdoor atmosphere, but a number of circumstances can cause combustion products to be emitted to the indoor atmosphere: improper installation (e.g., insufficient stack height), cracks or leaks in or poor fitting of stovepipe, negative air pressure indoors, downdrafts, and accidents, as when a log rolls out of the fireplace. Although much is known about the combustion products of fuels used for space heating, little is known about the impact of the emission from wood stoves and fireplaces on indoor air quality--a subject urgently in need of investigation. Combustion products of wood are highly irritating to the eyes, nose, and respiratory system and thus provide a warning to occupants that combustion products are present. Table IV-14, from the work of Duncan and co-workers, shows the types of pollutants associated with wood burning.[7]

A field monitoring program designed to compare indoor and outdoor pollution in 10 residences and two office buildings was undertaken in the Boston metropolitan area.[13] Three of the monitored residences used either a wood stove or a fireplace in the course of the sampling period. Increased indoor concentrations of total suspended particles (TSP), respirable particles, and benzo[a]pyrene were observed during periods of wood-burning. The average indoor TSP concentrations during days with wood-burning were about 3 times the corresponding concentrations during days without wood-burning. Indoor 24-h benzo[a]pyrene concentrations during days with wood-stove use were an average of 5 times higher than those during non-wood-burning periods. The authors concluded that the increased indoor concentrations of TSP,

TABLE IV-13

Pollutant-Gas Emission from Unvented Gas-Fired Space Heaters[a]

Operation	Heat Input, kcal/h	Pollutant Emission Factors, μg/kcal			Pollutant Emission Rates, mg/h		
		NO	NO_2	CO	NO	NO_2	CO
Low flame, steady state	2,800	76.4	46.4	632	214	130	1,770
High flame	6,200	135	43.8	319	837	272	1,982

[a]Based on Cote *et al.*[6]

TABLE IV-14

Emission from Residential Wood-Fired Stoves[a]

Substance Emitted	Emission, lb/cord[b] Range	Average
Criteria pollutants		
Particles	3-93	30.3
SO_x	0.5-1.5	0.7
NO_x	0.7-2.6	1.6
Hydrocarbons	1-146	41.6
Carbon monoxide	300-1,220	598.3
Noncriteria pollutants		
Polycyclic organic materials	0.6-1.22	0.9
Formaldehyde	0.3-1	0.8
Acetaldehyde	0.1-0.3	0.4
Phenols	0.3-8	3.3
Acetic acid	5-48	21.1
Aluminum	--	1.3
Calcium	--	10.2
Chlorine	--	0.1
Iron	--	0.7
Magnesium	--	2.0
Manganese	--	1.6
Phosphorus	--	1.0
Potassium	--	3.6
Silicon	--	1.6
Sodium	--	0.7
Titanium	--	0.02

[a]Based on Duncan *et al.*[7]

[b]The relationship used to convert from lb/ton to lb/cord was: 1 lb/ton = 1.65 lb/cord.

respirable particles, and benzo[a]pyrene attributed to wood-burning may have long-term health implications.[14]

Gas Stoves: Pollutant Emission Rates and Concentrations

The early pilot studies of Yocom and co-workers, reported in 1974, showed that, on the basis of relative indoor-outdoor concentrations of carbon monoxide, unvented gas stoves definitely contribute to the deterioration of indoor air quality.[28]

A brief study by the EPA showed that peak nitrogen dioxide concentrations up to 1 ppm (about 1,880 $\mu g/m^3$) and 1-h averages of 0.25-0.50 ppm (about 470-940 $\mu g/m^3$) are reached in a closed kitchen with no external ventilation.[8]

Wade and co-workers in 1973-1974 studied four homes equipped with gas stoves to determine the concentrations of nitric oxide, nitrogen dioxide and carbon monoxide and their impact on indoor air quality.[24] Sampling was carried out for 2-wk periods in each home, simultaneously at four sampling locations--three indoors and one outdoors. Table IV-15 presents the principal data from one home included in this study. The following were the main conclusions of the study:

- Emission from gas stoves contributes nitrogen dioxide, nitric oxide, and carbon monoxide to the indoor atmosphere of houses where such stoves are used. Kitchen concentrations of these gases responded rapidly to stove use and, for a given house during a given season, there was a rough correlation between average nitrogen dioxide concentrations and average stove use.
- Nitrogen dioxide and nitric oxide were produced in roughly equal amounts in the homes where testing was conducted. Indoor concentrations of these pollutants were invariably higher than those outside.
- Normal stove operations frequently resulted in nitrogen dioxide concentrations in the kitchens averaging over 100 $\mu g/m^3$ for the 2-wk sampling periods.
- Comparison of samplings carried out in the spring-summer of 1973 and the fall-winter of 1973-1974 showed that in the colder weather, when the house was closed up more often, pollutant concentrations were more uniformly distributed in the various rooms of the house than in the warmer months.
- A diffusion experiment conducted in one of the houses showed that the half-life of nitrogen dioxide was only one-third that of carbon monoxide and nitric oxide, indicating that nitrogen dioxide decays through reaction or adsorption, in addition to normal dilution from air exchange. This effect was observed in some of the other houses by comparing the relative concentrations of nitrogen dioxide and the other pollutants in various parts of the house.

Moschandreas and his associates carried out a 2-yr air sampling program to characterize the indoor residential air environment.[12] Indoor air quality was monitored for continuous periods of

TABLE IV-15

Summary of Indoor and Outdoor Concentrations of Pollutant Gases at a Suburban Home near Hartford, Connecticut[a]

Sampling Period	Pollutant Gas	Average Concentration, $\mu g/m^3$: Kitchen over Stove (1)	Kitchen 1 m from Stove (1A)	Living Room (2)	Bedroom (3)	Outside (4)	Indoor: Outdoor Ratio 1:4	2:4
Spring-summer, 1973	NO_2	--	100	61	52	44	--	1.39
	NO	--	102	64	65	26	--	2.46
	CO	--	4,490	4,070	4,170	3,480	--	1.17
Fall-winter, 1973-1974, first half	NO_2	67	60	55	--	50	1.34	1.10
	NO	136	134	94	--	63	2.16	1.49
	CO	4,190	3,520	3,230	--	1,670	2.51	1.93
Fall-winter, 1973-1974, second half	NO_2	110	67	--	49	46	2.39	--
	NO	134	131	--	102	65	2.06	--
	CO	4,790	4,210	--	3,820	2,310	2.07	--

[a]Based on Cote *et al.*[6] House specifications: split level, two bedrooms, centrally located, well-ventilated kitchen, 2,000-ft^2 living area, gas-fired stove and central heating system; occupied by two (smoking) adults and two teen-aged children; home 6 yr old at time of tests.

approximately 14 d in each of five detached dwellings, two semidetached dwellings (townhouses), six apartment units, two mobile homes, and one school. Three of the dwellings were referred to as "experimental," because they were designed to conserve energy. The remaining dwellings were referred to either as "conventional" or by structural type. In addition, the residences were divided according to their cooking and heating fuel. These structures are in five metropolitan areas: Baltimore, Washington, D.C., Chicago, Denver, and Pittsburgh. The dwellings in Baltimore, Washington, D.C., and Chicago were monitored twice to obtain seasonal variations. Conclusions from the study pertaining to pollutants associated with combustion products (carbon monoxide, nitric oxide, and nitrogen dioxide) follow:

> Indoor CO concentrations are generally higher than corresponding outdoor levels in all residences monitored. High indoor concentrations may be attributed to . . . indoor CO emission sources, such as gas-fired cooking appliances, attached garages, faulty furnaces, and cigarette smoking. . . .
>
> The complexity of the dynamics involved in the establishment of an indoor-outdoor relationship is clearly illustrated in the interpretation of the data base generated for NO. From the perspective of NO indoor variation and under real-life conditions, three types of indoor environments have emerged: 1) houses with electric cooking and heating appliances; 2) houses that are heated by gas furnaces, yet serviced by electric cooking appliances; and 3) houses that are furnished with gas cooking and heating equipment. In houses equipped with gas cooking appliances, observed indoor NO levels are consistently higher than observed outdoor levels. Houses with gas furnaces but electric cooking appliances display higher NO indoor levels than outdoor levels, most of the time. However, there are time intervals interspersed throughout the monitoring period during which the observed NO outdoor levels surpass corresponding indoor levels. Indoor NO concentrations in totally electric homes are almost always lower than corresponding outdoor concentrations. . . .
>
> The residential environment often provides a shelter from high outdoor NO_2 levels. The three classes of residences identified in the interpretation of the NO data also manifest themselves in the study of NO_2. The data base collected for this project indicates that the hourly average indoor concentrations of NO_2 are almost always lower than the corresponding ambient levels in totally electric houses. Houses equipped with gas furnaces and electric cooking appliances also shelter their occupants, but to a lesser extent during peak ambient NO_2 levels. Totally gas residences do not appear to provide such protection. . . .

With the large data base generated by the study, Moschandreas and Stark[11] formulated, documented, and validated a numerical prediction model (see Chapter VI). A series of numerical simulations with the model showed that, under some conditions involving gas appliances, indoor carbon monoxide, nitric oxide, and nitrogen dioxide concentrations increase substantially. Under these same conditions (with the oven in use for 2 consecutive hours), carbon monoxide may reach concentrations over 35 ppm.

The influence of gas stoves on indoor nitrogen dioxide content was confirmed by Palmes and associates, who used integrating "personal" monitors.[15] Melia and associates used this same type of sampler in an epidemiologic study to show that nitrogen dioxide concentrations were significantly higher in homes with gas stoves than in those with electric stoves.[10]

Puxbaum measured nitric oxide and nitrogen dioxide in a kitchen and an adjoining room in a home with a gas stove and gas hot-water appliance and compared indoor and outdoor concentrations. He also compared the burning of natural and "town" gas and found that the concentrations of both nitrogen dioxide and nitric oxide were significantly higher during the natural-gas runs than during the town-gas burning.[17]

In their 1973-1974 study, Cote and co-workers measured rates of pollutant emission from gas appliances.[6] Table IV-16 presents their data from various burner configurations and operating modes for a new and an old stove. (These stoves produce low but continuous emission from the pilot light, whereas current trends are toward low-heat-input gas pilot lights and non-gas ignition systems, which will reduce this source of indoor air pollution.)

More recently, Traynor and co-workers studied, in detail, emission from a gas-fired stove.[23] Table IV-17 summarizes some of the data from this work. In the case of nitrogen dioxide, there was generally good agreement between the Cote and Traynor studies. No laboratory tests on electric stoves were carried out for comparison in these studies, although, in an earlier field study, Hollowell *et al.* did observe small increases in kitchen ozone concentrations with electric-stove use.[9]

Both Cote *et al.* and Hollowell *et al.* measured gaseous pollutant emission from top burners while water-filled pans of different materials were being heated. Cote *et al.* noted some minor effects on gas-flame emission, especially with a low flame.[6] For example, carbon monoxide concentrations were higher when an aluminum pan was in place, and nitrogen dioxide concentrations were marginally lower with all types of pans under low-flame conditions. Hollowell and co-workers, however, could find no appreciable differences in gaseous emission, regardless of the types of pans used.

In the study by Cote and co-workers, the efficiency of an exhaust hood in removing stove-generated pollutants was determined.[6] Removal efficiencies varied from 4% with the hood fan off to 49% with the fan at its highest speed setting. (The low removal efficiency with the fan off was the result of natural ventilation through the hood.) A recirculating hood with activated charcoal, as might be expected, had

TABLE IV-16

Summary of Pollutant Emission of Gas Appliances for Typical Operating Conditions (No Pans in Place)[a]

Appliance	Operation	Heat Input Rate, kcal/h	Pollutant Emission Factors, μg/kcal			Pollutant Emission Rates, mg/h		
			NO	CO_2	CO	NO	NO_2	CO
Older gas stove with cast iron burners	Pilot lights	150	45.3	54.6	419	6.8	8.2	62.9
	1 burner, high flame	2,700	92.6	51.8	382	250	140	1,031
	3 burners, high flame	6,780	117.0	72.8	475	793	494	3,220
	Oven, steady state	2,200	91.4	73.1	530	201	161	1,166
Newer gas stove with pressed-steel burners	Pilot lights	100	4.7	18.6	842	0.5	1.9	84.2
	1 burner, high flame	3,500	130.0	79.0	510	455	277	1,795
	3 burners, high flame	10,200	138.0	65.6	315	1,408	669	3,213
	Oven, steady state	2,200	77.9	50.4	1,620	171	111	3,564

[2]Based on Cote *et al.*[6]

TABLE IV-17

Summary of Gas-Stove Emission[a]

Pollutant	Oven[b] Emission, μg/kcal[d]	Oven[b] No. Experimental Runs	Top Burner[c] Emission, μg/kcal[d]	Top Burner[c] No. Experimental Runs
Gases:				
CO	950 (650-1,600)	6	890 (720-1,090)	4
CO_2	200,000 (195,000-205,000)	6	205,000 (196,000-217,000)	3
NO	29 (14-50)	11	31 (21-47)	4
NO_2	62 (44-74)	11	85 (69-100)	4
SO_2	0.8 (0.5-1.0)	11	0.8 (0.6-0.9)	4
HCN	1.8 (1.6-2.3)	3	0.07	1
HCHO	11.4 (9.9-14.2)	5	5.2 (2.0-12.0)	5
Particles (<2.5 μm):				
Carbon	0.13 (0.05-0.24)	9	0.90 (0.86-0.96)	4
Sulfur (as SO_4)	0.01	9	0.05 (0.01-0.08)	4
Total respirable mass	--	--	1.7 (1.0-2.6)	3

[a]Based on Traynor et al.[23]

[b]Oven operated for 1 h at 180°C (350°F).

[c]Operated with water-filled cooking pots.

[d]Ranges in parentheses.

little effect on gaseous pollutants; however, such a device may partially control organic gases and vapors responsible for odors and aerosols released during cooking. The effect of kitchen ventilation on carbon monoxide, nitric oxide, and nitrogen dioxide was also included in the study by Traynor et al.,[23] and their data on nitrogen dioxide are presented in Figure IV-5.

Other Indoor Combustion Sources

Several other combustion sources can affect indoor air quality. Although they tend to be site-specific and not as common as domestic heating and gas-stove operation, several are worthy of mention.

Water Heaters and Clothes Dryers. In two homes with gas water heaters, Traynor observed that indoor nitrogen dioxide concentrations were greater than outdoor (G. W. Traynor, Lawrence Berkeley Laboratory, personal communication). Initially, passive monitors were used to measure average indoor concentrations over 1 wk. Both homes were found to have increased indoor concentrations and were investigated in more detail with a continuous analyzer. In both cases, nitrogen dioxide entered the living space from the flue collar on top of the water heater, despite the fact that the appliances were designed to vent the combustion products outside. Similar considerations apply to gas clothes dryers.

Automobiles. Driving a car into or out of a basement or attached garage can strongly affect indoor air quality, depending on the configuration of the house and the routes for entering and leaving the garage. In a number of documented cases, occupants have inadvertently left a car running in a basement garage from which the resulting carbon monoxide drifted into the associated house or apartment and caused sickness or death of occupants. The most critical situation would occur in a basement garage during cold weather, when the "stack effect" of a heated house tends to draw air in from the garage and distribute the pollutants captured in the garage throughout the house. A variant of this situation is an apartment house or office building with a basement garage that has stairwells or elevator shafts that can distribute pollutants throughout the building.

Little information is available on the impact on indoor air quality of automotive exhaust emitted in garages. Yocom and associates[28] sampled one house with an attached garage and concluded that the design of the house (split-level) caused automotive emission from the garage to have a greater impact on carbon monoxide concentration than the gas stove.

Charcoal Broiling. Charcoal cooking is usually an outdoor activity; however, it is sometimes done in a fireplace, and some fireplaces even include a charcoal cooker. Depending on how such cooking is carried out and how well the fireplace draws, the resulting emission can enter the indoor environment. There appear to be no data

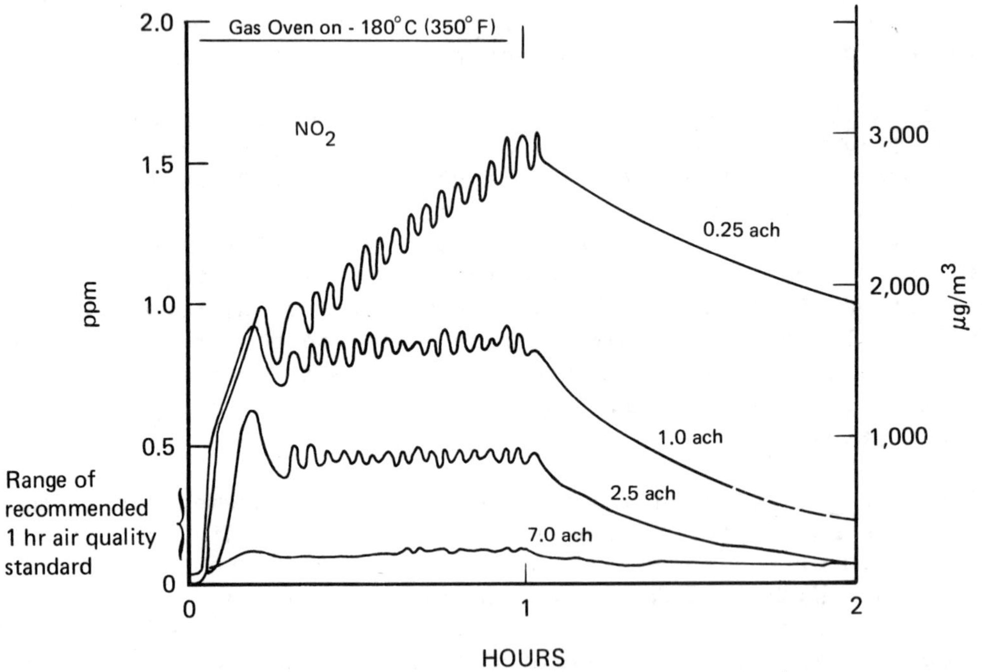

FIGURE IV-5 Nitrogen dioxide concentrations in test kitchens reported by Traynor *et al.*[23] as functions of use of gas oven with different kitchen exhaust rates.

on the direct impact of charcoal broilers on indoor air quality, but a recent study by Brookman and Birenzvige on exposure to air pollutants from "domestic combustion" sources provides some idea of how great these exposures might be.[5] They used personal samplers to measure hourly carbon monoxide and particulate matter in the outdoor environment while people were operating gas lawn mowers and chain saws and cooking with charcoal. A sampler adjacent to a charcoal cooker recorded that hourly carbon monoxide exposures ranged from 8 to 38 ppm. Analysis of the particulate filters for polynuclear aromatic organic matter showed that it was below the detection concentration of the screening method used.

Hobbies. A variety of hobbies involve combustion processes, for example, heating and soldering with an LPG torch and brazing or welding with an oxyacetylene torch. Depending on the location, extent, and duration of such activities, they could have an impact on indoor air quality. Under some circumstances, the exposure of those pursuing such hobbies to harmful contaminants could be substantial.

COMMERCIAL BUILDINGS

Only limited information is available on the effects of combustion sources on indoor air quality in commercial buildings. Nevertheless, any of the sources mentioned above can affect the quality of indoor air in commercial buildings as they do in residential buildings.

Spengler and co-workers determined carbon monoxide concetrations in several enclosed ice-skating rinks in Massachusetts[20] where gasoline-powered ice resurfacing machines are used; the concentrations were found to exceed 50 ppm regularly.

Large public garages where traffic congestion commonly occurs in confined spaces constitute another example of indoor air-quality problems in commercial buildings. There are many documented cases of drivers and passengers who experienced acute health effects (presumably from carbon monoxide) at times of heavy garage traffic. The most critical situation occurs at the end of the workday or at the completion of a major sporting event, when hundreds of cars line up with their engines running. Stankunas and associates carried out a study for the American Society of Heating, Refrigerating and Air-Conditioning Engineers (ASHRAE) to measure carbon monoxide in several parking garages.[21] They developed a model for calculating in-garage carbon monoxide concentrations on the basis of such variables as ventilation volume, outdoor ambient carbon monoxide concentration, initial carbon monoxide concentration in the garage, and garage volume.

REFERENCES

1. American Conference of Governmental Industrial Hygienists. TLVs. Threshold Limit Values for Chemical Substances in Workroom Air Adopted by ACGIH for 1979. Cincinnati: American Conference of Governmental Industrial Hygienists, 1979. 94 pp.

2. Amiro, A. Carbon monoxide presents public health problem. J. Environ. Health 32:83-88, 1969.
3. Biersteker, K., H. de Graaf, and C. A. G. Nass. Indoor air pollution in Rotterdam homes. Int. J. Air Water Pollut. 9:343-350, 1965.
4. Billings, C. E. Atmosphere, pp. 35-63. In J. F. Parker, Jr., and V. R. West, Eds. Bioastronautics Data Book. 2nd ed. National Aeronautics and Space Administration. Publication No. NASA SP-3006. Washington, D.C.: U.S. Government Printing Office, 1973.
5. Brookman, E. T., and A. Birenzvige. Exposure to Pollutants from Domestic Combustion Sources: A Preliminary Assessment. U.S. Environmental Protection Agency Report No. EPA-600/7-80-084. Washington, D.C.: U.S. Environmental Protection Agency, Office of Research and Development, 1980. 51 pp.
6. Cote, W. A., W. A. Wade, III, and J. E. Yocom. A Study of Indoor Air Quality. Final Report. U.S. Environmental Protection Agency Report No. EPA-650/4-74-042. Washington, D.C.: U.S. Environmental Protection Agency, 1974. 282 pp.
7. Duncan, J. R., K. M. Morkin, and M. P. Schmierbach. Air Quality Impact Potential from Residential Wood Burning Stoves. Paper 80-7.2, presented at 73rd Annual Meeting of the Air Pollution Control Association, Montreal, Quebec, June 22-27, 1980.
8. Eaton, W. C., J. N. Howard, Jr., R. M. Burton, F. Benson, and G. H. Ward. A Preliminary Study of Indoor Air Pollution in a Home Using a Gas Stove. Part I: Oxides of Nitrogen. U.S. Environmental Protection Agency, Human Studies Laboratory, 1972.
9. Hollowell, C. D., R. J. Budnitz, G. D. Case, and G. W. Traynor. Combustion-Generated Indoor Air Pollution. I. Field Measurements 8/75-10/75. Lawrence Berkeley Laboratory Report LBL-4416. Berkeley, Cal.: Lawrence Berkeley Laboratory, 1976. 25 pp. Available from National Technical Information Service, Springfield, Va., as LBL-4416.
10. Melia, R. J. W., C. Florey, S. C. Darby, E. D. Palmes, and B. D. Goldstein. Differences in NO_2 levels in kitchens with gas or electric cookers. Atmos. Environ. 12:1379-1381, 1978.
11. Moschandreas, D. J., and J. W. C. Stark. The Residential Environment and Energy Conservation: Predicting Indoor Air Quality. Paper 78-60.4, presented at the 71st Annual Meeting of the Air Pollution Control Association, Houston, Texas, June 1978.
12. Moschandreas, D. J., J. W. C. Stark, J. E. McFadden, and S. S. Morse. Indoor Air Pollution in the Residential Enviroment. Vol. 1. Data Collection, Analysis and Interpretation. U.S. Environmental Protection Agency Report No. EPA-600/7-78-229a. Research Triangle Park, N.C.: U.S. Environmental Protection Agency, Environmental Monitoring and Support Laboratory, 1978. 201 pp.
13. Moschandreas, D. J., J. Zabransky, and D. J. Pelton. Comparison of Indoor-Outdoor Concentrations of Atmospheric Pollutants. Final Report for the Electric Power Research Institute. GEOMET Technologies Inc., Contract No. EP 1301-1. May 1980.
14. Moschandreas, D. J., J. Zabransky, and H. E. Rector. The effects of woodburning on the indoor residential air quality. Environ. Int. (in press, 1981)

15. Palmes, E. D., C. Tomczyk, and J. DiMattio. Average NO_2 concentration in dwellings with gas or electric stoves. Atmos. Environ. 11:869-872, 1977.
16. Plotkin, S., and R. Kapplow. Food poisoning and carbon monoxide poisoning. N.Y. State J. Med. 52:2409-2411, 1952.
17. Puxbaum, H. Indoor Air Pollution by Combustion Sources--Influence of the Gas Type on NO_x Emissions. Paper presented at Euroanalyses III, Dublin, Ireland, August 20-25, 1978.
18. Rench, J., and E. P. Savage. Carbon monoxide in the home environment. J. Environ. Health 39:104-106, 1976.
19. Schaefer, K. E., Ed. Preventive Aspects of Submarine Medicine. Undersea Biomed. Res. 6(Suppl.):S-1--S-246, 1979.
20. Spengler, J. D., K. R. Stone, and F. W. Lilley. High carbon monoxide levels measured in enclosed skating rinks. J. Air Pollut. Control Assoc. 28:776-779, 1978.
21. Stankunas, A. R., P. T. Bartlett, and K. C. Tower. Contaminant Level Control in Parking Garages. Conference DV 80-5, No. 3, RP 223, presented at the Meeting of the American Society of Heating, Refrigerating and Air-Conditioning Engineers (ASHRAE), Denver, Colorado, June 23-27, 1980.
22. Tansey, W. A., J. M. Wilson, and K. E. Schaefer. Analysis of health data from 10 years of Polaris submarine patrols. Undersea Biomed. Res. 6(Suppl.):S-217--S-246, 1979.
23. Traynor, G. W., D. W. Anthon, and C. D. Hollowell. Indoor Air Quality: Gas Stove Emissions. Berkeley, Cal.: Lawrence Berkeley Laboratory, 1979.
24. Wade, W. A., III, W. A. Cote, and J. E. Yocom. A Study of Indoor Air Quality. J. Air Pollut. Control Assoc. 25:933-939, 1975.
25. Waligora, J. N., Ed. The Physiological Basis for Spacecraft Environmental Limits. National Aeronautics and Space Administration Reference Publication 1045. Washington, D.C.: National Aeronautics and Space Administration, 1979.
26. Yamanaka, S., H. Hirose, and S. Takada. Nitrogen oxides emissions for domestic kerosene-fired and gas-fired appliances. Atmos. Environ. 13:407-412, 1979.
27. Yates, M. W. A preliminary study of carbon monoxide gas in the home. J. Environ. Health 29:413-420, 1967.
28. Yocom, J., W. Coté, and W. Clink. A Study of Indoor-Outdoor Air Pollutant Relationships. Vol. 1. Summary Report. Washingon, D.C.: National Air Pollution Control Administration, 1974.
29. Yocom, J., W. Coté, and W. Clink. A Study of Indoor-Outdoor Air Pollutant Relationships. Vol. 2. Supplementary Study. Washington, D.C.: National Air Pollution Control Administration, 1974.

TOBACCO SMOKE

Nearly everyone is exposed to tobacco smoke at some time or other. Indirect exposure (i.e., exposure of nonsmokers) is referred to as "passive exposure," "passive smoking," or "involuntary smoking." The extent of the exposure is determined by the number of smokers one associates with, their smoking habits, and the characteristics of the environment in which exposure occurs. Actual population exposures to tobacco smoke are therefore quite variable. For some, passive exposure to cigarette, pipe, or cigar smoke may occur routinely at work, in transit, or at home. For others, passive exposure may result only from infrequent encounters with smokers in public facilities.

Over 2,000 compounds have been identified in cigarette smoke; many are established carcinogens that appear primarily in the particulate phase. It is reasonable to assume that passive tobacco-smoke exposure is many people's principal source of exposure to many of these compounds.

Despite the overwhelming evidence on health effects on smokers, the impact of tobacco-smoke exposure on nonsmokers is not well documented. (The health effects of gases and particles emanating from tobacco combustion are summarized in Chapter VII.) Tobacco smoke irritates the eyes, nose, and throat and is annoying to nonsmokers, even in the presence of "adequate" ventilation. Annoyance increases with increasing smoke contamination and increasing dryness of air. Aside from the irritation and annoyance that it causes, smoking in confined spaces increases annoyance from odors and particle accumulation. Aldehydes and ketones produced by burning tobacco give rise to odors. Particles that adsorb and release organic vapors can be odor sources long after the tobacco is extinguished, and the lingering odors can be smelled by those not desensitized.

This section discusses indoor exposure to pollution resulting from cigarette-smoking. The factors of concern here are the number of people exposed to cigarette smoke, the composition of the gases and particles emitted, and the concentrations of pollutants encountered.

BACKGROUND

The number of people exposed to passive smoking, who might also be termed "involuntary smokers," is not known. However, given the number of people who smoke in the United States, some involuntary inhalation of tobacco combustion products from smoke-contaminated atmospheres by nonsmokers is unavoidable. Passive exposure to tobacco smoke will inevitably occur in any number of public or private activities.

In 1978, an estimated 54 million persons smoked 615 billion cigarettes. The prevalence of regular cigarette-smoking in the adult population declined from 42% in 1964 to 33% in 1978. Figure IV-6 plots the annual consumption, from 1950 to 1978, of cigarettes and filter-tip cigarettes per person aged 18 and cver. The adult per capita consumption for 1978 is estimated at 3,965, which is the lowest recorded consumption since 1958. Surveys show that fewer men are smoking each year, but more women are smoking, particularly teen-agers. Figure IV-7 presents the proportion of U.S. men, women, and teen-agers that reported being regular smokers in 1974 and 1975. This figure implies that one of every three persons between the ages of 17 and 64 regularly smokes cigarettes. Association with adult males 35-44 yr old increases the likelihood that a person will be passively

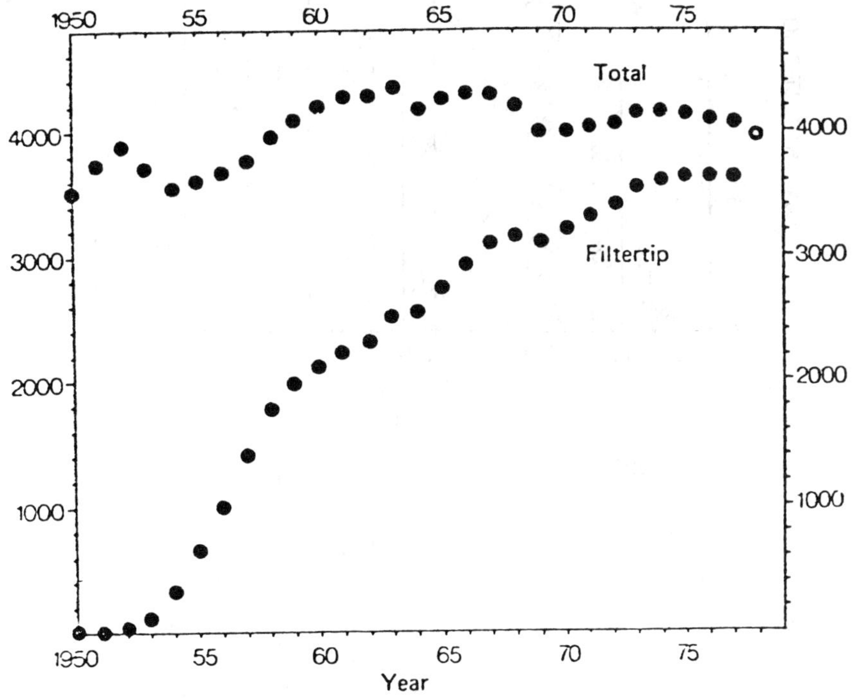

FIGURE IV-6 Annual consumption of cigarettes and filter-tip cigarettes per person aged 18 yr and over, 1950-1978. 1978 per capita consumption of cigarettes was 3,965, the lowest since 1958. Open circle, preliminary estimate. Reprinted from U.S. Department of Health, Education, and Welfare.[31(p.A-5)]

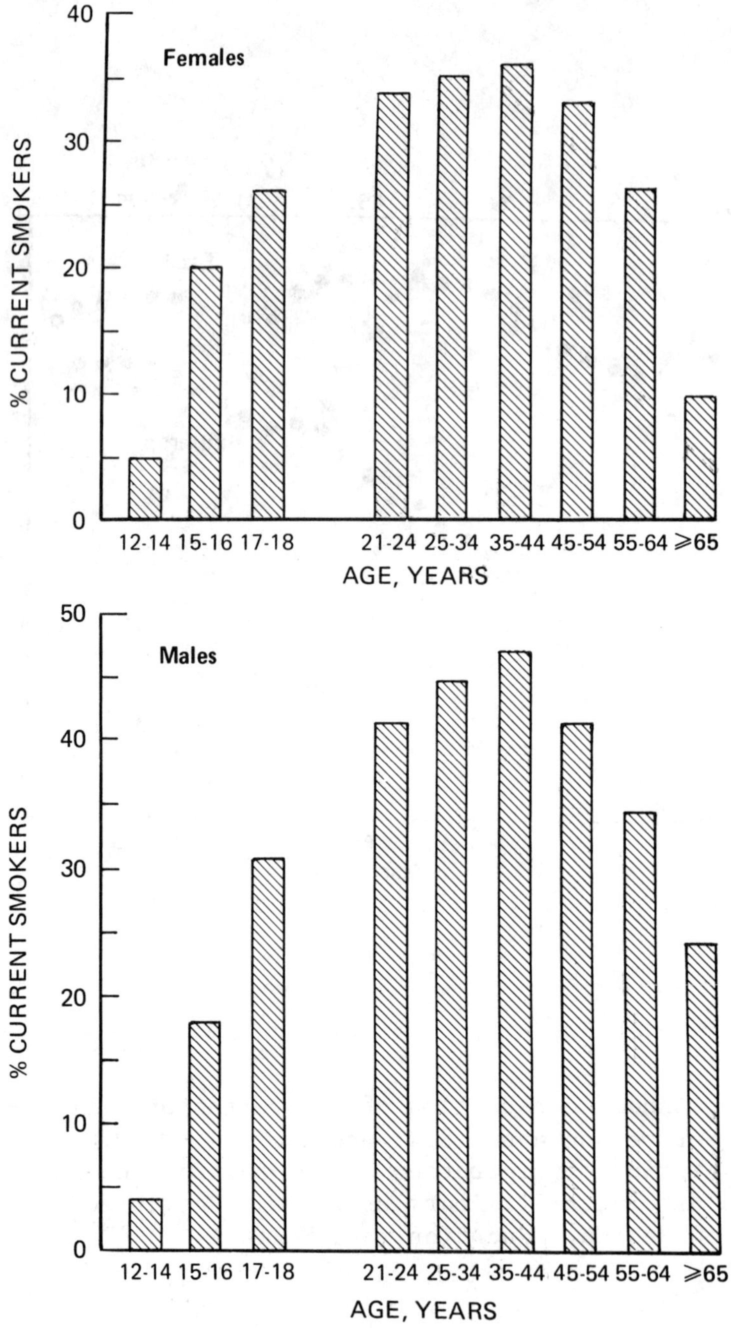

FIGURE IV-7 Percentage of regular smokers in population. Teenagers, 1974. Adults, 1975. Data from U.S. Department of Health, Education, and Welfare.[31(pp. A-10, A-14)]

exposed to cigarette smoke. In 1974, the percentages of girls and boys 15-16 yr old who smoked were essentially the same. The percentage of teen-aged boys smoking, however, has been dropping since 1970, whereas the percentage of teen-aged girls smoking has increased dramatically.

Because cigarette smoke is ubiquitous, passive exposure to it will be encountered in offices, industrial facilities, homes, public sporting events, restaurants, transportation facilities, and innumerable other locations. Professional and technical workers have the lowest percentage of cigarette-smokers; laborers, craftsmen, and other "blue-collar" workers have the highest. Of the males in "blue-collar" occupations, 47% were smokers in 1975, whereas 36% of the males in "white-collar" employment were smokers. Females showed the opposite relationship; 34% of the "white-collar" females and 32% of "blue-collar" female workers smoked. In the 1975 survey, 40% of the women in the sample worked outside the home; of these, 33% were cigarette-smokers, compared with 27% of the housewives.

Information on passive exposure of nonsmokers to tobacco smoke has never been systematically obtained, except in a limited number of epidemiologic investigations (see Chapter VII). Data on populations exposed at home, at work, or in other locations must be estimated from surveys on smoking prevalence and epidemiologic studies. The National Center on Smoking and Health (NCSH) has conducted the most extensive surveys on American smoking habits. Marital status, educational achievement, and income are demographic variables associated with differences in smoking prevalence. NCSH information indicates that divorced or separated persons smoke more than those in any other marital-status group. For men, the highest prevalence occurs in middle-income and high-school-educated groups. For women, there is a more direct relationship between income and smoking, with more smoking in the higher-income groups.

The demographics of smoking prevalence do not indicate exactly what groups are passively exposed to tobacco smoke or where or when they are likely to be exposed. We do not have information on the percentage of the 80 million workers in the United States who are employed in an environment free of tobacco smoke. We do not know what percentage of America's 80 million residential units have smokers, but the results of some surveys are summarized below.

Some 54 million adult Americans smoke (33%), so we can infer that the number of homes and other residential units with smokers is substantial. Summaries of prevalence of homes with smokers taken from a study on air-pollution health effects support this contention. Table IV-18 summarizes the response to a questionnaire in the children's study. On the basis of 8,493 questionnaire responses obtained across six cities (75% response rate), 70% of the homes reported having at least one smoker. The percentages reporting at least one smoker ranged from 63% in rural Portage, Wisconsin, to 76% in a middle-income community in St. Louis. Lebowitz and Burrows[13] reported that 53.8% of the children in their Tucson study had smokers in their homes. Schilling et al.[25] reported that an average of 63% of the sampled homes in two Connecticut towns had smokers. Substantial regional variations in the percentage of homes with smokers may be expected,

TABLE IV-18

Percentage of Homes Reporting One or More Smokers[a]

Location	No. Responses	Proportion of Homes with Smokers, %
St. Louis, Mo.	1,922	76.1
Steubenville, Oh.	1,808	74.2
Kingston-Harriman, Tenn.	810	71.8
Watertown, Mass.	838	69.8
Topeka, Kans.	1,663	63.3
Portage, Wis.	1,452	62.5
Tucson, Ariz.	676	53.8
Two towns, Conn.	376	63.3

[a]Data from Ferris _et_ _al_.,[8] Lebowitz and Burrows,[17] and Schilling _et_ _al_.[25]

owing to geographic variations in social, demographic, and religious variables that are associated with differences in smoking prevalence.

Further stratification of the responses to the question of parental smoking by level of parental education confirms earlier NCSH surveys. In the Harvard six-city study,[8] of the homes in which neither parent had a high-school education, over 80% had at least one smoker. In homes in which one parent had graduated from college, only 50% had smokers.

CONTAMINANTS IN SMOKE

A distinction can be made between mainstream smoke and sidestream smoke in the contaminants evolved from tobacco combustion. Both smokers and nonsmokers are, of course, exposed to sidestream smoke. Mainstream smoke is undiluted and is pulled through the tobacco into a smoker's lungs. Sidestream smoke is directly from the burning tobacco. Depending on smoking behavior, burning temperature, and type of filter, the composition of mainstream smoke exhaled by a smoker varies substantially. From 50% to more than 90% of water-soluble compounds are removed from the mainstream smoke by the smoker, depending on the depth of inhalation and the time of breath-holding. The insoluble compounds show equal variability over a range of 20-70%.

A typical cigarette-smoker inhales mainstream smoke 8-10 times, for a total of 24-30 s of a total 12-min burn time for a cigarette.[19] There is no dispute that the concentrations of almost all constituents are far greater in mainstream than in sidestream smoke. However, given approximately 24 to 1 disparity in burning time (i.e., the sidestream smoke is produced during 96% of the total smoking time) and the difference in combustion conditions, it is not surprising that sidestream smoke is more important to the involuntary, passive smoker. The sidestream smoke can also be enriched in many compounds. Sidestream smoke and mainstream smoke have been characterized by many investigators.[2-4,6,14,19] The mean particle size of fresh mainstream smoke is slightly greater than that of sidestream smoke. Exhaled-smoke particles are larger, on the average, than those in fresh mainstream or sidestream smoke, because of water absorption and coagulation. The same processes modify the size distribution of smoke particles as they age. If particles begin with a mass mean diameter less than 0.2 μm, they grow by agglomeration and water absorption to a mass mean diameter approaching 1 μm within minutes.

The passive smoker by no means receives a lung dose of smoke equivalent to that of the smoker. Several investigators have estimated the exposure of the nonsmoker in a smoke-filled environment as one-hundredth to one-tenth of the smoker's exposure.[12,13,22,23] These estimates were based primarily on measurements of carbon monoxide, suspended particulate matter, or nicotine concentration extrapolated from reported values for mainstream smoke. Substantial evidence indicates that many substances are increased in sidestream smoke. Therefore, direct comparison on the basis of particle mass or concentration of a specific gas may not adequately describe a passive smoker's exposure to cigarette smoke.

The sidestream-to-mainstream ratios (s:m) of specific compounds range from 0.7 to 46. The ratios of vapor-phase compounds vary more than those of particulate-phase compounds. Although the composition of mainstream smoke from nonfiltered cigarettes is quite different from that from filtered cigarettes, the compositions of their sidestream smoke are essentially the same. Table IV-19 presents a compilation of the concentrations of some substances found in cigarette smoke, with s:m ratios. Values are given in mass per cigarette. Unless otherwise noted, the values refer to nonfiltered cigarettes. Sidestream cigarette smoke, because of the length of the burn and the burn temperature, is a more important source of local air contamination with many substances--such as carbon monoxide, nicotine, ammonia, and aldehydes--than mainstream smoke.

INDOOR CONCENTRATIONS OF PARTICLES AND VAPORS FROM CIGARETTE SMOKE

Cigarette-smoking in enclosed areas increases concentrations of particles and gases. Increased concentrations of carbon monoxide, nicotine, nitrosamines, and benzopyrene are among the most frequent. Pollution measurements of air contaminated with tobacco smoke can be conveniently divided among controlled chamber experiments and actual exposure conditions. In many of the controlled-setting experiments, carbon monoxide concentrations exceeding the 1-h NAAQS of 35 ppm have been reported (see summary table on chamber studies of cigarette-smoke exposure in U.S. Surgeon General's Report, Smoking and Health[31]). Without ventilation, indoor carbon monoxide concentrations are proportional to the amount of tobacco burned and inversely proportional to room volume. These studies indicate that carbon monoxide concentrations of 50-100 ppm can be obtained and that increased ventilation substantially reduces concentrations (see Figure IV-8). Chapter 11 of the 1979 Surgeon General's report tabulates the results of many of the chamber experiments (Table IV-19).

The experiments performed in controlled environments generally involved heavier smoking than is normally encountered. Of more interest are the observations of carbon monoxide and other pollutants in normal indoor locations. Several studies are summarized in Table IV-20. In general, the carbon monoxide concentrations were less than those in the controlled experiments, probably because of the heavier smoking. Taverns, bars, nightclubs, and restaurants have been the more frequently reported locations in assessments of the air-quality impact of cigarette-smoking. Concentrations less than 35 ppm for an hourly average have been reported, primarily because of mechanical or natural ventilation. However, studies have indicated that the 8-h NAAQS of 9 ppm could be exceeded in public facilities that permit cigarette-smoking. Elliot and Rowe[7] reported that carbon monoxide concentrations in public assemblies of 2,000-14,000 people were between 9 and 25 ppm--up to 4 times higher than background.

As a more direct measure of exposure, several investigators have measured COHb in nonsmokers' blood after their exposure to cigarette smoke. The results were as expected: modest increases in COHb.

TABLE IV-19

Composition of Mainstream and Sidestream Smoke

	Concentration, mg/cigarette[a]			
Characteristic or Compound	Mainstream Smoke (1)	Sidestream Smoke (2)	Ratio, 2:1	Reference
General characteristics:				
Duration of smoke production, s	20	550	27.5	19
Tobacco burned	347	411	1.2	15
Particles, no. per cigarette	1.05×10^{12}	3.5×10^{12}	3.3	24
Particles:				
Tar (chloroform extract)	20.8	44.1	2.1	16
	10.2[b]	34.5[b]	3.4	16
Nicotine	0.92	1.69	1.8	16
	0.46[b]	1.27[b]	2.8	16
Benzo[a]pyrene	3.5×10^{-5}	1.35×10^{-4}	3.9	10
	4.4×10^{-5}	1.99×10^{-4}	4.5	18
Pyrene	1.3×10^{-4}	3.9×10^{-4}	3.0	10
	2.70×10^{-4}	1.011×10^{-3}	3.7	18
Fluoranthene	2.72×10^{-4}	1.255×10^{-3}	4.6	18
Benzo[a]fluorene	1.84×10^{-4}	7.51×10^{-4}	4.1	18
Benzo[b/c]fluorene	6.9×10^{-5}	2.51×10^{-4}	3.6	18
Chrysene, benz[a]anthracene	1.91×10^{-4}	1.224×10^{-3}	6.4	18
Benzo[b/k/j]fluoranthrene	4.9×10^{-5}	2.60×10^{-4}	5.3	18
Benzo[e]pyrene	2.5×10^{-5}	1.35×10^{-4}	5.4	18
Perylene	9.0×10^{-6}	3.9×10^{-5}	4.3	18
Dibenz[a,j]anthracene	1.1×10^{-5}	4.1×10^{-5}	3.7	18
Dibenz[a,h]anthracene, ideno-[2,3-ed]pyrene	3.1×10^{-5}	1.04×10^{-4}	3.4	18
Benzo[ghi]perylene	3.9×10^{-5}	9.8×10^{-5}	2.5	18
Anthanthrene	2.2×10^{-5}	3.9×10^{-5}	1.8	18
Phenols (total)	0.228	0.603	2.6	13
Cadmium	1.25×10^{-4}	4.5×10^{-4}	3.6	27
Gases and vapors:				
Water	7.5[c]	298[d]	39.7	13
Carbon monoxide	18.3	86.3	4.7	32
	--	72.6	--	26
Ammonia	0.16	7.4	46.3	20
Carbon dioxide	63.5	79.5	1.3	20
NO_x	0.014	0.051	3.6	20
Hydrogen cyanide	0.24	0.16	0.67	31
Acrolein	0.084	--	--	31
	--	0.825	--	26
Formaldehyde	--	1.44	--	26
Toluene	0.108	0.60	5.6	32
Acetone	0.578	1.45	2.5	32
Polonium-210, pCi	0.04-0.10	0.10-0.16	1-4	7

[a]Unless otherwise noted. [b]Filtered cigarettes.
[c]3.5 mg in particulate phase; rest in vapor phase.
[d]5.5 mg in particulate phase; rest in vapor phase.

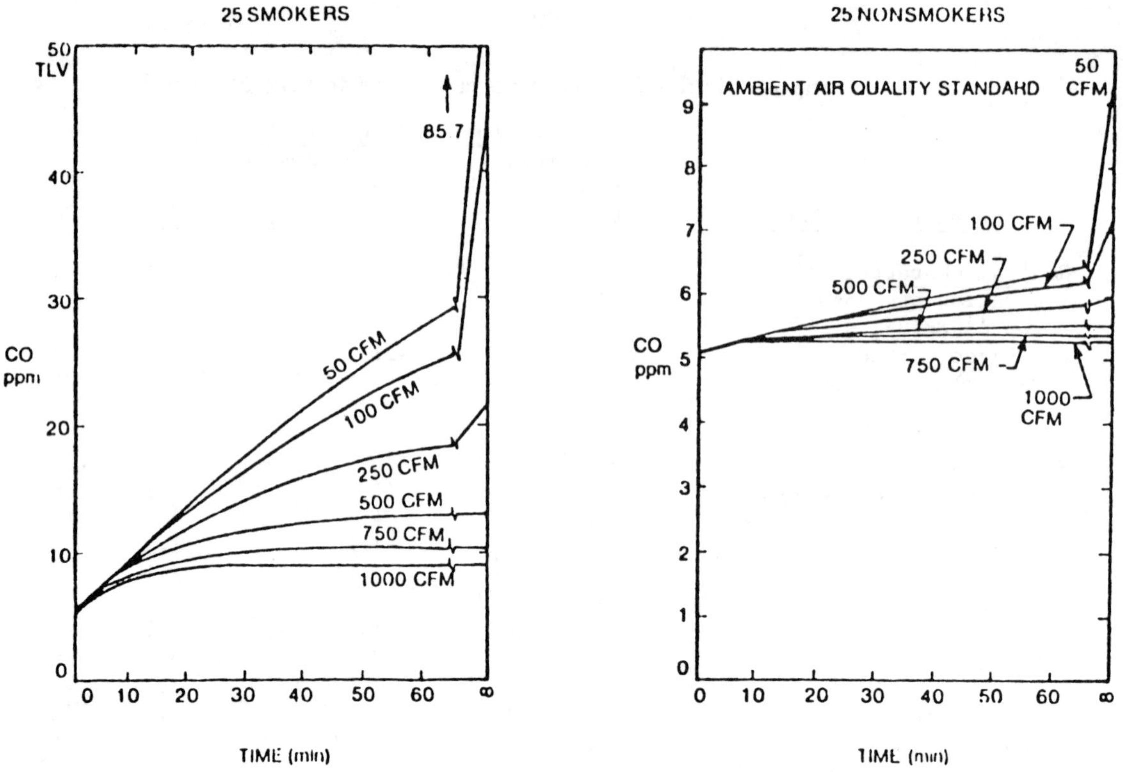

FIGURE IV-8 Calculated buildup of carbon monoxide under various conditions of ventilation and smoking. Calculated for a room of 3,000 ft^3 with 25 smokers on the left and for 25 nonsmokers on the right. TLV is the threshold limit value for carbon monoxide (50 ppm). CFM is ventilation in cubic feet per minute. Reprinted with permission from Galuškinová.[9]

Harke[11] showed a COHb increase from 0.9% to 2.1% after a 2-h exposure to carbon monoxide at 30 ppm. Aronow[1] reported similar results in patients exposed to smoke from 15 cigarettes smoked over 2 h in a 30.8-m^3 room without ventilation. It is possible that passive smoking was the reason that a portion of the nonsmoking population was reported by Stewart et al.[30] (in a national COHb survey) to have COHb of over 1.5%. But, given the range of carbon monoxide concentrations reported in smoke-filled indoor locations, it is unlikely that cigarettes alone contribute significantly to increased COHb in the nonsmoking population. Because carbon monoxide is slow to be removed from the environment, reducing fresh-air supplies to office buildings or public facilities that permit smoking would necessarily increase carbon monoxide concentrations.

Several other constituents of tobacco smoke have been measured indoors, including nicotine, acrolein, benzo[a]pyrene, nitrosamines, and aldehydes.

Under heavy-smoking conditions, acrolein is the only gaseous substance that has been shown to exceed threshold limit values established for industrial environments. Acrolein is found at 1-20 ppb in bars and restaurants; even to nonsensitive persons, these concentrations can cause annoying odors and eye and nose irritation.

The indoor concentrations of benzo[a]pyrene where smoking occurs are more ambiguous. Galuškinová[9] reported concentrations of 0.2-4.6 µg/m^3 in restaurants described as "smoky." These high concentrations may have been due to cooking; Elliot and Rowe[7] reported concentrations of 7-22 ng/m^3 in the presence of total suspended particles at 224-480 µg/m^3 in public arenas with smokers.

The nicotine concentration in air is an excellent indicator of cigarette smoke. A typical cigarette contains 2 mg of nicotine. Hinds and First[12] measured nicotine at 1-10.3 µg/m^3 in a number of public facilities, including cocktail lounges, transportation waiting rooms, trains, and buses. The submarine contains a unique environment for observing human exposure to cigarette smoke. Cano et al.[5] found nicotine in the urine of nonsmokers when the submarine environment had nicotine at 15-35 µg/m^3, and the nonsmokers' urinary nicotine concentration was only 1% of that of the smokers. It appears unlikely that the threshold limit value (500 µg/m^3) for exposure to nicotine in industrial environments would be exceeded in indoor locations with ventilation, although only a few studies have been reported.

Total suspended particles (TSP) and the fractions of respirable suspended particles (RSP) have been measured in the indoor environment in the presence of tobacco smoke. TSP concentrations of 50-400 µg/m^3 have been reported for integrated samples taken in public arenas, lounges, bus stations, and airplanes (see Table IV-20).

There are few reported measurements of RSP in the vicinity of tobacco-smoking. Repace and Lowrey[21] reported on 2-min RSP samples taken in 20 indoor environments where smoking is permitted (outdoor measurements were also reported). The indoor RSP concentrations ranged from 86 to 697 µg/m^3. These results are consistent with residential measurements of 24-h RSP concentrations reported by Spengler et al.[29] Daily indoor concentrations of RSP frequently

TABLE IV-20

Measurement of Constituents of Tobacco Smoke Under Natural Conditions[a]

Reference, Location, and Dimensions	Ventilation	Amount of Tobacco Burned	Level of Constituent: Smoking Section	Level of Constituent: Other Control Section
Brunnemann and Hoffmann			dimethylnitrosamine	
Train 1 (Bar Car)	--	--	.13 ng/l	--
Train 2 (Bar Car)	--	--	.11 ng/l	--
Bar	--	--	.24 ng/l	--
Nightclub	--	--	[.09 ng/l]	--
Cano, *et al.*				
Submarines 66 m^3	Yes	157 cig per day	<40 ppm CO, 32 $\mu g/m^3$ nicotine	--
		94-103 cig per day	<40 ppm CO, 15-35 $\mu g/m^3$ nicotine	--
Chappel and Parker				
General public places	--	--	3.5 ppm CO	2.0 ppm CO
Government offices	--	--	2.5 ppm CO	2.5 ppm CO
Restaurants	--	--	4.0 ppm CO	2.5 ppm CO
Night clubs and taverns	--	--	13.0 ppm CO	3.0 ppm CO
Cuddeback, *et al.*				
Tavern 1	6 air changes per hour	--	12.5 ppm CO .33 mg/m^3 TPM	--
Tavern 2	[natural (∿1-2/hr)]	--	17 ppm CO .98 mg/m^3 TPM	--
Elliott and Rowe				
Arenas	--	--	14.3 ppm CO .367 mg/m^3 TPM	3 ppm CO .068 mg/m^3 TPM

TABLE IV-20 (continued)

Reference, Location, and Dimensions	Ventilation	Amount of Tobacco Burned	Level of Constituent: Smoking Section	Level of Constituent: Other Control Section
Galuskinova				
Restaurant	--	--	.0002-.0046 mg/m^3 benzopyrene	--
Godin, *et al.*				
Ferry boat compartments	--	--	18.4 $\pm$ 8.7 ppm CO	3.0 $\pm$ 2.4 ppm C
Theater	--	--	3.4 $\pm$ 0.8 ppm CO	1.4 $\pm$ 0.8 ppm C
Harke				
Office building	air conditioned	--	<5 ppm CO	--
Office building	not air conditioned	--	<5 ppm CO	--
Room 78.3 m^3	--	3 smokers	15.6 ppm CO	--
Harke and Peters				
Automobile	35 km/hr speed, no ventilation.	4 cig	24.3 ppm CO	--
	80 km/hr speed, no ventilation.	4 cig	12.1 ppm CO	--
	30 km/hr speed, no ventilation.	4 cig	21.4 ppm CO	--
	30 km/hr speed, air jets open.	4 cig	15.7 ppm CO	--
	3 km/hr speed, air jets open & blower on.	4 cig	12.0 ppm CO	--

TABLE IV-20 (continued)

Reference, Location, and Dimensions	Ventilation	Amount of Tobacco Burned	Level of Constituent: Smoking Section	Level of Constituent: Other Control Section
Hinds and First			nicotine:	
Commuter train	--	--	.0049 mg/m^3	--
Commuter bus	--	--	.0063 mg/m^3	--
Bus waiting room	--	--	.001 mg/m^3	--
Airline waiting room	--	--	.0031 mg/m^3	--
Restaurant	--	--	.0052 mg/m^3	--
Cocktail lounge	--	--	.0103 mg/m^3	--
Student lounge	--	--	.0028 mg/m^3	--
Lefcoe and Inculet				
House	--	1 cig	48×10^6 particles per cubic foot	$.9 \times 10^6$ particles per cubic foot
Szadkowski, *et al.*				
Offices	--	--	2.7 ppm CO	--
Sebben, *et al.*				
Night clubs	--	--	13.4 ppm CO [6.5-41.9]	9.2 ppm CO
Restaurants	--	--	8-28 ppm CO	--
Bus	--	--	7.3 ppm CO [6-14]	6.2 ppm CO
Slavin and Hertz				
Conference room	8 air changes per hour	--	8 ppm CO	1-2 ppm CO
	6 air changes per hour	--	10 ppm CO	1-2 ppm CO

TABLE IV-20 (continued)

Reference, Location, and Dimensions	Ventilation	Amount of Tobacco Burned	Level of Constituent: Smoking Section	Level of Constituent: Other Control Section
Seiff				
Intercity bus	15 air changes per hour	23 cig burning continuously	33 ppm CO	--
		3 cig burning continuously	18 ppm CO	--
U.S. Dept. Transportation, *et al.*				
Airplane flights:				
Overseas--100% filled	15-20 air changes per hour	--	2-5 ppm CO, <.120 mg/m^3 TPM	--
Domestic--66% filled	--	--	2 ppm CO, <.120 mg/m^3 TPM	--

[a]Reprinted from U.S. Department of Health, Education, and Welfare.[31](pp. 11-16--11-20)
cig = cigarettes; -- = unknown; TPM = total particulate matter.

exceeded 200 μg/m^3 in homes with cigarette-smokers. Aggregating the data obtained from a study of 69 homes in six cities reveals that the indoor and outdoor concentrations in the 38 nonsmoking homes are essentially equivalent (24 μg/m^3 indoors versus 22 μg/m^3 outdoors). In the 22 homes with only one cigarette-smoker, the mean concentration indoors was 43 μg/m^3; the nine homes with two or more smokers had a mean concentration of 75 μg/m^3. These data, collected over a 3-yr period, are presented in Figure IV-9, which shows monthly mean RSP concentrations outside and inside homes without smokers, with one smoker, and with two or more smokers. The data clearly illustrate the contribution of cigarette-smoking to indoor particle concentrations. (The effects of pollution control on indoor concentrations of particles generated from tobacco-burning are discussed in Chapter IX.)

In recent investigations of personal exposures to respirable particles by Spengler et al.,[28] passive smoking was shown to be an important source. Volunteers in Topeka, Kansas, carried portable monitors for 12-h periods on 15 sampling days. The mean RSP concentration of samples where participants reported passive cigarette-smoke exposure for some time during the day was 40 μg/m^3. The nonsmoking, nonexposed participants had an overall mean concentration of approximately 22 μg/m^3, and the outdoor concentrations averaged less than 15 μg/m^3.

CONCLUSIONS

Tobacco-smoking indoors can contribute to or cause increased concentrations of respirable particles, nicotine, carbon monoxide, acrolein, and many other substances in the smoke. Many experimental and "real-life" measurements have demonstrated that where ventilation is low or nonexistent, indoor pollutant concentrations can exceed ambient-air quality standards and industrial standards. The indoor concentrations have been shown to depend on the number of smokers, how the tobacco is smoked (cigarette, pipe, or cigar), the room volume, the volume of fresh-air makeup, the efficiency of the air-cleaning apparatus, and the effectiveness of air mixing in the room. The absorbing characteristics of building and furnishing materials can affect the concentration. It has been demonstrated that the use of makeup air (for ventilation) that has lower concentrations of the contaminants effectively lowers the concentrations of carbon monoxide and other pollutants in tobacco smoke. Nevertheless, high respirable and total suspended particle concentrations have been noted even in the presence of "adequate" ventilation. Daily concentrations of respirable particles can exceed proposed ambient-air quality standards for TSP in homes with smokers.

Although it needs to be documented, tobacco smoke may be the most important source of exposure of nonsmoking populations to benzo[a]pyrene, nicotine, and other compounds in nonindustrialized areas. It must be noted, however, that there are other important sources, both indoors and outdoors, of many of the pollutants produced

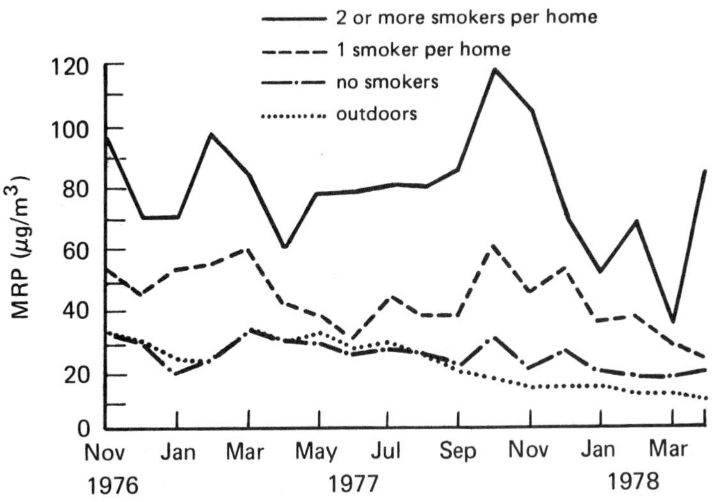

FIGURE IV-9 Monthly mean respirable particle concentrations. 0, outside. X, inside nonsmoker homes. 1, inside homes with one smoker. Solid circles, inside homes with two or more smokers. Sample represents 80 homes across six cities (approximately 10-15 homes per city). Reprinted with permission from Spengler _et al._[29]

by the burning of tobacco. And it is important to point out that direct exposure of a smoker is an order of magnitude greater than the passive exposure of a nonsmoker.

REFERENCES

1. Aronow, W. S. Effect of passive smoking on angina pectoris. N. Engl. J. Med. 299:21-24, 1978.
2. Brunnemann, K. D., J. D. Adams, D. P. S. Ho, and D. Hoffmann. The influence of tobacco smoke on indoor atmospheres. II. Volatile and tobacco specific nitrosamines in main- and sidestream smoke and their contribution to indoor pollution, pp. 876-880. In 4th Joint Conference on Sensing of Environmental Pollutants, New Orleans, Louisiana, November 6-11, 1977. Washington, D.C.: American Chemical Society, 1978.
3. Brunneman, K. D., and D. Hoffmann. Chemical studies on tobacco smoke. XXIV. A quantitative method for carbon monoxide and carbon dioxide in cigarette and cigar smoke. J. Chromat. Sci. 12:70-75, 1974.
4. Brunnemann, K. D., and D. Hoffmann. Chemical studies on tobacco smoke. LIX. Analysis of volatile nitrosamines in tobacco smoke and polluted indoor environments, pp. 343-356. In F. A. Walker, M. Castegnaro, L. Griciute, and R. E. Lyle, Eds. Environmental Aspects of N-Nitroso Compounds. IARC Scientific Publications No. 19. Lyon, France: International Agency for Research on Cancer, 1978.
5. Cano, J. P., J. Catalin, R. Badre, C. Dumas, A. Viala, and R. Guillerme. Determination de la nicotine par chromatagrophie en phase gazeuse. II. Applications. Ann. Pharm. Fr. 28(11):633-640, 1970.
6. Corn, M. Characteristics of tobacco sidestream smoke and factors influencing its concentration and distribution in occupied spaces. Scand. J. Respir. Dis. Suppl. 91:21-36, 1974.
7. Elliot, L. P., and D. R. Rowe. Air quality during public gatherings. J. Air Pollut. Control Assoc. 25:635-636, 1975.
8. Ferris, B. J., Jr., F. E. Spiezer, J. D. Spengler, D. Dockery, Y. M. M. Bishop, M. Wolfson, and C. Humble. Effects of sulfur oxides and respirable particles on human health. Methodology and demography of populations in study. Am. Rev. Respir. Dis. 120:767-779, 1979.
9. Galuškinová, V. 3,4-Benzpyrene determination in the smoky atmosphere of social meeting rooms and restaurants. A contribution to the problem of the noxiousness of so-called passive smoking. Neoplasma 11:465-468, 1964. (in Czech)
10. Grimmer, G., H. Böhnke, H. and H.-P. Harke. Passive smoking: Measuring of concentrations of polycyclic aromatic hydrocarbons in rooms after machine smoking of cigarettes. Int. Arch. Occup. Environ. Health 40:83-92, 1977. (in German; English summary)
11. Harke, H.-P. The problem of "passive smoking." Münch. Med. Wochenschr. 112:2328-2334, 1970. (in German; English summary)

12. Hinds, W. C., and M. W. First. Concentrations of nicotine and tobacco smoke in public places. N. Engl. J. Med. 292:844-845, 1975.
13. Hoegg, U. R. Cigarette smoke in closed spaces. Environ. Health Perspect. 2:117-128, 1972.
14. Johnson, W. R., R. W. Hale, J. W. Nedlock, H. J. Grubbs, and D. H. Powell. The distribution of products between mainstream and sidestream smoke. Tob. Sci. 175(21):43-46, October 12, 1973.
15. Keith, C. H., and J. C. Derrick. Measurement of the particle size distribution and concentration of cigarette smoke by the "conifuge." J. Colloid. Sci. 15:340-356, 1960.
16. Kotin, P., and H. L. Falk. The role and action of environmental agents in the pathogenesis of lung cancer. II. Cigarette smoke. Cancer 13:250-262, 1960.
17. Lebowitz, M. D., and B. Burrows. Respiratory symptoms related to smoking habits of family adults. Chest 69:48-50, 1976.
18. Neurath, G., and H. Ehmke. Apparatur zur Untersuchung des Nebenstromrauches. Beitr. Tabakforsch. 2:117-121, 1964. (in German; English summary)
19. Neurath, G., and H. Horstmann. Einfluss des Feuchtigkeitsgehaltes von Cigaretten auf die Zusammensetzung des Rauches und die Glutzonentemperatur. Beitr. Tabakforsch. 2:93-100, 1973. (in German; English summary)
20. Parfenov, Y. D. Polonium-210 in the environment and in the human organism. At. Energy Rev. 12/1:75-143, 1974.
21. Repace, J. L., and A. H. Lowrey. Indoor air pollution, tobacco smoke, and public health. Science 208:464-472, 1980.
22. Russell, M. A. H., P. V. Cole, and E. Brown. Absorption by non-smokers of carbon monoxide from room air polluted by tobacco smoke. Lancet 1:576-579, 1973.
23. Russell, M. A. H., and C. Feyerabend. Blood and urinary nicotine in non-smokers. Lancet 1:179-181, 1975.
24. Scassellati Sforzolini, G., and A. Savino. Evaluation of a rapid index of ambient contamination by cigarette smoke, in relation to the composition of the gas phases of the smoke. Riv. Ital. Ig. 28:43-55, 1968.
25. Schilling, R. S. F., A. D. Letai, S. L. Hui, G. J. Beck, J. B. Schoenberg, and A. Bouhuys. Lung function, respiratory disease, and smoking in families. Am. J. Epidemiol. 106:274-283, 1977.
26. Schmeltz, I., and D. Hoffmann. Chemical studies on tobacco smoke, XXXVIII. The physiochemical nature of cigarette smoke, pp. 13-34. In E. L. Wynder, D. Hoffmann, and G. B. Gori, Eds. Smoking and Health. 1. Modifying the Risk for the Smoker. Proceedings of the 3rd World Conference on Smoking and Health. Washington, D.C.: U.S. Government Printing Office, 1976.
27. Seehofer, F., D. Hanssen, H. Rabitz, and R. Schroder. Uber den Verbleib des Wassers beim Abrauchen. 2. Mitteilung. Beitr. Tabakforsch. 3:491-503, 1966. (in German; English summary)
28. Spengler, J. D., D. W. Dockery, M. P. Reed, T. Tosteson, and P. Quinlan. Personal Exposure to Respirable Particles. Paper 80-61.5b, presented at 73rd Annual Meeting of the Air Pollution Control Association, June 22-27, Montreal, Quebec, 1980.

29. Spengler, J. D., D. W. Dockery, W. A. Turner, J. M. Wolfson, and B. G. Ferris, Jr. Long-term measurements of respirable sulfates and particles inside and outside homes. Atmos. Environ. 15:23-30, 1981.
30. Stewart, R. D., E. D. Baretta, L. R. Platte, E. B. Stewart, J. H. Kalbfleisch, B. Van Yserloo, and A. A. Rimm. Carboxyhemoglobin levels in American blood donors. J. Am. Med. Assoc. 229:1187-1195, 1974.
31. U.S. Department of Health, Education, and Welfare, Public Health Service. Smoking and Health. A Report of the Surgeon General. DHEW Publication No. (PHS) 79-50066. Washington, D.C.: U.S. Government Printing Office, 1979. [1250] pp.
32. Weber-Tschopp, A., T. Fischer, and E. Grandjean. Physiological and psychological effects of passive smoking. Int. Arch. Occup. Environ. Health 37:277-288, 1976. (in German; English summary)

ODORS

Some substances in the indoor environment make their presence known primarily by their ability to evoke odor sensations. These substances generally arise from humans or their activities. A room full of people, for example, will invariably have "occupancy odor." A characteristic odor like this generally emerges from a mixture of many organic substances, each present at a low concentration. Some of the individual constituents might cause greater concern if present alone at much higher concentrations; as building blocks of a composite, relatively benign odor, they receive attention only as odorants. Thus, the smell of the indoor environment is a measure of environmental quality. The remarkable sensitivity of olfaction encourages this approach. That is, treating low-concentration organic contaminants on the basis of their olfactory impact usually places stringent requirements on the quality of the indoor air. With some notable exceptions (e.g., the presence of carbon monoxide or mercury vapor), an indoor environment that is odorless contains air of healthful quality.

Completely odorless conditions occur indoors only rarely. Weak odors may go unnoticed or may be tolerated, particularly after persons have had the opportunity to remain in a space for a while. The lability of olfaction, evident in this phenomenon of adaptation, may impair the credibility of the nose as an air-quality indicator. Its sensitivity, often as good as or better than that of the most sensitive instruments, offers some compensation for its functional instability.

This section deals with how odors arise, how human beings perceive them, and how their status as "perceived" contaminants determines the means to cope with them. The stringent air-quality standard effectively imposed by high olfactory sensitivity makes odorants a target of particular interest in any effort to keep indoor air quality high in an energy-conscious society.

SOURCES

Occupancy Odor

Most buildings exist to hold people, in some cases many people. With the exception of churches, structures built specifically to hold many people were relatively rare until the nineteenth century. Before then, people avoided crowded places because of possible contamination--it was believed that crowded places, with their odors, served as breeding grounds of disease. The odors were often blamed for the spread of infection.[27] This attitude makes some sense, in view of the imperceptibility of the actual agents of contagion. Even at the end of the nineteenth century, many people still found it difficult to accept the notion that contagion is primarily fingerborne, rather than airborne. Belief in the airborne route carried no evident penalty for the layman, merely some inconvenience. Hence, the notion that bad-smelling air indoors signaled unhealthful conditions could carry on undisturbed. Through the burning of incense, churches had long seemed to give credence to the idea that a good (i.e., pleasant) odor would purify the air and thereby protect against illness. Such a practice actually reflected a predominant view of pre-nineteenth-century medicine.[43]

Only in the second half of the nineteenth century did the toxicity of the body effluvia responsible for occupancy odor receive scientific attention. In a common type of laboratory experiment on the matter, animals breathed air previously breathed by other animals, or received liquid injections of condensed organic materials from previously breathed air. In the experiments on "rebreathing," the animals sometimes developed infections or other difficulties; but, despite such occurrences, experimenters could not point indisputably to any harmful effects of the organic materials in previously breathed air.[12] Experiments on injections of condensed materials yielded essentially the same result: no consistent hazard. At the close of the nineteenth century, the issue seemed more or less settled. On the basis of both a review of available literature and their own experiments, Billings _et al._[12] concluded that

> [it is] very improbable that the minute quantity of organic matter contained in the air expired from human lungs has any deleterious influence upon men who inhale it in ordinary rooms, and, hence, it is probably unnecessary to take this factor into account in providing for the ventilation of such rooms. . . . The discomfort produced by crowded, ill-ventilated rooms in persons not accustomed to them is not due to the excess of carbonic acid, nor to bacteria, nor, in most cases, to dusts of any kind. The two great causes of such discomfort, though not the only ones, are excessive temperature and unpleasant odors. . . . The cause of the unpleasant, musty odor which is perceptible to most persons on passing from the outer air into a crowded, unventilated room is unknown; it may, in part, be due to volatile products

> of decomposition contained in the expired air of persons having decayed teeth, foul mouths, or certain disorders of the digestive apparatus, and it is due, in part, to volatile fatty acids given off with, or produced from, the excretions of the skin, and from clothing soiled with such excretions. It may produce nausea and other disagreeable sensations in specially susceptible persons, but most men soon become accustomed to it, and cease to notice it, as they will do with regard to the odor of a smoking-car, or of a soap factory, after they have been for some time in the place. The direct and indirect effects of odors of various kinds upon the comfort, and perhaps also upon the health, of men are more considerable than would be indicated by any tests now known for determining the nature and quantity of the matters which give rise to them. (pp. 24, 26-27)

This statement would prompt little dispute today.

A quarter-century after the experiments of Billings and colleagues, a New York State commission focused on ventilation requirements for occupied classrooms.[66] Tests of such varied functions and indexes as comfort, body temperature, intellectual performance, motivation, respiration, metabolism, condition of the nasal mucosa, frequency of colds, blood pressure, hematocrit, appetite, and rate of physical work uncovered no cause for medical concern under normal conditions of occupancy. This 8-yr effort reinforced notions that control of occupancy odor should figure prominently in indoor-air quality control, but the justification had to rest on grounds of comfort, rather than on grounds of health. The first truly quantitative studies of ventilation requirements started with the premise that ventilation primarily must control occupancy odor.

Tobacco Smoking. Throughout the twentieth century, the air in occupied rooms has commonly been smoky. Mainstream cigarette smoke contains approximately 3,000 gaseous constituents,[76] and these and associated particulate matter may both constitute health hazards for occupants. To add insult to injury, tobacco smoke forms the most annoying and persistent indoor odor nuisance.[54] A survey of professional ventilating engineers placed it well ahead of the next two most disturbing indoor odorous contaminants, occupancy and cooking odors. Its severity as a nuisance derives from its properties: it is an apparent allergen for some persons; it is an eye, nose, and throat irritant for most persons; it is an odorant; it is a soiling agent; and it is a stimulus for chest discomfort in persons with angina pectoris.[57,71] Its "tar" content causes it to adsorb strongly to surfaces. After adsorption, it desorbs slowly and thereby promotes so-called secondary sources of odor. In general, such sources concentrate previously airborne odorants on their surfaces (e.g., air-conditioning coils) or in their interstices (e.g., fabric). When the adsorbed odorants desorb, they often have an odor character somewhat different from that of the parent contaminant, and this generally seems true of tobacco smoke.[54]

Tobacco-smoke odor increases in intensity and unpleasantness immediately after active smoking has ceased and after the particulate-vapor complex has adsorbed to surfaces.[79] Figure IV-10 demonstrates how odor increased in intensity during a period after cigarettes were extinguished in an unventilated room. During this time, the odor character changed from pungent and burnt to stale and sour. This presumably reflected some chemical instability of the airborne matter. The contaminants of mere occupancy also seem somewhat unstable; but, unlike tobacco odor, occupancy odor diminishes rapidly and dramatically with time (see Figure IV-10).

Cooking. One brand of cigarettes differs from another in type and blend of tobacco, type of additives sprayed on the tobacco or paper, and various other characteristics, such as porosity of paper and temperature of the ember.[42] In spite of these variations, all cigarettes give rise to an odor readily identifiable perceptually as tobacco odor.[19] In similar fashion, cooking gives rise to perceptually characteristic odors. These do not possess quite the simple integrity as a perceptual class as does tobacco-smoke odor. Some cooking odors constitute more serious nuisances than others. Some (e.g., cabbage odor) are generally considered disagreeable, whereas others (e.g., baking odors) are generally considered inoffensive. Some vapors (e.g., those from deep frying) adsorb tenaciously to surfaces and thereby become long-term contaminants. During initial generation, such vapors may evoke pungency, as well as odor; whether the pungency derives from organic gases generated by the reaction of the oil with the food, from inorganic gases generated by combustion, or from particulate matter remains unknown, but deserves attention.[40]

Bathroom and Waste Odors. In many buildings, exhaust hoods remove cooking odors at the site of generation. In some residences, cooking odors may have no such easy route of egress. Ductless range hoods, equipped with aluminum-mesh filters and carbon filters, sometimes serve as substitutes for exhaust ducts and true exhaust fans. Similarly, ductless bathroom "ventilators" have begun to see some use in homes. Nevertheless, most commercial, institutional, and industrial buildings exhaust bathroom air to the outside. In theory, recirculation of properly filtered bathroom air in nonresidential buildings offers an opportunity for great energy savings. In practice, the putative need to eliminate moisture before recirculation and the possible, if only occasional, breakthrough of contaminants through any filtration system limit the prospects for recirculation.[44] People will tolerate inadequate control of cooking odors (e.g., operation of a ductless range hood with a spent carbon filter) much more readily than inadequate control of bathroom odors.

Odor Control. Elimination of the odor source, local exhaust, general ventilation, and filtration (usually adsorption) are the principal ways of controlling indoor malodors physically. In some places, such as bathrooms and smoking areas, generation of malodor may exceed the limits of physical control and persons in charge of

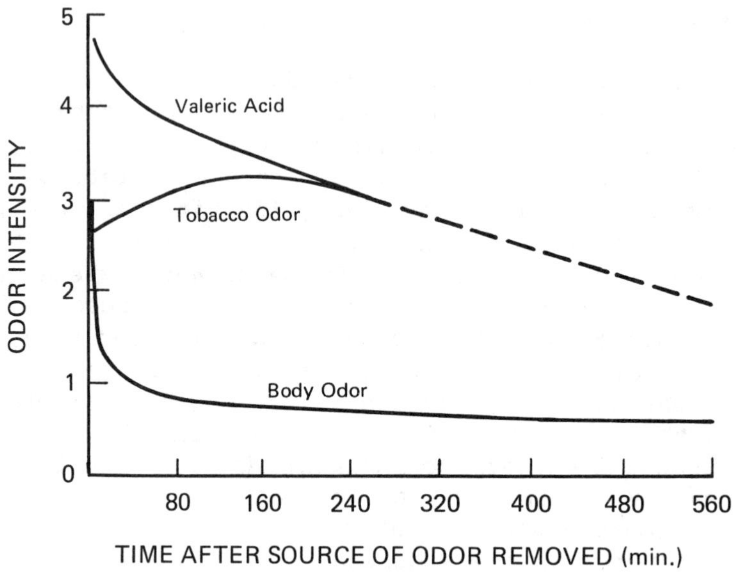

FIGURE IV-10 Decay of odor in still air in an unventilated chamber after an open flask of valeric acid had been removed, five cigarettes had been smoked, and a number of nonsmoking occupants had left the chamber ("body odor"). Odor judgments were made by observers who entered momentarily from time to time. Reprinted with permission from Yaglou and Witheridge.[79]

maintenance resort to commercially marketed odor counteractants.[59] Such products generally comprise a fragrance base made up of many aroma chemicals and possibly a single proprietary "active" ingredient. Typically, a manufacturer claims that the product has eliminated the objectionability or diminished the intensity of some standard malodor in laboratory tests. The claim invariably has some validity, because all odorants, including malodorants, can influence the perception of other odors. This rule forms the foundation of practical perfumery.[13] Basic research in olfaction supports the notion that a gas-phase mixture of components generally has an odor less intense than the total of the separate odors of the unmixed components,[20] perhaps less intense than some of its constituents alone--an extreme case of perceptual hypoadditivity.[10, 23] The matter of alterations in odor quality (character) has received little attention in the scientific laboratory. The perfumer knows through experience how to manipulate or blend malodorants to produce acceptability. Hence, some fruity-smelling natural essences (always mixtures of constituents) may contain some "subsurface" putrid-smelling constituents. The perfumer may therefore blend a "deodorizing" fragrance that will assimilate a putrid-smelling contaminant into a fruity complex. When unable to anticipate the particular malodorous quality of interest, the perfumer generally blends a fragrance with a nondescript, unnameable odor character. Such a broad-band masker may assimilate some malodors readily, but at the very least adds perceptual "noise" to the maldodorous environment. Eventually, the usual or frequent occupants of a space may come to smell the malodorant through the olfactory noise. Persons who ride airplanes often, for instance, eventually find that the smoking area smells strongly of both tobacco smoke <u>and</u> masking agent.

It might seem that deodorizing products are used commercially only in special locations (e.g., bathrooms) or only on the occasions of uncontrollable malodorous emission (e.g., in the case of water damage to upholstery). In fact, such products, under the generic name of "reodorants," appear in virtually every cleaning product (e.g., degreasers, detergents, soaps, and fabric shampoos) and in many other materials (e.g., fabrics, plastics, floor finishes, and carpets). Reodorants are sometimes used to cover up undesirable ambient odors (e.g., mildew in damp spaces) and often to cover up the intrinsic odors of manufactured products themselves (e.g., formaldehyde in permanent-press fabrics). No matter what their purpose, reodorants and odor counteractants have become permanent parts of the indoor environment.[49]

Building Materials and Furnishings

Almost any object indoors may serve as a primary or secondary source of odor. Accordingly, the list of indoor odorous contaminants could go on indefinitely. In fact, Jarke[48] identified over 200 organic constituents in residences under conditions designed to minimize active generation of contaminants. Mølhave and Møller[60] identified a similar number and noted that only six of 46 dwellings seemed odorless.

The identities of the more common indoor contaminants led Jarke to conclude that notable more or less permanent sources include food, plants, bodies, dry-cleaned clothes, cosmetics, household products, attached garages, heating systems, new furniture, carpeting, and redecorated surfaces. Chemical contamination is apprently much higher in new than in old homes.[60] New materials require a considerable amount of time to "cure" before the off-gassing of volatile odorous substances diminishes to an imperceptible point. One of the most notorious of the odorous contaminants in new buildings is formaldehyde.[8] It emanates from chipboard, panel adhesive, carpet backing, vinyl wall-covering, resin-treated fabrics, and urea-formaldehyde foam insulation. In some Danish homes, the concentration of gas-phase formaldehyde has exceeded threshold limit values for occupational exposure.[9]

Although formaldehyde has an odor, it has developed a reputation as an olfactory anesthetic.[52] The reason for its putative anesthetic properties has not been fully studied. Conceivably, its irritant properties play a role in the anesthetic phenomenon. Cain and Murphy[26] have reported that an irritant can immediately diminish and actually block olfaction. This effect seems to occur through interplay between sensory activity in the trigeminal nerve system and sensory activity in the olfactory nerve system. The trigeminal nerve mediates all cutaneous sensations of the face, as well as the pungency, irritation, warmth, cooling, and pain that can arise from chemical stimulation of mucosal tissue in the nose, mouth, and eyes. The evidence suggests that the interaction between odor and irritation takes place in the central nervous system and, hence, that it would hold true for virtually all combinations of odorous and irritating stimuli. It seems relevant that various successful deodorizers have contained irritants or pungent materials. For example, one commonly used deodorizer of the wick type contained formaldehyde, and other types of deodorizers contained other aldehydes. Ozone, a pungent gas, has long had a reputation as a deodorizer, even at concentrations too low to eliminate malodors through oxidation.[75] The "fresh-air" smell that deodorizers sometimes are claimed to produce can stem from the pungency produced by these substances at low concentrations in the product. Fresh Arctic air generally contains noticeable amounts of ozone. Hence, the association of pungency with "fresh air" has some basis in common experience. Nevertheless, the deliberate addition, even at low concentrations, of products that are irritating has questionable justification.

Odors That Enter from Outdoors

In theory, ventilation dilutes and displaces contaminants that are generated indoors. Bringing in odor-contaminated air for use in ventilation to reduce odors obviously can defeat the purpose. Odorous outside air is encountered in areas of great industrial pollution or where micrometeorologic conditions allow entrainment of emitted substances from local sources into intake vents. In recognition of the possibility that the air used for ventilation sometimes fails to meet normal standards of quality, the American Society of Heating, Refrigerating and Air-Conditioning Engineers (ASHRAE) specified, in Standards for Natural and Mechanical Ventilation,[6] that intake air should meet both objective and subjective criteria for cleanliness. A draft revision of this standard contains a list of notable outdoor contaminants, some odorous and some nonodorous, and concentrations not to be exceeded (see Table IV-21). Adherence to these criteria will not guarantee the odorlessness of air, but will probably minimize difficulties in polluted regions.

Odors may also be generated in the air delivery system itself. Berglund and Lindvall[11] discovered cases where the air supplied to a room had a higher degree of odorant contamination than the air already in the room. That can occur when air-to-air heat exchangers, installed to conserve energy, allow exchange of organic materials between exhaust and makeup air. Another source may be the accumulation of odorous materials on roughing filters and cooling coils and in humidifiers in the intake system.

MEASUREMENT OF ODOR

Odors have various attributes: intensity, quality (character), affective charge (acceptability-objectionability), and duration. Most environmental odorants are mixtures of substances. Hence, intensity and character generally represent the net action and nonlinear perceptual combination of various constituents. Figure IV-11 depicts a gas chromatogram of a sample of odorous air (perspiration odor). In this case, the odorous sample is split into two streams after it passes through the column of the chromatograph.[32] One stream goes to a flame ionization detector and the other to a sniffing port, where the experimenter can note the odor character associated with the various chromatogram peaks. Such an odor-annotated chromatogram, called an "odorogram," can help in deciding how many peaks (constituents) seem odor-relevant, whether a few constituents seem particularly redolent of the unfractionated sample, and whether there is a relationship between the height of a peak and the magnitude of its odor. Often, a barely detectable peak is related to a strong odor and an enormous peak to a barely detectable odor. This situation reflects the nonuniform sensitivity of the nose; some substances stimulate at much lower concentrations than others. Table IV-22 shows, for instance, that ethyl acrylate stimulates at a concentration six orders of magnitude below that of ethylene and five orders of magnitude below that of acetone.[46] Odor science has long sought, with limited success, to account for such large disparities in stimulating efficiency.[15,28]

TABLE IV-21

Ambient-Air Quality Standards for Notable Contaminants Unregulated by Federal Clean Air Act[a]

Contaminant[b]	Long-Term Standard Concentration	Long-Term Standard Period	Short-Term Standard Concentration	Short-Term Standard Period[c]
Acetone*	7 mg/m^3	24 h	24 mg/m^3	30 min
Acrolein*	--	--	25 μg/m^3	C
Ammonia*	0.5 mg/m^3	1 yr	7 mg/m^3	C
Beryllium	0.01 μg/m^3	30 d	--	--
Cadmium	2.0 μg/m^3	24 h	--	--
Calcium oxide (lime)	--	--	20-30 μg/m^3	C
Carbon disulfide*	0.15 mg/m^3	24 h	0.45 mg/m^3	30 min
Chlorine*	0.1 mg/m^3	24 h	0.3 mg/m^3	30 min
Chromium	1.5 μg/m^3	24 h	--	--
Cresol*	0.1 mg/m^3	24 h	--	--
Dichloroethane*	2.0 mg/m^3	24 h	6.0 mg/m^3	30 min
Ethyl acetate*	14 mg/m^3	24 h	42 mg/m^3	30 min
Formaldehyde[d]*	--	--	120 μg/m^3	C
Hydrochloric acid*	0.4 mg/m^3	--	3 mg/m^3	30 min
Hydrogen sulfide*	40-50 μg/m^3	24 h	42 μg/m^3	1 h
Mercaptans*	--	--	20 μg/m^3	1 h
Mercury	2 μg/m^3	24 h	--	--
Methyl alcohol*	1.5 mg/m^3	24 h	4.5 mg/m^3	30 min
Methylene chloride*	20 mg/m^3	1 yr	150 mg/m^3	30 min
	50 mg/m^3	24 h	--	--
Nickel	2 μg/m^3	24 h	--	--
Nitrogen monoxide	0.5 mg/m^3	24 h	1 mg/m^3	30 min
Phenol*	0.1 mg/m^3	24 h	--	--
Sulfates	4 μg/m^3	1 yr	--	--
	12 μg/m^3	24 h	--	--
Sulfuric acid*	50 μg/m^3	1 yr	200 μg/m^3	30 min
	100 μg/m^3	24 h	--	--
Trichloroethylene*	2 mg/m^3	1 yr	16 mg/m^3	30 min
	5 mg/m^3	24 h	--	--
Vanadium	2 μg/m^3	24 h	--	--
Zinc	50 μg/m^3	1 yr	--	--
	100 μg/m^3	24 h	--	--

[a]Reprinted with permission from ANSI/ASHRAE.[1] Concentrations listed should be corrected to standard conditions--25°C and 760 mm Hg.

[b]Contaminants marked with an asterisk have odors at concentrations sometimes found in outdoor air; concentrations listed do not necessarily result in absence of odor.

[c]C, ceiling (maximal allowable concentration).

[d]An industry organization has appealed the air quality limits of 120 μg/m^3 as shown in Tables 2 and 4 of Standard 62-1981. The appeal is under consideration. If any change in Standard 62-1981 results from the appeal, all original recipients will be informed by ASHRAE.

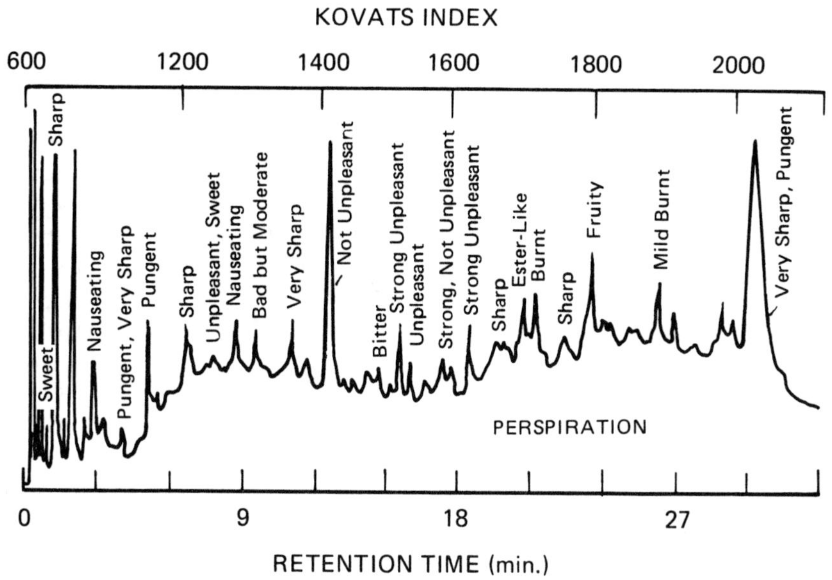

FIGURE IV-11 Odor-annotated gas chromatogram, called an "odorogram." Test vapor was human perspiration. Annotations refer to odor qualities noted by an observer when various constituents (represented by peaks) eluted from the chromatographic column. Reprinted with permission from Dravnieks.[32]

TABLE IV-22

Odor Threshold and Quality of Various Petrochemicals[a]

Compound	Threshold Concentration, ppm Absolute	50% Recognition	100% Recognition	Quality
Acetone	20.0	32.5	140	Sweet-fruity
2,6-Butanol	0.30	1.0	2.0	Rancid-sweet
Di-N-butylamine	0.08	0.27	0.48	Fishy-amine
Diethylamine	0.02	0.06	0.06	Musty-fishy-amine
Ethyl acetate	6.3	13.2	13.2	Sweet-ester
Ethyl acrylate	0.0002	0.00030	0.00036	Sour-pungent
Ethylene	260	400	700	Olefinic
N-Ethyl morpholine	0.08	0.25	0.25	Ammoniacal
Isobutyl acetate	0.35	0.50	0.50	Sweet-ester
Isobutyl acrylate	0.002	0.009	0.012	Sweet-musty
Methanol	4.26	53.3	53.3	Sour-sharp
Methylethylketone	2.0	5.5	6.0	Sweet-sharp
2-Methyl-5-ethyl pyridine	0.006	0.008	0.010	Sour-pungent
2,4-Pentanedione	0.01	0.020	0.024	Sour-rancid
Propanal	0.009	0.040	0.080	Sweet-ester
Propionic acid	0.028	0.034	0.034	Sour
Propylene	22.5	67.6	67.6	Sharp-amine

[a]Data from Hellman and Small.[46]

Sample Collection

Procedures used to collect environmental samples for odor analysis have begun to approach standardization. Figure IV-12 displays a collector that contains the porous polymer material Tenax GC.[35] Use of this adsorbent material avoids the need for solvents in collection and analysis of air samples. A 2-L sample drawn through the collector typically retains organic materials in a quantitatively faithful fashion if their molecules have more than about six carbon atoms. For analysis of its contents, the collector is connected to a gas chromatograph, where flash-heating desorbs the contents into a stream of inert carrier gas. New techniques, such as high-pressure capillary-column chromatography, allow greater resolution and sensitivity than could be achieved with packed columns of the sort used to analyze the sample in Figure IV-11. Unfortunately, there is some incompatibility between capillary injection and the odorogram procedure.

Psychophysical Analysis

The standardization that has evolved in the collection of samples has extended also to psychophysical procedures for evaluation of environmental odors. The choice of procedures obviously depends on the question of interest. A variety of techniques are used in the laboratory, and a considerable number of techniques compete for attention in the field. Nevertheless, techniques that have recently emerged from activities of the ASTM Committee on Sensory Evaluation seem reasonably stable and precise and therefore legitimate for use as standard techniques. This section highlights these methods for the assessment of important attributes of odors.

Odor Character. The odor of any substance can be described precisely by only one term--the name of the substance itself. Stated otherwise, the only precise name for the odor of a lemon is "lemon odor," for the odor of a rose, "rose odor," and for the odor of a goat, "goat odor." Unlike colors, tastes, and sounds, odors have never given rise to their own glossary.[17] Moreover, people often find it difficult to retrieve the names of even familiar odors, not to mention describing unfamiliar ones.[18] This situation motivated the derivation of a list of 146 descriptors to aid in the characterization of odors (Figure IV-13).[33] It seems necessary, if unwieldy, and has led to surprising reliability in a multilaboratory comparison.[36] Earlier systems of odor classification, generally hierarchic, with a small number of major categories and associated subdivisions never proved practical.[15]

Data derived from the list of 146 odor descriptors can have particular use in an effort to track down the source of a malodor, particularly an episodic malodor. The list enables persons influenced by the same malodor to express possible consensus regarding its character. Without such uniform terminology, even articulate persons often give such impoverished qualitative descriptions of malodors as "stinky," "rotten," "yucky," and "foul."

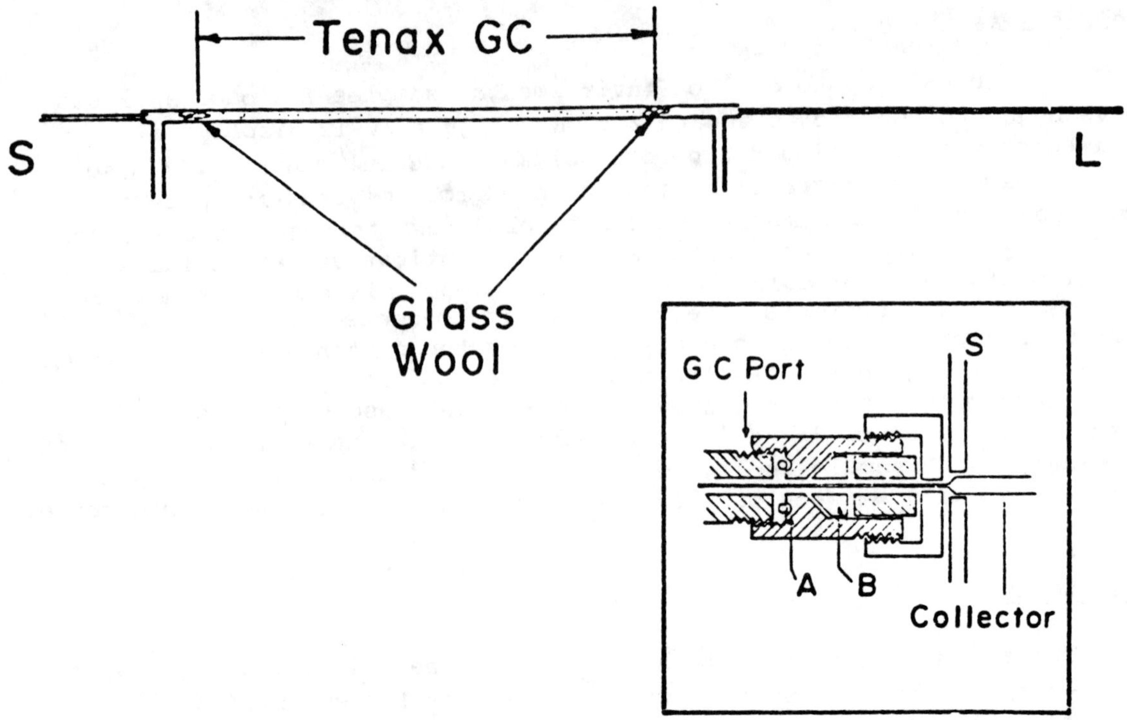

FIGURE IV-12 Tenax-filled collector for organic contaminants. Insert shows details of connecting the end of the collector to the injection port of a gas chromatograph. Length of collector, 200 mm; outside diameter, 3.1 mm. Reprinted with permission from Jarke.[48]

1	2	3	4	5	6	7	8	9	10	11	12
						1					

SAMPLE DESIGNATION — PANEL CODE — PANELIST
(Col. 1–6, 8–10: letters or numbers. Col. 11–12: numbers only.)

ODOR QUALITY EVALUATION

* SMELL SAMPLE. YOU CAN RE-SMELL IT AS NEEDED FOR EVALUATION.
** GO THROUGH LIST BELOW. FOR EACH DESCRIPTOR, *ENCIRCLE* THAT SCORE NUMBER WHICH BEST CHARACTERIZES THE *DEGREE OF PRESENCE* OF THAT ODOR NOTE IN THE SAMPLE ODOR. IF ABSENT, *DO NOT ENCIRCLE ZERO*
*** INITIAL OR SIGN, AND DATE.

Initial or Signature .... Date ....

MEANING OF THE ODOR QUALITY SCALE:

ABSENT	SLIGHTLY		MODERATELY		EXTREMELY
0	1	2	3	4	5

Index	DESCRIPTOR	SCORE
001	FRAGRANT	0 1 2 3 4 5
002	SWEATY	0 1 2 3 4 5
003	ALMOND-LIKE	0 1 2 3 4 5
004	BURNT, SMOKY	0 1 2 3 4 5
005	HERBAL, GREEN, CUT GRASS	0 1 2 3 4 5
006	ETHERISH, ANAESTHETIC	0 1 2 3 4 5
007	SOUR, ACID, VINEGAR	0 1 2 3 4 5
008	LIKE BLOOD, RAW MEAT	0 1 2 3 4 5
009	DRY, POWDERY	0 1 2 3 4 5
010	LIKE AMMONIA	0 1 2 3 4 5
011	DISINFECTANT, CARBOLIC	0 1 2 3 4 5
012	AROMATIC	0 1 2 3 4 5
013	MEATY (COOKED, GOOD)	0 1 2 3 4 5
014	SICKENING	0 1 2 3 4 5
015	MUSTY, EARTHY, MOLDY	0 1 2 3 4 5
016	SHARP, PUNGENT, ACID	0 1 2 3 4 5
017	CAMPHOR LIKE	0 1 2 3 4 5
018	LIGHT	0 1 2 3 4 5
019	HEAVY	0 1 2 3 4 5
020	COOL, COOLING	0 1 2 3 4 5
021	WARM	0 1 2 3 4 5
022	METALLIC	0 1 2 3 4 5
023	PERFUMERY	0 1 2 3 4 5
024	MALTY	0 1 2 3 4 5
025	CINNAMON	0 1 2 3 4 5
026	POPCORN	0 1 2 3 4 5
027	INCENSE	0 1 2 3 4 5
028	Cantaloupe, Honey Dew MELON	0 1 2 3 4 5
029	TAR-LIKE	0 1 2 3 4 5
030	EUCALYPTUS	0 1 2 3 4 5
031	OILY, FATTY	0 1 2 3 4 5
032	LIKE MOTHBALLS	0 1 2 3 4 5
033	LIKE GASOLINE, SOLVENT	0 1 2 3 4 5
034	COOKED VEGETABLES	0 1 2 3 4 5
035	SWEET	0 1 2 3 4 5
036	FISHY	0 1 2 3 4 5
037	SPICY	0 1 2 3 4 5
038	PAINT-LIKE	0 1 2 3 4 5
039	RANCID	0 1 2 3 4 5
040	MINTY, PEPPERMINT	0 1 2 3 4 5
041	SULPHIDIC	0 1 2 3 4 5
042	FRUITY (CITRUS)	0 1 2 3 4 5
043	FRUITY (OTHER)	0 1 2 3 4 5
044	PUTRID, FOUL, DECAYED	0 1 2 3 4 5
045	WOODY, RESINOUS	0 1 2 3 4 5
046	MUSK-LIKE	0 1 2 3 4 5
047	SOAPY	0 1 2 3 4 5
048	GARLIC, ONION	0 1 2 3 4 5
049	ANIMAL	0 1 2 3 4 5
050	VANILLA-LIKE	0 1 2 3 4 5
051	FECAL (LIKE MANURE)	0 1 2 3 4 5
052	FLORAL	0 1 2 3 4 5
053	YEASTY	0 1 2 3 4 5
054	CHEESY	0 1 2 3 4 5
055	HONEY-LIKE	0 1 2 3 4 5
056	ANISE (LICORICE)	0 1 2 3 4 5
057	TURPENTINE (PINE OIL)	0 1 2 3 4 5
058	FRESH GREEN VEGETABLES	0 1 2 3 4 5
059	MEDICINAL	0 1 2 3 4 5
060	ORANGE (FRUIT)	0 1 2 3 4 5
061	BUTTERY (FRESH)	0 1 2 3 4 5
062	LIKE BURNT PAPER	0 1 2 3 4 5
063	COLOGNE	0 1 2 3 4 5
064	CARAWAY	0 1 2 3 4 5
065	BARK-LIKE, BIRCH BARK	0 1 2 3 4 5
066	ROSE-LIKE	0 1 2 3 4 5
067	CELERY	0 1 2 3 4 5
068	BURNT CANDLE	0 1 2 3 4 5
069	MUSHROOM-LIKE	0 1 2 3 4 5
070	WET WOOL, WET DOG	0 1 2 3 4 5
071	CHALKY	0 1 2 3 4 5
072	LEATHER-LIKE	0 1 2 3 4 5
073	PEAR (FRUIT)	0 1 2 3 4 5
074	STALE TOBACCO SMOKE	0 1 2 3 4 5
075	RAW CUCUMBER-LIKE	0 1 2 3 4 5
076	RAW POTATO-LIKE	0 1 2 3 4 5
077	MOUSE-LIKE	0 1 2 3 4 5
078	BLACK PEPPER-LIKE	0 1 2 3 4 5
079	BEAN-LIKE	0 1 2 3 4 5
080	BANANA-LIKE	0 1 2 3 4 5
081	BURNT RUBBER-LIKE	0 1 2 3 4 5
082	GERANIUM LEAVES	0 1 2 3 4 5
083	URINE-LIKE	0 1 2 3 4 5
084	BEERY (BEER-LIKE)	0 1 2 3 4 5
085	CEDARWOOD-LIKE	0 1 2 3 4 5
086	COCONUT-LIKE	0 1 2 3 4 5
087	ROPE-LIKE	0 1 2 3 4 5
088	SEMINAL, SPERM-LIKE	0 1 2 3 4 5
089	LIKE CLEANING FLUID (Carbona)	0 1 2 3 4 5
090	CARDBOARD-LIKE	0 1 2 3 4 5
091	LEMON (FRUIT)	0 1 2 3 4 5
092	DIRTY LINEN-LIKE	0 1 2 3 4 5
093	KIPPERY (SMOKED FISH)	0 1 2 3 4 5
094	CARAMEL	0 1 2 3 4 5
095	SAUERKRAUT-LIKE	0 1 2 3 4 5
096	CRUSHED GRASS	0 1 2 3 4 5
097	CHOCOLATE	0 1 2 3 4 5
098	MOLASSES	0 1 2 3 4 5
099	ALCOHOL-LIKE	0 1 2 3 4 5
100	DILL-LIKE	0 1 2 3 4 5
101	CHEMICAL	0 1 2 3 4 5
102	CREOSOTE	0 1 2 3 4 5
103	GREEN PEPPER	0 1 2 3 4 5
104	HOUSEHOLD GAS	0 1 2 3 4 5
105	PEANUT BUTTER	0 1 2 3 4 5
106	VIOLETS	0 1 2 3 4 5
107	TEA-LEAVES-LIKE	0 1 2 3 4 5
108	STRAWBERRY-LIKE	0 1 2 3 4 5
109	STALE	0 1 2 3 4 5
110	CORK-LIKE	0 1 2 3 4 5
111	LAVENDER	0 1 2 3 4 5
112	CAT-URINE-LIKE	0 1 2 3 4 5
113	PINEAPPLE (FRUIT)	0 1 2 3 4 5
114	FRESH TOBACCO SMOKE	0 1 2 3 4 5
115	NUTTY (WALNUT, ETC.)	0 1 2 3 4 5
116	FRIED CHICKEN	0 1 2 3 4 5
117	WET PAPER-LIKE	0 1 2 3 4 5
118	COFFEE-LIKE	0 1 2 3 4 5
119	PEACH (FRUIT)	0 1 2 3 4 5
120	LAUREL LEAVES	0 1 2 3 4 5
121	BURNT MILK	0 1 2 3 4 5
122	SEWER ODOR	0 1 2 3 4 5
123	SOOTY	0 1 2 3 4 5
124	CRUSHED WEEDS	0 1 2 3 4 5
125	RUBBERY (NEW RUBBER)	0 1 2 3 4 5
126	BAKERY (FRESH BREAD)	0 1 2 3 4 5
127	OAK WOOD, COGNAC-LIKE	0 1 2 3 4 5
128	GRAPEFRUIT	0 1 2 3 4 5
129	GRAPE-JUICE-LIKE	0 1 2 3 4 5
130	EGGY (FRESH EGGS)	0 1 2 3 4 5
131	BITTER	0 1 2 3 4 5
132	CADAVEROUS, Like Dead Animal	0 1 2 3 4 5
133	MAPLE (AS IN SYRUP)	0 1 2 3 4 5
134	SEASONING (FOR MEAT)	0 1 2 3 4 5
135	APPLE (FRUIT)	0 1 2 3 4 5
136	SOUPY	0 1 2 3 4 5
137	GRAINY (AS GRAIN)	0 1 2 3 4 5
138	CLOVE-LIKE	0 1 2 3 4 5
139	RAISINS	0 1 2 3 4 5
140	HAY	0 1 2 3 4 5
141	KEROSENE	0 1 2 3 4 5
142	NAIL POLISH REMOVER	0 1 2 3 4 5
143	FERMENTED (Rotten) FRUIT	0 1 2 3 4 5
144	CHERRY (BERRY)	0 1 2 3 4 5
145	VARNISH	0 1 2 3 4 5
146	SOUR MILK	0 1 2 3 4 5

FIGURE IV-13 Data sheet for profiling of odor quality with 146 descriptors. Reprinted from National Research Council.[65]

Odor Intensity. From the standpoint of environmental engineering, intensity is the most important attribute of an odor. It also permits relatively precise measurement by matching. A recently adopted ASTM butanol reference scale has already served well in this capacity.[2] The reference scale entails the use of an olfactometer that sets up eight concentrations of butanol spanning a range of subjective intensity from very weak to very strong. As Figure IV-14 shows, an observer seeks to find a nozzle on a lazy Susan that matches the intensity of a test odor.[31] Figure IV-15 gives an example of some results obtained in this fashion.[37]

Intramodality matching, such as that used with the butanol reference scale, is the most fundamental and secure psychophysical operation. The widespread, successful use of intramodality matching in vision and hearing ensures its basic validity.[58] Nevertheless, its use requires some knowledge of psychophysics in general and olfactory psychophysics in particular. For example, adaptation is important in determining how many judgments a person can make in a given period.[24] Inattention to time-dependent changes in sensitivity can severely distort the outcome of the matching operation.

The NRC report Odors from Stationary and Mobile Sources[65] presented other techniques for assessing odor intensity. Some involve the use of a matching odorant other than butanol.[11, 29] Others involve only numerical judgments. The potential value of numerical judgements is derived from the possibility that they reflect the true form of the psychophysical function. Such judgments imply, for instance, that the perceived magnitude of butanol varies with concentration, not proportionally, but with approximately the 2/3 power of concentration. This relation, expressed in the equation $S = 0.261C^{0.66}$ (where S is perceived magnitude and C is concentration), can make it possible to convert a concentration of butanol chosen during matching to a numerical perceived magnitude.[61] The psychophysical technique of magnitude estimation used to derive the equation falls into a class of ratio scaling techniques. These commonly used techniques assume that subjects can judge ratio relations among sensations.[67]

Although the equation above has been proposed as a standard function, its parameters can vary systematically with conditions of stimulation, type of numerical scaling procedure, etc. Hence, the standard function requires specification of standard means of data acquisition. Some category scaling procedures--wherein subjects use a scale of five categories, seven categories, and so on--typically fail to produce a function that conforms to the equation. In principle, however, a category scale with a properly chosen number (and hence range) of categories should produce such a relation and might thereby bring the outcome of category scaling and ratio scaling into register. The issue of why these two classes of procedures generally produce different results has escaped simple resolution throughout sensory research.[58]

Odor Acceptability. The acceptability-objectionability or pleasantness-unpleasantness of an odor can be assessed by essentially

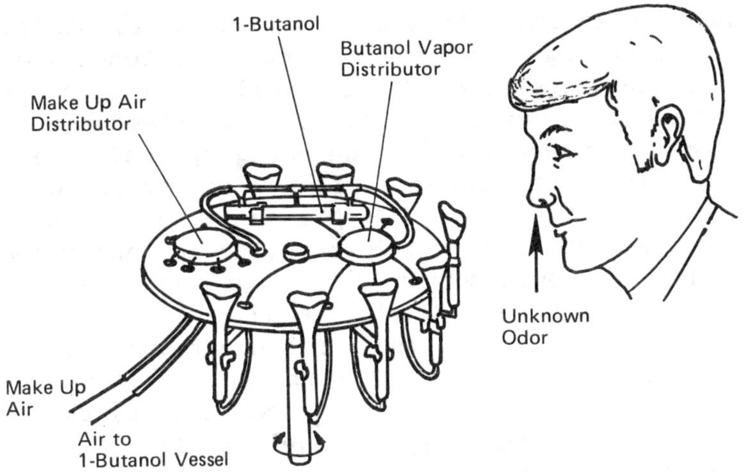

FIGURE IV-14 A subject using the butanol olfactometer (lazy Susan configuration) to find an odor that matches an unknown stimulus. As customarily arranged, the device delivers concentrations that range from 16 to 2,000 ppm. Concentration changes by a factor of 2 from port to port. Reprinted with permission from Flavor Quality: Objective Measurement.[31(p. 13)]

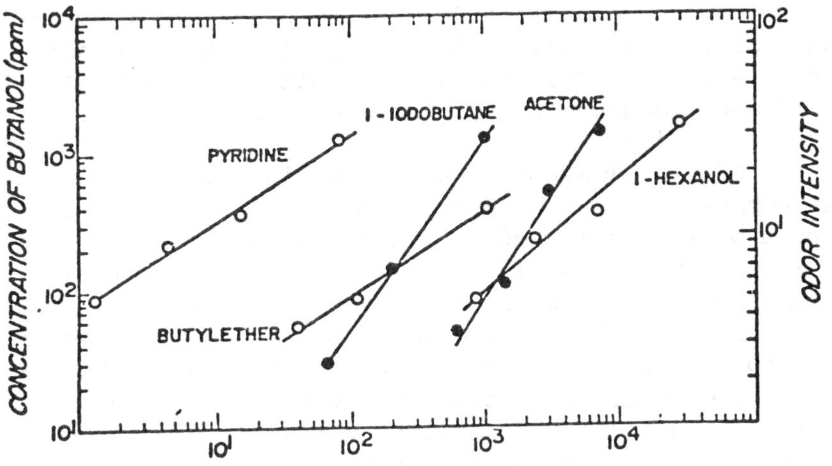

FIGURE IV-15 Psychophysical functions for five odorants. Functions were obtained by matching butanol (note left ordinate) to various concentrations of each odorant. Right ordinate shows odor intensity derived via the psychophysical function for butanol. Data from Dravnieks and Laffort.[37]

the same techniques as its intensity. For instance, Lindvall and Svensson[56] used matching to hydrogen sulfide (rotten-egg odor) to assess the unpleasantness of combustion-toilet emission. Many investigators have used numerical scaling, sometimes ratio scaling,[39] and sometimes category scaling.[30] The category scales have typically been bipolar, ranging, for instance, from -3 (very unpleasant) to +3 (very pleasant).[7] Unfortunately, such a scale seems often to misrepresent the variability of acceptability and objectionability in the extremes of the hedonic continuum. Accordingly, some persons have used a line-marking or line-producing technique.[62] This generally involves a bipolar scale, but without fixed categories, and seems to yield a more realistic picture of response variability than does category scaling.[25] Other techniques of continuous, rather than categoric, judgment may behave just as well as line-marking, with respect to response variability.

Odor Threshold. The measurement of odors has often comprised merely the measurement of the concentration (or dilution) necessary to achieve some criterion of detectability. The literature on olfaction contains well over 1,000 threshold values for odorants. All too commonly, these were obtained by techniques devised specifically for particular investigations;[41,72] for this reason, the threshold value for a particular odorant may vary widely, even by orders of magnitude, from one investigation to another. Specific factors contributing to this variability include the means of stimulus presentation to the odor-panel member, judgmental factors, and the implicit or explicit definition of threshold. Few experimenters have sought to verify the reliability of the vessels or olfactometers used to present odorous stimuli. Furthermore, each device customarily uses arbitrarily chosen conditions (e.g., flowrate, temperature, humidity, and solvent), and this often precludes comparison with results obtained with other devices or, more important, may preclude a comparison with typical environmental conditions.

The psychophysical means of eliciting information on detectability can alter apparent sensitivity by a factor of 100 or more.[55] Presenting a pair of stimulus samples--one containing an odorant at a fixed, weak concentration and the other containing only air--and forcing the subject to choose one of the two in each of scores or hundreds of trials will maximize detection. (This forced-choice procedure could also offer one odorous sample and two blanks, etc.) But presenting various concentrations randomly with no direct comparison stimulus ("blank") invites variability. One subject may set a low criterion and say "yes" to almost anything. Another subject may behave much more conservatively. Such differences between subjects have given incentive to erect a new psychophysics of signal detection.[38] Because of its time-consuming demands in data collection, the theory of signal detection is not popular with researchers. That theory challenges the concept of threshold and instead specifies, in probabilistic and relatively bias-free terms, the detectability of any given signal. In this respect, it highlights the probabilistic nature of all thresholds. Despite the layman's view, "threshold" hardly refers to a concentration below which a normal person can never detect odor. Even concentrations well below "threshold" may be detectable with sufficient frequency to cause concern about the validity of the threshold concentration.

Complications involved in the interpretation of existing threshold data and pitfalls in the collection of new data raise the question of whether "thresholds," or comparable indexes of detection, offer greater benefits than suprathreshold matching. Nevertheless, some persons will undoubtedly choose threshold measurement over matching. A relatively new ASTM procedure at least offers some standardization.[3] It recommends three-alternative forced-choice presentations wherein one nozzle of an air-dilution olfactometer presents an odorous stimulus at a given dilution and two companion nozzles present odorless air. Testing begins at a very low concentration and progresses to higher concentrations until the subject detects the stimulus reliably. Although the method contains arbitrary ingredients, it combines various important features designed to minimize response bias; and it allows a relatively speedy, if gross, estimate of the degree of dilution that a sample of odorous air can withstand before it ceases to be detected readily.

ODOR CONTROL

The control of odor should follow the same strategy as the control of virtually any other type of indoor contaminant. The first step should involve good housekeeping and prevention of the source. This will obviously fail in the many cases where the mere presence and normal activities of occupants inevitably give rise to odorous organic materials. A second step would eliminate airborne contaminants, such as kitchen odors, by local exhaust.

In principle, the removal of odorous contaminants can follow one or more of the classical strategies: oxidation (e.g., incineration), scrubbing (e.g., spray-washing), chemical conversion (e.g., chlorination), filtration, and dilution.[70] The first three of these are used almost exclusively in the control of industrial odorous emission. For odors in residential, commercial, and institutional spaces, filtration and dilution are the methods of choice. The use of deodorizers is justified only under unusual and temporary conditions.

Ventilation

Before the development of mechanical ventilation, the entrance of outside air through windows was the means of both thermal and contaminant control. The need for thermal control often dictated the demand for outside air.[21] Mechanical ventilation systems, however, allowed separate control of the temperature and contaminants. The amount of ventilation air necessary for control of contaminants historically has been a matter of contention. Some persons argued for ventilation rates that would render the air odorless. This strategy rested on the premise that odor-laden air necessarily contained harmful organic contaminants. Similarly, some argued for rates that would maintain carbon dioxide at a concentration only twice that of the ambient air.[51] Here again, the strategy rested on the notion that conservative control of a measurable (or, in the case of odors, perceptible) contaminant, even a rather innocuous one, would take care of unknown, but possibly harmful, airborne substances. This strategy led to unreasonably high ventilation rates. To achieve a criterion of approximate odorlessness, the ventilation rate must generally exceed 30 ft^3/min (about 14 L/s) per occupant, even during nonsmoking occupancy.

The New York State Commission on Ventilation found that the concentration of carbon dioxide in a normally ventilated schoolroom correlated only weakly with odor.[66] Hence, the use of one contaminant seemed unable to predict the concentrations of all contaminants. The correlation rule may actually work reasonably well under conditions of active control of the delivery of ventilation air.[53] Nevertheless, the New York State commission found that a rate of 10-15 ft^3/min (about 4.7-7.1 L/s) per student in a classroom with about 250 ft^3 (9 m^3) per student sufficed to control odor and carbon dioxide concentration reasonably well; furthermore, it seemed acceptable on the basis of criteria of comfort, health, and performance. In Winslow's[74] words:

> The chemical vitiation of the air of an occupied room (unless poisons or dusts from industrial processes or defective heating appliances are involved) is of relatively slight importance. The organic substances present, manifest as body odors, may exert a depressing effect upon inclination to work and upon appetite; therefore occupied rooms should be free from odors which are obvious to anyone entering from without. (Such odors are never perceived by those who have been in the room while they have been accumulating.) Objectionable effects of this sort have only been demonstrated, however, with a carbon dioxide content of over .2 per cent, which would correspond to an air change of less than 6 cubic feet per person per minute. (pp. 77-78)

As mechanical ventilation systems became more common, there was more interest in discovering how the odor of a room would vary with changes in the proportion of total supply air that consisted of ventilation (outdoor) air. Figure IV-16 depicts a functional relation that Houghten and colleagues[47] erected from judgments in a junior-high-school classroom. The function, derived from judgments of visitors who entered the occupied classroom briefly, intersects the line equal to a judgment of 2 ("noticeable [odor] but not objectionable") at a ventilation rate of about 11 ft^3/min (about 5.2 L/s) per student. This outcome seemed to confirm the findings of the New York State commission.

Experiments performed at the Harvard School of Public Health shortly after the study of Houghten and colleagues implied that ventilation requirements per occupant would vary with the amount of space (volume) available to each occupant. Yaglou and colleagues[78]

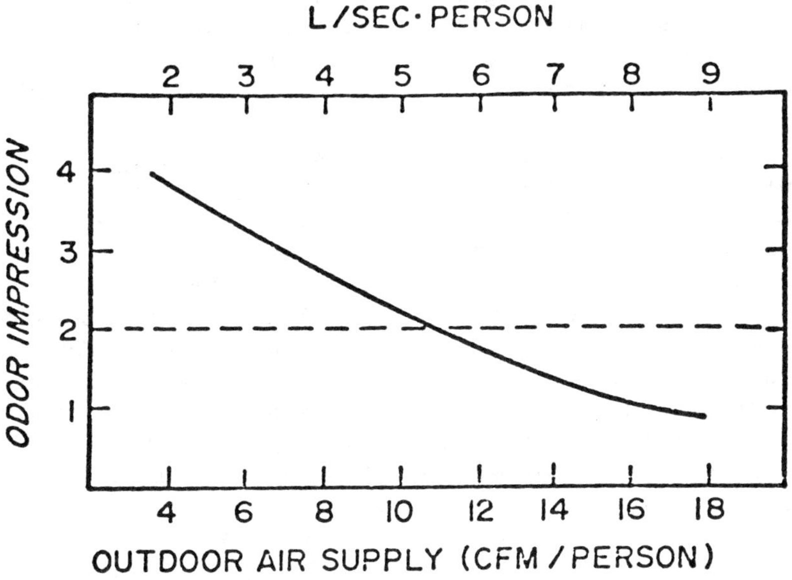

FIGURE IV-16 Relation between odor intensity and ventilation air in junior-high-school classrooms. Observers entered the occupied classroom from a relatively odor-free corridor. Adapted from Houghten *et al.*[47]

of Harvard, like Houghten et al., derived functions that related odor to ventilation rate and added such variables as relative crowding, hygiene, and age. Each variable had some influence, but crowding (i.e., occupant density in an experimental room) had the most noteworthy influence. Figure IV-17 depicts ventilation requirements decided by various criteria versus air space per occupant. Function C represents requirements according to an odor criterion of "moderate" (a rating of 2 on a scale of 0-5). It reflects the intersection of this perceived extent of odor with the various combinations of ventilation rate and air space per person in the functions for occupancy odor depicted in Figure IV-18.

Although admittedly incomplete, the work of Yaglou et al. stands as the most definitive investigation of ventilation requirements ever performed. It seemed to figure at least implicitly in the recommendations of ASHRAE Standard 62-73, a standard based on professional consensus.[6] As Figure IV-19 shows, the recommended ventilation rates for a diverse group of residential and commercial spaces follows much the same curvature as the function of Yaglou et al.[16] Almost all the recommended rates fall above the function--an unsurprising feature, inasmuch as most spaces have a higher odor load than that imposed merely by sedentary occupancy. Cigarette-smoking, for instance, leads to much higher ventilation requirements. Unfortunately, the study shown in Figure IV-20 proffered the only thorough look at the ventilation requirements necessary to control the odor of fresh tobacco smoke.[77] In many of the spaces represented in Figure IV-17, smoking or some physical activity might occur. Curve D in Figure IV-17 formed one attempt to account for the requirements in spaces with such activities. The function, a 50% upward transposition of curve C, apparently has no experimental justification, but may nevertheless serve well as a rough guide for ventilation requirements.[73]

See Chapter IX for more complete discussion of ventilation standards.

Air-Cleaning

The principles of contaminant dilution and displacement, achieved through ventilation, offer the simplest means of indoor contaminant control. In some instances, however, ventilation alone proves inefficient or ineffective. Other means of control can then assist. Outdoor air sometimes contains unwanted concentrations of contaminants, and recirculation of indoor air becomes desirable. For instance, the outdoor air may contain enough sulfur dioxide to damage sensitive electronic equipment. Filtration of one sort or another can reduce the concentrations of the contaminant in the incoming air or, sometimes more productively, can reduce the concentrations of the various contaminants generated indoors and thereby reduce reliance on the use of contaminated outdoor air.[50] In principle, filtration of recirculated air can achieve indoor air quality that exceeds that of outdoor air. But in practice, filtration leaves some contaminants (e.g., carbon monoxide and carbon dioxide) unattenuated. Only very expensive procedures will eliminate these contaminants. Thus, an

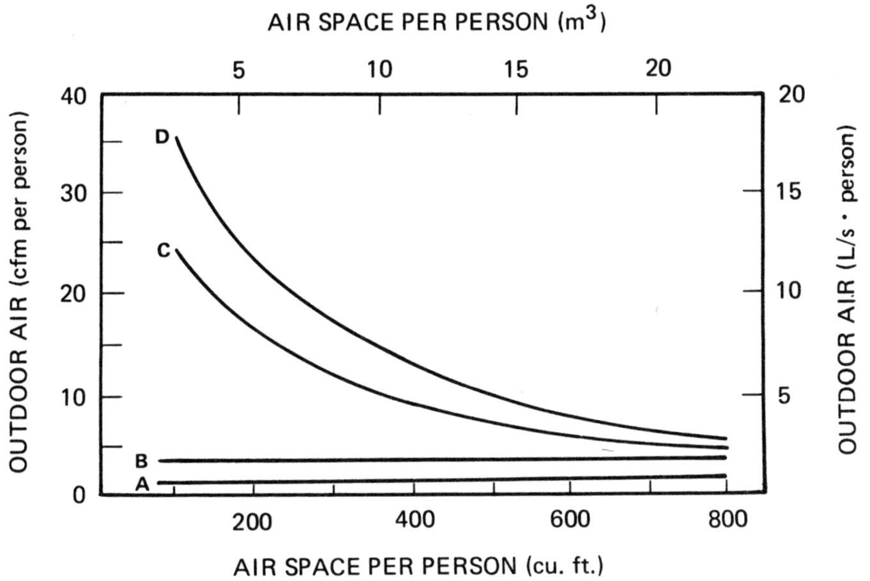

FIGURE IV-17 Relation between ventilation rate and air space per occupant according to four criteria: A, maintenance of oxygen; B, control of carbon dioxide (<0.6%); C, control of body odor under sedentary conditions (no smoking); and D, control of odor when occupants were slightly active and when smoking was permitted. Lower curve derived from data of Yaglou *et al*. Adapted from Viessman.[73]

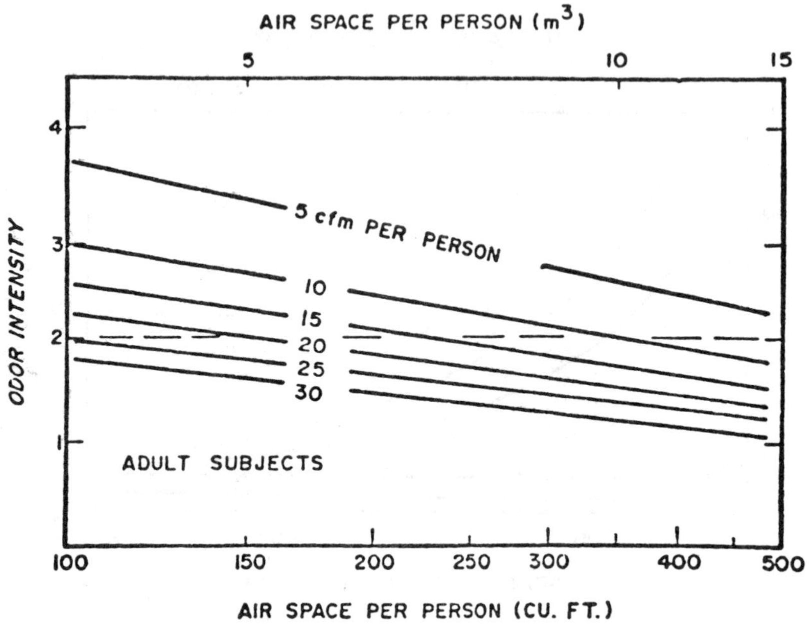

FIGURE IV-18 Odor intensity versus net air space per person for ventilation rates of 5-30 cfm (2.5-15 L/s). Adapted from Yaglou _et_ _al_.[78]

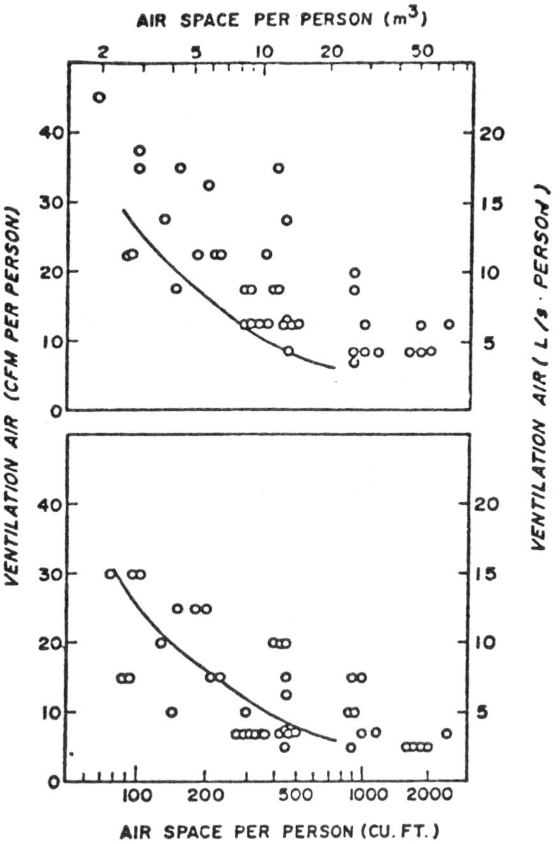

FIGURE IV-19 Top, points depict rates of ventilation recommended for various residential and commercial spaces by ASHRAE Standard 62-73 versus air space per person (logarithmic scale). Air space per person was derived from estimates of occupancy per 1,000 ft^2 (93 m^2) of floor area (incorporated into the standard) and from estimate of ceiling height. Excluded were spaces, such as theaters, where ceiling height might vary considerably from one space to another. Bottom, points depict rates of ventilation recommended for new buildings by ASHRAE Standard 90-75. The points represent the so-called minimal ventilation rates of Standard 62-73. Lines in top and bottom portions are function C from Figure IV-17.

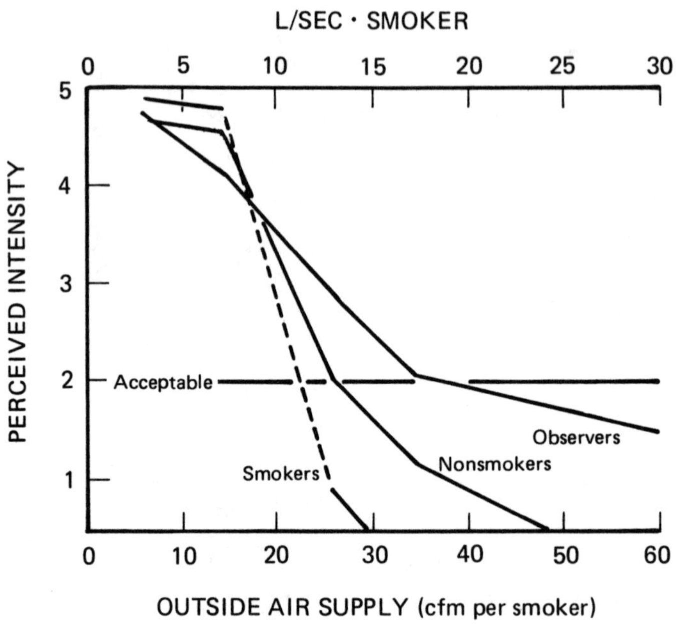

FIGURE IV-20 Variation in odor intensity with fresh air supply when nine persons, including six smokers, occupied Yaglou's chamber and smoked cigarettes at a rate of 24 per hour. Functions labeled "smokers" amd "nonsmokers" depict judgments of occupants. Function labeled "observers" depicts judgments of persons who entered briefly from an odor-free room. Adapted from Yaglou.[77]

engineer or designer cannot rely entirely on recirculation, but must deliver some minimal quantity of outdoor air. According to ASHRAE guidelines,[6] the minimum should equal 5 ft^3/min (about 2.4 L/s) per person when the filtration system has high efficiency. This quantity of air guarantees adequate control of carbon dioxide with a substantial margin of safety. In a space that might normally demand outdoor air at, say, 20 ft^3/min (about 9.4 L/s) per occupant to control the concentrations of organic materials generated during occupancy, use of a high-efficiency filtration system could save considerable amounts of the energy generally used to heat or cool ventilation air.

The characteristic way to clean indoor air involves the use of granular filter media and, if necessary, particle filters.[4] Particle filtration will prove necessary in the presence of cigarette-smoking. Some portion of tobacco-smoke odor is presumably eliminated by such filtration. Particle filters protect the granular filter.

Activated carbon is the most common type of granular medium for control of airborne organic matter.[69] It removes most odorous material and generally renders the air "fresh-smelling." Its adsorbent property leads to actual retention of the contaminant up to a saturation point, when the filter bed must be replaced. Activated carbon is available in many varieties, depending on starting materials and production conditions. Efficiency varies considerably, and life span may be unpredictable. Such technical vagaries detract from the use of activated carbon by nonspecialists, and it currently is little used in ordinary ventilation systems. Other adsorbent materials, such as porous polymers, have been developed for gas chromatography,[68] and perhaps these now-costly materials will become cheap enough for use in ventilation.

Finally, activated carbon or activated alumina can be impregnated with other materials to increase efficacy against particular contaminants. Activated alumina impregnated with postassium permanganate, developed specifically for odor control, offers the most readily available commercial alternative to activated charcoal.[45] It operates by adsorption and oxidation and thereby capitalizes on an empirically observed rule that a malodorant may lose its malodorous properties when oxidized.

RESEARCH NEEDS

Ventilation Codes

A survey of building codes in the 1960s uncovered a tenfold variation in the listed ventilation rates.[14] Such variation reflects in part the change in modes of thought regarding ventilation requirements through the years. Some codes apparently arose from local notions of "good engineering practice" and had seen little or no revision in decades. In the late nineteenth century, the American Society of Heating and Ventilating Engineers recommended 30 ft^3/min per occupant. As mentioned above, this value arose from the notion that ventilation should keep the indoor concentration of carbon dioxide at less than twice its outdoor concentration. Forty years later, the Society recommended a rate of 10 ft^3/min per occupant. Although the studies of the New York State commission and other researchers had made it clear that a rate as low as 10 ft^3/min per occupant would not lead to harm, the members of the Society's committee apparently felt the need for considerably more data. W. H. Driscoll, a member of the committee, noted:[78]

> When we finally decided that we would take 10 cfm per person as the minimum it was a sheer compromise, merely an attempt to finish the work of the Committee and get the report before the Society. There was a difference of opinion as to whether the 30 cu ft that have been set up as a standard since time immemorial should be adopted, or whether no cubic feet, for which there was very aggressive support, not necessarily within the Committee but from outside of the Committee, on the theory that no scientific studies had ever been made to support the necessity for the introduction of any outdoor air as a ventilation requirement. (p. 160)

The experiments of Yaglou _et al_.[78] added some critical information, but this did not necessarily become part of the thinking of code-makers. For instance, some codes still specify ventilation requirements in terms of air changes per hour, irrespective of density of occupancy. Yaglou _et al_. had suggested a need for great sensitivity regarding density of occupancy. Indeed, Yaglou and Witheridge's experiment prompted the comment:[79]

> One result of this investigation, for which I think we can be particularly grateful to the authors, is that it effectively explodes the myth of air change based on room volume as a measure of ventilation. I consider it a real step forward if the idea of air change on a room volume basis can be eliminated from our thinking. (p. 432)

Irrespective of the reasons for variation in ventilation codes, its mere existence implies that a reduction of rates in some of the localities now requiring high rates would probably arouse few complaints. Despite variation from one location to another, both code-makers and ventilating engineers have generally sought to supply ventilation air liberally. Most codes specify rates for conditions of full occupancy; but full occupancy may occur only rarely.

The increasing cost of energy has prompted an unprecedented desire for uniformity in American ventilation codes. The bottom part of Figure IV-19 shows the so-called minimal ventilation rates promulgated in ASHRAE Standard 62-73.[6] These values, like the recommended values shown above them, capture much of the curvature of the recommendations of Yaglou _et al_.[78] The minimal values actually cluster around the Yaglou _et al_. function. When energy first became particularly expensive, it seemed desirable to reduce the "recommended" ventilation rates to these minimal values, at least on a trial basis. These were therefore incorporated into the general standard, ASHRAE 90-75, _Energy_

Conservation in New Building Design.[5] The various model-code groups (Building Officials and Codes Administrators International, International Conference of Building Officials, and Southern Building Code Congress International) have since cooperated with the National Conference of States on Building Codes and Standards (NCSBCS) in drawing up an energy-conservation code, Code for Energy Conservation in New Building Construction.[63] It too incorporated the minimal rates of the ASHRAE ventilation standard.

An NCSBCS survey of building codes[64] conducted in May 1979 revealed widespread adoption of energy conservation in ventilation (Table IV-23). Virtually all large jurisdictions have conformed or will soon conform to conservation standards through adherence to one or another model code, a separate state code rooted in a model code, or ASHRAE Standard 90-75. At the time of the survey, only five jurisdictions had failed to take overt action toward the adoption of an energy-efficient code.

Energy Efficiency, Comfort, and Health

Both the recommended and minimal ventilation rates in ASHRAE Standard 62-73 comprise consensus values. Except, perhaps, for spaces with unexpectedly high rates of cigarette-smoking, the recommended values probably serve well. In some spaces with heavy smoking or cooking, the minimal values will probably serve very poorly. Without a foundation of modern research on a variety of contaminants, the justification of the minimal values will rest heavily on considerations of energy, rather than health. The research must seek to specify rates and strategies that will achieve both energy efficiency and healthful conditions. This will eventually require strict control over the internal atmosphere. Without guidance, some engineers will shut down intake-air dampers to save fuel and will thereafter rely on infiltration alone for fresh air. This strategy actually reduces, rather than increases, the engineer's active control over the building.

The dimensions of ventilation research are roughly the same now as in the time of the work of Yaglou *et al*. Ventilation dilutes contaminants. Its efficacy depends on the nature of indoor contaminants, the size of a space, environmental variables, furnishings, duration of occupancy, aesthetic standards of occupants of or visitors to the space, and sensitivity of occupants. Modern methods available for experimentation include the newly standarized psychophysical methods discussed here and monitoring equipment not available during earlier research on ventilation (e.g., continuous carbon monoxide analyzer, continuous carbon dioxide analyzer, gas chromatograph, mass spectrometer, particle-mass monitor, condensation nucleus counter, and electric aerosol analyzer). With these various tools, it is now possible to characterize the indoor environment in both a psychophysical and a physicochemical manner. Such characterization has now begun in North America and Europe.[22]

TABLE IV-23

NCSBCS Survey of Energy-Conservation Codes[a]

Jurisdiction	Energy-Conservation Code
Alabama	SBCCI
Alaska	(ASHRAE 90-75)
Arizona	(State code)*
Arkansas	(State code)*
California	State code
Colorado	MCEC
Connecticut	State code*
Delaware	(MCEC)
District of Columbia	(City code)
Florida	State code*
Georgia	MCEC
Hawaii	ICBO
Idaho	ICBO
Illinois	(State code)*
Indiana	MCEC
Iowa	MCEC
Kansas	State code
Kentucky	MCEC
Louisiana	(MCEC)
Maine	(State code)
Maryland	(ASHRAE 90-75)
Massachusetts	MCEC
Michigan	ASHRAE 90-75
Minnesota	ASHRAE 90-75
Mississippi	(MCEC)
Missouri	(MCEC)
Montana	MCEC
Nebraska	(ICBO)
Nevada	MCEC
New Hampshire	MCEC
New Jersey	BOCA
New Mexico	ICBO
New York	State code*
North Carolina	State code
North Dakota	ICBO
Ohio	MCEC
Oklahoma	None
Oregon	ICBO
Pennsylvania	(ASHRAE 90-75)
Rhode Island	ASHRAE 90-75
South Carolina	SBCCI
South Dakota	(MCEC)
Tennessee	MCEC
Texas	(MCEC)
Utah	MCEC
Vermont	(ASHRAE 90-75)
Virginia	BOCA, ASHRAE 90-75
Washington	State code
West Virginia	None
Wisconsin	State code*
Wyoming	ICBO
American Samoa	MCEC
Guam	ICBO
Puerto Rico	MCEC

[a]( ) denotes that legislation is pending.
ASHRAE 90-75 = ASHRAE STANDARD 90-75.
BOCA = Model Code, Building Officials & Code Administrators International, Inc.
ICBO = Model Code, International Conference of Building Officials.
MCEC = Model Code for Energy Conservation in New Building Construction.
SBCCI = Model Code, Southern Building Code Congress International Inc.
Asterisk denotes obvious incorporation of energy-conserving aspects of a model code or ASHRAE 90-75; codes or pending codes for California, Maine, and North Carolina also include some such aspects.

REFERENCES

1. American National Standards Institute, and Society of Heating, Refrigerating and Air-Conditioning Engineers, Inc. ANSI/ASHRAE Standard 62-1981. Ventilation for Acceptable Indoor Air Quality. New York: American Society of Heating, Refrigerating and Air-Conditioning Engineers, Inc., 1981. 48 pp.
2. American Society for Testing and Materials. ASTM E 544-75. Standard Recommended Practices for Referencing Suprathreshold Odor Intensity. Philadelphia: American Society for Testing and Materials, 1975.
3. American Society for Testing and Materials. ASTM D 1391. Standard Test Method for Measurement of Odor in Atmospheres (Dilution Method). Philadelphia: American Society for Testing and Materials, 1978.
4. American Society of Heating, Refrigerating and Air-Conditioning Engineers. Control of odors and gaseous contaminants, pp. 33.1-33.8. In ASHRAE Handbook and Product Directory. 1980 Systems. New York: American Society of Heating, Refrigerating and Air-Conditioning Engineers, Inc., 1980.
5. American Society of Heating, Refrigerating and Air-Conditioning Engineers. ASHRAE Standard 90-75. Energy Conservation in New Building Design. New York: American Society of Heating, Refrigerating and Air-Conditioning Engineers, Inc., 1975. 53 pp.
6. American Society of Heating, Refrigerating and Air-Conditioning Engineers. ASHRAE Standard 62-73. Standards for Natural and Mechanical Ventilation. New York: American Society of Heating, Refrigerating and Air-Conditioning Engineers, Inc., 1973. 17 pp.
7. Amerine, M. A., R. M. Pangborn, and E. B. Roessler. Principles of Sensory Evaluation of Food. New York: Academic Press, Inc., 1965. 602 pp.
8. Andersen, I. Formaldehyde in the indoor environment--Health implications and the setting of standards, pp. 65-87 (includes discussion). In P.O. Fanger and O. Valbjørn, Eds. Indoor Climate. Effects on Human Comfort, Performance, and Health in Residential, Commercial, and Light-Industry Buildings. Proceedings of the First International Indoor Climate Symposium, Copenhagen, August 30-September 1, 1978. Copenhagen: Danish Building Research Institute, 1979.
9. Andersen, I., G. R. Lundqvist, and L. Molhave. Indoor air pollution due to chipboard used as a construction material. Atmos. Environ. 9:1121-1127, 1975.
10. Berglund, B. Quantitative and qualitative analysis of industrial odors with human observers. Ann. N.Y. Acad. Sci. 237:35-51, 1974.
11. Berglund, B., and T. Lindvall. Olfactory evaluation of indoor air quality, pp. 141-157. In P.O. Fanger and O. Valbjørn, Eds. Indoor Climate. Effects on Human Comfort, Performance, and Health in Residential, Commercial, and Light-Industry Buildings. Proceedings of the First International Indoor Climate Symposium, Copenhagen, August 30-September 1, 1978. Copenhagen: Danish Building Research Institute, 1979.

12. Billings, J. S., S. W. Mitchell, and D. H. Bergey. The composition of expired air and its effects upon animal life. Smithsonian Contributions to Knowledge 29(989):1-81, 1895.
13. Billot, M., and F. V. Wells. Perfumery Technology. Art: Science: Industry. Chichester, England: Ellis Horwood Limited, 1975. 353 pp.
14. Brauer, R. L., and R. L. Kuehner. The Variability of Ventilation Codes. In Odors and Odorants: The Engineering View. ASHRAE Symposium Bulletin CH-62-2, January 27-30, 1969.
15. Cain, W. S. History of research on smell, pp. 197-229. In E. C. Carterette and M. P. Friedman, Eds. Handbook of Perception. Vol. 6A. Tasting and Smelling. New York: Academic Press, Inc., 1978.
16. Cain, W. S. Interactions among odors, environmental factors and ventilation, pp. 257-274. In P.O. Fanger and O. Valbjørn, Eds. Indoor Climate. Effects on Human Comfort, Performance, and Health in Residential, Commercial, and Light-Industry Buildings. Proceedings of the First International Indoor Climate Symposium, Copenhagen, August 30-September 1, 1978. Copenhagen: Danish Building Research Institute, 1979.
17. Cain, W. S. Physical and cognitive limitations on olfactory processing in human beings, pp. 287-302. In D. Müller-Schwarze and M. M. Mozell, Eds. Chemical Signals in Vertebrates. New York: Plenum Press, 1977.
18. Cain, W. S. To know with the nose: Keys to odor identification. Science 203:467-470, 1979.
19. Cain, W. S. Sensory attributes of cigarette smoking. In G. B. Gori and F. G. Bock, Eds. Banbury Report Three. A Safe Cigarette? Cold Spring Harbor, N.Y.: Cold Spring Harbor Laboratories, 1980.
20. Cain, W. S. The odoriferous environment and the application of olfactory research, pp. 277-304. In E. C. Carterette and M. P. Friedman, Eds. Handbook of Perception. Vol. 6A. Tasting and Smelling. New York: Academic Press, 1978.
21. Cain, W. S. Ventilation and odor control: Prospects for energy savings. ASHRAE Trans. 85(Pt. 1):784-793, 1979.
22. Cain, W. S., L. G. Berglund, R. A. Duffee, and A. Turk. Ventilation and Odor Control. Prospects for Energy Efficiency. Final Report of Phase I of Energy Efficient Ventilation Standards: Requirements for Odor Control. Lawrence Berkeley Laboratory Report LBL-9578. Berkeley, Cal.: Lawrence Berkeley Laboratory, Energy and Environment Division, 1979.
23. Cain, W. S., and M. Drexler. Scope and evaluation of odor counteraction and masking. Ann. N.Y. Acad. Sci. 237:427-439, 1974.
24. Cain, W. S., and T. Engen. Olfactory adaptation and the scaling of odor intensity, pp. 127-141. In C. Pfaffmann, Ed. Olfaction and Taste. Proceedings of the Third International Symposium. New York: The Rockefeller University Press, 1969.
25. Cain, W. S., and F. Johnson, Jr. Lability of odor pleasantness: Influence of mere exposure. Perception 7:459-465, 1978.
26. Cain, W. S., and C. L. Murphy. Interaction between chemoreceptive modalities of odor and irritation. Nature 284:255-257, 1980.
27. Chapin, C. V. The Sources and Modes of Infection. New York: John Wiley & Sons, 1910. 399 pp.

28. Davies, J. T. Olfactory theories, pp. 322-350. In L. M. Beidler, Ed. Handbook of Sensory Physiology. Vol. IV. Chemical Senses. Part 1. Olfaction. Berlin: Springer-Verlag, 1971.
29. Degobert, P. Hedonic and intensity ranking of different malodours by category estimation and paired comparison, pp. 107-121. In J. H. A. Kroeze, Ed. Preference Behaviour and Chemoreception. London: Information Retrieval Ltd., 1979.
30. Doty, R. L. An examination of relationships between the pleasantness, intensity, and concentration of 10 odorous stimuli. Percept. Psychophys. 17:492-496, 1975.
31. Dravnieks, A. Correlation of odor intensities and vapor pressures with structural properties of odorants, pp. 11-28. In R. A. Scanlan, Ed. Flavor Quality: Objective Measurement. American Chemical Society Symposium Series, No. 51. Washington, D.C.: American Chemical Society, 1977.
32. Dravnieks, A. Evaluation of human body odors: Methods and interpretations. J. Soc. Cosmet. Chem. 26:551-571, 1975.
33. Dravnieks, A. Fundamental considerations and methods for measuring air pollution odors, pp. 429-436. In J. LeMagnen and P. MacLeod, Eds. Olfaction and Taste. VI. London: Information Retrieval Ltd., 1978.
34. Dravnieks, A. Measurement of odors in an indoor environment, pp. 127-139. In P.O. Fanger, and O Valbjørn, Eds. Indoor Climate. Effects on Human Comfort, Performance, and Health in Residential, Commercial, and Light-Industry Buildings. Proceedings of the First International Indoor Climate Symposium, Copenhagen, August 30-September 1, 1978. Copenhagen: Danish Building Research Institute, 1979.
35. Dravnieks, A. Organic Contaminants in Indoor Air and Their Relationship to Outdoor Contaminants. Phase I, ASHRAE Research Project 183. IIT Research Institute. New York: American Society of Heating, Refrigerating and Air-Conditioning Engineers, Inc., February 1977. (unpublished)
36. Dravnieks, A., F. C. Bock, J. J. Powers, M. Tibbetts, and M. Ford. Comparison of odors directly and through profiling. Chem. Senses Flavor 3:191-225, 1978.
37. Dravnieks, A., and P. Laffort. Physico-chemical basis of quantitative and qualitative odor discrimination in humans, pp. 142-148. In D. Schneider, Ed. Olfaction and Taste. IV. Stuttgart: Wissenschaftliche Verlagsgesellschaft MBH, 1972.
38. Engen, T. Psychophysics. 1. Discrimination and detection, pp.11-46. In J. W. Kling and L. A. Riggs, Eds. Woodworth and Schlosberg's Experimental Psychology. 3rd ed. New York: Holt, Rinehart and Winston, Inc., 1972.
39. Engen, T., and D. H. McBurney. Magnitude and category scales of the pleasantness of odors. J. Exp. Psychol. 68:435-440, 1964.
40. Evans, C. D., K. Warner, G. R. List, and J. C. Cowan. Room odor evaluation of oils and cooking fats. J. Am. Oil Chem. Soc. 49:578-582, 1972.
41. Fazzalari, F. A., Ed. Compilation of Odor and Taste Threshold Values Data. Philadelphia: American Society for Testing and Materials, 1978. 497 pp.

42. Gori, G. B. Low-risk cigarettes: A prescription. Science 194:1243-1246, 1976.
43. Haggard, H. W. Devils, Drugs, and Doctors. The Story of the Science of Healing from Medicine-Man to Doctor. New York: Harper & Brothers, 1929. 405 pp.
44. Hanna, G. F., and R. L. Kuehner. Critical Factors in Odorant Measurement and Control. ASHRAE Symposium Bulletin CH-69-2:66-72, January 27-30, 1969.
45. Hanna, G. F., R. L. Kuehner, J. D. Karnes, and R. Garbowicz. A chemical method for odor control. Ann. N.Y. Acad. Sci. 116:663-675, 1964.
46. Hellman, T. M., and F. H. Small. Characterization of the odor properties of 101 petrochemicals using sensory methods. J. Air Pollut. Control Assoc. 24:979-982, 1974.
47. Houghten, F. C., H. H. Trimble, C. Gutberlet, and M. F. Lichtenfels. Classroom odors with reduced outside air supply. ASHVE Trans. 41:253-267, 1935.
48. Jarke, F. H. Organic Contaminants in Indoor Air and Their Relation to Outdoor Contaminants. Final Report of ASHRAE Research Project 183, December 1979.
49. Jellinek, J. S. The Use of Fragrance in Consumer Products. New York: John Wiley & Sons, Inc., 1975. 219 pp.
50. Kalika, P. W., J. K. Holcombe, and W. A. Cote. The re-use of interior air. ASHRAE J. 12(11):44-48, 1970.
51. Klauss, A. K., R. H. Tull, L. M. Roots, and J. R. Pfafflin. History of the changing concepts in ventilation requirements. ASHRAE J. 12(6):51-55, 1970.
52. Kulka, K. Odor control by modification. Ann. N.Y. Acad. Sci. 116:676-681, 1964.
53. Kusuda, T. Control of ventilation to conserve energy while maintaining acceptable indoor air quality. ASHRAE Trans. 82(Pt.1):1169-1181, 1976.
54. Leonardos, G., and D. A. Kendall. Questionnaire study on odor problems of enclosed space. ASHRAE Trans. 77(Pt. 1):101-112, 1971.
55. Lindvall, T. On sensory evaluation of odorous air pollutant intensities. Measurements of odor intensity in the laboratory and in the field with special reference to effluents of sulfate pulp factories. Nord. Hyg. Tidskr. Suppl. 2:1-181, 1970.
56. Lindvall, T., and L. T. Svensson. Equal unpleasantness matching of malodorous substances in the community. J. Appl. Psychol. 59:264-269, 1974.
57. Lundqvist, G. R. The effect of smoking on ventilation requirements, pp. 275-292 (includes discussion). In P.O. Fanger, and O. Valbjørn, Eds. Indoor Climate. Effects on Human Comfort, Performance, and Health in Residential, Commercial, and Light-Industry Buildings. Proceedings of the First International Indoor Climate Symposium, Copenhagen, August 30-September 1, 1978. Copenhagen: Danish Building Research Institute, 1979.
58. Marks, L. E. Sensory Processes. The New Psychophysics. New York: Academic Press, Inc., 1974.

59. McCord, C. P., and W. N. Witheridge. Odors. Physiology and control. New York: McGraw-Hill Book Company, Inc., 1949. 405 pp.
60. Mølhave, L., and J. Møller. The atmospheric environment in modern Danish dwellings--Measurements in 39 flats, pp. 171-186. In P. O. Fanger and O. Valbjørn, Eds. Indoor Climate. Effects on Human Comfort, Performance, and Health in Residential, Commercial, and Light-Industry Buildings. Proceedings of the First International Indoor Climate Symposium, Copenhagen, August 30-September 1, 1978. Copenhagen: Danish Building Research Institute, 1979
61. Moskowitz, H. R., A. Dravnieks, W. S. Cain, and A. Turk. Standardized procedure for expressing odor intensity. Chem. Senses Flavor 1:235-237, 1974.
62. Moskowitz, H. R., A. Dravnieks, and L. A. Klarman. Odor intensity and pleasantness for a diverse set of odorants. Percept. Psychophys. 19:122-128, 1976.
63. National Conference of States on Building Codes and Standards. Code for Energy Conservation in New Building Construction. McLean, Va.: National Conference of States on Building Codes and Standards, 1977.
64. National Conference of States on Building Codes and Standards. Survey of Energy Efficient Building Codes. McLean, Va.: National Conference of States on Building Codes and Standards, 1979.
65. National Research Council, Committee on Odors from Stationary and Mobile Sources. Odors from Stationary and Mobile Sources. Washington, D.C.: National Academy of Sciences, 1979. 491 pp.
66. New York State Commission on Ventilation. Ventilation. New York: E. P. Dutton & Company, 1923.
67. Stevens, S. S. Psychophysics: Introduction to Its Perceptual, Neural, and Social Prospects. New York: John Wiley & Sons, Inc., 1975. 329 pp.
68. Turk, A. Absorption, Chapter 8. In A. C. Stern, Ed. Air Pollution. 3rd ed. Vol. 4. Engineering Control of Air Pollution. New York: Academic Press, Inc., 1977.
69. Turk, A., and K. A. Bownes. Absorption can control odors. Chem. Eng. 58(5):156-158, 1951.
70. Turk, A., R. C. Haring, and R. W. Okey. Odor control technology. Environ. Sci. Technol. 6:602-607, 1972.
71. U.S. Department of Health, Education, and Welfare, Public Health Service. Smoking and Health. A Report of the Surgeon General. DHEW Publication No. (PHS) 79-50066. Washington, D.C.: U.S. Government Printing Office, 1979. [1250] pp.
72. van Gemert, L. J., and A. H. Nettenbreier, Eds. Compilation of Odour Threshold Values in Air and Water. Voorburg, Netherlands: National Institute for Water Supply, 1977.
73. Viessman, W. Ventilation control of odors. Ann. N.Y. Acad. Sci. 116:630-637, 1964.
74. Winslow, C.-E. A. Fresh Air and Ventilation. New York: E. P. Dutton & Company, 1926. 182 pp.
75. Witheridge, W. N., and C. P. Yaglou. Ozone in ventilation--Its possibilities and limitations. ASHVE Trans. 45:509-522, 1939. (includes discussion)
76. Wynder, E. L., and D. Hoffmann. Tobacco and health. A societal challenge. New Engl. J. Med. 300:894-903, 1979.

77. Yaglou, C. P. Ventilation requirements for cigarette smoke. ASHAE Trans. 61:25-32, 1955.
78. Yaglou, C. P., E. C. Riley, and D. I. Coggins. Ventilation requirements. ASHVE Trans. 42:133-162, 1936. (includes discussion)
79. Yaglou, C. P., and W. N. Witheridge. Ventilation requirements (Part 2). ASHVE Trans. 43:423-436, 1937. (includes discussion)

TEMPERATURE AND HUMIDITY

The atmosphere not only has an important role in respiratory gas exchange (supplying oxygen and accepting carbon dioxide), but also serves as the heat-exchange medium surrounding the human body. Atmospheric pressure, including its constituents and especially water-vapor pressure, environmental temperature, and the rate of air movement all affect rates of heat loss and heat gain. A change in the rate of heat loss or heat gain ultimately has an effect on body heat content and body temperature.

Human body temperatures are maintained within very narrow ranges either by involuntary physiologic responses controlled by the thermoregulatory system or by behavioral adjustments that modify the thermal environment toward thermal equilibrium. The physiologic responses are proportional to deviations from preferred body temperature, especially those associated with the hypothalamic region of the brain. Behavioral adjustments are proportional to deviations in thermal sensation and thermal comfort from thermoneutral and acceptable, respectively. In general, people prefer to use the behavioral adjustments, rather than having to rely on the physiologic responses.

The thermal state of the human body can be described by the heat balance equation:

$$S = M - E - W - R - C, \qquad (1)$$

in which S = rate of storage of body heat,
M = rate of metabolic heat production,
E = rate of evaporative heat loss,
W = rate of external work done,
R = rate of radiant heat loss, and
C = rate of convective heat loss.

All the above terms are usually expressed in watts per square meter of skin surface area. Body surface areas in adults range from 1.6 to 2.2 m^2 and can be estimated by Equation 2 from height and weight:

$$A_D = 0.202\ m^{0.425} H^{0.725}, \qquad (2)$$

in which A_D = body surface area, m^2,
m = body weight, kg, and
H = body height, m.

Except for short periods, Equation 1 should result in values of S very near zero. Any deviation from zero results in lowering or raising of body temperature (hypothermia or hyperthermia, respectively), with adverse consequences for health and well-being. The metabolic heat generated in the body varies from a minimum of 45 W/m^2 to a maximum that varies between 600 and 900 W/m^2 for short periods.

Heat losses in excess of 45 W/m^2 in the resting state tend to produce hypothermia in some people. Depending on the individual, evaporative heat loss must be at least 100-300 W/m^2 to prevent hyperthermia under conditions of sustained hard work. The range of deep-body temperatures that can be encountered is shown in diagrammatic form in Figure IV-21. A range of ambient temperatures with the appropriate responses is shown in Figure IV-22, and a range of metabolic rates associated with some typical activities is presented in Figure IV-23.

As is evident from the heat-balance equation, many factors determine the heat balance, and they must always be evaluated simultaneously. Energy-conservation strategies may involve lower or higher air temperatures, higher or lower vapor pressures, and higher or lower air velocities. Interruptions in energy supplies can result in sharply higher or lower air temperatures, which can have additional adverse health effects.

HEAT EXCHANGE WITH THE INDOOR ATMOSPHERE

A complete assessment of heat exchange between man and his thermal environment can be found in a recent review by Gagge and Nishi.[8] The following is a limited overview.

Whenever the value of S differs from zero in the heat-balance equation, body temperature will change. At an average body weight of 40 kg/m^2 and a weighted average specific heat of 36.5 $Wh \cdot {}^\circ C \cdot m^{-2}$, it follows that, for example, with S at 36.5 W/m^2 for 1 h, the average body temperature will rise by 1°C.

The rate of metabolic heat production, M, cannot be reduced to below about 40 W/m^2, but can rise to as high as 800 W/m^2 during maximal exercise. Heat exchange by convection (C), by radiation (R), and by evaporation (E) is affected by ambient temperature T_a, by air velocity, and by clothing insulation. For a nude person, convective heat exchange in still air (velocity, V, less than 1 m/s) is approximated by:

$$C = h_c(\bar{T}_s - T_a), \qquad (3)$$

in which C = convective heat exchange, W/m^2,

h_c = convective-heat-transfer coefficient, $W \cdot m^{-2} \cdot {}^\circ C^{-1}$,

$\bar{T}_s$ = mean weighted skin temperature, °C, and

T_a = ambient air temperature, °C.

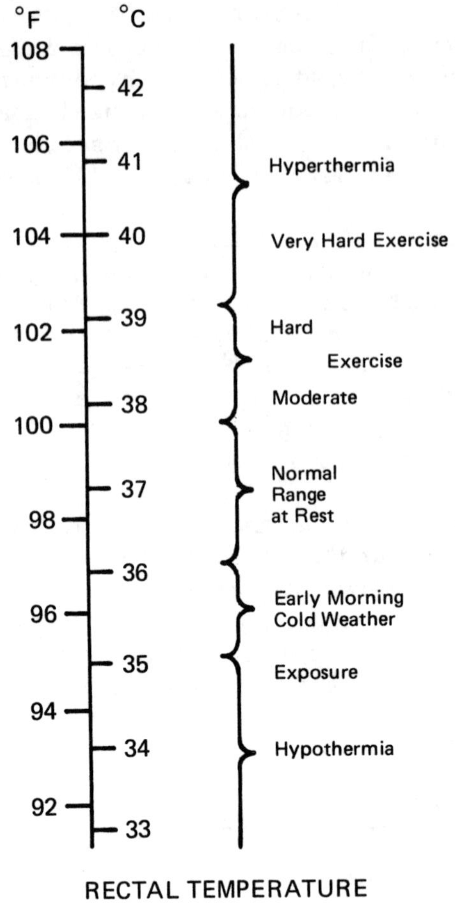

FIGURE IV-21 Range of rectal temperatures encountered in different conditions. Temperatures designated as "hyperthermia" and "hypothermia" are associated with increased risk to health and should be avoided.

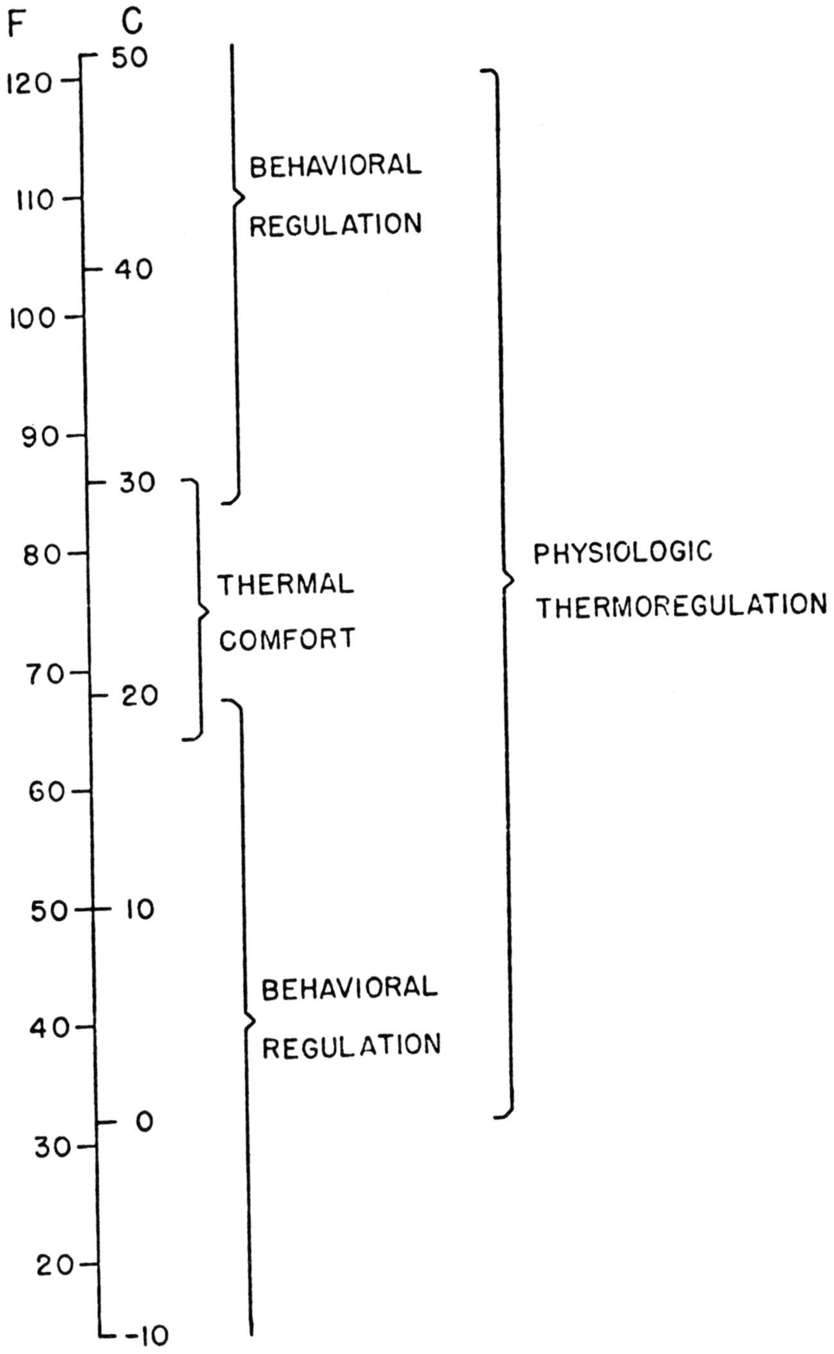

FIGURE IV-22 The range for thermal comfort assumes minimal clothing at the high end and substantial clothing at the low end and thus includes some behavioral regulation. The zones designated "behavioral regulation" require modification by means other than clothing. "Physiologic thermoregulation" indicates the limits of short-term physiologic regulation in a healthy person at rest with minimal clothing.

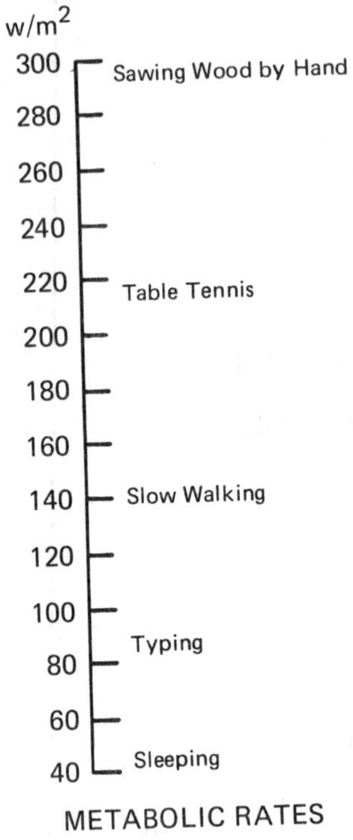

FIGURE IV-23 Metabolic heat production rates associated with various activities. Adapted from Berenson and Robertson.[1]

The convective-heat-transfer coefficient depends on position and posture and on air velocity. Reasonable estimates of h_c are about 4 $W \cdot m^{-2} \cdot {}^{\circ}C^{-1}$ in still air and 11.6 $V^{0.5}$ in air with a velocity of over 0.2 m/s.[11] The human body also exchanges heat by radiation. The radiant environment is usually characterized by a mean radiant temperature ($\overline{T}_r$) or by an effective radiant field (H_r) in watts per square meter. Equation 4 shows the conversion between these terms:

$$H_r = h_r(\overline{T}_r - T_a), \qquad (4)$$

in which h_r = radiant-heat-transfer coefficient.

A reasonable approximation of h_r is 4.5 $W \cdot m^{-2} \cdot {}^{\circ}C^{-1}$ at a skin temperature of 34°C and an ambient temperature of 29°C. It is often convenient, especially when $\overline{T}_r$ and T_a are relatively close together, to combine radiation and convection into an overall heat-transfer coefficient, h, and to use the operative temperature, T_o, for the thermal environment. Operative temperature is the temperature of an environment with uniform air and wall temperature that exchanges heat with the body at the same rate as with the complex environment that it describes. The combined heat-transfer coefficient (h) in still air is about 8-10 $W \cdot m^{-2} \cdot {}^{\circ}C^{-1}$.

The use of clothing reduces the heat exchange with the environment. This insulation, I_{clo}, is usually expressed in clo units; 1 clo unit corresponds to insulation at 0.155 $m^2 \cdot {}^{\circ}C \cdot W^{-1}$. The reciprocal of I_{clo} is h_{cl}, the conductance of clothing. If clothing efficiency is F_{cl}, then

$$F_{cl} = h_{cl}/(h_{cl} + h) = 1/(1 + 0.155hI_{clo}), \qquad (5)$$

and the total radiant and convective heat transfer with clothing becomes:

$$R + C = F_{cl}h(\overline{T}_s - T_o). \qquad (6)$$

F_{cl} varies from 1 for a nude man down to 0.25 for heavy winter clothing, including an overcoat. The temperature at which man feels thermoneutral at rest varies with clothing. At $F_{cl} = 1$, thermoneutrality occurs at 28°C; at $F_{cl} = 0.25$, it occurs at 15°C. Figure IV-24 further illustrates the relationship between clothing and temperature for thermoneutrality.

At temperatures above thermoneutrality, or when increased metabolic heat production exceeds loss of "dry" heat to the environment, sweat is secreted and the resulting evaporative heat loss (E) restores overall thermal equilibrium. For every gram of sweat that evaporates from the skin, 0.68 Wh of heat is lost from the skin. The maximal evaporative heat loss is limited by the maximal amount of sweat that can be secreted (about 1,500 g/h) and by the evaporative power of the environment. The maximal rate of evaporation, E_{max}, from a completely wetted skin is given in Equation 7:

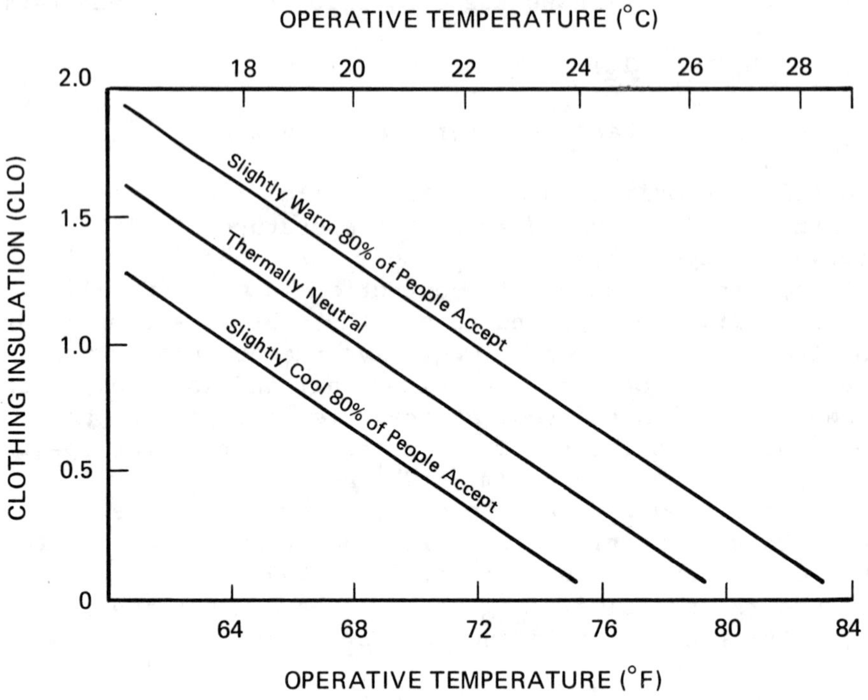

FIGURE IV-24 Effect of interaction of operative temperature and clothing insulation on thermoneutrality, and ambient temperatures accepted by 80% of people when sedentary at minimal air velocity.

$$E_{max} = 2.2h_c(P_{sk} - P_{dp})F_{pcl}, \quad (7)$$

in which

E_{max} = maximal evaporative rate, W/m^2,
2.2 = Lewis relation, °C/Torr,
h_c = convective heat transfer coefficient, $W \cdot m^{-2} \cdot °C^{-1}$,
P_{sk} = saturated water-vapor pressure at skin,
P_{dp} = water-vapor pressure in the atmosphere, Torr, and
F_{pcl} = Nishi permeation factor.[12]

For most types of porous clothing, Nishi and Gagge[12] have shown experimentally that

$$F_{pcl} = 1/(1 + 0.143h_cI_{clo}). \quad (8)$$

If the evaporative heat loss required for thermal equilibrium (E_{req}) is less than E_{max}, the skin will be less than 100% wet. If W is the percent of skin surface area wetted by sweat, then

$$W = E_{req}/E_{max}. \quad (9)$$

Increasing values of W produce increasing discomfort. The value of E_{max} is decreased by adding clothing, by increasing water-vapor pressure, and by lowering air velocity. Lowering the humidity, reducing clothing, and increasing air velocity reduce W and physiologic strain and increase thermal comfort.

PHYSIOLOGIC RESPONSES TO THE THERMAL ENVIRONMENT

In environments in which the body tends to lose or gain heat, body temperature changes. Such changes produce physiologic responses aimed at keeping body temperature constant. When body temperature falls, skin temperature tends to fall first, and then (more gradually) deep-body temperature. The first physiologic response is a reduction in blood flow to the extremities and to the skin in general. Peripheral vasoconstriction reduces the convective heat transfer between the skin and the trunk core and causes rapid lowering of the temperature of hands and feet while reducing the loss of heat from the core. Normal skin blood flow is at about 250 ml/min and can easily be reduced to 50 ml/min or less. If further reduction occurs in body temperature, the thermoregulatory system causes an increase in metabolic heat production through involuntary shivering, which can add 100-150 W/m^2 to the basal 40 W/m^2.

If, however, the thermal environment causes body temperature to rise, the initial physiologic response is an increase in blood flow to the extremities and the skin. Vasodilatation increases the convective heat transfer between the skin and the trunk core. The normal skin blood flow of 250 ml/min can be increased to as much as 3,000 ml/min. Effective overall thermal conductance between skin and core is about 18 $W \cdot m^{-2} \cdot °C^{-1}$, with a low of 6 $W \cdot m^{-2} \cdot °C^{-1}$ in the cold and a high of 100 $W \cdot m^{-2} \cdot {}^{\circ}C^{-1}$ in the heat. Increased body temperature also causes the secretion of sweat over most of the skin surface area at rates proportional to the body-temperature increase. Above the sweating threshold, a further 1°C rise in mean body temperature produces sweat secretion at 200-600 $g \cdot m^{-2} \cdot h^{-1}$, corresponding to evaporative heat loss at 150-400 W/m^2. For a more complete and quantitative review of human thermoregulation, the reader is referred to Stolwijk and Hardy.[15]

The involuntary physiologic responses just described are usually associated with feelings of thermal discomfort. Relationships among ambient temperature, thermal pleasantness, comfort, and temperature sensation are illustrated in Figure IV-25. Prediction of thermal sensation, thermal comfort, and thermal acceptability must take into account all the factors involved in heat production, heat loss, and heat transfer; and it must do it simultaneously. In a nearly steady state, this is fairly feasible, but it becomes more complex when body temperature or ambient temperature is changing.

Berglund and Gonzalez[2] evaluated the effect of slowly changing ambient temperature and water-vapor pressure and found that environmental temperature changes of 0.6°C/h from a 25°C starting point were quite acceptable, as were changes in the water-vapor pressure, as long as the instantaneous water-vapor pressure was kept below 16 Torr.

Internal body temperature has a considerable effect on comfort sensation. When the body interior is hyperthermic, as after vigorous exercise, a low air temperature even at high velocity (cold draft) is felt as comfortable and pleasant, although the same cold draft would be extremely uncomfortable to a person who is already slightly hypothermic from a previous exposure to cold. A hypothermic person would prefer very warm air, which in turn would be felt as very uncomfortable by someone in a hyperthermic state.

There is general agreement that age, gender, and physical fitness do not affect preferred ambient temperature, if metabolic heat production and clothing insulation are constant.

When people are exposed to thermal conditions outside the comfort range, the extent of their discomfort is affected by age, gender, and physical fitness, even if activity and clothing are controlled. The extent of discomfort is closely correlated with physiologic thermoregulatory responses. Those who have a vigorous involuntary response, such as sweat secretion on body warming, experience more intensive and earlier discomfort than those whose sweat secretion begins at higher body temperature.

When a space has thermal nonuniformities (such as vertical temperature gradients) or radiant nonuniformities (such as radiant temperature in one direction 10°C higher than in another direction), such nonuniformities can be perceived as uncomfortable, even if the average temperature is in the comfortable zone.

In a cool or thermoneutral environment, increased air movement is felt as uncomfortable; but at temperatures above thermoneutrality, increased air movement is desirable and extends the comfort zone to higher ambient temperatures.

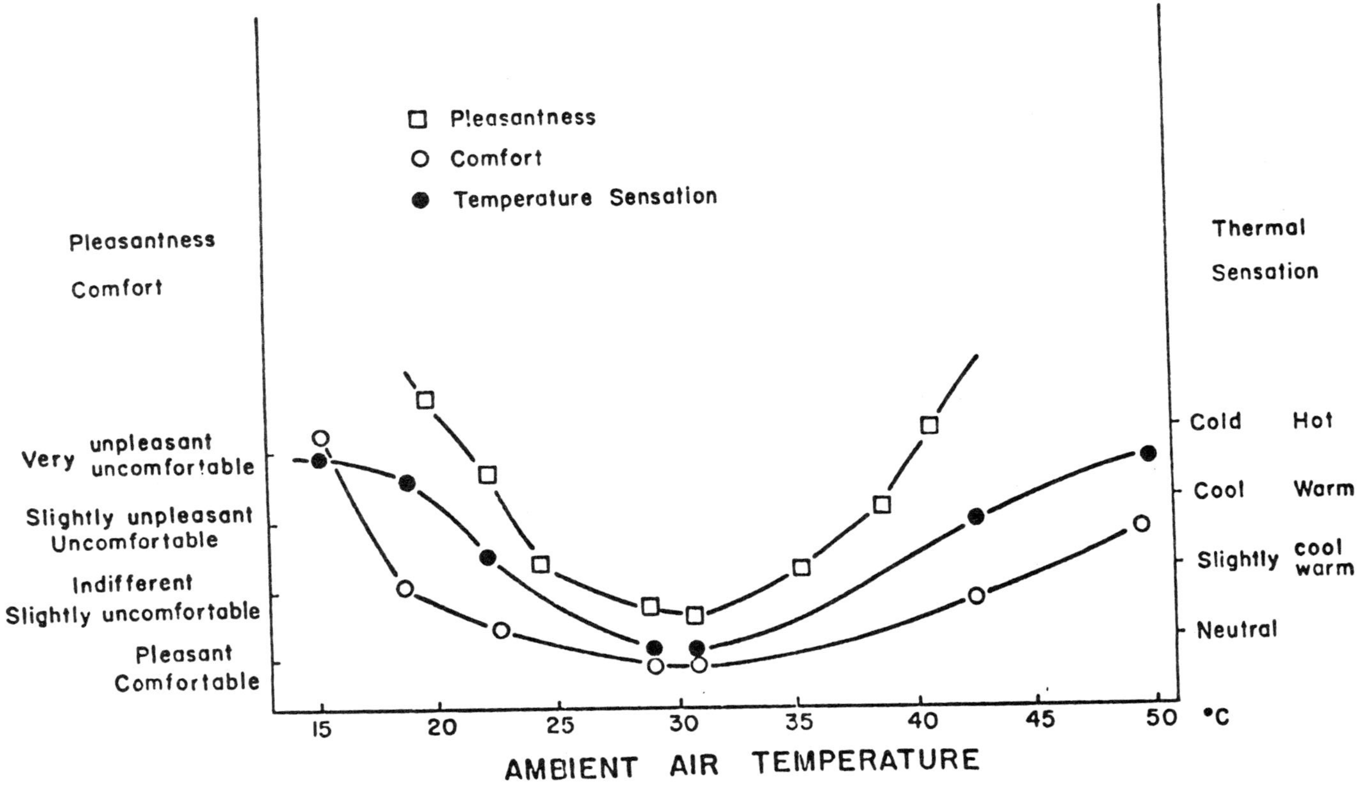

FIGURE IV-25 Reports of pleasantness, comfort sensation, and thermal sensation from sedentary subjects in minimal clothing in environments with low air velocities and uniform air and wall temperatures. Adapted from Hardy *et al.*[9]

HEALTH CONSEQUENCES OF EXTREMES OF TEMPERATURE AND HUMIDITY

Adverse health effects of extremes of temperature and humidity are in several categories. As extremes of temperature and humidity, we designate operative temperatures below 18°C and above 30°C and humidities below 4 Torr and above 16 Torr water-vapor pressure, for people in conventional clothing and at rest. Any temperature or water-vapor pressure that requires physiologic responses makes demands on the reserves of the systems that regulate cardiovascular function or fluid balance. In healthy people, these responses reduce such reserves, thus perhaps interfering temporarily with work capacity or limiting athletic achievement. In diseased persons, such as those with cardiovascular disease, such reserves are very small or nonexistent; such persons can be seriously threatened by even small excursions of temperature or water-vapor pressure.

Assessments of health consequences of extremes of temperature and humidity tend to be based on two kinds of observations: observations of acutely ill people whose status can be correlated with temperature measurements in their hospital rooms, and observations in epidemiologic studies of whole populations, with daily outdoor temperatures correlated with daily mortality or with the number of emergency-room visits. In the first case, the number of patients observed is relatively small, and each patient has a different disease state; it is difficult to draw very detailed conclusions from such observations. In the second case, very large populations can be observed, but it is difficult to evaluate actual exposure and to assess the pre-existing disease of those who have died.

Clinical reports of particular sensitivity of patients with congestive heart failure to hot and humid environments have been presented by Burch and co-workers.[3-6] Ellis,[7] Schuman,[14] and Oechsli and Buechley[13] have reported on excess mortality in populations exposed to unusually hot weather. This excess mortality occurred particularly among the aged, the hypertensives, diabetics, and patients with arteriosclerotic and other cardiovascular disease and chronic respiratory disease.

Exposure to low temperatures can produce accidental hypothermia, especially in the aged (Watts;[16] McNicol and Smith[10]). Low winter temperatures are associated with increased mortality,[7] although few studies have specifically addressed this association.

REFERENCES

1. Berenson, P. J., and W. G. Robertson. Temperature, pp. 65-148. In J. F. Parker, Jr. and V. R. West, Eds. Bioastronautics Data Book. 2nd ed. National Aeronautics and Space Administration Publication No. NASA SP-3006. Washington, D.C.: U.S. Government Printing Office, 1973.
2. Berglund, L. G., and R. R. Gonzalez. Application of acceptable temperature drifts to built environments as a mode of energy conservation. ASHRAE Trans. 84(1):110-121, 1978.

3. Burch, G. E. The influence of environmental temperature and relative humidity on the rate of water loss through the skin in congestive heart failure in a subtropical climate. Am. J. Med. Sci. 211:181-188, 1946.
4. Burch, G. E., and A. Ansari. Artificial acclimatization to heat in control subjects and patients with chronic congestive heart failure at bed rest. Am. J. Med. Sci. 256:180-194, 1968.
5. Burch, G. E., and N. DePasquale. Influence of air conditioning on hospitalized patients. J. Am. Med. Assoc. 170:160-163, 1959.
6. Burch, G. E., and G. C. Miller. The effects of warm, humid environment on patients with congestive heart failure. South. Med. J. 62:816-822, 1969.
7. Ellis, F. P. Mortality from heat illness and heat-aggravated illness in the United States. Environ. Res. 5:1-58, 1972.
8. Gagge, A. P., and Y. Nishi. Heat exchange between human skin surface and thermal environment, pp. 69-92. In D. H. K. Lee, H. L. Falk, S. D. Murphy, and S. R. Geiger, Eds. Handbook of Physiology. Section 9. Reactions to Environmental Agents. Bethesda, Md.: American Physiological Society, 1977.
9. Hardy, J. D., J. A. J. Stolwijk, and A. P. Gagge. Man, p. 342. In G. C. Whittow, Ed. Comparative Physiology of Thermoregulation. Vol. II. Mammals. New York: Academic Press, Inc., 1971.
10. McNicol, M. W., and R. Smith. Accidental hypothermia. Br. Med. J. 1:19-21, 1964.
11. Nishi, Y., and A. P. Gagge. Direct evaluation of convective heat transfer coefficient by naphthalene sublimation. J. Appl. Physiol. 29:830-838, 1970.
12. Nishi, Y., and A. P. Gagge. Moisture permeation of clothing--A factor governing thermal equilibrium and comfort. ASHRAE Trans. 76(Pt. 1):137-145, 1970.
13. Oechsli, F. W., and R. W. Buechley. Excess mortality associated with three Los Angeles September hot spells. Environ. Res. 3:277-284, 1970.
14. Schuman, S. H. Patterns of urban heat-wave deaths and implications for prevention: Data from New York and St. Louis during July, 1966. Environ. Res. 5:59-75, 1972.
15. Stolwijk, J. A. J., and J. D. Hardy. Control of body temperature, pp. 45-68. In D. H. K. Lee, H. L. Falk, S. D. Murphy, and S. R. Geiger, Eds. Handbook of Physiology. Section 9. Reactions to Environmental Agents. Bethesda, Md.: American Physiological Society, 1977.
16. Watts, A. J. Hypothermia in the aged: A study of the role of cold-sensitivity. Environ. Res. 5:119-126, 1972.

CHARACTERIZATION OF ADDITIONAL PHYSICAL INDOOR POLLUTANTS

The indoor pollutants thus far described have been defined primarily as airborne contaminants. Other forms of pollution do not depend on mass concentration, such as sound and electromagnetic radiation. Electromagnetic radiation occurs in the radiofrequency, infrared, visible light, ultraviolet, and x-ray portions of the electromagnetic spectrum. The frequencies of the electromagnetic radiation discussed here range from 10^4 Hz (radiofrequency) to 10^{15} Hz (ultraviolet). Table IV-24 summarizes the frequency distribution for each of these portions of the spectrum and of audible sound, which is transmitted via the vibration of air molecules.

SOUND AND NOISE

Physical Characteristics

The audibility of sound depends on intensity and frequency, with a maximal human response in the region near 3×10^3 Hz. A sound with predominant frequencies below 100 Hz or above 10^4 Hz may require a million times more energy to have the same audibility as a sound with a predominant frequency of 3×10^3 Hz. A method of weighting the pressure exerted by the sound waves at different frequencies has been developed to compensate for these variations. The decibel values (which constitute a logarithmic intensity scale) cited herein are measurements with level A weighting, the scale that most closely matches the response of the human ear. The differences in the treatment of the intensity content of a sound are slight and do not change substantially from one source to another.[14]

Sounds in the indoor environment are generated both outside and inside the occupied space. Table IV-25 gives examples of sound intensities in the outdoor environment. Table IV-26 lists sound intensities produced by typical household appliances in the indoor environment.

Sound intensities are usually measured by a meter satisfying the requirements of American National Standards Institute Specification SI.4-1971 (for sound meters).

Psychophysiologic Effects

The possible effects of sound include permanent and temporary loss of hearing, cardiovascular disease, sleep disruption, and psychologic effects.[14] The physiologic and psychologic responses to sound may be transitory;[14] however, there is insufficient information on the effects of sound by itself or in combination with other stressors. Sound at intensities that are found to be objectionable will affect productivity and decrease enjoyment of the environment.[14]

The EPA has identified sound intensities that, if not exceeded, should protect against some of the adverse effects of sound.[14] These values are expressed in terms of maximal 8-h (75 dB) and 24-h (70 dB) averages required to protect against hearing loss. There are also yearly average long-range environmental goals of 55 dB outdoors and 45 dB indoors, which are recommended to avoid activity interference or annoyance.[12] There is still debate in professional circles about the maximal intensities of short-duration environmental sound that can be

TABLE IV-24

Radiation Wavelengths and Frequencies

Type of Radiation	Wavelength	Frequency, Hz
Ultraviolet		
Ultraviolet C	0.19-0.28 μm	1.07×10^{15}-1.58×10^{15}
Ultraviolet B	0.28-0.315 μm	9.5×10^{14}-1.07×10^{15}
Ultraviolet A	0.315-0.4 μm	7.5×10^{14}-9.5×10^{14}
Visible light	0.4-0.7 μm	4.29×10^{14}-7.5×10^{14}
Infrared		
Near infrared	0.7-1.4 μm	2.14×10^{14}-4.29×10^{14}
Infrared	1.4-3 μm	1.00×10^{14}-2.14×10^{14}
Far infrared	3-1,000 μm	3×10^{11}-1.00×10^{14}
Radiofrequency		
Microwave	1-1,000 mm	3×10^{8}-3×10^{11}
Very high frequency	1-10 m	3×10^{7}-3×10^{8}
High frequency	10-100 m	3×10^{6}-3×10^{7}
Medium frequency	100-1,000 m	3×10^{5}-3×10^{6}
Low frequency	1,000-10,000 m	3×10^{4}-3×10^{5}
Very low frequency	10,000-30,000 m	1×10^{4}-3×10^{4}
Sound, audible	0.016-20.0 m	15-2×10^{4}

TABLE IV-25

Examples of Outdoor Day-Night Average Sound Intensities at Various Locations[a]

Location	Average Sound Intensity, dB(A)
Apartment next to freeway	88
Downtown with some construction	79
Urban high-density apartment	78
Urban row housing on major avenue	68
Old urban residential area	59
Wooded residential area	51
Agricultural cropland	44
Rural residential area	39
Wilderness ambient	35

[a]Data from Council on Environmental Quality.[4]

TABLE IV-26

Examples of Sound Intensities Generated Indoors by Household Appliances[a]

Appliance	Average Sound Intensity, dB(A)
Blender	80-90
Garbage disposer	80
Window air conditioner	60
Refrigerator	45
Vacuum cleaner	70-75
Hair dryer	78
Mixer	82

[a]Data from Jones.[8]

considered safe. However, above 110 dB, sound is so intense that most people experience pain or a tickle in their ears.[14]

Although it is difficult to determine the exact day and night indoor sound intensities, studies have indicated that an intensity of 60.4 dB with a standard deviation of 5.9 dB can be expected in a typical urban residential area, with instantaneous intensities exceeding 80 dB.[12] An expected intensity of 60.4 dB is below the 70 dB recommended by the EPA to prevent hearing loss, but it is well above the intensity recommended to avoid interference and annoyance. Therefore, day and night sound intensities in the 100-site EPA survey may contribute to speech interference, reduced worker productivity, and annoyance.[10(pp. 66-69)]

A high intensity of background noise in urban areas stemming primarily from transportation appears to affect the developing fetus. Women exposed to aircraft noise have a higher proportion of low-birthweight children, who are at higher risk of mortality and both physical and mental effects.[10(pp. 110-111)] This association cannot be separated from the social status of the women (a codetermining variable), inasmuch as many members of the lower social classes live in "noisy" areas.

Exposure to high intensities of sound affects communication and learning, including the acquisition of language.[10(p. 115)] Adaptation or resignation to annoyance may occur, and there do not appear to be groups of people that are particularly sensitive. After-effects of noise have been noted at home and at work, and noise appears to influence aggressiveness and minimize voluntary helping behavior.[10(pp. 120-121)]

RADIOFREQUENCY AND MICROWAVE RADIATION (10^4 to 3×10^{11} Hz)

Physical Characteristics

Although the physical characteristics of all electromagnetic radiation are similar, the frequency is inversely proportional to wavelength, and the effects of the longer wavelengths, such as radiofrequency radiation, are radically different from those of the shorter-wavelength ionizing radiation, such as x rays and gamma rays. The photon energy in radio waves is so small that there is no ionization when it is absorbed in an organism.[15]

Table IV-27 summarizes the radiation properties of some common nonionizing-radiation systems and their expected far-field power densities. Energy radiated by these systems can be additive, provided that the frequencies are within the same octave band.

For the purposes of this report, the densities of all radiofrequency energies generated outdoors are defined as "background power densities," and those of radiofrequencies generated indoors as "generated power densities."

Some radiofrequency energy is generated in the indoor environment. In general, all electric equipment produces some radiofrequency radiation. However, all but a few electric devices radiate energy at

TABLE IV-27

Comparison of Radiation Properties of Some Common Radiofrequency-Generating Systems

Radiating System	Frequency	Average Transmitter Power	Far-Field Power Density at Selected Distance
AM radio transmitter	500 kHz	50 kW (maximum allowed)	0.016 mW/cm^2 at 400 m
FM radio transmitter	88-108 MHz	100 kW	0.03 mW/cm^2 at 500 m
VHF TV	174-216 MHz	300 kW (maximum allowed)	0.04 mW/cm^2 at 500 m
UHF TV	470-890 MHz	5,000 kW (maximum allowed)	0.5 mW/cm^2 at 500 m
Acquisition radar[a] (ARSR)	1 GHz	20 kW	0.05 mW/cm^2 at 3,000 m (near field, 100 mW/cm^2 at 30 m)
Tracking radar:[a]			
Type A	10 GHz	5 kW	0.01 mW/cm^2 at 3,000 m (near field, 800 mW/cm^2 at 9 m)
Type B	1.3 GHz	150 kW	1 mW/cm^2 at 5,000 m (near field, 56 mW/cm^2 at 523 m)

[a]Radar power densities given along line of sight of transmitted power; off axis, power densities are generally 0.01 less than those along line of sight.

well below the American National Standards Institute recommended exposure limits, even in combination with one another. One major exception is the microwave oven. Under normal operating conditions, a residential microwave oven radiates approximately 1 mW/cm^2 at the seal on the door. However, if the door is defective, values in excess of 1 W/cm^2 can be achieved.

Psychophysiologic Effects

Effects of radiofrequency radiation can be divided into two major categories:[15] thermal effects (when the radio-wave energy is converted into heat) and nonthermal effects (which cannot be directly explained by thermal equivalents).

Biologic effects depend on the frequency and the intensity of the radiation; the duration of exposure; the dielectric constant, temperature, and thermal conductivity of the irradiated tissue; the ability of the tissue to dissipate heat; and the dimensions of the body. Absorption of microwave radiation by body tissues results in an increase in temperature, often producing internal burns due to local hot spots caused by nonuniformity in the field. The eyes and testes were found to be the most sensitive.[17]

Specific effects at the cellular or molecular level were postulated more than a decade ago without resolution of the importance of these effects with respect to biologic damage.[3] The possibilities of nonthermal effects, such as rearrangements within macromolecules and subcellular structures, have been under investigation for many years, but further studies will be necessary to clarify the issues. It is relatively clear that metabolic and functional disturbances at the cellular level can be caused by microwave radiation, but the mechanisms of these effects are not yet well understood.

Table IV-28 characterizes the relative rates of absorption by the human body; however, it is difficult to determine the exact effect of each frequency. Because the radiofrequency energy generated indoors is low, the major emphasis should be on outdoor sources. Indoor radiofrequency fields are generally lower than outdoor. Osepchuck has discussed sources of microwave and other forms of radiofrequency energy.[11]

FAR-INFRARED AND INFRARED RADIATION (3×10^{11} Hz to 4.3×10^{14} Hz)

Physical Characteristics

The infrared energy spectrum ranges from far-infrared (3×10^{11} Hz to 10^{14} Hz), through infrared ($1.0\text{-}2.14 \times 10^{14}$ Hz), to near-infrared ($2.14\text{-}4.29 \times 10^{14}$ Hz). Infrared radiation is produced naturally by the sun and by all common heating and artificial-light sources. The incandescent lamp is one of the major sources of infrared radiation and the most common artificial-light source in the indoor environment. Of the total input wattage of an incandescent lamp,

TABLE IV-28

Relative Absorption of Radiofrequencies by Human Body[a]

Frequency, 10^6 Hz	Maximal Absorption by Human Body, %
<400	<50
400-1,000	50-100
1,000-3,000	20-100
3,000-10,000	>50

[a]Data from U.S. Department of Health, Education, and Welfare.[13]

75-80% is converted to near-infrared and infrared radiation.[9] The ACGIH has adopted a TLV of 10 mW/cm^2 for infrared radiation in the workplace.[1] The power density 2 m from a 100-W lamp is approximately 0.6 mW/cm^2 for the total infrared spectrum. Sunlight on the earth's surface produces a flux of about 70 mW/cm^2, of which about half is infrared.

Psychophysiologic Effects

Depending on its wavelength, infrared is absorbed in the surface of the skin (wavelengths larger than 2 μm) or can penetrate several millimeters (wavelengths between 0.7 and 1.5 μm). Safety standards in industrial environments are based on the risk that infrared radiation may induce cataracts in the eyes of persons exposed to excessive infrared radiation, such as glassblowers or open-hearth steelworkers.[16]

Excessive infrared radiation is most easily controlled by shielding the source with reflecting metallic foils.

VISIBLE RADIATION

Physical Characteristics

Radiation in the near-infrared and visible spectrum is produced by many sources, both natural and artificial. Our sense of sight, feeling of well-being, and comfort are all, to a great extent, influenced by visible and near-infrared radiation.

Psychophysiologic Effects

Retinal burns from observation of the sun have been described throughout history. Chorioretinal burns rarely occur from exposure to artificial light, because the normal aversion to high-brightness light sources (the blink reflex) provides adequate protection, unless the exposure is hazardous within the duration of the blink reflex.

Many factors affect the usefulness of visible light. Among the most important are discomfort glare and disability glare. Light sources can cause a reduction in contrast of an image, owing to scattered visible radiation, by adding a uniform veil of luminance to the object. This effect, commonly called "veiling luminance," may cause a reduction in visual performance without physical damage.

Discomfort glare is a sensation of annoyance or pain caused by brightness in the field of view that is greater than that to which the eyes are adapted. It has been shown that the threshold of discomfort glare changes as a function of age.[2] Although discomfort glare does not necessarily interfere with visual performance, it can cause eye strain and contribute to fatigue. Disability glare and ocular stray light influence one's ability to perform a task by artificially veiling the contrast of the visual target. It is therefore a great contributor to eye fatigue.

ULTRAVIOLET RADIATION (0.75-1.58 x 10^{15} Hz; wavelength, 0.19-0.400 μm)

Physical Characterstics

Ultraviolet radiation is divided into three wavelength categories: ultraviolet-A (UV-A), 0.315-0.400 μm; ultraviolet-B (UV-B), 0.28-0.315 μm; and ultraviolet-C (UV-C), 0.19-0.28 μm. All fluorescent lamps emit UV-A, but not UV-B or UV-C. High-intensity discharge lamps produce UV-A, UV-B, and some UV-C. Incandescent lamps produce small amounts of UV-A, and essentially no UV-B and UV-C.

Ultraviolet radiation is measured with specialized radiometric photometers.

Psychophysiologic Effects

UV-B and UV-C are known photocarcinogens.[5] Doses of UV-B and UV-C 10 times the human minimal erythema dose (MED) have initiated squamous cell carcinomas, and chronic continuous exposure to UV-A can also have a carcinogenic effect.[5]

The ACGIH recommends limits on workplace ultraviolet exposure that depend on wavelength and on the duration of exposure.[1] For UV-A, the intensity should not exceed 1 mW/cm^2 for more than 1,000 s, nor should the dose exceed 1 J/cm^2 if given in less than 1,000 s. For UV-B and UV-C, the dose should not exceed about 3-10 mJ/cm^2 in any 8-h period. The degree of hazard seems to be associated with the erythemal efficiency of each frequency.[1,5]

SUMMARY

Ionizing and nonionizing electromagnetic radiation occurs in the indoor environment. This radiation can be harmful, and one cannot always sense its presence.

Sound can generally be heard and in some cases felt. Excessive sound can cause deterioration of hearing acuity and, if extremely intense or prolonged, cause deafness. Background sound in the urban residential environment can exceed the recommended intensities and result in interference and annoyance. Sound of 70-80 dB, commonly found in indoor environments, can inhibit task performance and possibly contribute to aggressive human behavior.[6]

Infrared, far-infrared, and radiofrequency radiation produce no visible or audible evidence of their presence. However, infrared radiation does provide sensory indication of its presence by heating of human tissue. Far-infrared and radiofrequency radiation, however, provide no indication of their presence, unless their power levels are so high as to increase skin temperature. Heating of human tissue occurs because of the infrared output of incandescent lamps. However, the detrimental effects of this heating have not been fully investigated. Surveys have shown that in several cities 98% of the people are exposed to less than 1 $\mu W/cm^2$ from broadcasting transmitters.[7] However, ultrahigh-frequency television transmitters can radiate radiofrequency pollution to adjacent buildings at 5-200 $\mu W/cm^2$.

Ultraviolet-A, visible light, and near-infrared radiation can produce surface heating of human and animal tissue. These frequencies are of concern because of their ability to affect human performance. The veiling reflections caused by most artificial lighting systems can have substantial influence on human visual performance. Reduction of veiling reflections can increase visual performance and decrease the energy consumed by lighting systems. Transient adaptation (dilation during or immediately after eye movements) is caused by sudden changes in the visual spectrum power. Transient adaptation contributes to eye fatigue and decreased visual performance.

REFERENCES

1. American Conference of Governmental Industrial Hygienists. TLVs. Threshold Limit Values for Chemical Substances in Workroom Air Adopted by ACGIH for 1980. Cincinnati: American Conference of Governmental Industrial Hygienists, 1980. 93 pp.
2. Bennet, H. J. Discomfort Glare: Demographic variables, p. 6. IERI Special Report No. 118, 1976.
3. Cleary, E. Biological Effects and Health Implications of Microwave Radiation. Symposium Proceedings. Richmond, Virginia, September 17-19, 1969. U.S. Department of Health, Education, and Welfare, Bureau of Radiological Health Publication No. BRH/DBE 70-2. Washington, D.C.: U.S. Government Printing Office, 1971. 265 pp.
4. Council on Environmental Quality. Noise, pp. 533-576. In Environmental Quality--1979. The Tenth Annual Report of the Council on Environmental Quality. Washington, D.C.: U.S. Government Printing Office, 1980.
5. Cunningham-Dunlop, S., and B. H. Kleinstein. Wavelength dependence, pp. 51-61. In Carcinogenic Properties of Ionizing and Nonionizing Radiation. Vol. I. Optical Radiation. DHEW (NIOSH) Publication No. 78-122. Washington, D.C.: U.S. Government Printing Office, 1977.
6. Geen, R. G., and E. C. O'Neal. Activation of the cue-elicited aggression by general arousal. J. Pers. Soc. Psychol. 11:289-292, 1969.
7. Janes, D. E., Jr. Radiation surveys--Measurement of leakage emissions and potential exposure fields. Bull. N.Y. Acad. Med. 55:1021-1041, 1979.
8. Jones, H. W. Noise in the Human Environment. Edmonton, Alberta: Environmental Council of Alberta, 1979.
9. Kaufman, J. E., and J. F. Christensen, Eds. IES Lighting Handbook. The Standard Lighting Guide. 5th ed. New York: Illuminating Engineering Society, 1972.

10. National Research Council, Committee on Appraisal of Societal Consequences of Transportation Noise Abatement. Noise Abatement: Policy Alternatives for Transportation. Washington, D.C.: National Academy of Sciences, 1977. 206 pp.
11. Osepchuck, J. M. Sources and basic characteristics of microwave/RF radiation. Bull. N.Y. Acad. Med. 55:976-998, 1979.
12. Schultz, T. J. Noise Assessment Guidelines. (Technical Background for Noise Abatement in HUD's Operating Programs.) U.S. Department of Housing and Urban Development Report No. TE/NA 172. Washington, D.C.: U.S. Government Printing Office, 1971. 210 pp.
13. Smith, S. W., and D. G. Brown. Radio Frequency and Radio Microwave Radiation Levels Resulting from Man-Made Sources in the Washington, D.C. Area, pp. 1-13. U.S. Department of Health, Education, and Welfare Pub. No. (FDA)72-8015. Washington, D.C.: U.S. Government Printing Office, 1972.
14. U.S. Environmental Protection Agency, Office of Noise Abatement and Control. Information on Levels of Environmental Noise Requisite to Protect Public Health and Welfare with an Adequate Margin of Safety. U.S. Environmental Protection Agency Report No. 550/9-74-004. Washington, D.C.: U.S. Government Printing Office, 1974. [214] pp.
15. Vogelman, J. H. Physical characteristics of microwave and other radiofrequency radiation, pp. 7-10. In S. F. Cleary, Ed. Biological Effects and Health Implications of Microwave Radiation. Symposium Proceedings. Richmond, Virginia, September 17-19, 1969. U.S. Department of Health, Education, and Welfare, Bureau of Radiological Health Publication No. BRH/DBE 70-2. Washington, D.C.: U.S. Government Printing Office, 1971.
16. Wallace, J., P. M. Sweetnam, C. G. Warner, P. A. Graham, and A. L. Cochrane. An epidemiological study of lens opacities among steel workers. Br. J. Ind. Med. 28:265-271, 1971.
17. World Health Organization. Health Hazards in the Human Environment. Geneva: World Health Organization, 1972. 387 pp.

V

FACTORS THAT INFLUENCE EXPOSURE TO INDOOR AIR POLLUTANTS

The types and quantities of pollutants found indoors vary temporally and spatially. Depending on the type of pollutant and its sources, sinks, and mixing conditions, its concentration can vary by a factor of 10 or more, even within a small area.

Human mobility constitutes an important kind of complexity in the determination of exposure to air pollutants. Human activity patterns differ between midweek and weekend, between one season and another, and between one part of one's lifetime and another. Activity patterns determine when and how long one is exposed to both indoor and outdoor pollutants. Therefore, in reviewing the factors that influence air-pollution exposures, we have specifically separated them into two major components: time (activity) and concentration (location).

Information on the time spent in various activities is summarized first, and then the variations in concentration encountered in different locations. Unfortunately, most of the studies discussed were not longitudinal and thus do not offer information on seasonal differences in time spent indoors and outdoors or on regional differences in activity patterns.

Outdoor concentrations of pollutants and rates of infiltration affect the concentrations to which people are exposed indoors. The emphasis of the second section of this chapter is on geographic variations in outdoor pollution and their impact on indoor pollution. Building construction techniques, as they vary geographically, and their effect on pollution infiltration rates are particularly important. But the measurement techniques available are limited; the need for additional studies is discussed. The rates of infiltration on a neighborhood scale have been studied by only a few researchers. Although their work has focused on energy conservation, their findings can easily be applied to the study of the impact on indoor pollution.

As shown in Chapter IV, there can be large indoor-outdoor differences in pollutant concentrations. Concentrations also vary among indoor locations and from one time to another. In determining total exposure to pollutants, therefore, both indoor and outdoor concentrations must be well characterized. The ways in which building characteristics affect indoor pollution vary with type of pollutant, type of building, building location and orientation, and even room use within a given building. Building characteristics are the subject of the final section of this chapter.

HUMAN ACTIVITIES

Patterns of human behavior and activity determine the time spent in any specific location, and thus knowlege of them is essential in estimating exposures of populations to pollutants. As indicated by Ott,[51] a large number and variety of studies in which data on human activities were collected from population samples have been completed over the last 50 yr.

When one examines the literature on human activities, the term "time budget" ("zeitbudget," "budget de temps") is encountered often. A time budget produces a systematic record of how time is spent by a person in some specified period, usually 24 h. It contains considerable detail on a person's activities, including the locations in which the activities take place.[46]

One way of obtaining time budget information from the populations surveyed is to ask each respondent to maintain a diary of his or her activities over a 24-h period or longer. In another approach, the so-called "yesterday" survey approach, the interviewer asks each respondent about his or her activities on the preceding day.

Several summaries of the historical development of time-budget research have been published.[11 16 52 70] Ott[51] discussed the literature on activity patterns in the context of estimation of exposure to air pollution.

The Multinational Comparative Time Budget Research Project, launched in September 1964 by a small group of social scientists from eastern and western countries, used common principles for sampling, interviewing, and data coding and tabulation on an international basis. The population sample consisted of nearly 30,000 persons in 12 countries (Belgium, Bulgaria, Czechoslovakia, France, East Germany, West Germany, Hungary, Peru, Poland, United States, Soviet Union, and Yugoslavia). A standardized coding sytem was developed for comparing activities in different countries. The multinational study developed a coding system with 100 categories of activities represented by a two-digit code (from 00 to 99). The activities represented by these codes can be grouped into 10 classes: working time and activities related to work, domestic house work, care of children, purchasing of goods and services, private needs (such as meals and sleep), adult education and professional training, civic and collective participation, sports and active leisure, passive leisure, and spectacles, entertainment, and social life.[69]

The Project yielded a rich data base that has been summarized in a number of tables, figures, and articles.[69] For example, the average time spent by employed men, employed women, and married housewives in various locations in 12 countries is shown in Table V-1. The data show that employed men in the 12 countries spend between 12 h (in Hungary) and 15.2 h (in Belgium) in their homes, whereas housewives spend

TABLE V-1

Time Spent in Various Locations in 12 Countries (Average Hours per Day)[a]

	Belgium	Kazanlik, Bulgaria	Olomouc, Czechoslovakia	Six cities, France	100 electoral districts, Fed. Rep. Germany	Osnabrück, Fed. Rep. Germany	Hoyerswerda, German Dem. Rep.	Győr, Hungary	Lima-Callao, Peru	Toruń, Poland	Forty-four cities, USA	Jackson, USA	Pskov, USSR	Kragujevac, Yugoslavia	Maribor, Yugoslavia
7-1.1 Employed men, all days															
inside one's home	15.2	12.5	14.3	13.6	13.6	14.2	13.8	12.0	12.9	14.0	13.4	13.6	13.4	12.9	13.0
just outside one's home	0.5	0.7	0.3	0.3	1.0	0.5	0.4	1.0	0.1	0.2	0.2	0.3	0.3	0.5	1.4
at one's workplace	5.0	7.7	5.9	7.2	5.4	5.1	6.8	7.5	6.4	7.0	6.7	6.5	6.8	7.1	6.1
in transit	1.5	2.1	1.6	1.5	1.7	2.2	1.7	2.0	2.5	1.7	1.6	1.5	2.0	1.8	2.2
in other people's home	0.5	0.2	0.3	0.5	0.5	0.6	0.3	0.3	0.5	0.5	0.5	0.6	0.2	0.7	0.5
in places of business	0.7	0.6	0.6	0.5	0.4	0.4	0.6	0.4	0.7	0.4	0.7	0.7	0.4	0.5	0.5
in restaurants and bars	0.2	0.0	0.1	0.2	0.5	0.4	0.1	0.2	0.3	0.0	0.4	0.4	0.2	0.2	0.0
in all other locations	0.4	0.2	0.9	0.2	0.9	0.6	0.3	0.6	0.6	0.2	0.5	0.4	0.7	0.3	0.3
total	24.0	24.0	24.0	24.0	24.0	24.0	24.0	24.0	24.0	24.0	24.0	24.0	24.0	24.0	24.0
7-1.2 Employed women, all days															
inside one's home	17.1	14.6	16.0	15.3	17.0	16.7	16.7	14.5	16.1	15.0	15.4	15.3	14.0	15.0	15.0
just outside one's home	0.1	0.3	0.2	0.0	0.7	0.2	0.2	0.3	0.4	0.1	0.0	0.1	0.1	0.3	0.4
at one's workplace	3.6	6.5	5.1	6.3	3.6	3.6	4.9	6.8	4.4	5.8	5.2	5.0	6.7	6.1	6.4
in transit	1.2	1.6	1.3	1.1	1.1	1.3	1.1	1.4	1.8	1.5	1.3	1.3	1.7	1.4	1.5
in other people's home	0.4	0.2	0.2	0.5	0.4	0.9	0.2	0.3	0.3	0.6	0.7	0.6	0.2	0.6	0.2
in places of business	1.0	0.6	0.8	0.6	0.6	0.8	0.7	0.5	0.7	0.8	0.9	1.1	0.6	0.4	0.4
in restaurants and bars	0.2	0.0	0.0	0.1	0.2	0.3	0.1	0.0	0.1	0.0	0.2	0.2	0.2	0.0	0.0
in all other locations	0.4	0.2	0.4	0.1	0.4	0.2	0.1	0.2	0.2	0.2	0.3	0.4	0.5	0.2	0.1
total	24.0	24.0	24.0	24.0	24.0	24.0	24.0	24.0	24.0	24.0	24.0	24.0	24.0	24.0	24.0
7-1.3 Housewives, all days (married only)															
inside one's home	21.6	20.4	20.9	21.7	20.4	20.5	21.3	19.7	21.0	20.9	20.5	20.9	19.6	20.5	19.7
just outside one's home	0.2	1.4	0.3	0.1	0.8	0.4	0.3	2.1	0.5	0.1	0.1	0.1	0.4	0.8	2.3
in transit	1.0	0.9	1.2	1.0	1.0	1.0	1.0	0.9	1.2	1.2	1.0	0.9	1.9	1.5	1.1
in other people's home	0.4	0.4	0.3	0.5	0.6	0.6	0.3	0.2	0.4	0.5	0.8	0.7	0.7	0.7	0.3
in places of business	0.5	0.7	1.1	0.6	0.7	1.1	0.9	0.9	0.7	1.2	1.2	1.1	1.1	0.4	0.5
in restaurants and bars	0.1	0.1	0.0	0.0	0.1	0.1	0.0	0.0	0.0	0.0	0.1	0.1	0.0	0.0	0.0
in all other locations	0.2	0.1	0.2	0.1	0.4	0.3	0.2	0.2	0.2	0.1	0.3	0.2	0.3	0.1	0.1
total	24.0	24.0	24.0	24.0	24.0	24.0	24.0	24.0	24.0	24.0	24.0	24.0	24.0	24.0	24.0

[a]Reprinted with permission from Szalai.[69] Data are weighted to ensure equality of days of the week and number of eligible respondents per household.

between 19.7 h (in Hungary) and 21.6 h (in Belgium) in their homes. In the 12 countries, therefore, employed men spend, on the average, 50-63% of the day in their homes, and housewives spend 82-90% of the day in their homes. It is difficult to determine the overall amount of time spent indoors from these data, because categories like "at one's workplace" do not distinguish between indoor and outdoor workplaces. Similarly, the categories "in places of business" and "in all other locations" do not specify whether they are indoors or outdoors. However, if one assumes that all "workplaces," "places of business," and "restaurants and bars" are indoors, along with the category "in other person's homes," and that the category "in all other locations" is assumed to be entirely outdoors, then it is possible to estimate the amount of time spent by respondents in three general categories: indoors, outdoors, and in transit (see Table V-2).[51]

With these assumptions and the restructured data shown in Table V-2, it is estimated that employed men in the 12 countries spend between 84% (in Maribor, Yugoslavia) and 92% (in France) indoors. It should be emphasized, however, that many of the entries in Table V-2 cannot be compared with each other on a statistical basis, because the numbers of respondents in the samples vary. Also, the representativeness varies, because some countries, such as the Soviet Union, are represented by a single city and its suburbs (Pskov, population 115,000), whereas others, such as the United States (44 cities), are represented by a national sample of metropolitan areas. Finally, some assumptions as to whether a location was indoors or outdoors need to be examined, because they may introduce error. However, the estimates in Table V-2 appear useful as rough approximations of the times spent by residents of 12 countries indoors, outdoors, and in transit.[51]

If only the data for the United States (44 cities) are considered, it appears that, on the average, employed men spend 90% of the day (21.7 h) indoors, whereas married housewives spend 95% of the day (22.8 h) indoors. Employed men in the United States are estimated to spend 2.9% of the day (0.7 h) outdoors, and housewives 1.7% (0.4 h).

Although the estimates in Tables V-1 and V-2 are useful for determining the total amount of time spent in various locations, they give little information about the time of day when persons are present in each location. Data from the multinational study can be displayed in a composite profile that shows the proportion of the population that are engaged during the day in selected activities, such as sleeping, eating, working, travel, and watching television (Figure V-1).

In addition to the studies of activities in the United States by Robinson,[57-59] activity-pattern studies have been carried out in Durham, N.C., by Chapin and Hightower,[13] on a sample of 43 Standard Metropolitan Statistical Areas (SMSAs) by Chapin and Brail,[12] on a followup U.S. national sample by Brail and Chapin,[8] and on the Washington, D.C., metropolitan area by Hammer and Chapin.[32] Information supplied in this section is limited to urban areas; this reflects the available published information. No comments are made on variations in numbers, because it is beyond the scope of this document to assess their reliability.

TABLE V-2

Estimated Time Spent in Three Environmental Categories[a]

Country	Average Time Spent, h/d					
	Employed Men			Housewives[b]		
	Indoors	Outdoors	In transit	Indoors	Outdoors	In transit
Belgium	21.6	0.9	1.5	23.2	0.4	0.4
Bulgaria (Kazanlik)	21.0	0.9	2.1	22.1	1.5	0.4
Czechoslovakia (Olomouc)	21.3	1.1	1.6	23.2	0.5	0.4
France (six cities)	22.0	0.5	1.5	23.3	0.2	0.5
West Germany (100 districts)	20.4	1.9	1.7	22.2	1.2	0.6
West Germany (Osnabruck)	20.7	1.1	2.2	22.7	0.7	0.6
East Germany (Hoyerswerda)	21.6	0.7	1.7	23.2	0.5	0.3
Hungary (Gyor)	20.4	1.6	2.0	21.5	2.3	0.2
Peru (Lima-Callao)	20.8	0.7	2.5	22.9	0.7	0.4
Poland (Torun)	21.9	0.4	1.7	23.3	0.2	0.5
United States (44 cities)	21.7	0.7	1.6	22.8	0.4	0.8
United States (Jackson, Mich.)	21.8	0.7	1.5	23.0	0.3	0.7
U.S.S.R. (Pskov)	21.0	1.0	2.0	22.6	0.7	0.7
Yugoslavia (Kragujevac)	21.4	0.8	1.8	22.4	0.9	0.7
Yugoslavia (Maribor)	20.1	1.7	2.2	21.3	2.4	0.3

[a]Reprinted from Ott.[51] Derived from data originally published in Szalai;[69(p. 795)] data are weighted to ensure equality of days of the week and number of eligible respondents per household.

[b]Married persons only.

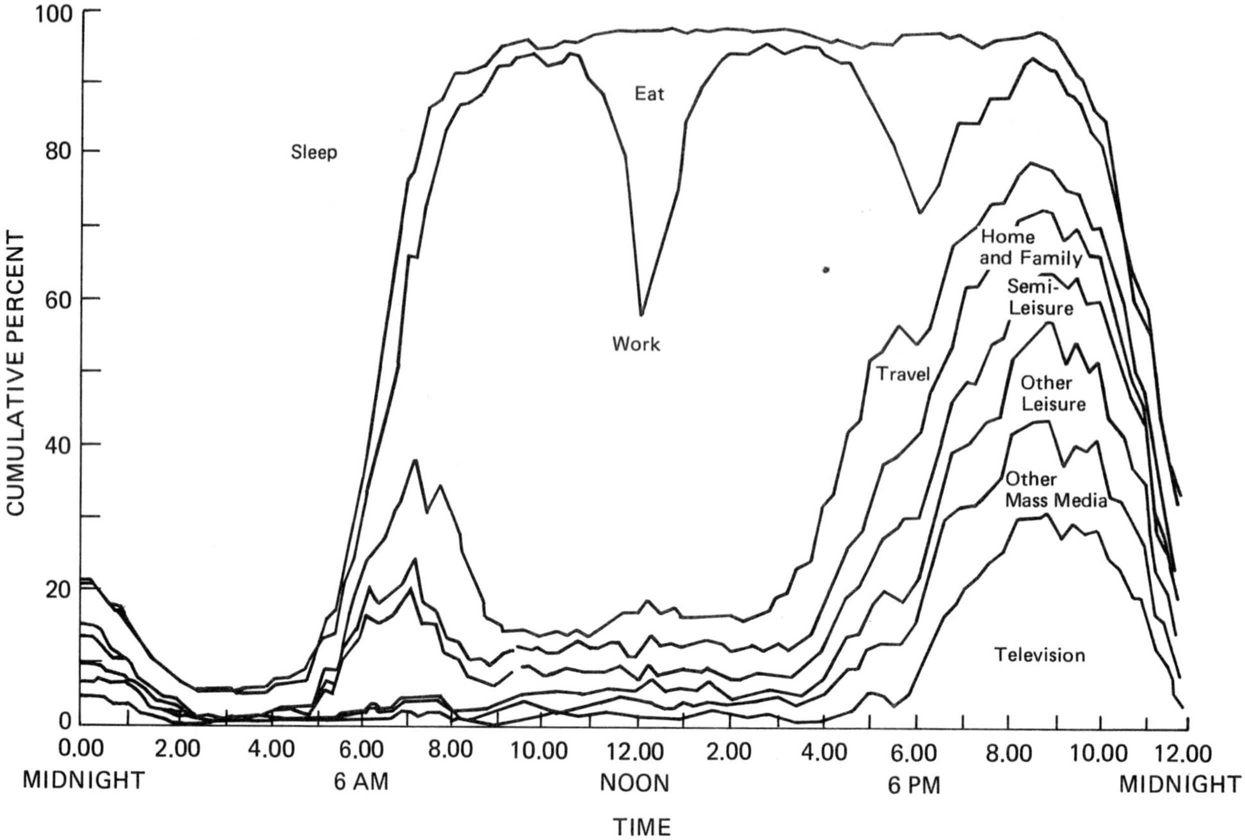

FIGURE V-1 Diurnal profiles showing percentage of employed men in United States (44 cities) engaged in nine types of activities as a function of time of day (weekdays only). Data weighted to ensure equality of days of the week and number of eligible respondents. Reprinted with permission from Szalai.[69]

In the United States, legislation passed in 1952 required urban areas to conduct metropolitan-area transportation studies as a prerequisite for receiving federal funds for highway construction.[64] As a result, transportation studies have been undertaken in 200 areas of the United States,[76] and these studies have usually involved collection of considerable detail about the transportation activities of the urban population, particularly in cities with populations in excess of 50,000. As reported by Robinson, Converse, and Szalai,[60] the multinational research project also collected information on the average time spent in commuting to and from work in various countries (see Table V-3). Most of the summaries of findings from time-budget studies have presented only average values and seldom given histograms or information on the variance of the time spent in various locations or activities.

In 1969-1970, the U.S. Department of Transportation made arrangements with the Bureau of the Census to carry out a nationwide study of the transportation-related activities of the U.S population. This study, called the Nationwide Personal Transportation Study, was based on home interviews and covered individual activities in considerable detail.[3,4,6,26,27,33,56,65-68] Figure V-2 shows a frequency distribution of the amount of time spent in commuting based on these data. Assuming two trips per day, the overall average of 22 min/trip compares reasonably well with the average of 46 min/d reported by Robinson, Converse, and Szalai.[60]

There is a need for a special-purpose activity-pattern study specifically tailored to the problem of estimating air-pollution exposures. Previous activity-pattern studies have not considered questions that apply to exposures to air pollutants. Such a survey should begin with a pilot study on a single city, to perfect the experimental design and data-collection methods, and should use personal monitoring instruments to measure exposures. Once the pilot study is completed and the results are evaluated, a large-scale research investigation could be carried out on a number of cities or on a national probability sample. The large-scale survey would use diaries and personal monitoring instruments to characterize the frequency distribution of air-pollution exposures of the population as a whole and in selected cities. Information from the diaries could be compared with the measurements of exposure to determine how different activities affect population exposure rates.

GEOGRAPHIC AND LOCAL VARIATIONS

The air quality of an indoor environment is often described on the basis of one 24-h average obtained from one indoor sampling location. The spatial distribution of indoor air pollutants within a structure is a little-studied subject. Therefore, recent or current unpublished works and technical papers related to environmental concerns, but not necessarily to indoor air quality, are incorporated in this review. Some of the associations made and conclusions reached are clearly based on explicitly stated assumptions, rather than on scientific documentation.

TABLE V-3

Average Time Spent Traveling To and From Work, by Mode of Transportation[a]

Country	Average Time Spent, min/d			
	Public Transport	Automobile	Walking	All Travel
Belgium	98	55	52	66
Bulgaria (Kazanlik)	93	73	47	57
Czechoslovakia (Olomouc)	73	62	46	59
France (six cities)	82	46	44	50
West Germany (100 districts)	--	--	--	40
West Germany (Osnabruck)	71	41	46	47
East Germany (Hoyerswerda	82	66	30	62
Hungary (Gyor)	104	48	40	64
Peru (Lima-Callao)	103	93	48	89
Poland (Torun)	71	50	41	60
United States (44 cities)	81	46	30	50
United States (Jackson, Mich.)	--	39	34	38
U.S.S.R. (Pskov)	67	--	32	--
Yugoslavia (Kragujevac)	70	53	47	51
Yugoslavia (Maribor)	71	44	40	51
Average	82.0	55.1	41.2	56.0
Standard deviation	13.3	15.1	7.2	12.7

[a]Reprinted with permission from Robinson *et al.*[60]

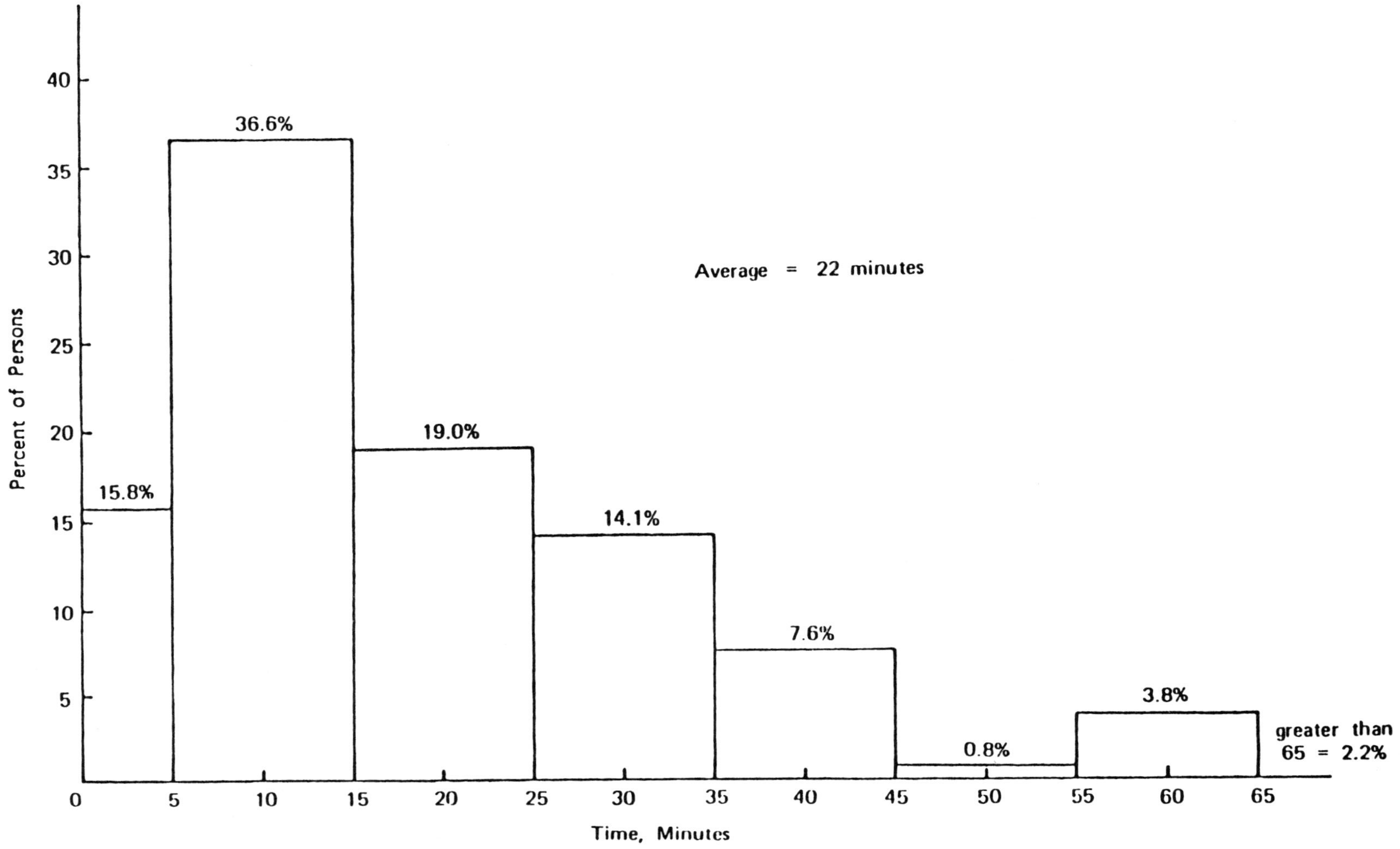

FIGURE V-2 Frequency distribution of home-to-work commuting times for employed persons in the United States (excludes persons who work at home or at no fixed address). Reprinted from Svercl and Asin.[68]

The types of pollutants and the concentrations of each type vary between locations within a structure, between structures within a geographic area, and between geographic areas. This section discusses some of these interrelationships.

GEOGRAPHIC VARIATIONS IN INDOOR AIR QUALITY

Owing to the small number of field monitoring studies, the geographic distribution of indoor air pollutants has not been determined. However, it is instructive to review the geographic distribution of the major factors that affect variations in the concentrations of pollutants and their impact on the quality of the indoor environment. Outdoor air quality, air-infiltration rates, and sources of emission of indoor pollutants are the major factors. Outdoor air quality has been studied with respect to some pollutants, and the geographic distribution of these few pollutants is well understood. Descriptive statistics published annually by EPA and state and local air-quality agencies furnish much scientific information useful in discerning regional and local differences in concentrations of carbon monoxide, total suspended particles, ozone, NO_x, sulfur dioxide, sulfates, and others. Clearly, it is beyond the scope of this document to summarize the existing information on geographic variations in the types and concentrations of all ambient pollutants. It should be noted that the geographic distribution of some criteria pollutants has been studied and is easily accessible from the literature; information on noncriteria pollutants is sparse and often collected and analyzed by questionable methods.

Concentrations of chemically nonreactive pollutants in residences generally correlate with those outdoors. Results from the six-city study,[63] which monitored indoor and outdoor environments for an extended period, clearly showed the influence of outdoor concentrations on the indoor environments (see Figure V-3). Another study,[47] [49] sponsored by EPA, supported the conjecture that indoor concentrations of inert gaseous contaminants correlate with outdoor concentrations. The available data base is not large enough to support statistical conclusions, but there is little doubt that the variations in indoor pollutant concentrations correlate with variations in outdoor concentrations. Thus, it is expected that a city with high outdoor pollutant concentrations will have high indoor concentrations, unless control strategies are used. Although this is a broad, general conclusion, it must be emphasized that the indoor concentration of a given pollutant is expected to vary widely among residences within one city. This variation may be sufficient to mask the impact of varying outdoor concentration.

Distribution of indoor air quality is extremely difficult to describe on a geographic scale, because indoor air quality is determined by complex dynamic relationships that depend heavily on occupant activity and highly variable structural characteristics. Weather, which has a regional character, influences indoor air concentrations of some chemicals, such as formaldehyde, and biologic

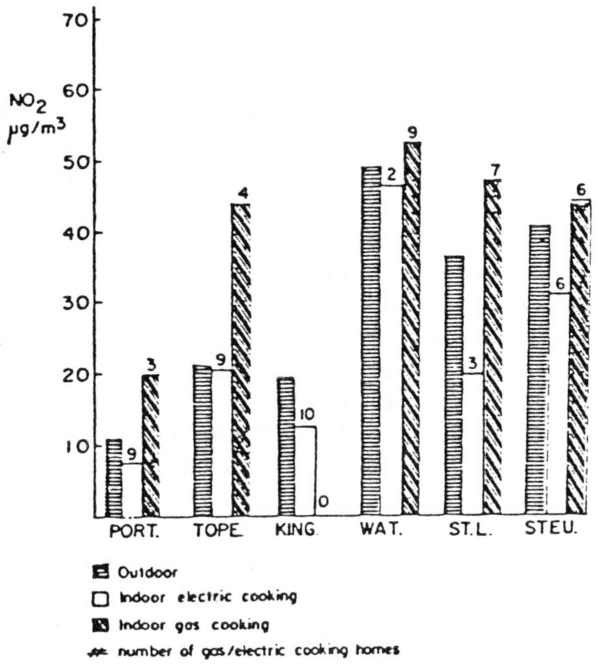

FIGURE V-3 Annual nitrogen dioxide concentration outside and inside electric- and gas-cooking homes, averaged across each community's indoor and outdoor network (May 1977-April 1978). Reprinted with permission from Spengler _et al._[63]

contaminants, such as bacteria and molds. Therefore, the influence of relative humidity and other weather-related conditions affecting indoor environmental quality needs to be studied geographically. Research specifically addressed to geographic distribution of indoor air quality is needed.

Other sections of this document address ventilation rates of large buildings, and the discussion of the geographic distribution of air-infiltration rates in this section focuses on residences. Typically, the air-infiltration rate for American residences is assumed to be 0.5-1.5 ach. This assumption is supported by the results of several energy and air-quality studies that experimentally determined the range of ventilation rates for typical residences to be between 0.7 and 1.1 ach (Moschandreas and Morse;[48] C. D. Hollowell, personal communication). Recently built or retrofitted residences had lower infiltration rates, between 0.5 and 0.8 ach (J. Woods, personal communication). However, the sample that yielded the data is small, and statistical documentation for such statements is not strong.

Only one experimental study appears to have been broad enough to allow generalizations on the spatial distribution of air-infiltration rates. A 1979 report[29] presented air-leakage characteristics of low-income residences 10-90 yr old in 14 cities in all climatic zones of the United States. Two measurement techniques were used: a tracer-gas decay technique with air bags to measure natural air infiltration,[28] and a fan-depressurization test that measures induced air-exchange rates.[39] Of the 266 low-income residences tested with these two techniques, 68% were frame buildings, 16% masonry, and 11% masonry-veneer. These proportions do not necessarily reflect those of the universe of low-income dwellings. Figure V-4 illustrates the findings of the study on air-infiltration rates from three cities. With the tracer-gas technique, it was found that 19% of the rates were below 0.5 ach; 40% were moderate, between 0.5 and 1.0 ach; 20% were high, between 1.0 and 1.5 ach; and 20% were very high, greater than 1.5 ach. This characterization of the rates as moderate, high, and very high was given by Grot and Clark and does not reflect a universally accepted nomenclature. Although their paper did not discuss the general geographic distribution of air-infiltration rates, investigations of the data base are continuing. They observed that, the higher the number of degree days, the lower the infiltration rate of the residences. This observation is preliminary (R. A. Grot, personal communication), and further work is needed to verify it. Furthermore, this was a study of low-income residences, not typical residences.

There have been studies that indicate the geographic distribution of residential indoor sources. The residential energy consumption in various geographic regions (Table V-4) shows the use of fuel types that are potential sources of high concentrations of nitric oxide, nitrogen dioxide, and carbon monoxide in residences with gas cooking and heating. A second example shows the number of mobile homes in each state (Table V-5)--an indoor environment with a reported potential for high formaldehyde concentrations. Finally, residences in Polk County, Florida, Grand Junction, Colorado, and Butte, Montana, are built with

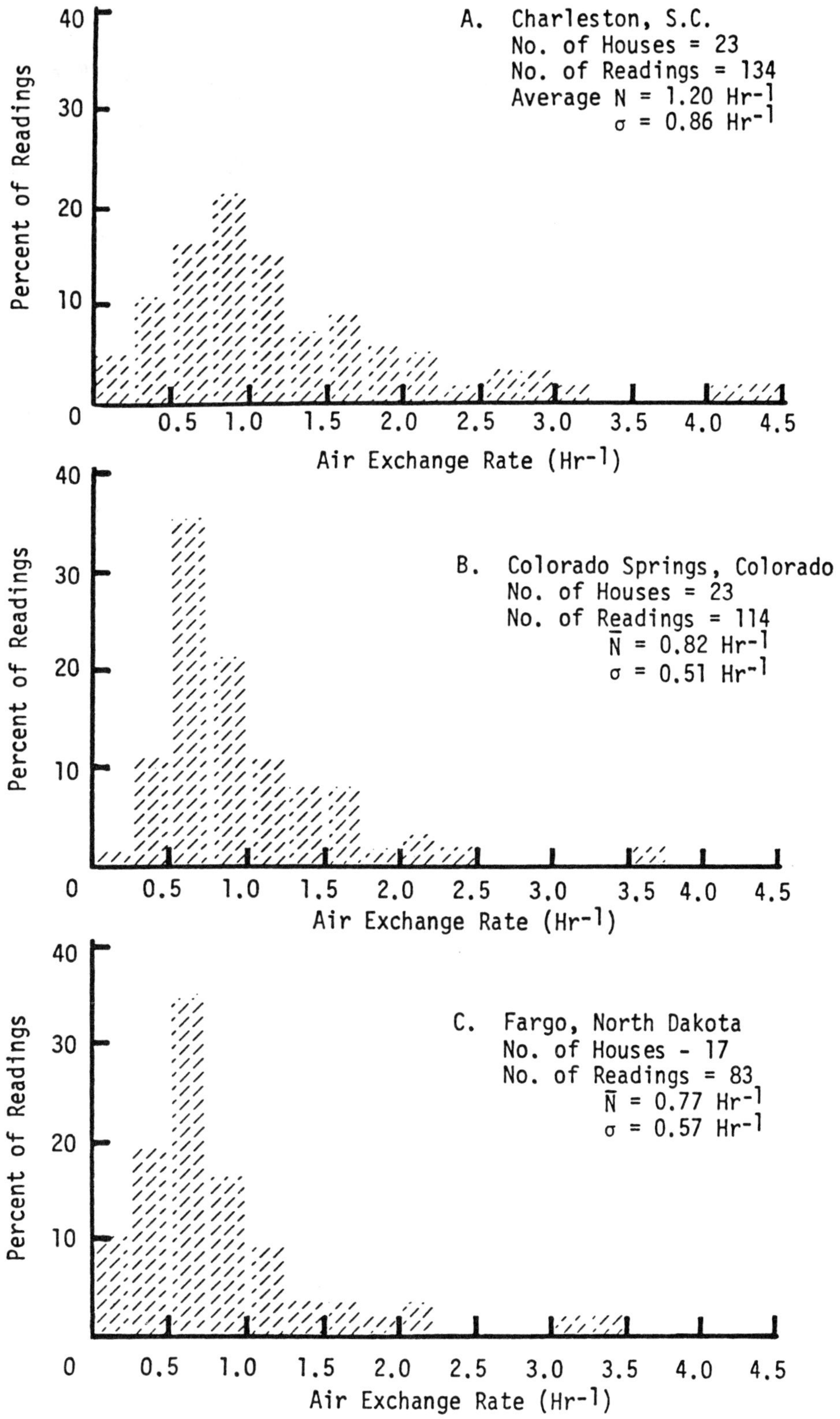

FIGURE V-4 Histograms of measured natural air-infiltration rates for three cities. Reprinted with permission from Grot and Clark.[29]

TABLE V-4

Distribution of Residential Energy Consumption by Fuel Type and Region, 1970 (Single-Family Detached Homes)[a]

Region	Distribution of Fuel Use, %			
	Gas	Oil	Electricity	Coal and Wood
Northeast				
New England	20	76	3	1
Middle Atlantic	46	45	3	6
North central				
East north central	71	23	2	4
West north central	76	20	2	2
South				
South Atlantic	41	39	13	7
East south central	60	4	20	16
West south central	93	--	4	3
West				
Rocky Mountain	81	8	5	6
Pacific	76	12	10	2

[a]Reprinted from Keyes.[37]

TABLE V-5

Estimated Stock of Year-Round Occupied Mobile Homes at End of 1977

State	No. Mobile Homes	% of Total	State	No. Mobile Homes	% of Total
Alabama	104,272	2.80	Montana	30,363	0.82
Alaska	14,215	0.38	Nebraska	25,957	0.70
Arizona	104,711	2.81	Nevada	28,937	0.78
Arkansas	57,616	1.55	New Hampshire	17,853	0.48
California	286,888	7.71	New Jersey	20,387	0.55
Colorado	54,245	1.46	New Mexico	44,930	1.21
Connecticut	10,017	0.27	New York	104,216	2.80
Delaware	17,661	0.47	North Carolina	192,893	5.18
Florida	317,708	8.54	North Dakota	18,561	0.50
Georgia	153,349	4.12	Ohio	135,374	3.64
Hawaii	231	0.01	Oklahoma	53,121	1.43
Idaho	34,599	0.93	Oregon	85,431	2.30
Illinois	106,125	2.85	Pennsylvania	150,838	4.05
Indiana	104,601	2.81	Rhode Island	2,842	0.08
Iowa	38,280	1.03	South Carolina	103,071	2.77
Kansas	48,396	1.30	South Dakota	19,641	0.53
Kentucky	83,360	2.24	Tennessee	92,750	2.49
Louisiana	83,682	2.25	Texas	232,550	6.25
Maine	26,675	0.72	Utah	20,520	0.55
Maryland	26,724	0.72	Vermont	11,049	0.30
Massachusetts	13,558	0.36	Virginia	83,576	2.25
Michigan	134,353	3.61	Washington	88,330	2.37
Minnesota	58,103	1.56	West Virginia	53,758	1.44
Mississippi	69,961	1.88	Wisconsin	53,737	1.44
Missouri	84,993	2.28	Wyoming	16,988	0.46
			Total	3,721,996	100

or on radon-emitting materials and have high indoor radon concentrations. The geographic distribution of radon-emitting materials is being generated by DOE and should be available soon. Spatial distributions of this kind have not been generalized to document variations in indoor air quality.

URBAN, SUBURBAN, AND NEIGHBORHOOD VARIATIONS IN INDOOR AIR QUALITY

The quality of indoor air is a function of outdoor air quality, emission from indoor sources, air-infiltration rates, and occupant activity and is likely to vary within each metropolitan and suburban area, and indeed within each neighborhood. Within a metropolitan area, it has been shown that an urban complex leads to the so-called urban heat reservoir.[2] Urban characteristics--such as city size, density of buildings, and population--correlate with such meteorologic factors as temperature, pressure, and wind velocity.[25] The urban heat island affects both urban pollution patterns and meteorologic characteristics that affect the infiltration rates of buildings. Thus, although the exact nature of the impact on indoor air quality is not known, it is fair to expect the heat island to have an impact on the indoor environment that is likely to be adverse. Also, the variations due to mechanical ventilation, structural differences, and air infiltration may vary within a neighborhood as a function of such factors as house orientation, tree barriers, and terrain roughness.

Occupant activity, air-infiltration rates, the indoor sources of pollutants, and their chemical natures are some of the factors that cause variations within a city. A recent study[50] in the Boston metropolitan area obtained indoor air samples from 14 residences under occupied "real-life" conditions for 2 wk each. As illustrated in Figure V-5, the indoor-air character not only was driven by outdoor concentrations, but was greatly affected by other factors, such as indoor activities.

Air-infiltration rates may be estimated by many dynamic models.[5,14,34,36,40,41,44,71,72] Network computer models[7] are also available. For tall buildings, there are methods for calculating infiltration rates on an overall and floor-by-floor basis.[73-75] The models vary in complexity and applicability. Their operational use also varies considerably, and only a few have been experimentally verified. Each of these models requires a number of input parameters for estimating the air infiltration: number of exterior walls and windows; use of each room; wind speed, direction, and pressure differentials; indoor-outdoor temperature difference; heating, ventilating, and air-conditioning (HVAC) systems; structural characteristics; and terrain characteristics.

Wind speed, temperature difference, pressure differential, terrain characteristics (roughness and barriers, such as trees and fences), building orientation, and structure characteristics may be affected by the location of one residence relative to another within a neighborhood. Energy-consumption patterns in residences were the subject of a 3-yr program at Princeton University's Center for

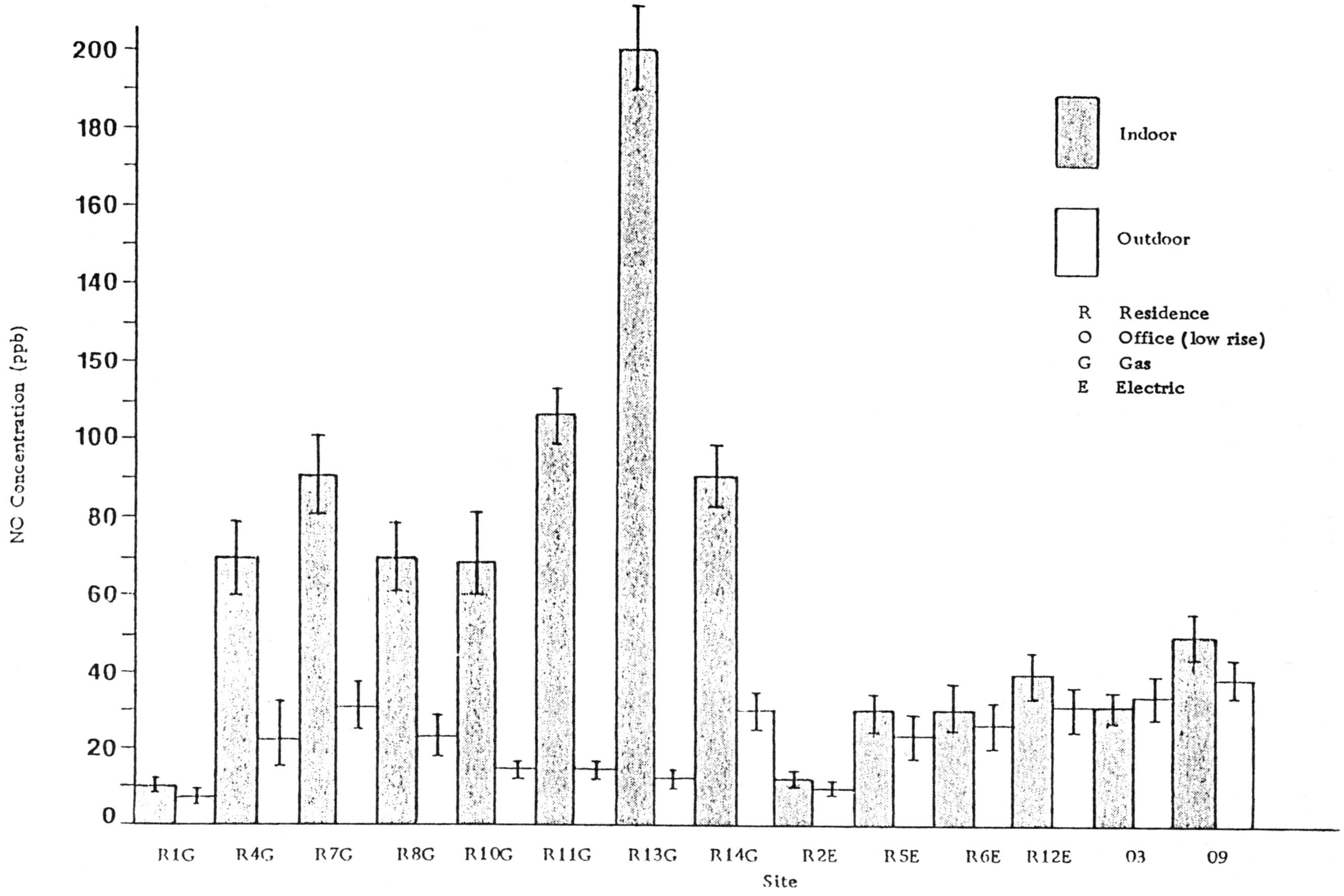

FIGURE V-5 Mean of hourly indoor and outdoor nitric oxide concentrations (ppb) in Boston metropolitan area with 95% confidence limits. Reprinted with permission from Moschandreas *et al.*[50]

Environmental Studies. An important segment of this work was to determine the effects of barriers to prevailing winds on air infiltration in residences. A series of wind-tunnel tests were used to validate the air-infiltration model reported by Princeton researchers.[44] Figure V-6 describes quantitative results for end units and interior units with and without tree sheltering. The "X" shows the unit tested in the configuration of "townhouses" oriented to the wind. The three wind speeds tested were 92 ft/s (28 m/s), 120 ft/s (37 m/s), and 140 ft/s (43 m/s). The figures shown for air infiltration were averaged for the three wind speeds. The authors reached the following conclusions relevant to this section:

- Significant differences occur for the variety of house/wind configurations tested. Depending upon the particular house/wind configurations, end units experienced about 20 to 30% more infiltration than interior units due to the large expanse of side wall exposed to the wind.
- Sheltering effects, such as small, solid fences and tall evergreen trees were found to significantly reduce wind-produced air infiltration losses.
- The most effective windbreak tested in the present quantitative study is that of a straight row of tall evergreen trees. This type of windbreak produced a reduction in air infiltration of 40% compared to the unsheltered interior house. The combination of trees and fences results in a reduction of 60%, compared to the case without trees.

A followup study collected data on the air infiltration in an occupied residence before, during, and after a temporary tree windbreak was installed. The benefits of tree-row sheltering were amply illustrated with a wind speed of 5.6 m/s (12.6 mph) and air temperature of 18°C (32.5°F). The air infiltration was reduced from 1.13 to 0.66 ach, a 42% reduction. Analysis of the weather conditions prevailing during the heating season led to the conclusion that the tree barrier would cause an overall reduction in air infiltration. A final report[7] documented that the wind-velocity profile varies with the roughness of terrain; the wind pressure distribution is changed and the absolute pressure on a building is decreased by the presence of obstacles within a few building lengths.

In conclusion, the studies done at Princeton showed that the location and orientation of a residence within a neighborhood and the terrain and barriers surrounding it do affect the rate of air infiltration. The difference between urban and suburban surroundings also contributes to the complexity of determining the effects of location on rates of air infiltration in residences. The functional relationship between air infiltration and indoor air quality has not been fully established, nor have the distribution patterns of indoor air quality in the urban, suburban, and neighborhood areas. Further research is warranted to study the cause-and-effect relationship between air infiltration and air quality and to formulate the best

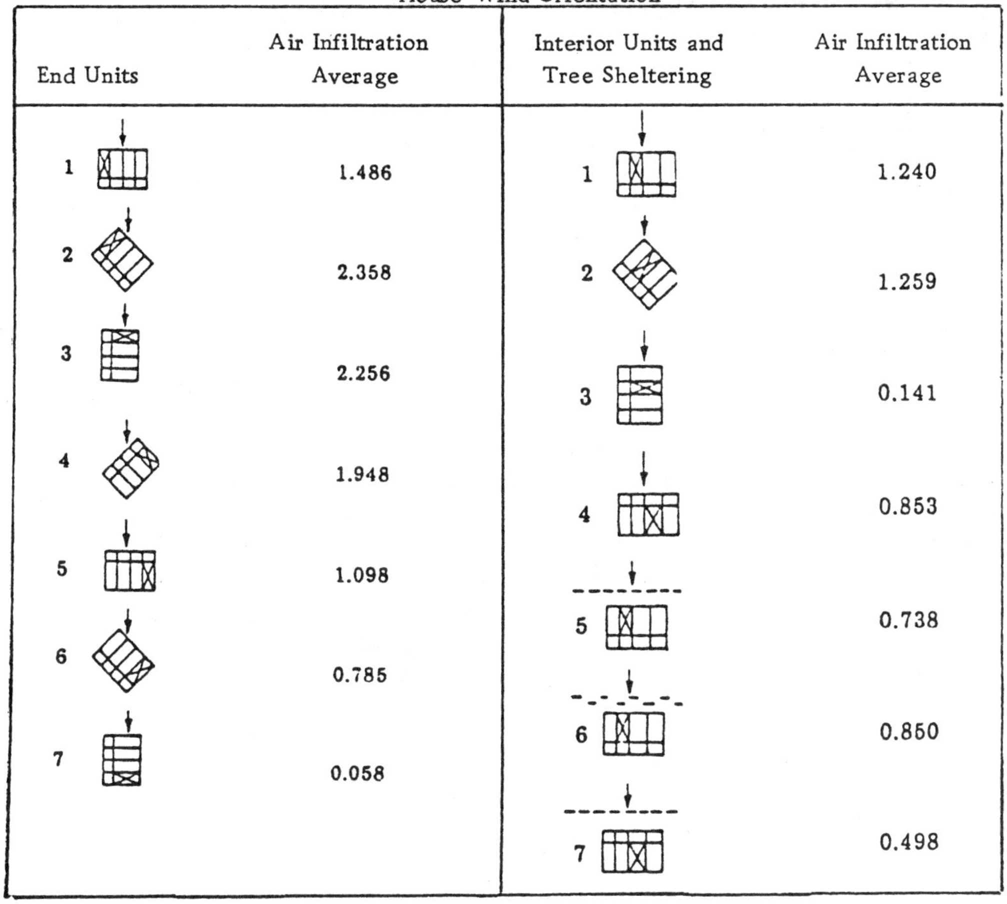

House-Wind Orientation

End Units	Air Infiltration Average	Interior Units and Tree Sheltering	Air Infiltration Average
1	1.486	1	1.240
2	2.358	2	1.259
3	2.256	3	0.141
4	1.948	4	0.853
5	1.098	5	0.738
6	0.785	6	0.860
7	0.058	7	0.498

FIGURE V-6 Effects of orientation and wind barriers on air infiltration in townhouse units. Arrow indicates wind direction. X indicates unit tested. Dashed lines indicate wind breaks. Infiltration average in ach. Reprinted with permission from Mattingly and Peters.[44]

balance between energy conservation and indoor air quality in commercial, as well as residential buildings.

A recent HUD publication,[73-75] entitled Air Quality Considerations in Residential Planning, is designed for routine use by HUD staff to determine air quality at potential housing sites. A special HUD environmental clearance rating is used to assess the relationship between estimated air quality and ambient-air standards and to determine whether the potential project should be rejected, proceed with standard construction practices, or proceed after mitigating steps are imposed. The mitigating steps include setting residences away from major roads and specifying air-infiltration rates and the use of pollutant control devices.

In a business district, ambient air quality and, correspondingly, indoor air pollution are affected by the amount of local automobile traffic. The General Electric Co.[24] investigated the indoor-pollution variations caused by traffic in the area of two complex high-rise buildings. One of the buildings was an air-rights apartment building that straddles the Trans Manhattan Expressway; the second was a more conventional high-rise structure on a canyon-like street in midtown Manhattan. Several observations in this study are pertinent when one is considering the distribution patterns of outdoor and indoor pollutant concentrations as they are affected by site configuration.

- The vertical wind profile was different at the two sites. At the air-rights building, a wind vortex was present at times; at the canyon structure, road winds were limited to particular directions only.
- The traffic flow rate and wind direction between the street level and the third-floor level of the air-rights building resulted in a random relationship, but "significantly lower" carbon monoxide concentrations at the third-floor level. However, at the other site, a linear relationship was observed between street- and third-floor-level concentrations, regardless of wind direction. This resulted in third-floor carbon monoxide concentrations that were only slightly lower than those observed at street level.
- It was found that "pollutants generated at road level diffuse as a function of vertical distance." Specifically, "typical exponential reductions in CO concentrations from the bottom to top floors" were observed outdoors at both sites. This elevation-related reduction in pollutant concentration was also observed indoors, but it is less pronounced.

In addition, the General Electric study made a number of recommendations that are relevant to air quality, its relationship to neighborhood building planning, and specific efforts to reduce indoor pollutant concentrations: lower floors of high-rise buildings must be specially sealed from traffic-generated pollutants; building entrances should be placed so that prevailing road winds are parallel to them; convection paths inside buildings should be minimized; elevator control rooms at roof level should be force-ventilated to reduce pollutant entrapment; and a parking garage in a large complex building should be force-ventilated outside the building at a point that will minimize reintroduction of the exhaust into the structure.

VARIATIONS IN INDOOR AIR QUALITY IN BUILDINGS

The indoor air quality of an individual building is often characterized by the 24-h average for the concentration of one pollutant measured at one sampling location. Because the activity patterns of persons are such that more time is spent in some indoor areas than in others, the question arises:[49] "Do indoor zones (independent areas) with distinct pollutant patterns exist?" At issue here is whether sampling from one monitoring zone is sufficient to characterize the air quality of an entire building. Moschandreas and co-workers tried to answer this question with data obtained from 24 residences monitored under "real-life" conditions. Four-minute average pollutant concentrations were obtained sequentially from four sampling sites (kitchen, bedroom, living room, and one outdoor site). Hourly averages of concentration were calculated from the 4-min information. Corresponding hourly average concentrations of pollutants at the three indoor sampling sites were not always equal. Indoor nitrogen dioxide concentrations (Figure V-7) from two residences, one with gas cooking facilities and the other with electric cooking facilities, illustrate a more pronounced increase in the room with an indoor pollutant-generating source.[50] Statistical analysis did show significant differences in the concentrations of pollutants between different locations within a residence. Air-quality measurements are made to determine the concentrations to which people are exposed. If there are indoor zones where concentrations of pollutants are high and where people spend substantial amounts of their indoor residence time, their calculated exposure could be very different from that calculated on the basis of a single measurement for an entire building.

The null hypothesis, tested by a two-tailed, paired t-test, was that the mean of the differences between corresponding hourly average pollutant concentrations at two indoor sites is equal to zero. This null hypothesis was rejected in more cases than it was accepted in. Comparison of the observed range and calculated differences led to the conclusion that, although corresponding hourly indoor pollutant concentrations are not uniform throughout a residence, the differences between sampled sites are small and probably of minimal health importance.

In an extensive analytic study of indoor air quality, Shair and Heitner[61] assumed that there are no pollutant gradients in the indoor environment. The experimental data base of Moschandreas and co-workers[50] verified that the gradients in concentrations of several gaseous pollutants in the residential environment are negligible. J.D. Spengler, R.E. Letz, J.B. Ferris, Jr., T.Tibbets, and C. Duffy reported (at the annual meeting of the Air Pollution Control Association, in June 1981) on weekly nitrogen dioxide measurements in 135 homes in Portage, Wisconsin. On the average, kitchen concentrations were twice

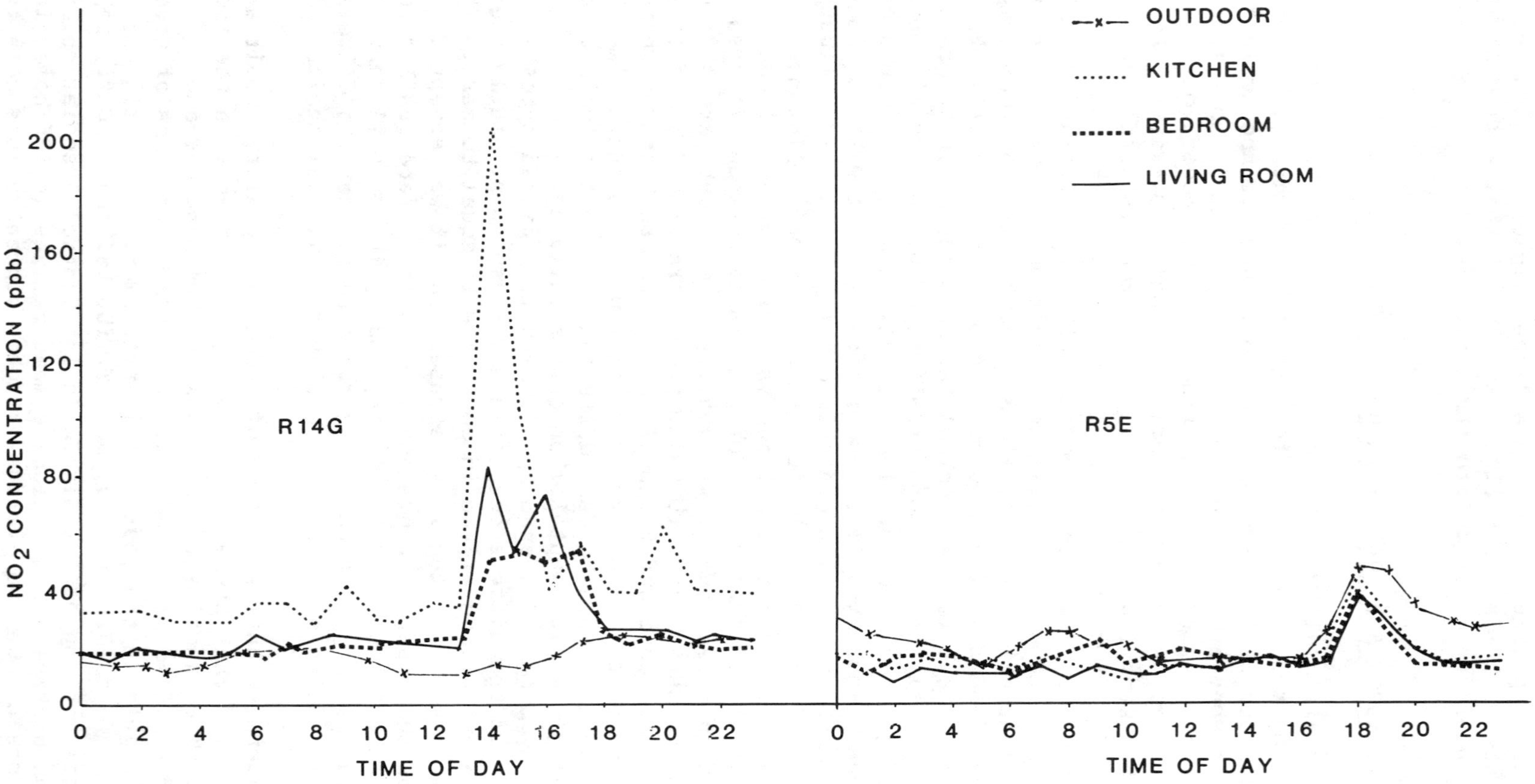

FIGURE V-7 Hourly concentrations of nitrogen dioxide at three indoor locations in residences with gas (R14G) and electric (R5E) cooking facilities. Reprinted with permission from Moschandreas _et al_.[50]

those in bedrooms in homes that had gas stoves. A study of the air quality in a scientific laboratory by West[78] showed an almost uniform distribution of an inert tracer continuously released in the room. Similar experiments performed by Moschandreas *et al.* in residential environments showed that equilibrium is reached throughout a house within an hour. Episodic release of sulfur hexafluoride tracer gas also illustrates this point. Figure V-8 shows the measured sulfur hexafluoride concentrations plotted against time. The source location was the living room; adjacent locations were the kitchen and the hall. Episodic release of this inert gas in 24 residences was followed by uniform indoor distributions within 30 min.[47 49] The one-zone concept does not require instantaneous mixing, because it is based on the behavior of hourly average pollutant concentrations.

Moschandreas and associates[50] used a different data base derived from the monitoring of 14 indoor environments in the Boston metropolitan area. Analysis of variance was used to reach the following conclusions:

- Pollutants (ozone and sulfur dioxide) generated principally outdoors have little or no interzonal statistical difference indoors.
- Pollutants with strong indoor generation have interzonal statistical differences in residences with gas facilities and offices, but not in electric-cooking residences. In general, the observed differences are not large, and the health differences are not expected to be serious.
- Depending on indoor activity and outdoor episodic pollutant activity, the indoor arithmetic 24-h average may or may not adequately represent the variation of hourly indoor concentrations.
- Although more than one zone would be preferable, hourly pollutant concentrations obtained from one indoor zone adequately characterize the indoor environment.

These conclusions were arrived at in a particular investigation. A properly designed, much larger experimental study is required to determine the general significance and applicability of these findings.

The above conclusions are not applicable to short-lived pollutants. Contaminants associated with tobacco smoke, bathroom odors, allergens, and other pollutants related to dust are expected to vary considerably in a given residence. Additional documentation is needed to determine the extent of this variation.

BUILDING FACTORS

Site characteristics, design, operation, and occupancy all may affect indoor air quality. Each affects the adequacy of a building's systems for controlling environmental quality. They are components of an environmental control system with the following functions: to mitigate adverse ambient conditions, to provide for variations in the intended occupant activity, to sustain the integrity of the structure of the building, and to support continuous operation over the

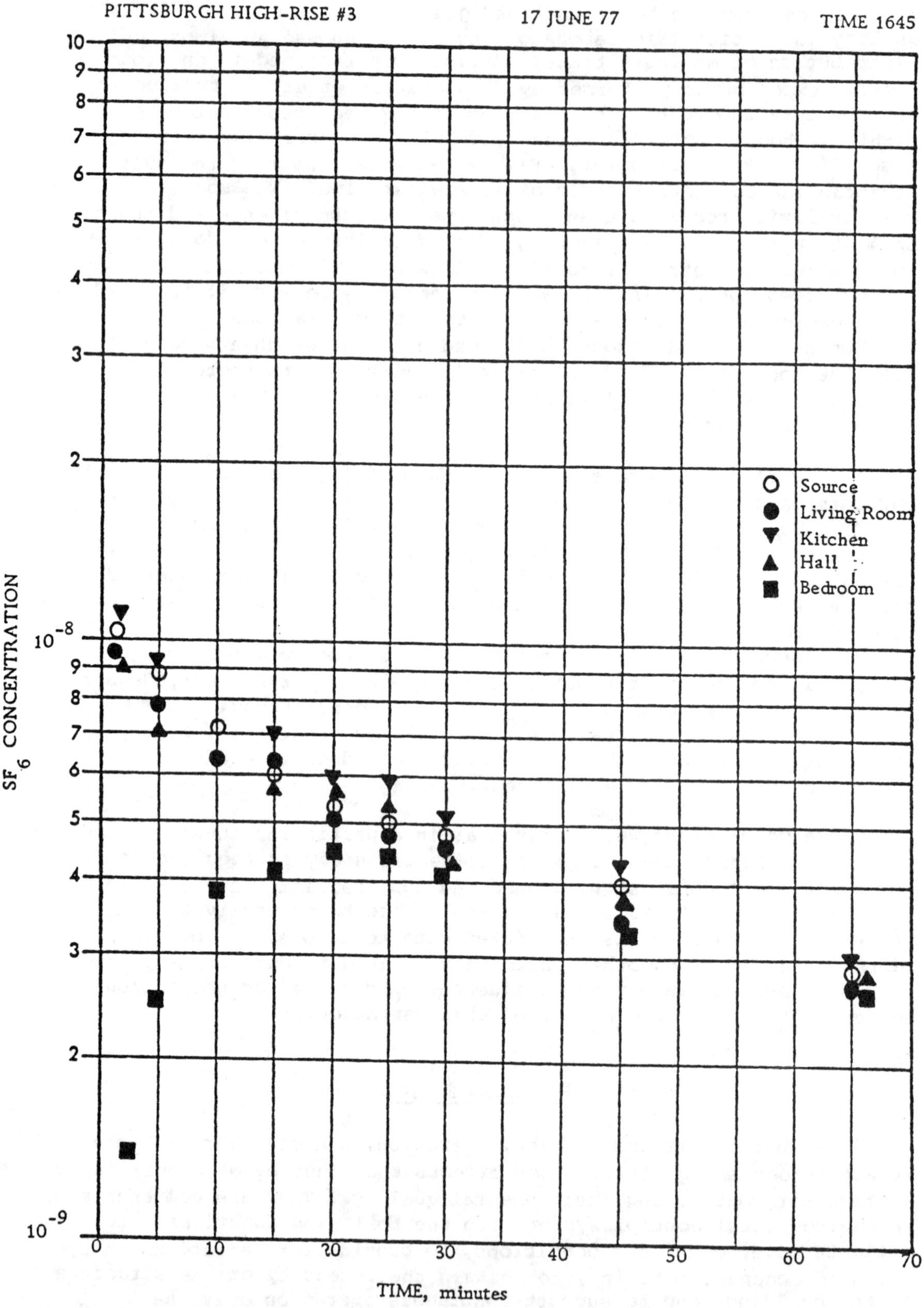

FIGURE V-8 Episodic release of sulfur hexafluoride gas. Reprinted with permission from Moschandreas et al.[50]

building's service life. The relationship between building factors and indoor air quality has not been quantified, or for that matter extensively studied, but it is important. Designing and controlling building factors may prove to be an effective mechanism for achieving desirable indoor air quality.

SITE CHARACTERISTICS

The characteristics of a building site that influence indoor air quality are addressed as three related subjects: air flow around buildings, proximity to major sources of outdoor pollution, and type of utility service available.

The air flow around a building has been shown to be determined by the local characteristics of the geometry of surrounding buildings,[54] the location and type of surrounding vegetation,[79] the terrain,[22] and the size and shape of the building itself.[31] Pollutants can be transported by the air flow from street level, over the facade of the building, and onto the roof.[10,20,35] Field tests of isolated buildings have been used to develop scaling coefficients for both isothermal and stratified cases of surface wind pressures, turbulence, and dispersion.[17,53] Air flow around a building creates low pressure on the leeward side and/or the sides adjacent to the windward face, as well as the roof.[30] Air pollutants released from stacks, flues, vents, and cooling towers in the region can reenter the building through makeup-air intakes for ventilation.[10]

Trees and forests have been generally studied as shelter belts in an agricultural context. Shelter belts affect air flow around buildings. When an air current reaches a shelter belt, part of it is deflected upward with only a slight change in velocity, part passes through the crowns of the trees with very low velocity, and part is deflected beneath the canopy with rapidly decreasing velocity.[21,38] The changes in velocity of air flow outside may change the infiltration rate and thus affect indoor air quality.

The location of a building relative to a major outdoor pollution source can affect indoor air quality. For example, buildings near major streets or highways often have high carbon monoxide and lead concentrations, owing to the infiltration of these pollutants.[15,18,23]

The type of utility service available is also related to the siting of a building and may affect the character of its indoor environment. The availability of particular fuels (e.g., natural gas and oil) influences the types and concentrations of pollutants (e.g., combustion products) emitted by space- and water-heating. Service moratoria, development timing, and development scale are institutional elements that contribute to the variability of utility services and thus can affect indoor air quality.

OCCUPANCY

Occupancy factors that affect indoor air quality include the type and intensity of human activity, spatial characteristics of a given activity, and the operation schedule of a building.

Several human activities--such as smoking, cleaning, and cooking--generate gaseous and particulate contaminants indoors. The number of occupants of a space and the degree of their physical activity (i.e., metabolic rate at rest or under intense activity) are related to the production of various pollutants, such as carbon dioxide, water vapor, and biologic agents. If the only source of indoor carbon dioxide production is that caused by occupants, ventilation rates may be proportional to the number of people and their metabolic rates.[45] Although studies have shown no constant relationship between carbon dioxide concentrations and the concentrations of other pollutants, carbon dioxide concentration is often used as a general indicator of the adequacy of ventilation in an occupied space.

Building occupancy is often expressed as occupant density and the ratio of building volume to floor area. The importance of occupancy in indoor air quality is illustrated by the fact that the choice of natural or mechanical ventilation is based on occupant density and the spatial characteristics of the buiding under consideration. The use of Yaglou's early work on the relationship between occupant density and detectable body odor in determining necessary ventilation rates is discussed elsewhere.

Occupancy schedules and associated building use may affect the type, concentration, and time and space distribution of indoor pollutants. Because most buildings are unoccupied for substantial portions of each day, the manipulation of "operating schedule" is a means of controlling energy use.[1] Efforts to conserve energy through the design of ventilation systems can result in the degradation of indoor air quality.[55] However, detailed studies relating ventilation capacity, occupancy schedules, energy requirements, and indoor air quality have only recently been implemented.

DESIGN

Elements of building design that affect the indoor environment include interior-space design (space planning), envelope design, and selection of materials.

The evolution of space planning in many building types has resulted in flexibility in assigning functions to specific locations. However, this flexibility is accompanied by a decrease in the ability to predict exposure to air pollutants. In particular, "open-plan" offices and schools have serious technical problems of redundant service distribution, limited acoustic control, incomplete air diffusion, and incomplete pollutant dispersion indoors, compared with "fixed-plan" floor layouts.

Evaluation of the success of a floor plan in achieving space efficiency, structural economy, and energy efficiency is usually in terms of net area per occupant and ratio of net usable area to total area. Explicit planning for environmental quality must be included to ensure that spatial arrangements are acceptable to the occupants.

A building's structural envelope consists of both primary elements--foundations, floors, walls, and roofs--and secondary "skin" elements--facings, claddings, and sheathing. To various degrees, the function of these is to maintain the integrity of the structure under the stresses caused by structural load, wind pressure, thermal expansion, precipitation, earth movement, and fire. The integrity of the building envelope is a major consideration in uncontrolled air movement into and out of a building--usually referred to as "infiltration." This is a major factor in indoor air quality. There has been no systematic survey of infiltration rates of buildings in the United States. The dominant factor in determining a building's infiltration rate is the total area of effective leakage, as measured with fan pressurization. Following the leakage area in importance are the terrain and shielding near the building, the mean climatic conditions during heating (or cooling) periods, and the building height.[62] There is much evidence,[19] both in the United States and in Europe, that houses in mild climates are "very leaky," whereas houses in severe climates are "tight."

Greater height of a building increases the "stack effect," or updrafting, and exposes the building to higher wind speeds. Thus, higher wind pressures drive air through existing openings, referred to as "leakage," increasing the infiltration rate.[43,62]

The dominant building factors that determine infiltration have not been identified, but a catalog of leakage openings found in typical structures is as follows:

- Walls: Leakage around sill plates (the openings at the bottom of wallboard), electric outlets, plumbing penetrations, and headers in attics for both interior and exterior walls.
- Windows and doors: Window type is more important than manufacturer in determining window leakage.[77] This source of leakage tends to be overrated; it contributes only about 20% of the total leakage of a house.[9,71]
- Fireplaces: This includes dampers, glass screens, and fireplace caps.
- Heating and cooling systems: The variables include combustion air for furnaces, dampers for stack air draft, air-conditioning units, and location of ductwork.
- Vapor barrier and insulation penetrations.
- Utility accesses: This includes recessed lighting and plumbing and electric penetrations leading to attic or outside.
- Terminal devices in conditioned space: This includes leakage of dampers, especially those for large air-handling systems.
- Structural types: Examples are drop ceilings above cupboards or bathtubs, prism-shaped enclosures over staircases in two-story houses, and elevator and utility shafts that lead from basement to attic.

Wall and ceiling materials and floor finishes are the constituents of the building interior. Modular components, weight, strength, thermal insulation, thermal stability, sound insulation, fire resistance, ease and speed of installation, and ease of maintenance are among the criteria considered in the selection of materials for walls, ceilings, and floors. But emphasis on first cost, ease of installation, maintenance, and long service life has also led to the use of materials that may be sources of indoor contaminants, as mentioned in Chapter IV.

OPERATIONS

Depending on the type of ownership (owner-occupied or developer-owned), building operation may vary considerably, and this variation may have an impact on indoor air quality. "Building operation" pertains to the following elements of a building:[42] the building envelope, service and plant, building facilities, equipment, and landscaping. Cleaning, preventive maintenance, and replacement and repair of defects are also included in building operation. The staff responsible for building operation include management, engineering, and custodial personnel. The care responsibilities are operation of the heating, ventilation, and air-conditioning systems and building services, such as hot water, lighting, and power distribution. Building operation has an impact on indoor air quality in numerous ways, but the magnitude of this impact is not known.

SUMMARY AND RECOMMENDATIONS

The nearness of a building to pollution sources and its orientation with respect to wind affect the impact of airborne pollutants within its envelope and the performance of its HVAC system. Air flow around buildings, protective building placement, and landscaping at the site and on an urban scale are useful in mitigating indoor contamination. The magnitude and duration of activity in a building affect the generation and dispersion of pollutants. Building classifications that specify occupancy limits for safety and fire protection can also be used to determine its environmental control requirements. The control of indoor pollutants depends on floor layout, pollutant concentrations, emission rates of sources, and type of ventilation system. These factors vary with the age, region, and type of construction of the buildings.

A systematic formulation of interactions of air turbulence, stratification, and pressure distribution between buildings needs to be developed to predict the effect of site conditions and design measures for buildings. Also, objective measurements of concentrations of contaminants for major classes of buildings need to be made for use in predicting the effects of building factors on the requirements for pollution mitigation within buildings. The measurement of dispersion characteristics for basic floor layouts and systems should be undertaken to identify methods of dilution or masking of pollutants. The amount of energy required for mitigation with various control strategies should be studied to optimize energy efficiency and indoor air quality. The lifetime costs of various mitigation strategies should be measured to identify promising first-cost and annual-cost alternatives both for the design of new buildings and for the redesign of existing buildings.

REFERENCES

1. American Institute of Architects Research Corporation. Phase Two Report for the Development of Energy Performance Standards for New Buildings. Report to U.S. Department of Housing and Urban Development and U.S. Department of Energy, 1979. 197 pp.
2. American Society of Heating, Refrigerating and Air-Conditioning Engineers. ASHRAE Handbook of Fundamentals. New York: American Society of Heating, Refrigerating and Air-Conditioning Engineers, Inc., 1972. 688 pp.
3. Asin, R. H. Nationwide Personal Transportation Study. Purposes of Automobile Trips and Travel. Report No. 10. Washington, D.C.: U.S. Department of Transportation, Federal Highway Administration, Office of Highway Planning, 1974. 99 pp.
4. Asin, R. H., and P. V. Svercl. Nationwide Personal Transportation Study. Automobile Ownership. Report No. 11. Washington, D.C.: U.S. Department of Transportation, Federal Highway Administration, Office of Highway Planning, 1974. 74 pp.
5. Bahnfleth, D. R., T. D. Mosley, and W. S. Harris. Measurement of infiltration in two residences. Part I: Technique and measured infiltration. ASHRAE Trans. 63:439-452, 1957.
6. Beschen, D. A., Jr. Nationwide Personal Transportation Study. Transportation Characteristics of School Children. Report No. 4. Washington, D.C.: U.S. Department of Transportation, Federal Highway Administration, 1972. 32 pp.
7. Blomsterberg, A. K., and D. T. Harrje. Approaches to evaluation of air infiltration energy losses in buildings. ASHRAE Trans. 85(Pt. 1):797-815, 1979.
8. Brail, R. K., and F. S. Chapin, Jr. Activity patterns of urban residents. Environ. Behav. 5:163-191, 1973.
9. Caffey, G. E. Residential air infiltration. ASHRAE Trans. 85(Pt. 1):41-57, 1979.
10. Cermak, J. E. Nature of air flow around buildings. ASHRAE Trans. 82(Pt. 1):1044-1054, 1976.
11. Chapin, F. S., Jr. Human Activity Patterns in the City: Things People Do in Time and in Space. New York: John Wiley & Sons, Inc., 1974. 272 pp.
12. Chapin, F. S., Jr., and R. K. Brail. Human activity systems in the metropolitan United States. Environ. Behav. 1:107-130, 1969.
13. Chapin, F. S., Jr., and H. C. Hightower. Household activity patterns and land use. J. Am. Inst. Planners 31(3):222-231, 1965.
14. Coblentz, C. W., and P. R. Achenbach. Field measurements of air infiltration in ten electrically-heated houses. ASHRAE Trans. 69:358-365, 1963.
15. Cohen, A. S., and K. Granauf. Summary and Conclusions: Carbon Monoxide Monitoring Programme, Naperville, Illinois. Technical Report of the Environmental Pollutants and the Urban Economy Project. Argonne, Ill.: Argonne National Laboratory, 1974.
16. Converse, P. E. Time budgets. pp. 42-47. In D. L. Sills, Ed. International Encyclopedia of the Social Sciences. Vol. 16. New York: The Macmillan Company and the Free Press, 1968.

17. Davenport, A. G. A rationale for determination of design wind velocities. Proceedings of the ASCE Journal of the Structures Division 86:39-66, 1960.
18. Department of Environmental Control, Chicago. Indoor-Outdoor Carbon Monoxide Concentration Survey within the City of Chicago's Central Business District. Chicago: Department of Environmental Control, 1973.
19. Dickerhoff, D., D. T. Grinsrud, and B. Shohl. Infiltration and Air Conditioning: A Case Study. Lawrence Berkeley Laboratory Report LBL-11674. Berkeley, Cal.: Lawrence Berkeley Laboratory, 1980.
20. Evans, B. H. Natural Air Flow around Buildings. Texas Engineering Experimental Station Research Report 59, College Station, Texas, 1957.
21. Federer, C. A. Effect of trees in modifying urban microclimate, pp. 23-28. In S. Little, and J. H. Noyes, Eds. Trees and Forests in an Urbanizing Environment. Amherst, Mass.: University of Massachusetts Cooperative Extension Service, 1971.
22. Geiger, R. The Climate near the Ground. Cambridge, Mass.: Harvard University Press, 1965. 611 pp.
23. General Electric Company. Final Report on Study of Air Pollution Aspects of Various Roadway Configurations. New York: New York City Department of Air Resources, 1971.
24. General Electric Company. Indoor-Outdoor Carbon Monoxide Pollution Study. U.S. Environmental Protection Agency Report No. EPA-R4-73-020. Washington, D.C.: U.S. Government Printing Office, 1973. [448] pp.
25. Gibson, U. E., and R. E. Cawley. The heat pump solar collector interface--A practical experiment. Appliance Eng. 11(4):68-71, 1977.
26. Gish, R. E. Nationwide Transportation Study. Characteristics of Licensed Drivers. Report No. 6. Washington, D.C.: U.S. Department of Transportation, Federal Highway Administration, Office of Highway Planning, 1973. 36 pp.
27. Goley, B. T., G. Brown, and E. Samson. Nationwide Personal Transportation Study. Household Travel in the United States. Report No. 7. Washington, D.C.: U.S. Department of Transportation, Federal Highway Administration, Office of Highway Planning, 1972. 40 pp.
28. Grot, R. A. A Low-Cost Method for Measuring Air Infiltration Rates in a Large Sample of Dwellings. National Bureau of Standards Report No. NBSIR 79-1728. Washington, D.C.: U.S. Department of Commerce, National Bureau of Standards, 1979. 10 pp.
29. Grot, R. A., and R. E. Clark. Air Leakage Characteristics and Weatherization Techniques for Low-Income Housing. Presented at DOE/ASHRAE Conference on Thermal Performance of Exterior Envelopes of Buildings, Orlando, Florida, December, 1979.
30. Halitsky, J. Air flow and pressures near exterior building surfaces. ASHRAE J. 7(7):37-38, 1965.
31. Halitsky, J. Gas diffusion near buildings. ASHRAE Trans. 69:464-484, 1963.
32. Hammer, P. G., Jr., and F. S. Chapin, Jr. Human Time Allocation: A Case Study of Washington, D.C. Technical Monograph. Chapel Hill: University of North Carolina, Center for Urban and Regional Studies, 1972. 242 pp.

33. Hatley, R. M. Nationwide Personal Transportation Study. Availability of Public Transportation and Shopping Characteristics of SMSA Households. Report No. 5. Washington, D.C.: U.S. Department of Transportation, Federal Highway Administration, Office of Highway Planning, 1972. 36 pp.
34. Hittman Associates, Inc. Residential Energy Consumption in Single-Family Housing. March 1973. U.S. Department of Housing and Urban Development Publication No. HUD-PDR-29-2. Washington, D.C.: U.S. Government Printing Office, 1974. 174 pp.
35. Jensen, M. The model-law for phenomena in natural wind. Ingeniøren 2:121-128, 1958.
36. Jordan, R. C., G. A Erickson, and R. R. Leonard. Infiltration measurements in two research houses. ASHRAE Trans. 69:344-350, 1963.
37. Keyes, D. L. Population redistribution: Implications for environmental quality and natural resource consumption, p. 213. In B. J. L. Berry and L. P. Silverman, Eds. Population Redistribution and Public Policy. Washington, D.C.: National Academy of Sciences, 1980.
38. Kittredge, J. Forest Influences. The Effects of Woody Vegetation on Climate, Water, and Soil, with Applications to the Conservation of Water and the Control of Floods and Erosion. 1st ed. New York: McGraw-Hill Book Company, Inc., 1948. 394 pp.
38. Kronvall, J. Testing of houses for air leakage using a pressure method. ASHRAE Trans. 84(Pt. 1):72-79, 1978.
40. Laschober, R. R., and J. H. Healy. Statistical analyses of air leakage in split-level residences. ASHRAE Trans. 70:364-374, 1964.
41. Malik, N. Field studies of dependence of air infiltration on outside temperature and wind. Energy Build. 1:281-292, 1978.
42. Manasseh, L., and R. Cunliffe. Office Buildings. New York: Reinhold Publishing Corporation, 1962. 208 pp.
43. Marin, A. Influence of stack effect on the heat loss in tall buildings. ASHVE Trans. 40:377-386, 1934.
44. Mattingly, G. E., and E. F. Peters. Wind and trees: Air infiltration effects on energy in housing. J. Ind. Aerodyn. 2:1-19, 1977.
45. McIntyre, D. A. Indoor Climate. London: Applied Science Publishers Ltd., 1980. 443 pp.
46. Michelson, W. Time-budgets in environmental research: Some introductory considerations, pp. 262-268. In W. F. E. Preiser, Ed. Environmental Design Research. Vol. 2. Symposia and Workshops. Fourth International EDRA Conference. Stroudsburg, Pa.: Dowden, Hutchinson, and Ross, Inc., 1973.
47. Moschandreas, D. J., Ed. Indoor Air Pollution in the Residential Environment. Vol. II. Field Monitoring Protocol, Indoor Episodic Pollutant Release Experiments and Numerical Analyses. U.S. Environmental Protection Agency Report No. EPA-600/7-78-229b. Research Triangle Park, N.C.: U.S. Environmental Protection Agency, Environmental Monitoring and Support Laboratory, 1978. 240 pp.
48. Moschandreas, D. J. and S. S. Morse. The Relationship between Energy Conservation Measures and Exposure to Indoor Toxic Pollutants. Paper No. 27b, presented at the American Institute of

Chemical Engineers 87th National Meeting, Boston, Massachusetts, August 19-22, 1979.
49. Moschandreas, D. J., J. W. C. Stark, J. E. McFadden, and S. S. Morse. Indoor Air Pollution in the Residential Environment. Vol. I. Data Collection, Analysis and Interpretation. U.S. Environmental Protection Agency Report No. EPA-600/7-78-229a. Research Triangle Park: U.S. Environmental Protection Agency, Environmental Monitoring and Support Laboratory, 1978. 201 pp.
50. Moschandreas, D. J., J. Zabransky, and D. J. Pelton. Comparison of Indoor-Outdoor Concentrations of Atmospheric Pollutants. GEOMET Report No. ES-823. Palo Alto, Cal.: Electric Power Research Institute, 1980.
51. Ott, W. R. Human Activity Patterns: A Review of the Literature for Air Pollution Exposure Estimation. SIMS Technical Report. Stanford, Cal.: Stanford University, Department of Statistics. (to be published)
52. Ottensmann, J. R. Systems of Urban Activities and Time: An Interpretive Review of the Literature. An Urban Studies Research Paper. Chapel Hill, N.C.: University of North Carolina, Center for Urban and Regional Studies, 1972. 45 pp.
53. Panofsky, H. A. The atmospheric boundary layer below 150 meters. Ann. Rev. Fluid Mechanics 6:147-177, 1974.
54. Peterka, J. A., and J. E. Cermak. Turbulence in building wakes, pp. 447-463. In K. J. Eaton, Ed. Proceedings of the Fourth International Conference on Wind Effects on Buildings and Structures, Heathrow, 1975. New York: Cambridge University Press, 1977.
55. Rand, G. H. Caution: The office environment may be hazardous to your health. AIA J. 68(12):38-41, 78, 1979.
56. Randill, A., H. Greenhalgh, and E. Samson. Nationwide Personal Transportation Study. Mode of Transportation and Personal Characteristics of Tripmakers. Report No. 9. U.S. Department of Transportation, Federal Highway Administration, Office of Highway Planning. Washington, D.C.: U.S. Government Printing Office, 1973. 49 pp.
57. Robinson, J. P. Changes in Americans' Use of Time: 1965-1975: A Progress Report. Cleveland, Ohio: Cleveland State University, Communications Research Center, 1977.
58. Robinson, J. P. How Americans Used Time in 1965. Ann Arbor: University of Michigan, University Microfilms International, 1977.
59. Robinson, J. P. How Americans Use Time: A Social-Psychological Analysis of Everyday Behavior. New York: Praeger Publishers, Praeger Special Studies, 1977. 209 pp.
60. Robinson, J. P., P. E. Converse, and A. Szalai. Everyday life in twelve countries, pp. 113-144. In A. Szalai, Ed. The Use of Time: Daily Activities of Urban and Suburban Populations in Twelve Countries. The Hague: Mouton & Co., 1972.
61. Shair, F. H., and K. L. Heitner. Theoretical model for relating indoor pollution concentrations to those outside. Environ. Sci. Technol. 8:444-451, 1974.
62. Sherman, M. H. Air Infiltration in Buildings. Berkeley, Cal.: University of California, Ph.D. Dissertation, 1981. (to be published as Lawrence Berkeley Laboratory Report LBL-10712)

63. Spengler, J. D., B. G. Ferris, Jr., and D. W. Dockery. Sulfur dioxide and nitrogen dioxide levels inside and outside homes and the implications on health effects research. Environ. Sci. Technol. 13:1276-1280, 1979.
64. Stopher, P. R., and A. H. Meyburg. Urban Transportation Modeling and Planning. Lexington, Mass.: D.C. Heath and Co., Lexington Books, 1975. 345 pp.
65. Strate, H. E. Nationwide Personal Transportation Study. Annual Miles of Automobile Travel. Report No. 2. Washington, D.C.: U.S. Department of Transportation, Federal Highway Administration, Office of Highway Planning, 1972. 32 pp.
66. Strate, H. E. Nationwide Personal Transportation Study. Automobile Occupancy. Report No. 1. Washington, D.C.: U.S. Department of Transportation, Federal Highway Administration, Office of Highway Planning, 1972. 32 pp.
67. Strate, H. E. National Personal Transportation Study. Seasonal Variations of Automobile Trips and Travel. Report No. 3. U.S. Department of Transportation, Federal Highway Administration. Washington, D.C.: U.S. Government Printing Office, 1972. 28 pp.
68. Svercl, P. V., and R. H. Asin. Nationwide Personal Transportation Study. Home-to-Work Trips and Travel. Report No. 8. U.S. Department of Transportation, Federal Highway Administration, Office of Highway Planning. Washington, D.C.: U.S. Government Printing Office, 1973. 104 pp.
69. Szalai, A., Ed. The Use of Time. Daily Activities of Urban and Suburban Populations in Twelve Countries. The Hague: Mouton & Co., 1972. 868 pp.
70. Szalai, A. Trends in comparative time-budget research. Am. Behav. Scientist 9(9):3-8, 1966.
71. Tamura, G. T. Measurement of air leakage characteristics of house enclosures. ASHRAE Trans. 81(Pt. 1):202-211, 1975.
72. Tamura, G. T., and A. G. Wilson. Air leakage and pressure measurements on two occupied houses. ASHRAE Trans. 70:110-119, 1964.
73. Thuillier, R. H. Air Quality Considerations in Residential Planning. Vol. 1. Guide for Rapid Assessment of Air Quality at Housing Sites. Final Report. May 1978. U.S. Department of Housing and Urban Development (Office of Policy Development and Research) Report No. HUD-PDR-524-1. Washington, D.C.: U.S. Government Printing Office, 1980.
74. Thuillier, R. H. Air Quality Considerations in Residential Planning. Vol. 2. Manual for Air Quality Considerations in Residential Location, Design and Construction, Final Report. May 1978. U.S. Department of Housing and Urban Development (Office of Policy Development and Research) Report No. HUD-PDR-524-2. Washington, D.C.: U.S. Government Printing Office, 1980.
75. Thuillier, R. H. Air Quality Considerations in Residential Planning. Vol. 3. Scientific Support and Documentation. Final Report 1978. U.S. Department of Housing and Urban Development (Office of Policy Development and Research) Report No. HUD-PDR-524-3. Washington, D.C.: U.S. Government Printing Office, 1980.

76. U.S. Department of Transportation, Federal Highway Administration. Urban Origin-Destination Surveys, Dwelling Unit Survey, Truck and Taxi Surveys, External Survey. Washington, D.C.: U.S. Government Printing Office, 1975. 309 pp.
77. Weidt, J. L., J. Weidt, and S. Selkowitz. Field Air Leakage of Newly Installed Residential Windows. Lawrence Berkeley Laboratory Report LBL-9937. Berkeley, Cal.: Lawrence Berkeley Laboratory, 1979. 17 pp.
78. West, D. L. Contaminant dispersion and dilution in a ventilated space. ASHRAE Trans. 83(Pt. 1):125-140, 1977.
79. White, R. F. Landscape development and natural ventilation. Effect of moving air on buildings and adjacent areas. Landscape Archit. 45:72-81, 1955.

VI

MONITORING AND MODELING OF INDOOR AIR POLLUTION

This chapter discusses the research tools required for measuring or estimating indoor pollution and exposure to it.

Techniques and instruments used for the measurement of outdoor pollution may be modified for the sampling of indoor environments. Several problems emerge with such modifications, and those problems are discussed here, as well as instruments designed specifically for the sampling of indoor air.

Personal monitors are increasingly recognized as powerful scientific tools for determining individual and population exposure to air pollutants. Although they are still in the early stages of application, it is clear the personal monitors can yield data that are useful in associating human activities with exposure to air pollution. The benefits and deficiencies of personal monitors are discussed in a separate section of this chapter.

The extent of indoor air pollution can be estimated with numerical models; mass-balance equations are used to estimate concentrations of indoor pollutants as fractions of outdoor concentrations and to estimate infiltration rates, indoor source strengths, pollutant decay rates, and mixing factors. Several models have been developed, but few have been validated against data obtained from measurements.

In estimating the total exposure of humans to pollutants (exposure to pollutants encountered indoor and outdoors, in industrial sites and other workplaces, etc.), it is essential to know not only the pollutant concentrations, but also individual patterns of mobility and use of time. The available information pertinent to the last two characteristics has been gathered mostly by social scientists and, although interesting, does not meet the information needs for assessing exposure to air pollution. The final section of this chapter discusses the idea of total exposure and what knowledge is needed to measure it.

FIXED-STATION SAMPLING AND MONITORING

There is an extensive data base on outdoor air quality, and much of the knowledge gained from studies of outdoor air quality is applicable to the characterization of nonindustrial indoor environments. However, the characteristics of indoor air quality in residential and commercial buildings and at other indoor sites can be quite different from those of outdoor and heavy-industrial environments. Thus, a number of special problems arise: the quality of indoor air is affected by a broad spectrum of pollutants from both outdoor and indoor sources; measurements of indoor air concentrations may require sampling instruments considerably different from those used in the outdoor or industrial environment; and the air volume inside a building is finite, and the rate of air exchange (especially in residential units) may be very low, and therefore, when air samples are drawn from an occupied space by external samplers, the sampling flow rate must be so low as to have only a negligible effect on indoor air movement and on the air-exchange rate.

Because of the effects of equipment heat and noise, as well as occupant inconvenience, sampling and monitoring equipment should (and usually can) be placed in remote locations outside the building being evaluated. Thus, it is common practice is to locate the instruments outside the building space and draw air-sample streams to them.

Sampling techniques fall into the following broad categories:

- Continuous sampling: Provides "real-time" sampling; required to observe temporal fluctuations in concentration over short periods.
- Integrated or continuous sampling: Provides an average sampling over a specified period; used when the mean concentration is either desirable or adequate for the purpose.
- Grab or spot sampling: Provides single samples taken at specified intervals; typically consists of admitting an air sample into a previously evacuated vessel, drawing a sample into a deflated bag for later analysis, or drawing (by mechanical pump) a sample through a sample collector to extract a contaminant from the air; suitable when "spot" samples are adequate for the measurement of a pollutant and knowledge of temporal concentration variation over short periods is not important.

Some instruments sample and measure pollutants directly, and others sample for later laboratory analysis. The direct-reading instruments required for continuous monitoring use various types of physicochemical detectors that can measure the concentrations of pollutants *in situ*. Integrated or grab-sampling methods are used when there is no suitable concentration sensor available, when the pollutants of interest are present at concentrations too low to permit use of direct-reading instruments, or when sampling sites are inaccessible to bulky instruments. Further information on sampling and measurement methods for air pollutants is available elsewhere.[2,53,103]

CONTINUOUS MONITORING

Continuous monitoring is a technique for sampling and measuring the real-time concentration of pollutants. Indoor air quality is subject to both temporal and spatial variations, and data on these variations would be needed to determine the concentrations to which occupants are exposed or to model indoor air pollution. The choice of monitoring techniques must be consistent with the types of information desired and the resources and manpower available. Although continuous monitoring has numerous benefits, it also has a number of disadvantages.

Two positive features of continuous monitoring are that peak short-term concentrations can be determined, in addition to average concentrations calculated over any period, and that concentration variation as a function of time can be correlated with source generation, infiltration-ventilation, and other characteristics.

The availability of continuous-monitoring instrumentation depends on many factors, including the chemical properties of the pollutant and the range of concentrations to be measured. Continuous monitors are commercially available for all the gaseous pollutants that are designated "criteria" pollutants by the EPA--carbon monoxide, sulfur dioxide, nitrogen dioxide, ozone, and total nonmethane hydrocarbons. The EPA has specified performance criteria for the instruments used to measure each of these pollutants, and all analyzers that meet these specifications in performance tests are designated "EPA-approved."

Continuous-monitoring systems, even with high-quality instrumentation, are not trouble-free. For example, continuous monitors are expensive and require frequent calibration and routine maintenance. In addition, they have their own power and ambient-temperature requirements and can create safety, heat, and noise problems if they are placed at the sampling points. For these reasons, monitoring systems are generally designed to have all equipment for continuous analysis and recording at a single remote site, often a mobile laboratory. Such a laboratory usually contains facilities for calibration and maintenance, and it may also provide electric power and suitable environmental conditions for the equipment. If sampling lines made of flexible fluorocarbon tubing or other nonreactive materials are used, air from several sampling points can be drawn into the laboratory for analysis. One set of continuous monitoring equipment can be shared by several sampling sites if the individual lines are sequentially sampled.[71] In this scheme, all instruments obtain air samples from a common manifold, which, in turn, is supplied with air from one of four sampling sites (one of which is usually outdoors) or from a calibration system.[71] The length of the sampling interval for each site can be determined by the response times of the individual instruments, the actual transit time in the sampling line, and the details of temporal information required at each sampling site.

Continuous monitoring requires highly trained field personnel, rigorous quality-control (calibration) procedures, and provisions for quality assurance (independent performance audits of routine monitoring and data-handling operations). Securing electric power and a suitable location for a mobile laboratory equipped with sampling lines and cables can require long-term planning and entail considerable expense. This type of fixed-station monitoring is not suitable for large-scale surveys, because of these time and cost considerations. For large-scale survey work, integrated sampling and grab-sampling techniques are generally more appropriate.

INTEGRATED SAMPLING

Integrated sampling, in which a known sampling rate is maintained over some period, is commonly used for pollutants that must be accumulated to permit analysis. The period of collection may be several minutes or several weeks or months. Analysis may be performed at the collection site or in a laboratory. The data resulting from analysis of integrated samples are expressed as an average concentration over the sampling period. A variety of particles, gases, and vapors are sampled by this technique.

Particles can be collected on filter media for later gravimetric and chemical analyses. Size-selective particle samplers, such as various dichotomous air samplers[58] and portable cyclone samplers,[18 102] are used in indoor aerosol sampling when it is desirable to determine the concentration of fine particles (less than 2.5 µm in aerodynamic diameter) or respirable particles (less than 3.5 µm in aerodynamic diameter). The samples can be analyzed by beta gauge or gravimetric techniques capable of determining mass concentration; by x-ray fluorescence, neutron-activation analysis, etc., to determine their elemental composition; and by a variety of separation and analytic techniques to determine chemical composition. Aerosol samplers must be placed directly at the sampling sites, to avoid the particle losses that occur when air is drawn through sampling lines. The sophistication of particle samplers ranges from hand-held units that require manual operation to fully automated units that can be programed to operate unattended for several weeks.

Gaseous substances can be collected by both passive (diffusion-controlled) and active (powered bulk air-flow) samplers. Soluble vapors, such as formaldehyde and ammonia, can be collected by liquid gas washers and bubblers. Air sampling with bubblers, as well as with other accumulating sample collectors (such as adsorbers and condensation traps), requires that the total volume of the air sample be accurately known. This can be accomplished with dry- or wet-test meters, which measure sampled volume directly, or by measuring or controlling the sampling rate and time.

Many techniques have been used to measure the concentrations of radon and radon daughters. Because of the low level of radioactivity usually found in buildings, integrated measurements are often necessary. Passive devices that use sensitive thermoluminescent dosimeter (TLD) chips,[29] passive film, or track-etch techniques[25] can record alpha decay over periods of weeks or months to determine average radon concentrations. Radon-daughter concentrations can be determined by passing a known volume of air through a filter paper (typically for 10 min) and then measuring total alpha activity on the filter with an alpha-decay ratemeter.

Integrated sampling techniques have several advantages: they are less expensive and require less manpower than continuous monitors, they can be used to measure concentrations that are too low to be measured directly, samples can often be analyzed later at a more convenient time or place, and average concentrations over long periods are easily obtained. But they also have some disadvantages: short-term temporal information is lost; samples must be taken frequently if temporal variability in concentration is to be assessed; transporting the sample to its point of analysis may require special handling, special environmental conditions, or rapid delivery to avoid deterioration; and quality control may be more difficult to implement.

GRAB SAMPLING

In grab sampling, one sample is collected over a very short period. Grab samples have to be taken frequently, if temporal variability in pollutant concentrations is to be assessed. It is usually the least expensive technique for field sampling, unless very frequent samples are required. It may simply involve filling a container with an air sample and transporting it to a laboratory for analysis, or it may involve extractive sampling, as in colorimetric detector tubes. It is most useful when the laboratory equipment required for analysis is at a remote location, when a very large number of samples are required, or when manpower and equipment are limited. Sampling vessels commonly used include plastic bottles, glass tubes filled with adsorbent, stainless-steel containers, and bags of aluminum polyester (Mylar), PVC film, and fluoroplastic film.[93] Grab sampling has been used to estimate concentrations of radon, tracer gases, and organic compounds.

Grab sampling can be used to measure radon concentrations by pumping a known volume of air through a filter into a Tedlar[24] bag, which is impervious to radon. The time at which the sample is taken must be recorded, and, because of the decay properties of radon gas, analysis must be performed within a few days. If necessary, the sample can be concentrated with a cryogenic trap or transferred directly into a zinc sulfide scintillation chamber[59] for alpha-counting. The bags are inexpensive and can be mailed, with manual pumps, to field sites. Similarly, air samples can be qualitatively and quantitatively analyzed for organic compounds with gas-chromatographic techniques.

Grab sampling, with its low cost and minimal manpower requirement, is suitable for large-scale survey work. However, a number of problems are associated with this technique. No information other than an "instantaneous" concentration can be obtained, and this value could be greatly affected by something as simple as the opening of a door or window. Sampled volumes are relatively small, and the laboratory measurement technique must be sensitive enough to determine ambient concentrations directly. Inward and outward diffusion of various gases has been observed for many materials used in collection bags, and leaks in the containers and connectors are common. Particular attention must be given to degradation, adsorption, contamination, transformation, and the possible formation of artifact pollutants.[48] The expeditious transport of grab samples with reference to time, temperature, sealing, and handling is important. Quality control is difficult to maintain, but must be established before this technique can be used with confidence.

MONITORING OF VENTILATION RATE

Indoor air quality is directly affected by the rate at which outdoor air enters a building. Ventilation can be used to maintain low concentrations of indoor-generated pollutants. In turn, human comfort conditions, such as temperature and relative humidity, can be the determining factors in setting ventilation rates. Measurement of the infiltration-ventilation rate, the meteorologic factors that affect it (outdoor temperature, wind speed, and wind direction), and comfort factors (temperature and relative humidity) can be an integral part of fixed-location field monitoring.

Ventilation systems vary considerably. Detached single residential units are ventilated primarily by infiltration--the uncontrolled leakage of air through cracks in the building envelope (around doors and windows, through walls and floor joints, etc.)--and by the controlled opening of windows and doors. Large buildings are usually ventilated by mechanical systems of varied complexity.

So-called fresh air enters detached residential structures by infiltration; the term "air changes per hour" (ach) is routinely used for this source of ventilation--"1 ach" means that a volume of outdoor air equal to the volume of the interior building space "leaks" inside each hour. That does not imply that the incoming air drives out or displaces the old air as it enters; rather, it is assumed that perfect mixing takes place. In practice, however, perfect mixing is impossible to achieve. Therefore, an estimate of outdoor-air flow rate is based on the assumption of perfect mixing and homogeneity of indoor air to facilitate calculating infiltration rates.

By far the most commonly used method of estimating air-exchange rates is the tracer-gas decay technique.[40] In this method, a tracer gas is released into the building space at one or more points, possibly with the use of fans. In this way, an attempt is made to produce a uniform concentration throughout the building space. If homogeneity is maintained, the decay of the tracer gas is exponential, and the infiltration rates can be determined by sampling the air at several times. The air-exchange rate can be obtained from the slope of a semilogarithmic plot of the natural logarithms of the pollutant concentration versus time.

In a similar method, the equilibrium-concentration method, a tracer gas is released at a constant rate into the building space.[40] In the steady-state condition with perfect mixing, the indoor concentration will reach a steady-state value. From this and the injection rate, the infiltration rate can be calculated. With this technique, although it is simple to perform, it often takes many hours to reach a steady-state equilibrium.

More complex tracer-gas systems can measure infiltration rates on a semicontinuous or continuous basis.[47] Many gases have been used for tracer-gas measurements. Some of the properties that such a gas should have are easy measurement at low concentrations, minimal interference from other air constituents, chemical stability, nonreactivity, lack of absorption by building contents, a density comparable with that of air, safety for humans, lack of explosiveness and flammability, absence of other interior or exterior sources, low cost, and ready availability. Some gases commonly used are sulfur hexafluoride, nitrous oxide, and ethane.[40]

Mechanical ventilation systems vary considerably in design and complexity, and methods chosen to estimate ventilation rates must be suitable for the systems under consideration. The methods commonly used to estimate the ventilation rate for systems that use recirculation include pressure-measuring devices (such as inclined manometers and U-tubes), velocity meters (such as pitot tubes, hot-wire flowmeters, heated-thermistor flowmeters, and heated-thermocouple flowmeters), mechanical gas-flow indicators (such as rotating and deflecting-vane anemometers), tracer-gas techniques, and heat-balance techniques.[53] Care must be taken to distinguish between the total rate at which air enters a particular zone and the rate at which outside air enters the zone.

Temperature-measuring devices suitable for continuous monitoring include thermocouples, semiconductors, and thermistors. Typical indoor temperatures range from 15 to 40°C. Thermocouples present problems with low voltage outputs near 0°C and have nonlinear characteristics, but only when the cold junction is at 0°C. Semiconductor temperature sensors that use integrated circuits and have a voltage output linear with temperature are suitable for continuous recording. The most common temperature probe for measuring temperatures in this range is probably the thermistor, because of its high resistance ratio (which yields large voltage changes for small changes in temperature), its linearized output, and its wide operating range. Temperature gradients can be large in a building and even in an individual room. Probes should be placed where they will sense the temperature experienced by the occupants. Temperature probes should be calibrated against mercury thermometers that meet the specifications of the American Society for Testing and Materials (ASTM).

Relative humidity can be measured with sensors of the "human hair" type, which expand and contract with changes in humidity, or with dewpoint-measuring devices. Commercially available dewpoint hygrometers, based on the principle that the vapor pressure of water is decreased by the pressure of an inorganic salt, are well suited to continuous monitoring. Relative humidity can be readily calculated from dry-bulb and dewpoint temperatures. Relative-humidity measuring devices can be calibrated with the aid of sling psychrometers.

PERSONAL MONITORS

Over the last 2 decades, a wide variety of miniaturized air samplers have become available that collect gaseous and particulate samples from the immediate vicinity of people, even as they conduct their normal activities. The initial devices used battery-powered samplers, defined as "nonpassive." Although widely used, these devices are often larger and heavier than desirable. More recently, a variety of diffusion- and permeation-controlled samplers have become available. These "passive" devices are applicable solely to gas- and vapor-sampling and are very small and light. They all use sensitive chemical or physical analytic methods.

Three recent workshops reviewed candidate technologies for personal sampling and monitoring of air-pollution exposure. A Brookhaven group[67] identified potential methods for gas- and particle-monitoring. An EPA feasibility study[38] identified useful methods for monitoring sulfur dioxide, nitrogen dioxide, and ozone. Another EPA symposium[60] explored the use of available technology for health-effects studies and other uses.

Blood carboxyhemoglobin (COHb) can be used as a measure of the actual dose of carbon monoxide received by a person. Respirable-particle concentrations are also of prime concern in health-effects studies; some of the factors involved in obtaining reliable data have been evaluated.[91,102]

PERSONAL SAMPLING DEVICES

Gas-Sampling

The major techniques developed for sampling gaseous pollutants are passive (based on membrane permeation or diffusion through a geometrically defined air space) and nonpassive (in which air-pumping devices draw defined air volumes through devices of known collection efficiency).

Passive Samplers. Passive samplers use the kinetic energy of gas molecules and the efficiency of the adsorbent collector to extract pollutant molecules from the air at a known rate. The sampler must be placed at the collection site, but has no requirement for a pump, flow regulator, or batteries. Such samplers therefore have major advantages with respect to weight, cost, and maintenance. There are two basic types: diffusion and permeation. Their use is limited by the rate and amount of gaseous diffusion through a geometrically controlled air space or by transport through a permeable membrane that is specific for the pollutants being sampled.

The choice and use of the diffusion-collector technique require knowledge of the coefficient of diffusion of the pollutant to be sampled in air under conditions similar to those normally encountered. Humidity effects have been encountered; these are most probably caused by changing absorbent efficiency.

One diffusion sampler has been developed[83-85] for the measurement of ambient nitrogen dioxide. It uses the principle of diffusion through the bore of an open tube that defines the rate of transport to the collector. The quantity of nitrogen dioxide diffused from the open end of the tube to the collector surface (triethanolamine) is calculable by Fick's first law of diffusion, which may be expressed as: $Q_{NO_2} = D(A/L)Ct$, where Q_{NO_2} = number of moles of nitrogen dioxide transferred during time t, A = cross-sectional area of tube (cm^2), L = distance from open end to collector surface (cm), C = concentration difference between tube entrance and closed end (mol/cm^3), and D = coefficient of diffusion of nitrogen dioxide in air (cm^2/s). The required nitrogen dioxide absorbed[82] is given as a time-weighted average concentration for the sampling period.

Substituting typical values of the parameters--D = 0.154 cm^2/s, A = 0.71 cm^2, and L = 7.1 cm--simplifies the expression to: $Q_{NO_2} = 2.3\ (\text{ppm-h}) \times 10^{-9}$.

In principle, this method is applicable to determination of any gaseous air pollutant for which an efficient, selective absorbent is available and for which an appropriate analytic chemical procedure may be devised. The size of the sampler can be varied, but attention must be given to scaling factors.[41,80] Diffusion samplers have reportedly been used for water vapor and sulfur dioxide,[83] nitric oxide,[81,85] aniline,[13] benzene,[4,35] ammonia,[62] carbon monoxide,[81] and NO_x.[85] An activated-carbon element has been used as the collector in a badge that has an open grid to define the geometry of the gaseous diffusion port.[32] This method of sampling requires use of gas-chromatographic (GC) analysis for measurement of the specific gaseous pollutant absorbed. A variety of organic compounds can be sampled and measured by this technique.

A large variety of permeation samplers used for monitor systems are available commercially. All use membranes fabricated and calibrated to control the rate of permeation of the pollutant to the collector, which may be a solid medium, such as charcoal or Tenax GC for specific chemicals. Processing of the collected sample varies widely; chromatographic or colorimetric procedures are commonly applied for measurement.

Transport of a gaseous pollutant across a membrane resembles the diffusion process.[107] However, permeation involves solution of the gaseous species in the membrane. Specific interaction between the gas and the polymer matrix introduces variables. As in a diffusion collector, the concentration of the gas approaches zero on the side of the membrane next to the collector, causing a gradient that results in flow from the ambient-air side.

The permeability constant, P, of a membrane is defined by the equation $N = PA(C_1 - C_2)/S$, where N = rate of transport across the membrane (mol/s), P = permeability constant (cm^2/s), A = cross-sectional area (cm^2), S = membrane thickness (cm), C_1 = concentration of gas on ambient side of membrane ($\mu g/m^3$), and C_2 = concentration of gas on collector side of membrane ($\mu g/m^3$). As a function of time, $Nt = PAC_1t/S$. If W = Nt, the amount of gas that passes through the membrane in time t, then, because P, A, and s are constant for any device and gas, $W = C_1t/k$, where k = s/PA, or $C_1 = Wk/t$.

Thus, the amount of a gaseous pollutant, W, trapped in the collector medium is proportional to its concentration, C_1, in the air. The value of k must be determined by calibration.

Membranes used in permeation collectors are made from polymeric materials, such as dimethylsilicones, silicone polycarbonate, silicones, cellulose acetate, TFE Teflon, FEP Teflon, Mylar, polyvinyl fluoride, Iolon, and Silastic. Thicknesses of 2.5-25 x 10^{-3} cm have been tested.[88]

The permeation collector with activated absorbents is particularly useful for organic pollutants when GC analysis is applied. Commercial monitors are available from the 3M Corporation[35] and Du Pont that allow determination of more than 80 compounds by this method. Sulfur dioxide,[63 88 107] chlorine,[37 107] vinyl chloride,[75 105 107] nitrogen dioxide,[106] alkyl lead,[107] and benzene[107] have been determined. The utility of a spectrum of absorbents in these passive collectors has been tested.[9]

Nonpassive Systems. A considerable variety of sampling systems using pumps to move the air have come into use over the last 2 decades, including impinger systems and solid adsorbers for gases and impingers, filters, and impactors for solid particles. Directly indicating devices using impregnated papers, chalks, and crayons and stain-detector tubes have also been used. These techniques have been reviewed in detail by Linch[55] and Saltzman.[91]

A recent development in personal monitors involves a pump that is positioned next to the wearer's diaphragm by a light harness. The volume of air pumped by the motion of the thoracic cavity is recorded by an electronic package, which may be checked by a detached readout system. This system samples air for gases through coated diffusion-tube collectors or for particles through small filters at flow rates of 75-500 ml/min, depending on the wearer's breathing rate. The complete apparatus weighs 590 g. In conjunction with spirometer calibration, actual exposures to measured pollutants may be calculated.

Particle-Sampling

Particulate samples are collected by using the same principles used for large-scale samplers: filtration, impaction, and liquid impingement. The separation of the respirable fraction of particles is of considerable importance for personal monitors, and collection of adequate numbers of samples for analysis is critical. All particle-samplers use some device for moving the air sample and for separating the respirable-particle fraction. Collection is preceded by such a device as a cyclone presampler. The relatively low power available to drive the air-sampling pump usually limits particle collection by personal monitors to filtration, either in a single stage or in a second stage that follows a cyclone that collects the large, nonrespirable particles. For respirable-particle-sampling, the air-flow rate must be precisely controlled.

The theory of aerosol collection by filtration has been extensively reviewed by Dorman,[21] Pich,[87] Fuchs,[26] Green and Lane,[36] Liu and Lee,[57] and Lippmann.[56] Small-scale impactors suitable for respirable-dust-sampling with a personal monitor have been described by Marple[61] and Willeke.[108]

Biologic Monitoring

Measurements made on the human body and its excretions constitute an alternative way of measuring exposure to environmental pollutants. They include measurements of blood, urine, feces, and hair. The methods that can be used to relate environmental pollutant exposure to human composition have been reviewed at length in other NRC reports (on carbon monoxide, nitrogen oxides, and various trace metals).

USE OF PERSONAL MONITORS IN EXPOSURE STUDIES

Exposures to air pollution usually vary with a person's mobility patterns and activities. Therefore, estimating the total exposure of a person from one or a few air-pollution measurements at stationary locations cannot properly characterize the variation in a population's or a person's actual exposures. To evaluate health effects, it is necessary to know actual personal exposures and the distribution of those exposures in a population. The need for direct measurement of personal exposure to pollution has been noted by several authors.[49,67]

Personal monitors for various pollutants are commercially available. The Brookhaven workshop identified four basic experimental designs:[68]

> 1. Use of Individual Air Pollution Monitors for Direct Determination of Exposure. Each person in the study population would wear or carry an individual air pollution monitor during the course of the study. The same individuals would also be subjected to continuous or periodic evaluation of health responses. Individual exposure and response would thus be measured. . . . Because of economic constraints, only relatively small populations could be studied by this direct approach.
>
> 2. Use of Individual Air Pollution Monitors to Adjust Results from Fixed Stations. As previously indicated, there can be substantial variations between area level measurements and personal exposure measurements. By monitoring exposure of individuals with individual air pollution monitors in areas also monitored with fixed stations, one would obtain the distribution of individual exposures in relation to measurements obtained at the fixed stations. If one or several relatively constant relations were found in various areas, fixed-station data would then be corrected for use in estimating population exposures.
>
> 3. Use of Representative Sampling to Determine Subgroup Exposure. A carefully selected sample of the study population would be asked to wear or carry individual air pollution monitors. The sample would be stratified, grouping those expected to have similar exposures (e.g., office

workers or street workers). The measured exposure of each subgroup in the sample could be used as representative of the entire group.

4. Use of Individual Air Pollution Monitors to Calibrate Personal Activity Models. Activity models have been developed that describe how and where people spend their time. . . . These models could prove to be useful in estimating population exposure. They have not been applied in air pollution epidemiology except in a very limited way, and they could be best calibrated or verified through experiments using individual air pollution monitors. In such an experiment, a carefully selected sample of the study population would be asked to wear individual air pollution monitors, and their measured exposure would be compared with the estimated exposure of the activity model.

Gaseous Pollutants

Carbon Monoxide. Ott and Mage[79] collected 425 integrated carbon monoxide samples over 21 d in November 1970 and January 1971 in downtown San Jose, California. Breathing-zone samples collected while the subjects walked typical pedestrian routes were compared with those measured at the fixed monitoring stations. The mean pedestrian exposure was 1.6 times the mean concentration measured by the fixed monitors, but individual measurements of exposure varied from those measured at the fixed stations by a factor of up to 10. Ott and Mage concluded that the fixed-station monitoring data "provide a relatively poor measure of the true exposure of members of the general public to air pollutants."

Wright *et al.*[111] sampled exposures in Toronto with portable carbon monoxide monitors. They demonstrated a "substantial discrepancy between the carbon monoxide concentrations detected by the provincial network of fixed-site sampling stations and the much higher concentrations commonly met by people living and working in a large metropolitan area such as Toronto."

Cortese and Spengler[16] measured exposure of Boston commuters equipped with portable carbon monoxide instruments (Ecolyzers). They reported that 1-h exposures exceeded fixed-monitor measurements by a factor of 1.3-2.1; 8-h mean exposures were considerably below the 8-h mean from the fixed monitor.

Wallace[104] carried a carbon monoxide dosimeter during 30 commuting trips by bus to his office in Washington, D.C. Concentrations inside the vehicles were typically 2-4 times those continuously measured in the central city at the fixed monitoring station (Figure VI-1). There was no correlation between the ambient and personal in-vehicle measurements.

Nitrogen Dioxide. In a personal-monitoring study of children in Ansonia, Connecticut,[6] nitrogen dioxide and sulfur dioxide were

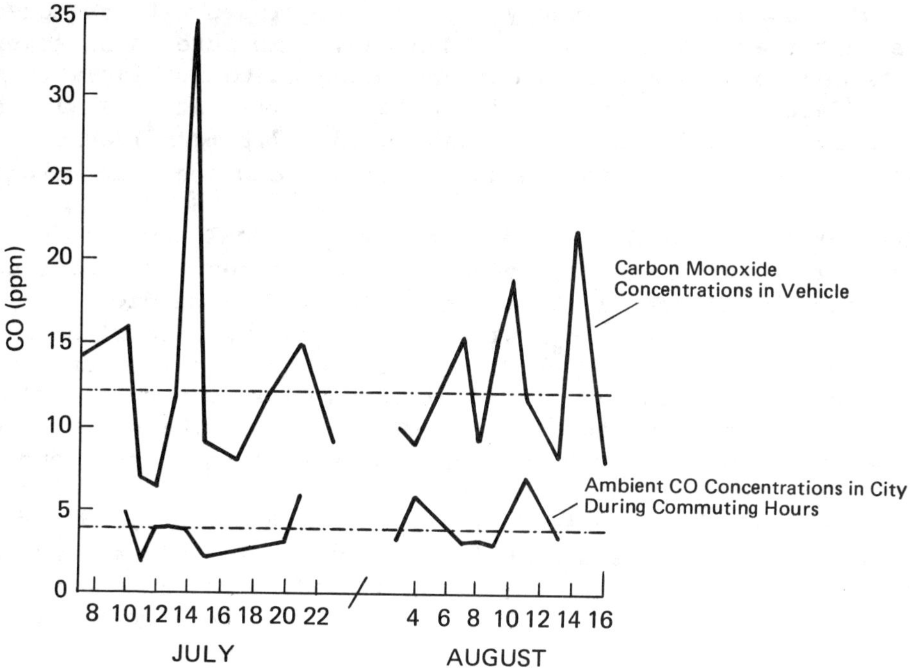

FIGURE VI-1 Carbon monoxide concentrations in vehicle, compared with ambient concentrations in city. Reprinted from Wallace.[104]

measured with bubblers.[10] Twenty boys were equipped with suitcase samplers that they carried for one 24-h day. Exposure to nitrogen dioxide tended to be greater in children exposed to smoking at home, but the differences were not statistically significant. As with sulfur dioxide, mean personal exposure values (61.3 ± 7.2 μg/m^3) were significantly lower than mean outdoor nitrogen dioxide concentrations (100.1 ± 9.0 μg/m^3).

Palmes et al.[84] has described a passive personal sampler for nitrogen dioxide suitable for occupational exposures. The same device has been used to estimate 1-wk average indoor nitrogen dioxide concentrations in 109 dwellings with gas stoves and nine with electric stoves in metropolitan New York.[86] It was found that the homes with gas stoves had significantly higher nitrogen dioxide concentrations than those with electric stoves. Average values in the kitchens with gas stoves approached the U.S. primary ambient-air quality standard of 50 ppb (annual average).

These dosimeters were used in a personal-monitoring study of five families with gas stoves and four families with electric stoves in Topeka, Kansas.[20] In each family, the husband, wife, and one child wore the dosimeters for four 1-wk samples. Dosimeters were also placed outside, in the kitchen, and in the bedroom. No significant differences were found between the personal-monitor and outdoor measurements for the families with electric stoves (Figure VI-2). For the families with gas stoves, personal exposures were significantly higher than outdoor values and correlated best with the fixed dosimeters in the bedroom. No significant differences in exposures were found between family members.

Sulfur Dioxide. Exposure to sulfur dioxide has been estimated in several studies by the calculation of a time-weighted average exposure from the time spent and average concentrations in various places.[14,27] Sulfur dioxide personal monitors have not been extensively used in field studies.

In the personal-monitoring study of children in Ansonia, Connecticut, no significant difference was found among the sulfur dioxide personal exposure measurements of the boys.[6] The personal samplers had a mean of 5.5 ± 0.07 μg/m^3, which was significantly lower than the outdoor mean of 12.0 ± 2.2 μg/m^3. The reports of daily activities showed that the children were indoors between 60% and 80% of the day.

Passive personal monitors using the collection principle of gas permeation through polymer membranes have been shown to be sensitive to 24-h average concentrations of sulfur dioxide down to 0.01 ppm.[1,107] Sampling at ambient concentrations with treated filters also appears feasible.[97] However, no personal-monitoring results with these techniques have been reported.

Organic Substances. Several passive monitors have been developed that may have the sensitivity to measure mean personal exposure to organic substances for sampling periods of 1 d to 1 wk.[9,32,33,35] Their use has yet to be demonstrated in personal-monitoring programs in nonindustrial environments.

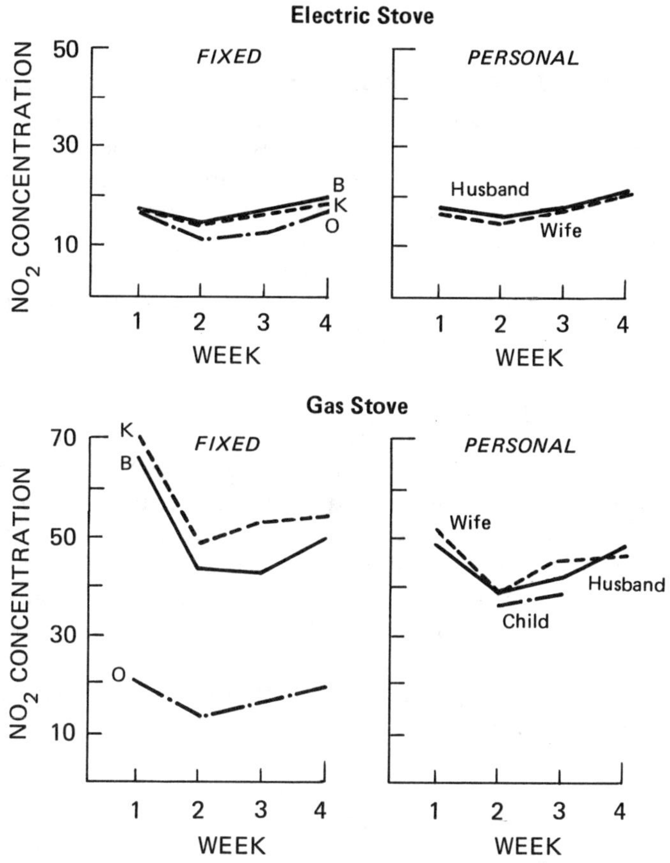

FIGURE VI-2 Week-long nitrogen dioxide concentrations, μg/m^3, for an electric-cooking and a gas-cooking family in Topeka, Kansas, May-June 1979. B, bedroom; K, kitchen; O, outdoor. Reprinted with permission from Dockery _et al._[20]

Particles

Respirable Particles. In the Ansonia study,[6] personal exposures to respirable particles were significantly higher among children who lived with one or more smokers. The mean personal exposure, 114.5 ± 9.0 μg/m^3, was significantly higher than the mean outdoor concentration, 58.4 ± 5.9 μg/m^3. The outdoor high-volume samplers, however, collected both respirable and nonrespirable particles. The principal conclusion of this study was that a child's exposure "load" of air pollutants, especially respirable particles, is determined primarily by indoor exposures.

Personal exposure to respirable particles and sulfates has been measured in two cities as part of the Harvard six-city study,[18] in which 37 people carried personal respirable-particle monitors with them during the day. Fixed-station monitors were run simultaneously in the main activity room of each home and at several locations outside. There were at least three complete sample days for each person.

Mean personal exposures to respirable particles and sulfates for each city were determined on the basis of mean outdoor concentrations. In each city, there were significant differences in results between individuals, as determined by their activities. A linear increase in personal exposure to respirable particles with the number of smokers in the home and workplace was found.

Lead. Berlandi *et al.*[5] have reported personal lead-exposure measurements from 2 d of sampling in metropolitan Boston. Samples collected while subjects were driving into Boston had a time-weighted average of 4.5 μg/m^3 the first day and 3.7 μg/m^3 the second; indoor personal samples were all less than 1 μg/m^3. Fugaš *et al.*[28] estimated average air lead exposure of an office worker in Zagreb. Air lead was measured with a personal monitor inside and outside her home and her office and at other sites. Table VI-1 shows that the average concentrations were highest in association with outdoor activities--6.3 μg/m^3--with only 1.7 μg/m^3 or less from indoor sampling. A time-weighted average exposure was then calculated on the basis of her activities each week (Table VI-1). The average weekly exposure was estimated to be 1.1 μg/m^3, compared with the average measured value of 0.72 μg/m^3 outside the subject's home.

Fugaš[27] extended this method of estimating exposure to lead in air to a middle-sized industrial town in Yugoslavia. Air lead was measured at various locations indoors and outdoors during the winter of 1972-1973. Estimated exposures were considerably higher than the average urban monitoring-station value of 0.9 μg/m^3.

Lead concentrations up to 10 times that found in buildings may be found in vehicles or outdoors. The contribution of outdoor and vehicular exposure to mean personal exposures may be small, because of the comparatively short exposure times.

TABLE VI-1

Calculation of Time-Weighted Weekly Average Exposure of Office Worker in Zagreb to Airborne Lead[a]

Location or Activity	Average Lead Concentration, $\mu g/m^3$	Duration of Exposure, h/wk	Integrated Exposure, $\mu g\text{-}h/m^3$
Workplace	1.2	42	50.4
Outdoor activities	6.3	14	88.2
Recreation	0.2	6	1.2
At home:			
Rest of day	0.7	22	15.4
Night	0.3	48	14.4
Weekends	0.5	36	18.0
Total	--	168	187.6

Weighted-average weekly exposure: 1.1 $\mu g/m^3$

[a]Data from Fugaš *et al.*[28]

Biologic Indicators

Direct measurement of individual dose by biologic means is possible for several pollutants. This method intrinsically compensates for different rates of uptake by different persons, as well as for differences in exposure.

Carbon Monoxide. Actual carbon monoxide dose received can be measured directly by measuring blood carboxyhemoglobin.

Stewart *et al.*[99 100] measured COHb of blood donors in 26 American cities. They reported COHb concentrations higher than those expected from fixed-station monitoring data. Goldmuntz[34] compared COHb measurements for nonsmokers from 1969 through 1972 with carbon monoxide measurements from 37 EPA fixed-station monitors in 1973. Goldmuntz argued that the measurements from some of these fixed stations may be inappropriately high, because of their siting.

Morgan and Morris[68] calculated COHb concentrations that would be expected if the population were in equilibrium with the measured fixed-station carbon monoxide concentrations by the relation, % COHb = 0.16[CO] + 0.3, where [CO] is the carbon monoxide concentration (ppm). Comparing these calculations with nonsmoker COHb measurements, they concluded that "the average dose indicated by COHb levels exceeded that predicted from the fixed-station data by a factor of 2."

Stewart *et al.*,[100] in a similar analysis of data obtained in Chicago in 1970 and 1974, reported COHb concentrations close to, but consistently higher than, those predicted from fixed-station measurements. They noted that the use of fixed-station measurements to define population exposure may not reflect worst-exposure situations, indoor exposures to carbon monoxide from cigarette smoke, or exposures from faulty heating systems. Kahn *et al.*[50] demonstrated that COHb concentrations among nonsmokers in the St. Louis population are strongly affected by occupational exposure and by exposure to smokers.

Lead. There is little doubt of a correlation between the high exposures to lead in the air of industrial areas and indexes of lead absorption, such as blood lead, urinary lead, delta-aminolevulinic acid dehydrase (ALAD), and delta-aminolevulinic acid (DALA). The relationship between these measures of lead dose and the lower concentrations of air lead characteristic of the community--i.e., less than 10 $\mu g/m^3$--is not well established. A National Research Council committee has stated that "more precise studies are needed of the relation between atmospheric lead exposure in the urban environment and the concentration of lead in the blood, perhaps by the use of personal monitors."[74]

In an attempt to characterize this relationship, Azar *et al.*[3] compared average exposures to lead in air measured by personal monitoring with the biologic indexes of lead absorption. Two groups of 30 taxi drivers in two cities and three groups of 30 Du Pont employees in three cities carried personal particle monitors with them for 2-4 wk. Their exposure to air lead was calculated as a time-weighted mean for their exposures at home and at work. Blood lead was determined weekly, and urinary lead daily. Different relationships were found between average exposure to airborne lead and the logarithm of the blood lead concentration in each city. The plots of the data for all five groups, however, had a similar slope, with different intercepts. The authors suggested that the different intercepts indicate that variables other than airborne lead, presumably ingested lead, are affecting blood lead content. No comparisons with fixed monitoring were made.

MODELING OF INDOOR AIR QUALITY

The value of an indoor air pollution model is twofold. First, it provides a framework for interpreting experimental results and for planning new experiments. Specifically, a model is useful in relating indoor pollutant concentrations to various geometric, ventilation, source, and sink parameters. Modeling can be used to determine the accuracy and precision to which various quantities must be measured if the desired accuracy of prediction is to be achieved. It can also be used in sorting out trends in the experimental data.

Second, and more important, a model provides a means to predict accurately some desired function of concentration (such as peak concentration or dosage) for places and conditions other than those tested experimentally.

In epidemiologic studies, it is important to consider the quality of the air to which subjects are actually exposed; in many cases, the air quality associated with the home, the mode of transportation, and the workplace should not be taken to be the same as that associated with the outside.

Indoor-air-quality models are developed to aid in understanding and predicting indoor air-pollutant concentrations and dosages as functions of outdoor air-pollutant concentrations, indoor-outdoor air-exchange rates, and indoor air-pollutant sources and sinks.

Air pollution indoors may be of outdoor or indoor origin. Outdoor pollutants may enter a structure through infiltration or ventilation. Pollutants of indoor origin may arise from point or diffuse sources. Regardless of their source, air pollutants may be transported and dispersed throughout various regions of the enclosure. Some pollutants may be removed by filters through which the makeup air or the recirculated air flows, by exfiltration or ventilation to the outdoors, and by chemical change. In the case of particles, surface removal and generation are often important.

Given familiarity with the system to be described and with the purpose of developing and using an indoor air pollution model, the starting point in developing a model is usually a statement of the mass balance concerning the pollutant of interest. For example, consider a structure of volume V, in which makeup air enters from the outside and passes through a filter at a rate q_0. Part of the building air is recirculated through another filter at a rate q_1, and air infiltrates the structure at a rate q_2. Each filter is characterized by a factor $F \equiv (C_{inlet} - C_{outlet})/C_{inlet}$. In this example, the pollutant concentration is assumed to be uniform throughout the structure. The indoor and outdoor pollutant concentrations at time t are C and C_0, respectively. The rate at which the pollutant is added to the indoor air owing to internal sources is S. The rate at which the pollutant is removed from the air owing to internal sinks is R. In this case, the appropriate starting equation is:

$$V \frac{dC}{dt} = \underbrace{q_0 C_0 (1 - F_0) + q_1 C (1 - F_1) + qC_0}_{\text{Input rate due to makeup, recirculated, and infiltrated air}} - \underbrace{(Q_0 + q_1 + q_2)C}_{\text{Output rate}} + \underbrace{S}_{\text{Source rate}} - \underbrace{R}_{\text{Sink rate}}. \quad (1)$$

The decay rate is a function of C; however, in modeling indoor air quality, the sink rate is often considered constant. As indicated in Equation 1, solutions to mass-balance equations invariably contain parameters that must be evaluated independently. Geometric parameters, such as volumes and surface areas, can be measured directly or obtained from blueprints. Accurate values of the ventilation parameters are usually more difficult to determine. Experimental techniques for obtaining net exchange rates between indoor and outdoor air have been reviewed by Georgii[30] and Gilath.[31] The use of sulfur hexafluoride, SF_6, as a tracer for air-exchange-rate studies appears to be increasing; air samples can be collected by hand in the region of interest,[22] remotely,[45] or with automated instruments.[39] Without forced ventilation, air exchange is due primarily to infiltration; in most forced-ventilation systems (which are balanced so that negative pressures are not created inside the building), the rate of infiltration is negligible, compared with the forced-ventilation rate.

Although ventilation parameters are often hard to obtain, the most difficult to evaluate are usually those associated with the rate at which the pollutant is being released or being removed (i.e., the strengths of the sources and sinks).

In view of the uncertainties associated with many of the parameter values and the difficulty of doing otherwise, compartments have been widely used in modeling indoor air quality. Traditionally, a compartment is defined as a region within which spatial variations in pollutant concentrations can be neglected over the time scale of interest. At any given instant, the concentration of a pollutant might vary substantially throughout the region of interest. However, if dosages are similar throughout the region of interest over intervals that are shorter than the time during which receptors (people, plants, equipment, etc.) are exposed, and if the damage is primarily a function of the dose, then the region may be treated as a compartment.

Depending on ventilation conditions, a single room, a floor, or a whole building may be adequately approximated as a single compartment. However, when either sources or sinks are not uniformly distributed throughout the region of interest, and the rate of mixing throughout the region of interest is low, compared with the characteristic residence time, then the single-compartment model may not provide an adequate description. For example, stratification in a room cannot be neglected when one is describing the movement of smoke and toxic gases associated with building fires;[112] however, even in such a case (where intense stratification is to be expected), only two compartments (coupled) were needed to obtain a satisfactory description.

Sulfur hexafluoride tracer experiments, conducted with average-sized rooms (20 x 20 x 8 ft) in which one or more persons were moving and in which the air was being exchanged about 3 times per hour, have suggested that associated eddy diffusivities are around 10^3 cm^2/s (D. D. Reible and F. H. Shair, personal communication); thus, about 5 min after an instantaneous point-source release, tracer concentrations (although decreasing) were about equal throughout most of a room.

The solution to the two-compartment model with constant coefficients is presented below, after a brief general discussion of multicompartment models. Examples of two-compartment and single-compartment models of indoor air quality are also discussed.

MULTICOMPARTMENT MODELS

Most n-compartment models have been (or probably will be) described by n coupled first-order linear ordinary differential equations of the form:

$$\frac{dx_1}{dt} + a_1X_1 = a_2X_2 + a_3X_3 + \quad a_nX_n + a_n + 1$$

$$\frac{dx_2}{dt} + b_1X_2 = b_2X_1 + b_3X_3 + \quad b_nX_n + b_n + 1 \qquad (2)$$

$$\frac{dx_n}{dt} + \ldots$$

In general, the terms a_1, b_1, etc., represent the sum of first-order losses from the compartment due to exhaust streams, filtration of any recirculating streams, and sources and sinks due to first-order chemical reactions. In most cases, the sources and sinks due to chemical reactions may be simulated as pseudo first-order, because of the low concentrations (parts per million, or less) of the pollutant. In cases where higher-order chemical reactions are important, the model equations will be nonlinear and generally will have to be solved numerically. In the case of particulate pollutants, the parameters a_1, b_1, etc., will probably contain loss terms, owing to surface deposition. The coefficients $a_2 \ldots a_n$, $b_2 \ldots b_n$, etc., represent the gain of pollutants in various compartments that may result from the intrusion of air from other compartments. The terms a_{n+1}, b_{n+1}, etc., represent the sums of the zeroth-order source and sink terms associated with each of the compartments.

As indicated by Equations 2, an n-compartment model will contain n(n + 1) parameters, whose values should be determined independently. Any temptation merely to fit the data through blind adjustment of the values of the parameters should be resisted, if the model is to be of broad value. The aim of any model should be to explain (and predict) as many data as possible with the smallest possible number of "adjustable parameters."

Because it is always possible to define X_n as the concentration of the pollutant in the nth compartment at any time minus the initial concentration, the initial conditions for Equations 2 may be taken as $X_1(0) = X_2(0) = \ldots X_n(0) = 0$.

TWO-COMPARTMENT MODELS

In general, the equations that describe two-compartment models are of the form:

$$\frac{dX_1}{dt} + a_1X_1 = a_2X_2 + a_3. \qquad (3)$$

and

$$\frac{dX_2}{dt} + b_1X_2 = b_2X_1 + b_3. \qquad (4)$$

The initial conditions are:

$$X_1 = X_2 = 0 \qquad \text{at } t = 0. \qquad (5)$$

When the coefficients of Equations 3 and 4 are constant,

$$X_1 = \frac{a_3b_1 + a_2b_3}{a_1b_1 - a_2b_2} + \left(\frac{a_3b_1 + a_2b_3}{a_1b_1 + a_2b_2}\right)\left(\frac{\beta e^{-\alpha t} - \alpha e^{-\beta t}}{[(a_1 - b_1)^2 + 4a_2b_2]^{1/2}}\right) - a_3\left(\frac{e^{-\alpha t} - e^{-\beta t}}{[(a_1 - b_1)^2 + 4a_2b_2]^{1/2}}\right) \qquad (6)$$

and

$$X_2 = \frac{a_1 b_3 + a_3 b_2}{a_1 b_1 - a_2 b_2} + \left(\frac{a_1 b_3 + a_3 b_2}{a_1 b_1 - a_2 b_2}\right) \left(\frac{\beta e^{-\alpha t} - \alpha e^{-\beta t}}{[(a_1 - b_1)^2 + 4a_2 b_2]^{1/2}}\right) - b_3 \left(\frac{e^{-\alpha t} - e^{-\beta t}}{[(a_1 - b_1)^2 + 4a_2 b_2]^{1/2}}\right) \quad (7)$$

where

$$\alpha \equiv (a_1 + b_1)/2 + [(a_1 - b_1)^2 + 4a_2 b_2]^{1/2}/2 \quad (8)$$

and

$$\beta \equiv (a_1 + b_1)/2 - [(a_1 - b_1)^2 + 4a_2 b_2]^{1/2}/2. \quad (9)$$

In Equations 6 and 7, the first term on the right side represents the steady-state solutions that are reached after the transient terms decay.

Woods et al.[110] used a two-compartment model in their analysis of thermal and ventilation requirements for laboratory-animal cage environments (see also Woods[109]). The two compartments were the room and the animal cage. Mass or energy balances for each compartment were coupled by both free convection and forced circulation of room air through the cage. Their models permit estimation of dry-bulb and dew-point temperatures and concentrations of gaseous particulate contaminants in cages, as well as in a laboratory room. Such models can be used to determine an acceptable means of safely reducing room ventilation rates with implications of reduced energy consumption and operational costs.

Miller[64] has used a two-compartment model in his description of the reentry of the exhausts from laboratory fume hoods. The two compartments were the building and the building wake (from which the makeup air is drawn). Miller[64] and Sasaki et al.[92] have shown that the reentry of fume-hood exhaust is a much more pervasive problem than is commonly recognized. Sulfur hexafluoride tracer experiments have shown that reentry of a portion of the fume-hood exhaust is usually the dominant factor in determining the concentrations of the pollutants to which all persons are exposed in the laboratory building. Indoor concentrations of fume-hood exhausts, normalized to the source strengths, range from about 1 to about 350 ppb per mole released per hour. Although the chemical nature of the fume-hood emission is of prime importance, persons typically complain often when they are in buildings whose normalized indoor concentrations are above 100 ppb per mole released per hour from fume hoods.

SINGLE-COMPARTMENT MODELS

Lidwell and Lovelock[54] were apparently among the first to compare concentrations of a pollutant with a mass-balance model. Their model involved the instantaneous introduction of a nonreactive pollutant into a room and considered the dilution resulting from a constant ventilation rate in which the input air was pollutant-free. They noted that, when the air in the room was not well mixed, the dilution by ventilation air was not necessarily exponential; nor were the rates of dilution the same in all parts of the room. A portion of the inlet air stream often tends to bypass part of the room. For instance, when both the inlet and exhaust ducts are on the ceiling, the lower half of the room (and especially the corners) is apparently bypassed and the air in it is diluted more slowly than expected.

Brief[8] suggested the use of a mixing factor (a constant, usually ranging in value between 1/3 and 1/10, that multiplies the ventilation rate) to account for dilution rates that are lower than would exist if the room air were continually well mixed. Constance[15] also recommended the use of mixing factors. Drivas et al.[22] derived mixing factors ranging in value between 0.3 and 0.7, except when fans were used; with fans, the characteristic time for mixing the air throughout the room was short, compared with the characteristic residence time, and the mixing factors were close to unity.

Milly[65] used a single-compartment model involving the instantaneous introduction of a nonreactive contaminant with a pollutant-free input air stream in his discussion of chemical attack of tanks and fortifications. Calder[11] used a single-compartment model in his analysis of the protection afforded by buildings against biologic-warfare aerosol attack; he permitted the outside concentration to vary with time and took into account the surface removal of aerosol by means of a first-order sink term; his results can also be used to describe doses associated with radioactive or chemical contaminants. Calder[12] also used a single-compartment model to calculate dosages associated with the penetration of a forest canopy by aerosols. Milly and Thayer[66] developed a technique for predicting indoor dosages of pollutants generated outdoors, on the basis of a single-compartment model.

Turk[101] presented a detailed analysis of the transient behavior of a single-compartment model involving a constant generation term, and a constant outside concentration; he then considered several special cases during his analysis of the measurement of odorous vapors in test chambers.

Hunt[44] used a single-compartment model with a constant internal source and a first-order sink term to interpret data regarding airborne dust in post-office facilities. Hunt *et al.*[46] and Cote and Holcombe[17] used a single-compartment model in their investigations of nonreactive gaseous pollutants indoors. Bridge and Corn[7] used a single-compartment model to predict concentrations of carbon monoxide and particles associated with smoking of cigarettes and cigars; their results were in good agreement with measured values. Sabersky *et al.*[90] reported that a single-compartment model involving an outdoor concentration that varied sinusoidally in time and a first-order heterogeneous (surface) decomposition term gave qualitative agreement with data for indoor concentrations of ozone. Shair and Heitner[95] started with a single-compartment model to develop a "linear-dynamic" model by which the indoor concentrations of ozone can be related to those outside by means of a simple expression. Tests conducted with 24 forced-ventilation systems in 13 laboratory-office buildings yielded values of k (the heterogeneous-loss constant) of 0.02-0.08 cm^3/cm^2-s, with an average of 0.04 cm^3/cm^2-s.[94]

To save energy, the makeup-air flow rate in buildings has been reduced. In one case, the reduction in the makeup-air flow rate was sufficient to permit economical selective filtering of the makeup airstream (with activated charcoal) during the times when the outdoor air quality was relatively poor (see Figure VI-3); use of the activated-charcoal filters only when needed and replacement of inexpensive prefilters every couple of months extended the life of the activated-charcoal filters to about 3 yr. This system was designed with the aid of the "linear-dynamic" model.[94] Kusuda[51] used a single-compartment model to examine the feasibility of intermittent operation of mechanical ventilation systems with an eye to conserving energy while maintaining acceptable indoor air quality.

Moschandreas (personal communication) used a single-compartment model of air pollution in nonworkplace indoor environments in Baltimore, Washington, D.C., Pittsburgh, Chicago, and Denver. He monitored carbon monoxide, nitric oxide, nitrogen dioxide, sulfur dioxide, ozone, methane, total hydrocarbons, and carbon dioxide continuously for periods of approximately 14 d in each of five detached dwellings (townhouses), six apartment units, two mobile homes, and one school. In addition, there was a 5-d period of monitoring in one hospital. The model discussed by Moschandreas *et al.*[72,73] explicitly included a chemical-decay term that was validated with the data base just mentioned. Numerical predictions of hourly carbon monoxide, nitric oxide, nitrogen dioxide, carbon dioxide, and nonmethane-hydrocarbon concentrations were found to be within 20% of the observed values 80% of the time. The model did less well in predicting the indoor sulfur dioxide and ozone concentrations; this was attributed to the chemical reactivity of the pollutants. A study was implemented to rank the sensitivity in magnitude changes in output of the model caused by perturbation of an input parameter. The ranking of input parameters, in ascending order of sensitivity, is as follows: initial condition and volume of the structure, indoor source, and pollutant decay and air-infiltration rate of the structure.[72]

Shair *et al.*[96] used a single-compartment model to describe the moisture content of bathroom air during and after the use of a shower;

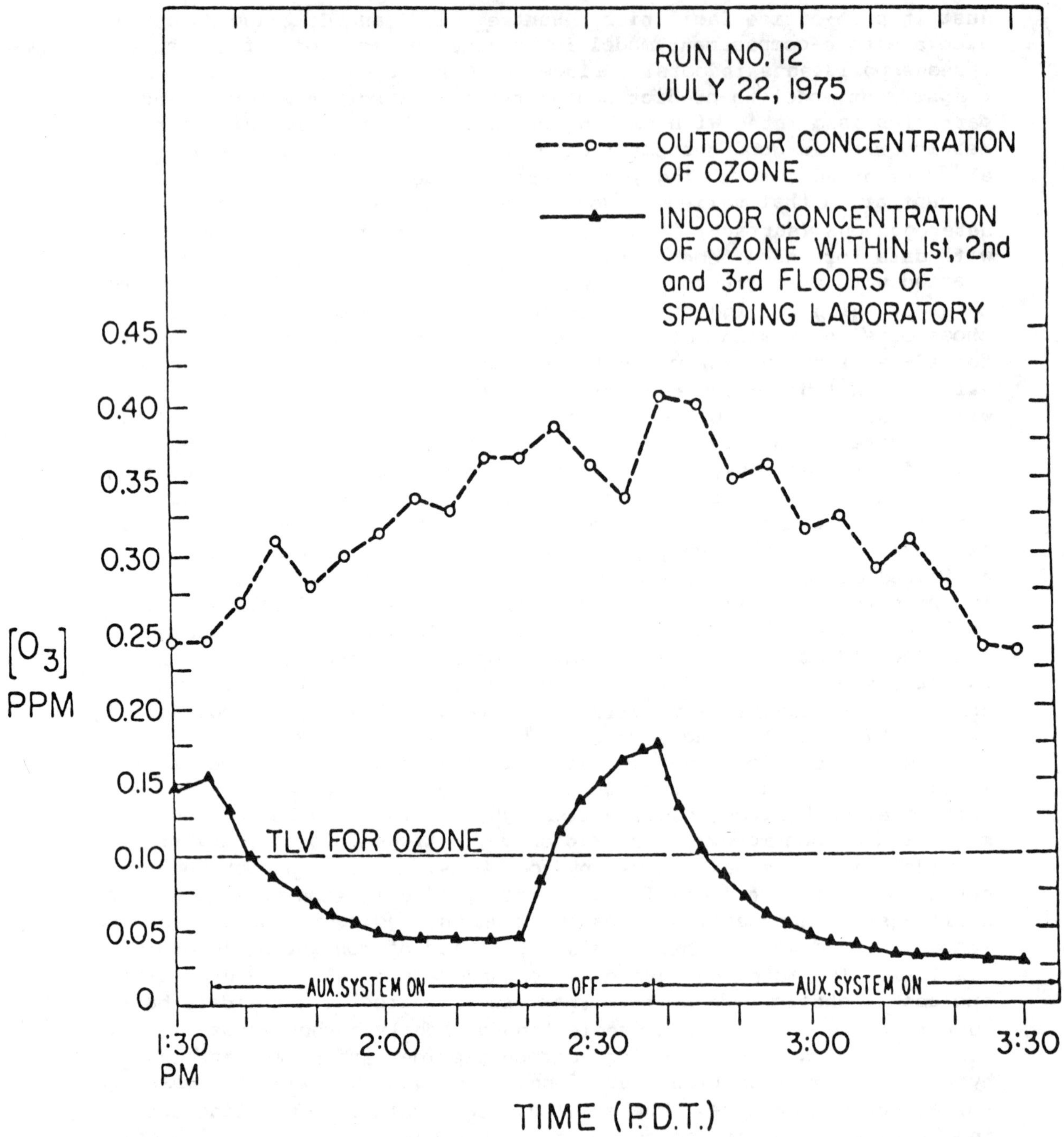

Figure VI-3 Relative indoor and outdoor ozone concentrations. From F. H. Shair (personal communication).

they found good agreement with experimental results by considering the solid surfaces in the bathroom to be a sink while the shower is on and a source shortly after the shower is turned off.

Repace and Lowrey[89] have developed a one-compartment model that describes growth equilibrium and decay of tobacco-smoke aerosol under different room mixing conditions.

Hollowell et al.[42] have discussed the impact of radon on indoor air quality. Kusuda et al.[53] used the available indoor-radon data to develop a single-compartment model with a first-order radioactive-decay term and a constant generation rate:

$$V\frac{dC}{dt} = qC_0 - \lambda VC + VS, \tag{10}$$

in which V is volume, C_0 is outdoor concentration, and C is indoor concentration. In addition to terms previously defined, there is the radon-decay constant $\lambda = 1.258 \times 10^{-4}$/min and the average source strength per unit volume of air, S. Setting the left-hand side of the above equation to zero and solving for the air-exchange rate yields:

$$q/V = (S - \lambda C)/(C - C_0). \tag{11}$$

To facilitate determination of effective radon source strengths, future measurements of indoor-radon concentrations should be accompanied by corresponding outdoor measurements and air-exchange rates, as determined, for example, by a tracer-dilution technique.

SUMMARY AND CONCLUSIONS

The main purpose of an indoor air quality model is to show the relationships of indoor pollutant concentrations to those outside, to geometric and ventilation characteristics of a structure, and to internal sources and sinks.

When the characteristic time for mixing throughout the region of interest is short, compared with the characteristic residence time, the region can be considered as a well-mixed "compartment." Even when that criterion is not met, the uncertainty in the values of the ventilation, source, and sink parameters (with the difficulty of doing otherwise) usually does not justify the development of a more sophisticated model.

Consequently, the starting point for essentially all indoor air pollution models has so far been a first-order differential equation representing a pollutant mass balance in a compartment. In many cases, only the steady-state solution is needed, and the model reduces to an algebraic equation. In a few cases, two or more compartments (usually coupled by internal ventilation streams) have been used to develop a model of indoor air pollution. The main difference between various models arises from choosing different source or sink mechanisms.

To obtain as much general information as possible, the researcher should include the following in the model: corresponding outside concentrations, appropriate geometric parameters, reasonably accurate ventilation rates, and, if possible, sink and source strengths. Such information is required for any model based on a pollutant mass balance.

ESTIMATION OF TOTAL EXPOSURE TO AIR POLLUTION

Today's data on urban air quality come mostly from measurements at fixed monitoring stations. Such data probably show accurately the exposure of a hypothetical person who spends all his time at the station's intake probe. However, people are in constant motion in urban areas, moving from residential areas to places of work to commercial areas, etc. To determine individual human exposure to air pollution accurately, it is necessary to find some means to measure and correlate the movement of individuals in a population and the spatial variation in concentrations of pollutants, whether indoors or outdoors.

One way to estimate better the total individual exposure to environmental pollutants is to equip a large number of persons with monitoring instruments and allow them to go about their daily activities in a normal manner. However, no large-scale personal monitoring studies have been done or are in progress, at least partly because the development of total-exposure monitoring is still in an early stage. Although no large-scale national program to develop personal monitors has evolved, limited funds from federal agencies have resulted in the development of specific monitors, and private companies have developed some instruments that are portable, small, and reasonably priced.[19 69 76]

Other approaches have used theoretical analyses and models for estimating total exposure. Fugas[27] made one of the first attempts to compute total exposure from experimental data; her approach was intended only as an illustrative example. She obtained measurements of average concentrations of lead, manganese, and sulfur dioxide during the winter of 1972-1973 from official air-monitoring stations in the city. The measurements were taken at the breathing zone in several streets during business hours, indoors close to the streets during business hours, and in the countryside. By estimating the time spent by inhabitants of the city in five locations--home, work, street 1, street 2, and the countryside--Fugaš calculated the "weighted weekly exposure" (WWE) for each of these air pollutants (see Table VI-2). An intermediate computation is the "integrated exposure," which is the product of the average concentration and the time during which pollution occurs. To calculate the WWE for sulfur dioxide, for example, we note that a person spent an average of 110 h/wk at home, where the average concentration was 89 $\mu g/m^3$, for an integrated exposure of 9,790 $\mu g\text{-}h/m^3$ for the time spent at home. By adding all the integrated exposure components, Fugaš obtained the total of 16,896 $\mu g\text{-}h/m^3$ for the week. The WWE to sulfur dioxide was then obtained by dividing by the number of hours in a week: 16,896/168 = 101 $\mu g/m^3$.[27]

Duan[23] has modified Fugaš's approach by substituting the term "microenvironment types" for the "locations" used to compute WWE. In Duan's model, a person's integrated exposure over some period (for

TABLE VI-2

Example by Fugaš Illustrating Computation of Weighted Weekly Exposure[a]

Type of Exposure	Duration of Exposure, h/wk	Sulfur Dioxide C	Sulfur Dioxide Ct	Lead C	Lead Ct	Manganese C	Manganese Ct
Home	110	89	9,790	2.5	275	0.04	4.4
Work	42	8	336	0.3	12.6	0.02	0.84
Street 1	10	600	6,000	6.0	60	0.80	8.0
Street 2	4	180	720	3.5	14	0.12	0.48
Countryside	2	25	50	0.1	0.2	0.01	0.02
Total	168	--	16,896	--	361.8	--	13.74
Weighted weekly exposure	--	--	101	--	2.2	--	0.08

[a]Data from Fugaš.[27] Values for C (concentration) are expressed in $\mu g/m^3$, and values for Ct (integrated exposure) are expressed in $\mu g\text{-}h/m^3$.

example, a week) is computed as a weighted average of the exposures from various microenvironment types, weighted by the proportion of time spent in each microenvironment type:

$$E_{ij} = \sum_{k=1}^{K} C_{ijk} t_{ijk} ,$$

where E_{ij} is the integrated exposure of the ith individual during the jth period, c_{ijk} is the average concentration in the kth microenvironment type during the jth period, and t_{ijk} is the activity-pattern coefficient denoting the time the ith individual spent in the kth microenvironment type during the jth period. Duan[23] suggested that a microenvironment type should be defined "finely" enough to be homogeneous; that is, the concentration coefficients should not vary appreciably over the individuals. However, the microenvironment types have to be somewhat "coarse," so that the analyst will not have too many types to deal with. Some types might be "rush-hour highway commuting," "daytime urban office with air-conditioning," and "weekend daytime outdoors in the park."[23]

Moschandreas and Morse[70] have suggested an analogous approach for computing air-pollution exposure and have applied it to real data. They introduced the application of "mobility patterns," which are designed to capture "the daily movements of individuals as they move to and from work, from home to points of amusements, adventure, business, and so on."[70] By examining the literature on activity patterns and "time budgets," they arrived at the estimated time that all persons--all races, ages, socioeconomic groups, workers, students, etc.--spend in various "environmental modes" (which are analogous to the "microenvironment types" of Duan[23] and the "locations" of Fugaš[27]). The population spends 72.8% of its time inside homes, but the figure is different for different population subgroups (workers, children, the elderly, etc.).

Because of the importance of considering the mobility patterns of population subgroups, Moschandreas and Morse[21] examined U.S. census data to define the percentage of each of six subgroups in the total population: housewives, office personnel, industrial workers, outdoor workers, elderly and infirm people, and students. However, students are not considered in the overall model, because few studies have been made of their mobility patterns. The model can be viewed as a three-dimensional drawing in which persons move through time occupying different environmental modes, thereby exposing themselves to the particular concentration that is associated with each environmental mode and period. On the basis of data on typical diurnal ozone concentrations for three environmental modes (residential, office, and indoors) in the Boston area and estimates of the percentage of the population in each environmental mode as a function of time, Moschandreas and Morse[70] estimated that 21% of the population is exposed to ozone at 80 ppb or more. At the time, the federal NAAQS for ozone (1-h average) was 80 ppb, but it has since been raised. Among individual population subgroups, Moschandreas and Morse estimated that outdoor workers are exposed to high ozone concentrations (over 80 ppb) for about 6 h and industrial workers for the 1-h period between 4:00 and 5:00 p.m. If we take a "snapshot" of the estimated population exposures at some particular point in time, such as 3:00 p.m. (that is, 1500; see Figure VI-4), some 25% of the population (3.9% in outdoor activities and 21.3% in transit) is exposed to ozone at 80 ppb or more.

Although these are excellent recent examples of ways to estimate population exposure to air pollution, they have several limitations. Health-related air quality standards usually do not have weekly averaging times. Thus, the value of the WWE calculated by Fugaš[27] cannot be compared directly with existing air quality standards. Duan's formulation[23] was intended to be more flexible, allowing any averaging period (for example, a week or a month) to be used. The computation of exposure by Moschandreas and Morse[70] did not cause difficulties with averaging periods, because ozone has a 1-h NAAQS, and the population is assumed to spend 1-h increments (or multiples of 1 h), in each environmental mode.

Each of the above approaches for calculating exposure gives the population's <u>average exposure</u> over some specified period, and a problem arises from the emphasis on the arithmetic mean. In any given period, some people will be involved in combinations of activities that can result in exposures much higher and much lower than the mean. In addition, the time spent in each activity varies from day to day and from person to person. Thus, the mean value for exposure concentration is not adequate to characterize the highest concentrations to which members of the population are exposed, and the variance of exposures also must be considered. Ideally, there would be an effective technique for determining the entire frequency distribution of exposures of the population to air pollution.

As discussed above, two approaches are used for estimating the frequency distribution of human air-pollution exposures: modeling, which relates the activities of persons as a function of time and the concentrations to which they are exposed; and field studies, which use personal monitors to cover a large enough population sample (or a stratified sample) to represent statistically the distribution of exposures. Although a large-scale field study of exposure has not been completed, efforts are under way to use computer simulation to model human exposures to air pollution.

Ott[77,78] has developed a computer-simulation model of human exposure to air pollution that includes the movements of individual people in a metropolitan area as a series of transitions from one "microenvironment" to another. Ott's computer program, "Simulation of Human Air Pollution Exposures" (SHAPE), uses probability distributions of the time that people spend in each microenvironment--distributions derived from studies of human activity patterns. In each microenvironment, the concentration to which an individual is exposed is treated stochastically, with distributional models that are based on field studies of air-pollutant concentrations reported in the research literature. In the simulation, the computer keeps track of the exposure received by each person as he or she moves forward in time and occupies successive microenvironments.

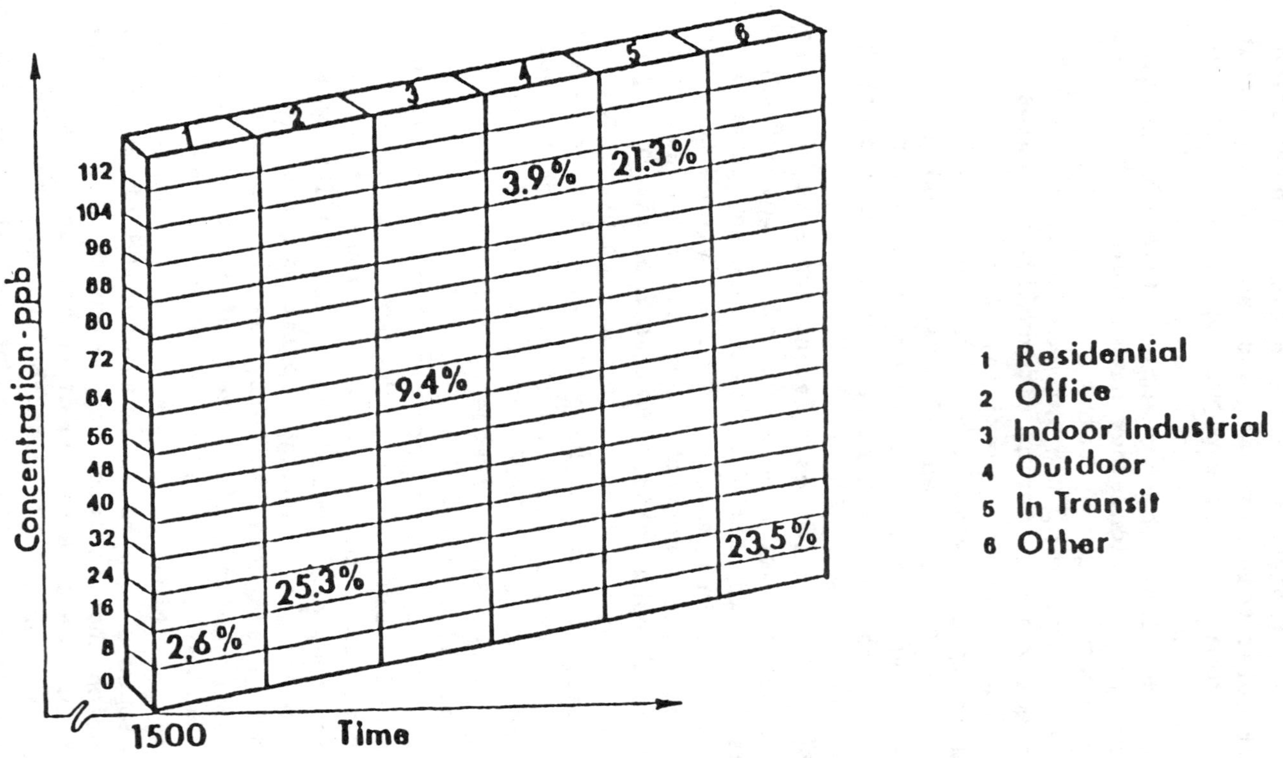

FIGURE VI-4 Example showing proportion of population exposed to hourly ozone concentrations during the 1500 hour (3:00 p.m.) in each environmental mode (numbered above). Reprinted with permission from Moschandreas and Morse.[70]

Ott[77] defines an "exposure" of person i to concentration c statistically as the joint occurrence of two independent events: person i is present in microenvironment j, and the concentration C(j) = c occurs in microenvironment j. C(j) denotes the probability distribution of the concentrations associated with the microenvironment and is based on field monitoring data. Then, the integrated exposure ε_i of person i is computed as the sum of the products of the concentrations encountered in each microenvironment and the time spent there:

$$\varepsilon_i = \sum_{j=1}^{J} c_j t_{ij} ,$$

where c_j is the concentration associated with microenvironment j, t_{ij} is the time spent by person i in microenvironment j, and J is the total number of microenvironments occupied by person i in some period of interest. Conceptually, this model can be represented by a three-dimensional array (see Figure VI-5) that is similar to a three-dimensional space developed by Moschandreas and Morse.[70] However, the computer simulation follows one person at a time through his or her daily activities, and the resulting distribution of exposures is obtained by considering those of all persons in the simulation. The times t_{ij} spent in the microenvironments are variable and do not need to be integer multiples of 1 h.

REFERENCES

1. Amass, C. E. Passive membrane-limited dosimeters using specific ion electrode analysis, pp. 437-460. In D. T. Mage, and L. Wallace, Eds. Proceedings of the Symposium on the Development and Usage of Personal Monitors for Exposure and Health Effect Studies. U.S. Environmental Protection Agency (Environmental Monitoring and Support Laboratory, and Health Effects Research Laboratory) Report No. EPA-600/9-79-032. Washington, D.C.: U.S. Government Printing Office, 1979.
2. American Conference of Governmental Industrial Hygienists. Air Sampling Instruments for Evaluation of Atmospheric Contaminants. 5th ed. Cincinnati: American Conference of Governmental Industrial Hygienists, 1978.
3. Azar, A., R. D. Snee, and K. Habibi. Relationship of community levels of air lead and indices of lead absorption, pp. 581-594. In D. Barth, A. Berlin, R. Engel, P. Recht, and J. Smeets, Eds. Proceedings. International Symposium. Environmental Health Aspects of Lead, Amsterdam, October 2-6, 1972.
4. Bamberger, R. L., G. G. Esposito, B. W. Jacobs, G. E. Podolak, and J. F. Mazur. A new personal sampler for organic vapors. Am. Ind. Hyg. Assoc. J. 39:701-708, 1978.
5. Berlandi, F. J., G. R. Dulude, R. M. Griffin, and E. R. Zink. Electrochemical air lead analysis for personal environmental

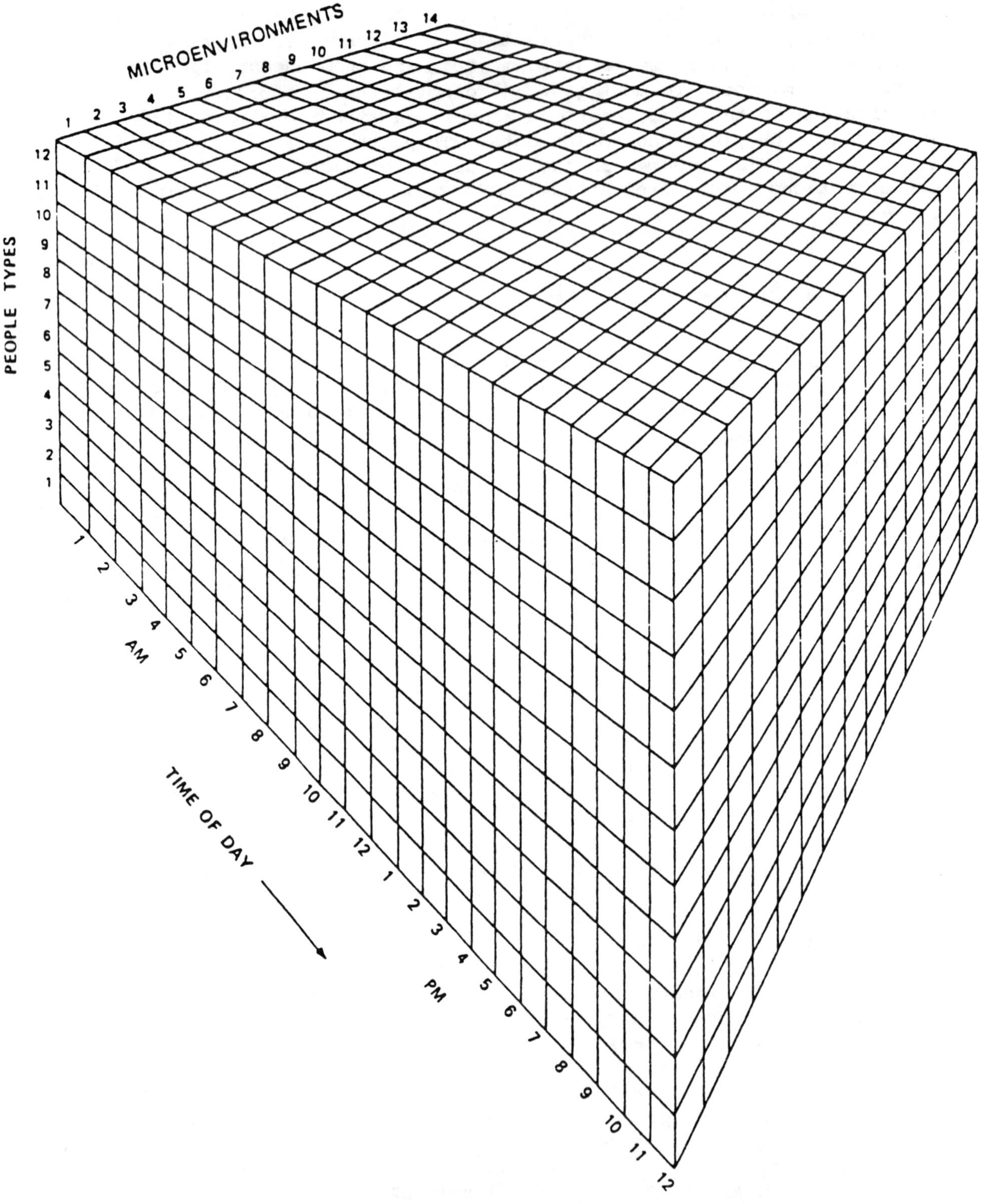

FIGURE VI-5 Graphic representation of person-environment-time array in computer simulation model of exposure to air pollution suggested by Ott.[77]

surveys, pp. 161-172. In D. T. Mage, and L. Wallace, Eds. Proceedings of the Symposium on the Development and Usage of Personal Monitors for Exposure and Health Effect Studies. U.S. Environmental Protection Agency (Environmental Monitoring and Support Laboratory, and Health Effects Research Laboratory) Report No. EPA-600/9-79-032. Washington, D.C.: U.S. Government Printing Office, 1979.

6. Binder, R. E., C. A. Mitchell, H. R. Hosein, and A. Bouhuys. Importance of the indoor environment in air pollution exposure. Arch. Environ. Health 31:277-279, 1976.

7. Bridge, D. P., and M. Corn. Contribution to the assessment of exposure of non-smokers to air pollution from cigarette and cigar smoke in occupied spaces. Environ. Res. 5:192-209, 1972.

8. Brief, R. S. Simple way to determine air contaminants. Air Eng. 2:39-41, 1960.

9. Brooks, J. J., D. S. West, D. J. David, and J. D. Mulik. A combination sorbent system for broad range organic sampling in air, pp. 383-412. In D. T. Mage, and L. Wallace, Eds. Proceedings of the Symposium on the Development and Usage of Personal Monitors for Exposure and Health Effect Studies. U.S. Environmental Protection Agency (Environmental Monitoring and Support Laboratory, and Health Effects Research Laboratory) Report No. EPA-600/9-79-032. Washington, D.C.: U.S. Government Printing Office, 1979.

10. Burgess, W., L. DiBerardinis, and F. E. Speizer. Exposure to automobile exhaust. III. An environmental assessment. Arch. Environ. Health 26:325-329, 1973.

11. Calder, K. L. A Numerical Analysis of the Protection Afforded by Buildings against BW Aerosol Attack. Biological Warfare Laboratories Technical Study No. 2. Fort Detrick, Md.: Office of the Deputy Commander for Scientific Activities, 1957. 24 pp.

12. Calder, K. L. A Simple Mathematical Model for the Penetration of Forest Canopy by Aerosols. Biological Laboratories Technical Study No. 37. Fort Detrick, Md.: Office of the Scientific Director, 1961.

13. Campbell, J. E. The development of a passive dosimeter for airborne aniline vapors. Am. Ind. Hyg. Assoc. J. 41:180-184, 1980.

14. Carnow, B. W., M. H. Lepper, R. B. Shekelle, and J. Stamler. Chicago air pollution study. SO_2 levels and acute illness in patients with chronic bronchopulmonary disease. Arch. Environ. Health 18:768-776, 1969.

15. Constance, J. D. Mixing factor is guide to ventilation. Power 114(2):56-57, 1970.

16. Cortese, A. D., and J. D. Spengler. Ability of fixed monitoring stations to represent carbon monoxide exposure. J. Air Pollut. Control Assoc. 26:1144-1150, 1976.

17. Cote, W. A., and K. Holcombe. The influence of air conditioning systems on indoor pollutant levels, pp. 1-20, Session 5, Paper 3. In J. W. White, Ed. Proceedings of the First Conference on Natural Gas Research and Technology. Chicago: Institute of Gas Technology, 1971.

18. Dockery, D. W, and J. D. Spengler. Personal exposure to respirable particulates and sulfates: Measurement and prediction, pp. 111-129 (includes discussion). In D. T. Mage and L. Wallace, Eds. Proceedings of the Symposium on the Development and Usage of Personal Monitors for Exposure and Health Effect Studies. U.S. Environmental Protection Agency (Environmental Monitoring and Support Laboratory, and Health Effects Research Laboratory) Report No. EPA-600/9-79-032. Washington, D.C.: U.S. Government Printing Office, 1979.

19. Dockery, D. W., and J. D. Spengler. Personal exposure to respirable particulates and sulfates versus ambient measurements. Paper No. 77-44.6 presented at the 70th Annual Meeting of the Air Pollution Control Association, June 20-24, 1977, Toronto, Canada.

20. Dockery, D. W., J. D. Spengler, M. P. Reed, and J. Ware. Relationships Among Personal, Indoor and Outdoor NO_2 Measurements. Paper presented at the 73rd Annual Meeting of the Air Pollution Control Association, Montreal, Canada, June 23-27, 1980.

21. Dorman, R. G. Filtration, pp. 195-222. In C. N. Davies, Ed. Aerosol Science. New York: Academic Press, Inc., 1966.

22. Drivas, P. J., P. G. Simmonds, and F. H. Shair. Experimentation characterization of ventilation systems in buildings. Environ. Sci. Technol. 6:609-614, 1972.

23. Duan, N. Microenvironment Types: A Model for Human Exposures to Air Pollution. SIMS Technical Report. Stanford, Cal.: Stanford University, Department of Statistics, 1981. (in preparation)

24. Environmental Measurements Incorporated. Radon Sampling, pp. 6-8. Brochure No. 2, Annapolis, Md.: Environmental Measurements Incorporated, 1978.

25. Fleischer, R. L., W. A. Giard, A. Mogro-Campero, L. G. Turner, H. W. Alter, and J. E. Gingrich. Dosimetry of environmental radon: Methods and theory for low-dose, integrated measurements. Health Physics 39:957-962, 1980.

26. Fuchs, N. A. The Mechanics of Aerosols. New York: Pergamon Press, 1964. 408 pp.

27. Fugaš, M. Assessment of total exposure to an air pollutant. Paper no. 38.5 in Proceedings of the International Conference on Environmental Sensing and Assessment, Vol. 2, September 14-19, 1975, Las Vegas, Nevada. Institute of Electrical and Electronics Engineers Publication No. #75-CH 1004-1 ICESA.

28. Fugaš, M., B. Wilder, R. Pauković, J. Hršak, and D. Steiner-Škreb. Concentration levels and particle size distribution of lead in the air of an urban and an industrial area as a basis for the calculation of population exposure, pp. 961-968. In D. Barth, A. Berlin, R. Engel, P. Recht, and J. Smeets, Eds. Proceedings. International Symposium on Environmental Health Aspects of Lead, Amsterdam, October 2-6, 1972. Luxembourg: Commission of the European Communities, Center for Information and Documentation, 1973.

29. George, A. C. A passive environmental radon-monitor, pp. 25-30. In A. J. Breslin, Ed. Radon Workshop, February 1977. Energy Research and Development Administration, Health and Safety

Laboratory Report HASL-325 (CONF. 770231). New York: U.S. Energy Research and Development Administration, 1977.

30. Georgii, H.-W. Investigation of the air exchange between rooms and the air outside. Arch. Meteorol. Geophys. Bioklimatol. (Ser. B) 5:191-214, 1954. (in German; English summary)

31. Gilath, C. Ventilation and air pollution studies using radioactive tracers. A critical review. Int. J. Appl. Radiat. Isot. 28:847-854, 1977.

32. Gillespie, J. C., and L. B. Daniel. A new sampling tool for monitoring exposures to toxic gases and vapors, pp. 479-491. In D. T. Mage and L. Wallace, Eds. Proceedings of the Symposium on the Development and Usage of Personal Monitors for Exposure and Health Effect Studies. U.S. Environmental Protection Agency (Environmental Monitoring and Support Laboratory, and Health Effects Research Laboratory) Report No. EPA 600/9-79-032. Washington, D.C.: U.S. Government Printing Office, 1979.

33. Gold, A., T. J. Smith, C. E. Dube, and J. J. Cafarella. Solid sorbent for acrolein and formaldehyde in air, pp. 425-435. In D. T. Mage and L. Wallace, Eds. Proceedings of the Symposium on the Development and Usage of Personal Monitors for Exposure and Health Effect Studies. U.S. Environmental Protection Agency (Environmental Monitoring and Support Laboratory, and Health Effects Research Laboratory) Report No. EPA-600/9-79-032. Washington, D.C.: U.S. Government Printing Office, 1979.

34. Goldmuntz, L. A. Extent to Which EPA Monitoring Station Measurements Reflect Human Exposure to Carbon Monoxide. National Science Foundation Grant No. STP 75-21384. Washington, D.C.: Economics & Science Planning, Inc., 1976.

35. Gosselink, D. W., D. L. Braun, H. E. Mullins, and S. T. Rodriguez. A new personal organic vapor monitor with in situ sample elution, pp. 365-382. In D. T. Mage and L. Wallace, Eds. Proceedings of the Symposium on the Development and Usage of Personal Monitors for Exposure and Health Effect Studies. U.S. Environmental Protection Agency (Environmental Monitoring and Support Laboratory, and Health Effects Research Laboratory) Report No. EPA-600/9-79-032. Washington, D.C.: U.S. Government Printing Office, 1979.

36. Green, H. L., and W. R. Lane. Particulate Clouds: Dusts, Smokes and Mists. 2nd ed. Belfast: Universities Press, 1964. 479 pp.

37. Hardy, J. K., P. K. Dasgupta, K. D. Reiszner, and P. W. West. A personal chlorine monitor utilizing permeation sampling. Environ. Sci. Technol. 13:1090-1093, 1979.

38. Harrison, O. Development Strategy for Pollutant Dosimetry. U.S. Environmental Protection Agency Report No. 600/2-76-034. Research Triangle Park, N.C.: U.S. Environmental Protection Agency, 1976.

39. Harrje, D. T., C. M. Hunt, S. J. Tredo, and N. J. Malik. Automated Instrumentation for an Infiltration Measurement in Buildings. Center for Environmental Studies Report No. 13. Princeton, N. J.: Princeton University, 1975.

40. Hitchin, E. R., and C. B. Wilson. A review of experimental techniques for the investigation of natural ventilation in buildings. Build. Sci. 2:59-82, 1967.

41. Hollingdale-Smith, P. A. Authors' reply. Ann. Occup. Hyg. 22:86, 1979.
42. Hollowell, C. D., J. V. Berk, and G. W. Traynor. Impact of reduced infiltration and ventilation on indoor air quality. ASHRAE J. 21(7):49-53, 1979.
43. Horiuchi, H. Radon-222 and its daughters in buildings at Uranium City, Saskatchewan. In J. E. Turner, C. F. Holoway, and A. S. Loebl, Eds. Radon Workshop on Dosimetry for Radon and Radon Daughters. Oak Ridge National Laboratories Report No. ORNL-4348.
44. Hunt, C. M. The Control of Airborne Dust in Post Office Facilities. Final Report. NBS Project 4213417. Bureau of Research and Engineering. Washington, D.C.: Post Office Department, 1970.
45. Hunt, C. M., and D. M. Burch. Air infiltration measurements in a four-bedroom townhouse using sulfur hexafluoride as a tracer gas. ASHRAE Trans. 81 (Pt. 1):186-201, 1975. (includes discussion)
46. Hunt, C. M., B. C. Cadoff, and F. J. Powell. Indoor Air Pollution Status Report. National Bureau of Standards Report 10 591. Washington, D.C.: U.S. Department of Commerce, National Bureau of Standards, Building Research Division, 1971.
47. International Energy Agency. Methods of measuring infiltration and building tightness, pp. 22-48. In Draft Program Plan. Air Infiltration in Buildings. U.S. Department of Energy Report No. DOE/CS-0099-D. Washington, D.C.: U.S. Department of Energy, Division of Buildings and Community Systems, 1979.
48. Intersociety Committee. Methods of Air Sampling and Analysis, pp. 117-129. Washington, D.C.: American Public Health Association, 1972.
49. Jackson, D. L., and V. A. Newill. The strengths and weaknesses of population studies in assessing environmental health effects, pp. 161-175. In Proceedings. International Symposium. Recent Advances in the Assessment of the Health Effects of Environmental Pollution, Paris, (France), June 24-28, 1974. Vol. I. Luxembourg: Commission of the European Communities, Directorate General Scientific and Technical Information and Information Management, 1975.
50. Kahn, A., R. B. Rutledge, G. L. Davis, J. A. Altes, G. E. Gantner, C. A. Thornton, and N. D. Wallace. Carboxyhemoglobin sources in the metropolitan St. Louis population. Arch. Environ. Health 29:127-135, 1974.
51. Kusuda, T. Control of ventilation to conserve energy while maintaining acceptable indoor air quality. ASHRAE Trans. 82(Pt. 1):1169-1181, 1976.
52. Kusuda, T., C. M. Hunt, and P. E. McNall. Radioactivity (radon and daughter products) as a potential factor in building ventilation. ASHRAE J. 21(7):30-34, 1979.
53. Lawrence Berkeley Laboratory, Environmental Instrumentation Group. Gases, pp. 1-51. In Instrumentation for Environmental Monitoring. Vol. I. Air. Lawrence Berkeley Laboratory Report LBL-1. Berkeley, Cal.: Lawrence Berkeley Laboratory, 1976.
54. Lidwell, O. M., and J. E. Lovelock. Some methods of measuring ventilation. J. Hyg. 44:326-332, 1946.

55. Linch, A. L. Evaluation of Ambient Air Quality by Personnel Monitoring. Cleveland, Ohio: CRC Press, Inc., 1974. 226 pp.
56. Lippmann, M. "Respirable" dust sampling, pp. G-1--G-23. In Air Sampling Instruments for Evaluation of Atmospheric Contaminants. 5th ed. Cincinnati: American Conference of Governmental Industrial Hygienists, 1978.
57. Liu, B. Y. H., and K .W. Lee. Efficiency of membrane and nucleopore filters for submicrometer aerosols. Environ. Sci. Technol. 10:345-350, 1976.
58. Loo, B. W., R. S. Adachi, C. P. Cork, F. S. Goulding, J. M. Jaklevic, D. A. Landis, and W. L. Searles. A Second Generation Dichotomous Sampler for Large-Scale Monitoring of Airborne Particulate Matter. Lawrence Berkeley Laboratory Report No. LBL-8725. Berkeley, Cal.: Lawrence Berkeley Laboratory, 1979. 18 pp.
59. Lucas, H. F. Alpha scintillation radon counting, pp. 69-156. In E. D. Harward, Ed. Workshop on Methods for Measuring Radiation in and around Uranium Mills, May 23-26, Albuquerque, New Mexico. Program Report 3-9. Washington, D.C.: Atomic Industrial Forum, 1977.
60. Mage, D. T., and L. Wallace, Eds. Proceedings of the Symposium on the Development and Usage of Personal Monitors for Exposure and Health Effect Studies, January 22-24, 1979, Chapel Hill, N.C. U.S. Environmental Protection Agency (Environmental Monitoring and Support Laboratory, and Health Effects Research Laboratory) Report No. EPA-600/9-79-032. Washington, D.C.: U.S. Government Printing Office, 1979. 525 pp.
61. Marple, V. A. Simlation of respirable penetration characteristics by inertial impaction. J. Aerosol Sci. 9:125-134, 1978.
62. Mazur, J. F., R. L. Bamberger, G. E. Podolak, and G. G. Esposito. Development and evaluation of an ammonia dosimeter. Am. Ind. Hyg. Assoc. J. 39:749-753, 1978.
63. McDermott, D. L., K. D. Reiszner, and P. W. West. Development of long-term sulfur dioxide monitor using permeation sampling. Environ. Sci. Technol. 13:1087-1090, 1979.
64. Miller, M. E. Laboratory Fumehoods and the Exhaust Gas Re-entry Problem. Pasadena: California Institute of Technology, Department of Chemical Engineering, Master of Science thesis, 1979.
65. Milly, G. H. A Theory of Chemical Attack of Tanks and Enclosed Fortifications. Report for the Chemical Corps. Edgewood Arsenal, Md.: Army Chemical Center, Chemical and Radiological Laboratories, 1953.
66. Milly, G. H., and S. D. Thayer. Techniques for Dosage Prediction. Vol. I. Development of Techniques. Travelers Research Center Report No. TRC247, 1967.
67. Morgan, M. G., and S. C. Morris. Individual Air Pollution Monitors: An Assessment of National Research Needs. Brookhaven National Laboratory Report No. 50482. Upton, N.Y.: Brookhaven National Laboratory, 1976. 35 pp.
68. Morgan, M. G., and S. C. Morris. Individual Air Pollution Monitors. 2. Examination of Some Nonoccupational Research and

Regulatory Uses and Needs. Brookhaven National Laboratory Report No. 50637. Upton, N.Y.: Brookhaven National Laboratory, 1977. 11 pp.

69. Moschandreas, D. J., Ed. Indoor Air Pollution in the Residential Environment. Volume II. Field Monitoring Protocol, Indoor Episodic Pollutant Release Experiments and Numerical Analyses. U.S. Environmental Protection Agency (Environmental Monitoring and Support Laboratory) Report No. EPA-600/7-78-229b. Research Triangle Park, N.C.: U.S. Environmental Protection Agency, 1978.

70. Moschandreas, D. J., and S. S. Morse. Exposure estimation and mobility patterns. Paper No. 79-14.4 presented at the 72nd Annual Meeting of the Air Pollution Control Association, Cincinnati, Ohio, June 24-29, 1979.

71. Moschandreas, D. J., D. J. Pelton, D. J. Sibbett, J. W. C. Stark, and J. E. McFadden. Comparison of Indoor-Outdoor Concentrations of Atmospheric Pollutants. Field Monitoring Protocol. Scientific Report RP1309. GEOMET Report No. E-721. Gaithersburg, Md.: GEOMET, Inc., July 1978. 103 pp.

72. Moschandreas, D. J., and J. W. C. Stark. The GEOMET Indoor-Outdoor Air Pollution Model. U.S. Environmental Protection Agency Report No. EPA-600/7-78-106. Research Triangle Park, N.C.: U.S. Environmental Protection Agency, Environmental Monitoring and Support Laboratory, 1978. 75 pp.

73. Moschandreas, D. J., J. W. C. Stark, J. E. McFadden, and S. S. Morse. Indoor Air Pollution in the Residential Environment. Final Report for U.S. Environmental Protection Agency, Environmental Research Center and U.S. Department of Housing and Urban Development, Office of Policy Development and Research under EPA Contract No. 68-02-2294, GEOMET Report No. EF-688. Gaithersburg, Md.: GEOMET, Inc., 1978.

74. National Research Council, Committee on Biologic Effects of Atmospheric Pollutants. Lead. Airborne Lead in Perspective, p. 216. Washington, D.C.: National Academy of Sciences, 1972.

75. Nelms, L. H., K. D. Reiszner, and P. W. West. Personal vinyl chloride monitoring device with permeation technique for sampling. Anal. Chem. 49:994-998, 1977.

76. Ott, W. R. A Field Quality Assurance Evaluation of Two Personal Monitors for Carbon Monoxide. Stanford, Cal.: Stanford University, Department of Statistics. (unpublished report)

77. Ott, W. R. Concepts of Human Exposure to Environmental Pollution. SIMS Technical Report No. 32. Stanford, Cal.: Stanford University, Department of Statistics, 1980.

78. Ott, W. R. Development of Activity Pattern Models for Human Exposure Monitoring. Innovative Research Program Proposal to U.S. Environmental Protection Agency, July 1, 1979.

79. Ott, W. R., and D. T. Mage. A method for simulating the true human exposure of critical population groups to air pollutants, pp. 2097-2107. In Proceedings. International Symposium. Recent Advances in the Assessment of the Health Effects of Environmental Pollution. Paris, (France), June 24-28, 1974. Vol. IV. Luxembourg: Commission of the European Communities, Directorate General Scientific and Technical Information and Information Management, 1975.

80. Palmes, E. D. Effect of wind on diffusion samplers. Ann. Occup. Hyg. 22:85, 1979.

81. Palmes, E. D. Personal Samplers for CO, NO and NO_2 in Air. U.S. Bureau of Mines Open File Report 92-77. Washington, D.C.: U.S. Department of Interior, Bureau of Mines, 1977. 22 pp.

82. Palmes, E. D. Personal sampler for measurement of ambient levels of NO_2, pp. 57-64. In D. T. Mage and L. Wallace, Eds. Proceedings of the Symposium on the Development and Usage of Personal Monitors for Exposure and Health Effect Studies. U.S. Environmental Protection Agency (Environmental Monitoring and Support Laboratory, and Health Effects Research Laboratory) Report No. EPA-600/9-79-032. Washingon, D.C.: U.S. Government Printing Office, 1979.

83. Palmes, E. D., and A. F. Gunnison. Personal monitoring device for gaseous contaminants. Am. Ind. Hyg. Assoc. J. 34:78-81, 1973.

84. Palmes, E. D. A. F. Gunnison, J. DiMattio, and C. Tomczyk. Personal sampler for nitrogen dioxide. Am. Ind. Hyg. Assoc. J. 37:570-577, 1976.

85. Palmes, E. D., and C. Tomczyk. Personal sampler for NO_x. Am. Ind. Hyg. Assoc. J. 40:588-591, 1979.

86. Palmes, E. D., C. Tomczyk, and J. DiMattio. Average NO_2 concentrations in dwellings with gas or electric stoves. Atmos. Environ. 11:869-872, 1977.

87. Pich, J. Theory of aerosol filtration by fibrous and membrane filters, pp. 223-285. In C. N. Davies, Ed. Aerosol Science. New York: Academic Press, Inc., 1966.

88. Reiszner, K. D., and P. W. West. Collection and determination of sulfur dioxide incorporating permeation and West-Gaeke procedure. Environ. Sci. Technol. 7:526-532, 1973.

89. Repace, J. L., and A. H. Lowrey. Indoor air pollution, tobacco smoke, and public health. Science 208:464-472, 1980.

90. Sabersky, R. H., D. A. Sinema, and F. H. Shair. Concentrations, decay rates, and removal of ozone and their relation to establishing clean indoor air. Environ. Sci. Technol. 7:347-353, 1973.

91. Saltzman, B. E. Direct reading colorimetric indicators, pp. S-1--S-39. In Air Sampling Instruments for Evaluation of Atmospheric Contaminants. 5th ed. Cincinnati: American Conference of Governmental Industrial Hygienists, 1978.

92. Sasaki, E., J. Schienle, T. Shibata, and F. H. Shair. Influence of meteorology upon the reentry of fumehood exhaust. Pasadena: California Institute of Technology, Department of Chemical Engineering, 1981. (unpublished)

93. Schuette, F. J. Plastic bags for collection of gas samples. Atmos. Environ. 1:515-519, 1967.

94. Shair, F. H. Relating Indoor Pollutant Concentrations of Ozone and Sulfur Dioxide to Those Outside: A. Determination of the Rate Constants Associated with the Heterogeneous Losses of Ozone and Sulfur Dioxide inside Laboratory-Office Buildings, B. Economic Reduction of Indoor Concentrations of Smog Pollutants through Selective Filtration of the Make-up Airstream. Research Project

142 Financial Report to ASHRAE. American Society of Heating, Refrigerating and Air-Conditioning Engineers Technical Paper No. 2618. Chicago: American Society of Heating, Refrigerating and Air-Conditioning Engineers, Inc., 1979. 16 pp. plus appendices. ASHRAE Trans. (in press, 1981)

95. Shair, F. H., and K. L. Heitner. A theoretical model for relating indoor pollutant concentrations to those outside. Environ. Sci. Technol. 8:444-451, 1974.

96. Shair, F. H., D. W. Wolbrink, L. O. Bowen, C. E. Neelley, and K. E. Sampsel. Influence of mechanical ventilation on moisture content of bathroom air. ASHRAE J. 21(7):54-60, 1979.

97. Shaw, R. W. and R. K. Stevens. The tandem filter package, pp. 131-143. In D. T. Mage and L. Wallace, Eds. Proceedings of the Symposium on the Development and Usage of Personal Monitors for Exposure and Health Effect Studies. U.S. Environmental Protection Agency (Environmental Monitoring and Support Laboratory, and Health Effects Research Laboratory) Report No. EPA-600/9-79-032. Washington, D.C.: U.S. Government Printing Office, 1979.

98. Spitz, H. B., and M. E. Wrenn. The Diurnal Variation of the Radon-222 Concentrations in Residential Structures in Grand Junction, Colorado. Second Workshop on the Natural Radiation Environment, 1974.

99. Stewart, R. D., E. D. Baretta, L. R. Platte, E. B. Stewart, J. H. Kalbfleisch, B. Van Yserloo, and A. R. Rimm. Carboxyhemoglobin levels in American blood donors. J. Am. Med. Assoc. 229:1187-1195, 1974.

100. Stewart, R. D., C. L. Hake, A. Wu, T. A. Stewart, and J. H. Kalbfleisch. Carboxyhemoglobin trend in Chicago blood donors, 1970-1974. Arch. Environ. Health 31:280-286, 1976.

101. Turk, A. Measurements of odorous vapors in test chambers: Theoretical. ASHRAE J. 5(10):55-58, 1963.

102. Turner, W. A., J. D. Spengler, D. W. Dockery, and S.D. Colome. Design and performance of a reliable personal monitoring system for respirable particulates, pp. 101-109. In D. T. Mage and L. Wallace, Eds. Proceedings of the Symposium on the Development and Usage of Personal Monitors for Exposure and Health Effect Studies. U.S. Environmental Protection Agency (Environmental Monitoring and Support Laboratory, and Health Effects Research Laboratory) Report No. EPA 600/9-79-032. Washington, D.C.: U.S. Government Printing Office, 1979.

103. U.S. Environmental Protection Agency, Environmental Monitoring Systems Laboratory. List of Designated Reference and Equivalent Methods. Research Triangle Park: U.S. Environmental Protection Agency, February 15, 1980. 22 pp.

104. Wallace, L. Personal air quality monitors: Uses in studies of human exposure, pp. 7-18. In D. T. Mage and L. Wallace, Eds. Proceedings of the Symposium on the Development and Usage of Personal Monitors for Exposure and Health Effect Studies. U.S. Environmental Protection Agency (Environmental Monitoring and Support Laboratory, and Health Effects Research Laboratory) Report No. EPA-600/9-79-032. Washington, D.C.: U.S. Government Printing Office, 1979.

105. West, P. W., and K. D. Reiszner. Field tests of a permeation-type personal monitor for vinyl chloride. Am. Ind. Hyg. Assoc. J. 39:645-650, 1978.

106. West, P. W., and K. D. Reiszner. Personal Monitor for Nitrogen Dioxide. U.S. Environmental Protection Agency Report No. EPA-600/2-78-001. Research Triangle Park, N.C.: U.S. Environmental Protection Agency, 1978. 33 pp.

107. West, P. W., and K. D. Reiszner. Personal monitoring by means of gas permeation, pp. 461-471. In D. T. Mage, and L. Wallace, Eds. Proceedings of the Symposium on the Development and Usage of Personal Monitors for Exposure and Health Effect Studies. U.S. Environmental Protection Agency (Environmental Monitoring and Support Laboratory, and Health Effects Research Laboratory) Report No. EPA 600/9-79-032. Washington, D.C.: U.S. Government Printing Office, 1979.

108. Willeke, K. Selection and design of an aerosol sampler simulating respirable penetration. Am. Ind. Hyg. Assoc. J. 39:317-321, 1978.

109. Woods, J. E. Ventilation, health and energy consumption: A status report. ASHRAE J. 21(7):23-27, 1979.

110. Woods, J. E., R. G. Nevins, and E. L. Besch. Analysis of thermal and ventilation requirements for laboratory animal cage environments. ASHRAE Trans. 81(Pt. 1):45-66, 1975.

111. Wright, G. R., S. Jewczyk, J. Onrot, P. Tomlinson, and R. J. Shephard. Carbon monoxide in the urban atmosphere. Hazards to the pedestrian and the street-worker. Arch. Environ. Health 30:123-129, 1975.

112. Zukoski, E. E., and T. Kubota. Two-layer modeling of smoke movement in building fires. Fire Mater. 4:17-27, 1980.

VII

HEALTH EFFECTS OF INDOOR POLLUTION

INTRODUCTION

The Committee, charged with characterizing the quality of the indoor environment and determining the potential adverse health effects of pollutants in that environment, selected the following pollutants for detailed discussion: radon and radon progeny, formaldehyde and other organic substances, fibrous building materials, combustion products (resulting from combustion of fuels in space-heating, water-heating, cooking, hobbies and crafts, etc.), involuntary smoking, airborne agents of contagion, and airborne allergens. These are obviously only examples of hazardous pollutants. They were chosen because there was a large volume of published material available on the sources of their presence indoors that could be used to document the adverse effects of human exposure to them. The sources of these and other pollutants are described in Chapter IV; the biologic responses to the selected pollutants are discussed here.

It is beyond the scope of this report to list all the pollutants found indoors that are hazardous to human health. Some pollutant sources have been known for a long time but only recently recognized as important. Cigarette-smoking is an example; although the smoke components that cause adverse health effects need more study, considerable progress has been made, as reported in this chapter. The examples given in this chapter make it plain that humans are exposed to a variety of potentially hazardous indoor pollutants from diverse sources. It is hoped that this report will encourage researchers to broaden the list of hazardous indoor pollutants and to characterize the hazards, so that the general public and those responsible for pollution control and abatement can be informed.

Throughout this report, pollutants are mentioned without discussion of their health effects. This does not constitute an oversight on the part of the Committee, but rather reflects a decision that the discussion here be adequate to show that there are indoor pollutants that cause adverse health effects in humans. The reader's attention is directed to Chapter III, which offers some recommendations for further health research with respect to these pollutants, for further exposure studies, and for public education about effective ways of reducing exposure to many contaminants encountered indoors.

Pollutants are inhaled, ingested, and absorbed. They may have effects at their first point of contact with the body, or they may affect internal organs. They may be changed physically or chemically (metabolically) in the process of exerting their effects, or they may undergo intermediate physical or metabolic changes before exerting an effect. They may be stored in tissue for a time and be released later; many of them are eventually excreted. Their own behavior helps to shape the mechanisms of their effects. Pollutants may act independently, antagonistically, or synergistically.

Inhalation is generally the most important route by which toxic substances enter the body. Inhaled substances may exert their effects in the lungs, or they may pass from the lungs to other organ systems in blood, lymph, etc. Ingestion is far less common than inhalation as a route of exposure, but is important for some toxic substances, such as lead, arsenic, and mercury. In addition to the direct physical or chemical effect of ingested substances in the gastrointestinal tract, they may pass through the tract into the blood and be distributed to other organs. Liquid and vapor-phase pollutants may be absorbed through the skin and affect the skin, pass through the skin and then conjugate with tissue protein, or enter the bloodstream and be distributed further.[13,21,24,25]

Environmental agents may exert their effects either by physical or by chemical-physiologic (enzymatic) means. The full toxic potential of most substances is usually not expressed in normal healthy people, because of the body's defense mechanisms and mechanisms of elimination or because the substances are sequestered in inactive forms at various tissue sites (bone, skin, hair, and nails). However, impairment of the body's defensive processes may lead to increased toxicity, owing to the higher concentrations of the substances that build up when the usual means of elimination or reduction are blocked. Effects can occur metabolically at the cell or organ level. Various trace substances (e.g., halogenated hydrocarbons and trace metals) can have their effects at both levels.[13,21,24,25]

Some physical signs give evidence of primary toxicity, such as contact with substances that produce irritation, inflammation, or contraction. Some gases, such as carbon monoxide and nitrogen dioxide, when inhaled can affect the body's capacity to absorb oxygen.

Secondary mechanisms of toxicity include metabolic alteration of the substance and accumulation of the byproducts from the initial action of the pollutant. Some substances are detoxified by metabolic processes (oxidation, reduction, and synthesis), and the detoxification mechanisms may themselves cause damage, as in the oxidation of alcohol to formaldehyde and the reduction of arsenic or manganese, which may produce more toxic forms.

Respiratory effects can be directly attributed to only a few pollutants encountered at high concentrations indoors: nitrogen dioxide, carbon monoxide, formaldehyde, and probably particles are important in this regard.

Physical factors (such as temperature, humidity, noise, nonionizing radiation, and light) and their effects are discussed in Chapters IV and VIII; knowledge of their effects in the indoor environment is sketchy and difficult to assess. Information on the health effects of pollution due to consumer products in general has the same limitations and is treated in the same way.

A variety of trace metals may be present indoors as a result of filtration of outdoor air and as a result of indoor sources of pollutants. These trace metals are also found in the domestic water and in the diet. Some of them, especially lead and mercury, have adverse health effects.[5,9,15] Exposure to mercury indoors may result from spills of liquid mercury and deterioration of paint. Mercury vapor is quickly and efficiently absorbed by the lung and may be absorbed through the skin.[22] Although much of the body burden of lead may come from the diet, the combined effects of air, soil, house dust, and water as sources of indoor lead are appreciable.[1,17,26]

The effects of lead and mercury on the brain are well known.[1,3,8,10,17-20,22,23,26-28] Behavioral dysfunctions caused by lead may occur through modification of the enzymatic response to a wide variety of toxic agents and through interference with neuromuscular and ganglionic transmission.[3,10]

Gastrointestinal symptoms may be produced by inhalation of toxic substances, such as lead and mercury, that reach the gastrointestinal tract through the bile duct.[4,22] Organic mercury is also hepatotoxic and may cause kidney damage by destroying cells in the tubular system.[27] Lead and arsenic deposited in the kidney at low concentrations may produce sensitization to damage by endotoxins or exotoxins, such as analgesics and bacteria, although this is still debatable.[16] Mercuric chloride may produce acute renal failure.[22] Mercury has toxic effects on the thyroid and therefore may have further systemic effects.[11] Cadmium interacts with other nutrients and may be stored in the kidney and damage capillaries there.[7,19,27,28] It also accumulates in the liver at concentrations that depend on age and smoking habits.[7] Lead can inhibit heme synthesis,[13] especially in school-age children. Lead, zinc, and delta-aminolevulinic acid (ALA-D) interact, and porphyrins (free erythrocyte porphyrins and zinc protoporphyrins) are active in the blood; that activity determines the influence of lead on heme synthesis.[27(pp.217-232)] Lead may increase the inhibition of ALA-D in erythrocytes, shorten erythrocyte life span, and produce reticulocytosis or anemia.[4,8,11] It may also increase hypertension and vascular disease.[11]

Lead is stored in the body and has effects related to its storage or its release.[12] Deposition occurs in soft tissue and bone tissue--predominantly in the latter. Effects may occur in those tissues, but often occur systemically on release of deposited lead or when the body burden becomes too great.[11-13] Release may be caused by acidosis or fractures. The lead in the soft tissues causes enzyme inhibition,[12] which in turn can lead to interactions of toxins.[6]

Mercury is a general sensory irritant. It may produce skin burns,[2] rash,[22] excessive perspiration, easy blushing, partial loss of scalp hair,[22] or a decrease in hearing.[11] It can affect taste, and it produces irritation in the mouth.[11 22] Mercury poisoning may affect the sense of touch, owing to the swelling of all extremities, including ears and nose.[14] Except for spills of inorganic mercury and excessive use of mercury-based paint, it is debatable whether indoor concentrations of mercury are ever high enough to produce those effects.[15]

This chapter deals with the biologic responses to specific pollutants and biologic agents. The pollutants discussed are sidestream cigarette smoke, radon progeny, mineral and vitreous fibers, formaldehyde, and products of indoor combustion (predominantly carbon monoxide and nitrogen oxides). Gases not usually found indoors in moderate or high concentrations--such as sulfur oxides and ozone--are not discussed at length. Nor are sources like cooking, which may produce some particles or hydrocarbons, but about which little is known. For information on substances that are known to have adverse effects in the occupational environment or on solvents, dusts, etc., which have been reviewed thoroughly, the reader is referred to the published literature (e.g., reports issued by FDA and CPSC). Environmental factors that are not known to have adverse biologic impact are not discussed here; rather, there are appropriate references to other chapters.

REFERENCES

1. Angle, C. R., and M. S. McIntire. Environmental lead and children: The Omaha study. J. Toxicol. Environ. Health 5:855-870, 1979.
2. Berkout, P. G., N. J. Paterson, A. C. Ladd, and L. J. Goldwater. Treatment of skin burns due to alkyl mercury compounds. Arch. Environ. Health 3:592-593, 1961.
3. Bull, R. J. Effects of trace metals and their derivatives on the control of brain energy metabolism, pp. 425-440. In S. D. Lee, Ed. Biochemical Effects of Environmental Pollutants. Ann Arbor, Mich: Ann Arbor Science Publishers, Inc., 1977.
4. Dahlgren, J. Abdominal pain in lead workers. Arch. Environ. Health 33:156-159, 1978.
5. Daines, R. H., D. W. Smith, A. Feliciano, and J. R. Trout. Air levels of lead inside and outside of homes. Ind. Med. Surg. 41(10):26-28, 1972.
6. DuBois, K. P. Interactions of chemicals as a result of enzyme inhibition, pp. 95-107. In D. H. K. Lee, and P. Kotin, Eds. Multiple Factors in the Causation of Environmentally Induced Disease. New York: Academic Press, Inc., 1972.
7. Elinder, C.-G., T. Kjellström, L. Friberg, B. Lind, and L. Linnman. Cadmium in kidney cortex, liver, and pancreas from Swedish autopsies. Arch. Environ. Health 31:292-302, 1976.
8. Finelli, V. N . Lead, zinc, and δ-aminolevulinate dehydratase, pp. 351-363. In S. D. Lee, Ed. Biochemical Effects of Environmental Pollutants. Ann Arbor, Mich.: Ann Arbor Science Publishers, Inc., 1977.
9. Foote, R. S. Mercury vapor concentrations inside buildings. Science 177:513-514, 1972.

10. Goldberg, A. M. Neurotransmitter mechanisms in inorganic lead poisoning, pp. 413-423. In S. D. Lee, Ed. Biochemical Effects of Environmental Pollutants. Ann Arbor, Mich.: Ann Arbor Science Publishers, Inc., 1977.
11. Goldsmith, J. R., and L. T. Friberg. Effects of air pollution on human health, pp. 457-610. In A. C. Stern, Ed. Air Pollution. 3rd ed. Vol. II. The Effects of Air Pollution. New York: Academic Press, Inc., 1977.
12. Hayes, W. J., Jr., R. A. Neal, and H. H. Sandstead. Role of body stores in environmentally induced disease - DDT and lead, pp. 136-164. In D. H. K. Lee and P. Kotin, Eds. Multiple Factors in the Causation of Environmentally Induced Disease. New York: Academic Press, Inc., 1972.
13. Hernberg, S. Lead, pp. 715-769. In C. Zenz, Ed., In Occupational Medicine. Principles and Practical Applications. Chicago: Year Book Medical Publishers, Inc., 1977.
14. Hirschman, S. Z., M. Feingold, and G. Boylen. Mercury in house paint as a cause of acrodynia. Effect of therapy with N-acetyl-D,L-penicillamine. N. Engl. J. Med. 269:889-893, 1963.
15. Joselow, M. M. Indoor air pollution by mercury. Ann. Intern. Med. 78:449-450, 1973.
16. Kass, E. H. Multiple factors in the causation of renal disease, pp. 83-91. In D. H. K. Lee and P. Kotin, Eds. Multiple Factors in the Causation of Environmentally Induced Disease. New York: Academic Press Inc., 1972.
17. Morse, D. L., W. N. Watson, J. Housworth, L. E. Witherell, and P. J. Landrigan. Exposure of children to lead in drinking water. Am. J. Public Health 69:711-712, 1979.
18. Needleman, H. L., C. Gunnoe, A. Leviton, R. Reed, H. Peresie, C. Maher, and P. Barrett. Deficits in psychologic and classroom performance of children with elevated dentine lead levels. N. Engl. J. Med. 300:689-695, 1979.
19. Petering, H. G., L. Murthy, and F. L. Cerklewski. Role of nutrition in heavy metal toxicity, pp. 365-376. In S. D. Lee, Ed. Biochemical Effects of Environmental Pollutants. Ann Arbor, Mich.: Ann Arbor Science Publishers, Inc., 1977.
20. Roels, H., J.-P. Buchet, R. Lauwerys, G. Hubermont, P. Bruaux, F. Claeys-Thoreau, A. Lafontaine, and J. Van Overschelde. Impact of air pollution by lead on the heme biosynthetic pathway in school-age children. Arch. Environ. Health 31:310-316, 1976.
21. Schanker, L. S. Flow of environmental agents in reaching their site of action, pp. 6-14. In D. H. K. Lee, and D. Minard, Eds. Physiology, Environment, and Man. New York: Academic Press Inc., 1970.
22. Sexton, D. J., K. E. Powell, J. Liddle, A. Smrek, J. C. Smith, and T. W. Clarkson. A nonoccupational outbreak of inorganic mercury vapor poisoning. Arch. Environ. Health 33:186-191, 1978.
23. Shy, C., J. Goldsmith, J. Hackney, M. D. Lebowitz, and D. Menzel. Statement on the Health Effects of Air Pollution: ATS News 4:22-62, Spring, 1978.

24. Stokinger, H. E. Means of contact and entry of toxic agents, pp. 7-11. In W. M. Gafafer, Ed. Occupational Diseases: A Guide to Their Recognition. U.S. Department of Health, Education, and Welfare, Public Health Service Publication No. 1097. Washington, D.C.: U.S. Government Printing Office, 1964.
25. Stokinger, H. E. Mode of action of toxic substances, pp. 13-26. In W. M. Gafafer, Ed. Occupational Diseases: A Guide to Their Recognition. U.S. Department of Health, Education, and Welfare, Public Health Service Publication No. 1097. Washington, D.C.: U.S. Government Printing Office, 1964.
26. Ter Haar, G. An investigation of elevated blood lead levels in Detroit children. Arch. Environ. Health 34:145-150, 1979.
27. Waldbott, G. L. Health Effects of Environmental Pollutants. Saint Louis: The C. V. Mosby Company, 1973. 316 pp.
28. World Health Organization. Health Hazards of the Human Environment. Geneva: World Health Organization, 1972. 387 pp.

RADON AND RADON PROGENY

The physical, chemical, and radiologic properties of radon-222 (referred to as radon), radon-220 (thoron), and their progeny and the principles of dosimetry are summarized in Chapter IV.

The unit of exposure of man is the working level (WL), defined as the quantity of short-lived progeny that will result in 1.3×10^5 MeV of potential alpha energy per liter of air. This is equivalent to a concentration of short-lived radon progeny in complete equilibrium with radon-222 at 100 pCi/L in air. The working-level month (WLM) is a term defined originally for occupational exposure, and 1 WLM is exposure at 1 WL for 170 h. Thus, the working-level month is a measure of cumulative exposure.

The working level is a measure of exposure rate; it has been widely assumed that, over a 70-yr lifetime, typical total-lifetime background exposures are in the range of 5-20 WLM. However, the average and distribution in the United States are not well studied. Some restrictions on the use of the working level must be noted. First, it is not useful for thoron progeny, because the dose delivered to the bronchial epithelium for the same amount of potential alpha energy (1.3×10^5 MeV) per liter of air can be much higher than that of radon progeny. Second, characterization of the dose to lung airways based solely on the working level involves a degree of uncertainty: the distribution of the lung dose depends on the unattached fraction, the particle size distribution of the aerosol to which the radon progeny are attached, lung morphometry, breathing rate, etc. Even with a general knowledge of the physical factors, other uncertainties in calculating dose are sufficiently great that characterization of the exposure atmosphere in terms of any measure more precise than working level is inappropriate for dose approximations. The difficulties in characterizing dose and relating it to effects have been reviewed recently by Cross _et al._[14] It should be noted that deviations in the exposure environment from reference conditions may result in actual lung doses that differ from those expected on the basis of the reference conditions assumed.

REVIEW OF DOSE AND EXPOSURE CALCULATIONS

The inhalation of radon progeny leads to a very inhomogeneous alpha dose to the human lung. For a variety of reasons--including preferential deposition, mucociliary clearance of aerosols deposited on conductive airways, and the observed tumor sites and types--it is believed (but by no means certain) that the radiation from the alpha-particle irradiation of the basal cells of the upper bronchial epithelium is the exposure characteristic most closely relatable to carcinogenic risk. However, it is difficult to determine the alpha-particle dose, because of the intractable difficulty of measuring it *in vivo*. Hence, dose calculations have been based on physical and biologic models. Dosimetric models have been developed for adults and have been summarized in several recent reports.[30 31 45] Recently, an age-dependent model was developed by Hofmann *et al*.[21] Moreover, the reference atmosphere is important for dose calculations, which are influenced by the fraction of unattached progeny and the particle size distribution of the progeny. Breathing rate, mucociliary clearance, lung morphometry, age, and sex must also be considered.

Depending on assumptions about the equilibrium, unattached fraction of progeny, carrier aerosol distribution, and the locus of target cells chosen for the estimates, calculated dose estimates per working-level month can vary by up to a factor of 100. A comprehensive evaluation of the dose through the various regions of the lung, taking into account attached and unattached fractions and particle size distributions, has recently been published.[22]

The table of background dose rates cited in Chapter IV is taken from National Commission on Radiation Protection and Measurements (NCRP) Report 45, which assumes that the reference exposure atmosphere for the United States is at about the concentration found in outdoor air, assumed to be radon-222 at 150 pCi/m^3 in equilibrium with the progeny. George and Breslin[18] measured radon working levels in cellars, first-floor spaces, and outdoors for 21 houses in New York and New Jersey and found the ratio of first-floor to outdoor average annual radon content to be 4.6, with median outdoor content of 180 pCi/m^3. The first-floor-to-outdoor working-level ratio was lower, 2.6; that suggests a reduced equilibrium indoors, as might be expected. The annual mean on the first floor was 0.004 WL. How representative these are of the metropolitan New York area or other areas is not known.

On the assumption that there was an 80% occupancy factor in the houses, with the 20% balance spent outdoors, the annual weighted estimate for the New York-New Jersey study was 0.11 WLM/yr. Over a 70-yr life, that would produce roughly 8 WLM.

BIOLOGIC EFFECTS

This section deals with the estimation of potential risk to man from inhalation exposure to radon progeny, the basis for the estimates of risk, and the shortcomings in our knowledge related to the exposures normally encountered.

Underground Miners

Much of our knowledge about the human health effects of radon and its progeny is based on the experience of underground miners whose exposures must be characterized, in relation to environmental characteristics, as having high dose rates (working levels) and high cumulative doses (working-level months). Table VII-1 shows representative values for underground mines and typical indoor measurements in houses, to provide perspective on the use of the term "high."

In the general population, exposure to radon progeny occurs under conditions rather different from those in underground mines, and it is therefore necessary to consider the extent to which epidemiologic studies in miners are germane to the general population. The feasibility of conducting epidemiologic studies of nonmining populations has recently been examined, and populations of health-spa workers were identified as promising.[47] There have been five major reviews of results of studies on underground miners. The analysis here draws partly on those and on the reports cited in them. All five reviews dealt with underground-mining experience and with miners who were, for the most part, adult males.

Conclusions patterned after those of Seltser derived from those studies are as follows:

- There is no reason to doubt an excessive lung-cancer risk among the early Bohemian uranium-miners in Schneeberg and Joachimstal,[16] the U.S. uranium-miners at the Colorado Plateau,[27] and Czechoslovakian uranium-miners.[24,41] In addition, there were increased occupational lung-cancer rates, relative to those of equivalent smoking groups in the general population, among underground miners with large exposures to radon and progeny in hematite, fluorspar, and zinc mines in several countries.[30]
- It is clear that the respiratory tracts of the uranium-miners received massive exposure from the alpha-emitting progeny, which are responsible for much more of the radiation exposure than the parent radon itself.
- There appears to be no convincing evidence that there are any other components of the mine environment that are responsible for the excess lung-cancer risk. Conversely, there is no evidence to rule out a contributory role of other components of this unusual environment, i.e., respirable silica dust and variable background dust concentrations and size distributions.

TABLE VII-1

Representative Exposures to Radon-222 Progeny

Subjects or Location	WL[a]	WLM[a,b]
Uranium miners[c]	1-20	100-10,000
Outdoors	<0.001	--
Indoors	<0.01	10

[a]To nearest order of magnitude.

[b]Lifetime or duration of exposure.

[c]Includes exposure before mid-1960s.

• There has been no definitive study in which a valid comparison group for the highly selected occupational populations was used. The observed-to-expected ratios have generally been expressed in relation to the general population or to a selected portion of the general population, and not to other underground miners. Such a comparison may be difficult to obtain, because most underground mining involves exposure to radon progeny at a higher-than-background concentration.

• Cigarette-smoking is clearly important, but not essential, in the induction of lung cancer. Lung cancer is greatly increased in these studies among uranium-miners who smoke, but is also higher among non-cigarette-smoking miners. Inferences from both the human epidemiologic work and the animal toxicologic studies are contradictory: in each case, one can cite opposite conclusions on the importance of smoking.

Fundamentally, the existing information is insufficient for a decision of whether radiation exposure multiplies the risk of lung cancer associated with other factors, such as smoking, or whether it produces a cancer risk that is proportional to the radiation exposure and merely additive to these other risks. In this review, a model based on the latter idea, the "absolute-risk model," has been adopted, although it must be kept in mind that it may not represent the true situation.[17]

Epidemiologic studies of carcinogenesis may be considered complete if all the population at risk has died and the followup is complete. Thus what is usually measured is some cumulative tumor incidence in the population up to the time of the analysis, which is lower than the lifetime excess risk.

For such data, risks may be defined as cumulative incidence to time t from exposures X. Or one may try to express the risk in terms of appearance per unit time (usually years), being careful to define the period over which tumors appear. One must distinguish between latent period and followup time of the study group. Sometimes, average risk per year is found by dividing cumulative incidence to time t by the followup time (i.e., as is done by UNSCEAR[45]); but recently the National Research Council Committee on the Biological Effects of Ionizing Radiations (the BEIR Committee) excluded the latent period to define the risk per year. Thus, risk estimates (in cancers/10^6 person-years per WLM) should not be directly compared with other dimensionally equivalent risk estimates found for a different period.

The method chosen here uses the cumulative incidence divided by the followup time. In any event, the time over which a time-dependent risk estimate is derived is always specified.

The results of studies of lung cancer in underground uranium-miners in the United States and Czechoslovakia and non-uranium-miners in Sweden, Canada, and the United Kingdom, analyzed as a linear, no-threshold phenomenon, are summarized in Table VII-2. The first column shows the excess risk in terms of the number of expected lung-cancer cases per working-level month per year; these range between 2.2×10^{-6} and 8×10^{-6}. Column 1 is obtained by dividing the observed number of lung cancers in the study group by the followup time

TABLE VII-2

Summary of Lung-Cancer Risk Estimates

Humans

Lung-Cancer Risk, no./WLM/yr	Followup Time, yr	Cumulative Excess Risk, no./WLM	Type of Mine	Exposure, WLM Average	Exposure, WLM Range	Location	Reference
2.1×10^{-6} to 2.9×10^{-6}	--	--	Uranium	740	0-10,000	Colorado plateau	45
8×10^{-6}	19-23	170×10^{-6}	Uranium	--	--	Czechoslovakia	41
6×10^{-6} to 16×10^{-6}	21-26	230×10^{-6}	--	--	--	--	24
3.4×10^{-6}	--	--	Nonuranium	--	--	Sweden	45
2.2×10^{-6}	--	--	Fluorspar	--	--	Canada	45
7×10^{-6}	--	--	Iron	--	--	U.K.	45

Experimental Animals

Lung-Tumor Risk, no./WLM	Animal	Exposure, WLM	Other Exposures	Reference
280×10^{-6}	Rat	170	None	11
9×10^{-6}	Beagle	11,000	Ore dust, smoking	15
33×10^{-6}	Beagle	12,000	Ore dust	15

in years and the average exposure of the population. The second column shows the followup time of the study in years. These studies in general have not followed the exposed populations until the end of life. Therefore, a valid estimate of the cumulative risk to the end of life cannot be derived from these studies. Column 3 is an underestimate of lung-cancer risk to the end of life. Indeed, UNSCEAR[45] has suggested that, beyond 15 yr of exposure, the risk per working-level month per year decreases from 4 x 10^{-6} to 2.2 x 10^{-6} up to 25 yr of exposure. UNSCEAR has suggested that the cumulative lifetime lung-cancer risk among these miners might be as high as 200 x 10^{-6} or 450 x 10^{-6} per working-level month.

In the study of Czechoslovakian miners, Kunz et al.[24] found that the proportion of cigarette-smokers in the underground miners was about 70%, roughly equal to the proportion in their general male population, and concluded that the risk increases with age at onset of exposure. A more recent reanalysis of the data from the uranium-miners in the Colorado Plateau done by Lundin et al.[26] concluded that the risk increased progressively from nonsmokers through light to heavy cigarette-smokers, but the excess risk over that to nonminers with equivalent smoking was much less for nonsmokers and was about the same for light and heavy smokers. Lundin et al. further suggested that smoking promotes the appearance of lung tumors and thus that lung cancers appear earlier in smokers than in nonsmokers. If true, this would effectively result in a lower cumulative incidence in nonsmokers.

Saccomanno et al. (personal communication), in analyzing ages of smokers and nonsmokers among 302 Colorado Plateau miners with lung cancer, found the average age at death to be earlier in the nonsmokers (48.5 yr) than in the smokers (56.5 yr).

Thus, the risk estimates shown in Table VII-2 seem to cluster in two groups, 2.2 x 10^{-6} to 3.4 x 10^{-6} WLM/yr in the U.S. uranium-miners and the Swedes and fluorspar-miners and 3-4 times higher in the Czechoslovakians and iron-miners.

The physical measurements of mine atmospheres in the U.S. and Czechoslovakian studies have not been compared, although systematic errors exceeding 50% seem unlikely (Domanski, personal communication).

The reason for the higher risk among the Czechoslovakian miners is not clear. Lundin et al.[26] suggested that they may be higher because Czechoslovakian miners started mining at younger ages. This hypothesis is not consistent with the observed dependence on age for equal doses in animals.

Thus, the risk estimates differ between the various studies by a factor of 3 or more for linear, absolute-risk estimates. However, greater uncertainties are involved in extrapolating to lower exposures in the range below 100 WLM, and further uncertainties are involved in extrapolating to populations with a wide distribution in ages. The question of the influence of age and gender on effect is not yet settled in either experimental animals or man.

Archer[1] has recently plotted his evaluation of the excess annual risk per working-level month (i.e., that above the risk established for the control population) for lung cancer versus the estimate of cumulative working-level months. This is reproduced in slightly modified form in Figure VII-1. The risk per working-level month increases monotonically down to several hundred working-level months. Although the reliability of the individual points is poor, the fact that all are well below the maximum is highly suggestive.

Histologic and Cytogenetic Studies in Man--Lung Cancer

Exposure to radon and thoron progeny has been shown to be correlated with chromosomal aberrations measured in peripheral blood lymphocytes.[12] [33] In uranium-miners, there appears to be a good correspondence between increased prevalence of chromosomal aberrations and cytologic sputum changes characteristic of markedly atypical cells and carcinomas *in situ*.[9]

On the basis of pathologic examination of 52 lung cancers in uranium-miners in the Colorado Plateau, Saccomanno *et al.* in 1964 reported that there was a predominance of small-cell undifferentiated carcinoma,[38] and in 1971 150 cases were evaluated by a panel of pathologists, who came to the same conclusion.[35] An interesting conclusion of this group was that the mean latent period for the small-cell types in miners with fewer than 700 WLM was 9.4 yr, compared with mean latent periods of 17.8 yr for all cancer types for exposure greater than 700 WLM.

The prevalence of the small-cell type seems to be decreasing with time, from 76% (13 of 17) in the period 1954-1959 to 22% (13 of 58) in the period 1975-1979 (Saccomanno, personal communication). The meaning of this is unclear, although the small-cell type of cancer may be associated with the higher dose rates (in working levels) of the 1950s and early 1960s in uranium mines in the Colorado Plateau. Archer *et al.*[3] analyzed lung cancer by type in a study group of 3,366 uranium-miners between 1950 and 1970 and found 66 small-cell cancers, compared with two expected. Saccomanno *et al.*[37] have shown that an early indicator of both precancerous lesions and carcinoma of the lung can be obtained by cytologic examination of cells from the sputum; this offers the opportunity to identify persons at risk early and to provide early treatment.

Auerbach *et al.*[7] have made detailed histologic examinations of lungs of uranium-miners and nonmining controls matched for age and smoking history obtained at autopsy. They concluded "that the synergistic effect of the exposure to uranium dust along with cigarette smoking increases the risk of lung cancer and that in addition to a main tumor mass, other sites of tissue alterations leading to tumor development are frequently already present in the lung."

In epidemiologic studies in nonminers, Hess *et al.*[20] have demonstrated a correlation between high radon concentrations in well water in Maine and mortality from lung cancer and all cancers, on the basis of available vital statistics. There were no controls for smoking. The lung-cancer incidence in Maine is lower than the national incidence, and a definitive analysis of the relationship of radon in water and long-term radon concentration in indoor air was not presented. These studies should be followed up with detailed

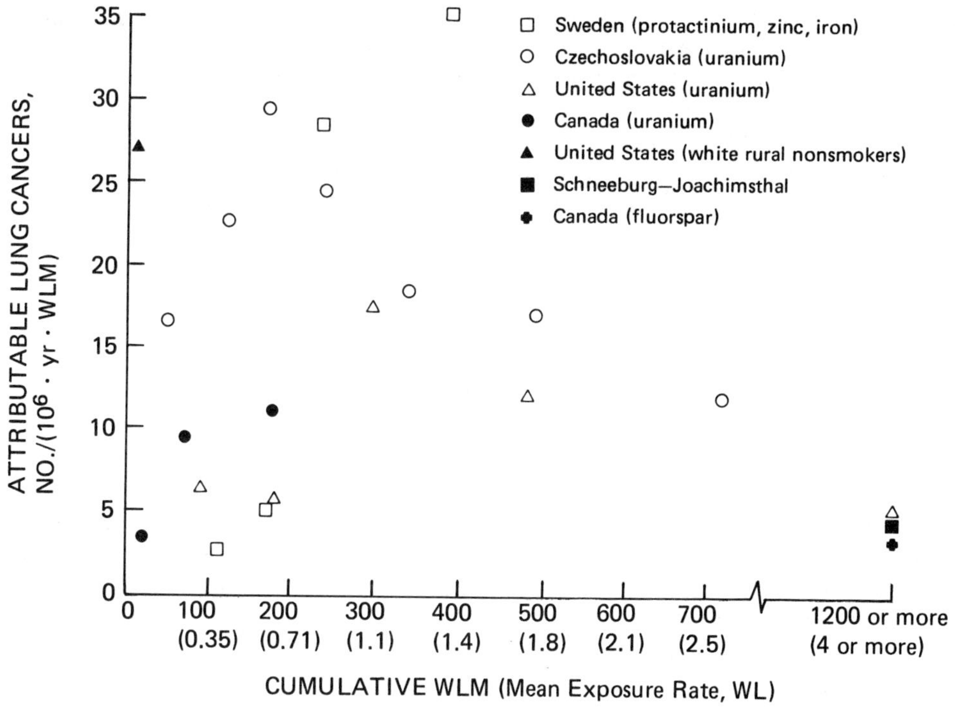

FIGURE VII-1 Attributable lung cancer per unit of radiation from radon progeny as function of cumulative exposure and exposure rate. Adapted from Archer.[5]

case-control studies in which the long-term radon concentrations in air are carefully evaluated.

Axelson et al.[8] conducted a case-control study based on death certificates and housing types in Sweden and concluded that there may be a relationship between housing type and lung cancer. Residence in stone houses was more common among those with lung cancer than residence in wooden houses. No exposure data on radon or radon progeny were correlated, nor was information available on heating and cooking practices.

Although both studies suggest a link between radon exposure and lung cancer, neither is sufficient to suggest causality.

Inhalation-Toxicology Studies in Animals

Experimental studies in animals offer the opportunity to test hypotheses related to the effects of smoking on animals and whether smoking is additive or synergistic in inducing or accelerating the appearance of lung cancer. Until recently, lung cancers had not been produced with radon progeny in animals. Early experiments with very large exposures showed no short-term effects on the lung, and, although dogs were exposed, the negative short-term pathologic results discouraged the pursuit of such studies in the United States until recently.

Studies of carcinogenesis of radon progeny are being conducted in France by researchers at CEA[11] and in the United States at the Battelle Pacific Northwest Laboratory.[14] Short-term exposures conducted at the University of Rochester showed little pathologic change in lungs of dogs for short followup times (D. Morken, personal communication). There has also been work on lung carcinogenesis after intratracheal instillation or inhalation of long-lived alpha-emitters. Those studies in which exposures are primarily of the pulmonary regions of the lung, rather than of the bronchial airways, are not considered relevant.

The combined effects of smoking and inhalation of uranium-ore dust, radon progeny, and diesel-engine exhaust have been studied in Syrian hamsters and dogs.[15] One conclusion was that the Syrian hamster was not an appropriate animal model for the study of pulmonary carcinogenesis, inasmuch as it appeared to be resistant to carcinoma induction by realistic exposures in lifespan exposure studies. Information obtained with beagles was useful, although limited. Four groups of 20 dogs were studied. All animals with lung tumors had cumulative exposures greater than 13,000 WLM. It was observed that smoking decreased the number of observed tumors. The tumors in uranium-miners are in the smaller bronchi, whereas in animals they are bronchioalveolar and in the nasal epithelium. It was suggested from this study that the human and animal data are not directly comparable. However, it is to be expected that the morphometric differences between animals and man and differences in exposure conditions will change the relative sites of particle deposition, the cell types being irradiated most heavily, and the loci and cellular origin of tumors.

Complementary studies were done in France with rodents exposed to radon progeny with and without concurrent exposure to other substances.[11] This very extensive program of studies on lung tumorigenesis in the rat demonstrates that it is a suitable model for some aspects of radon-progeny tumorigenesis in the lung. The tumors that are examined in histopathologic studies at death appear very late in life, grow slowly (relative to the rat's life span), and are rarely fatal.[28]

The data on rodents indicate a cumulative incidence per working-level month that is rather similar to that estimated for man, even though the tumor types and sites in rodents differ considerably from those found in humans. The pathologic basis for this and the comparability have been discussed elsewhere.[11] Similar studies of rats are underway at Battelle-Pacific Northwest. These are not yet complete; exposure began later than in the French studies. Because lung tumors in rodents show up after the median life span and are therefore restricted to less than half the population at risk, it is premature to speculate on the eventual results of the studies[14] and their relationship to the French studies.

SUMMARY AND CONCLUSIONS

There is no doubt that radon and its progeny in sufficient doses can produce lung cancer in man.

It is also generally believed, on the basis of dosimetric considerations, that the short-lived progeny are responsible for most of the lung-cancer risk.

Some epidemiologic studies of underground miners have suggested that smokers seem to be at higher risk of lung cancer than nonsmokers, but the risk to nonsmokers is also increased. Whether the risks are additive or synergistic is not yet clear.

The cumulative exposures (in working-level months) at which human and animal carcinogenesis has been observed are generally higher by an order of magnitude or more than those characteristic of the normal indoor environment. Excess cancers have been seen in association with exposures that were 2-3 orders of magnitude greater than those found in normal indoor environments.

Thus, to predict the results of the effects of decreased indoor ventilation on exposure to radon progeny, it is necessary to extrapolate beyond the range of exposures for which effects have clearly been documented.

Although the generally accepted linear dose-response function does not fit the available data very well, there is no established alternative dose-response function. We can conclude that dose rate or fractionation has an effect that cannot yet be adequately described.

Only by a combination of human, animal, and cellular studies will it be possible to estimate with any confidence the risk coefficients for natural indoor exposures in the range of interest, as enhanced by low air turnover associated with energy-conservation efforts.

Although there are legitimate reasons to criticize it, the working-level month is probably the best available measure of potential dose as related to likely biologic effect. Its use should be judicious.

REFERENCES

1. Archer, V. E. Effects of Low Levels of Radon on Man. Paper presented at the Specialist Meeting on Assessment of Radon and Daughter Exposure to Man and Related Biological Effects, Rome, 1980.
2. Archer, V. E. Factors in exposure-response relationships of radon daughter injury, pp. 324-367. In Conference/Workshop on Lung Cancer Epidemiology and Industrial Applications of Sputum Cytology. Golden, Colo: Colorado School of Mines Press, 1979.
3. Archer, V. E., G. Saccomanno, and J. H. Jones. Frequency of different histologic types of bronchogenic carcinoma as related to radiation exposure. Cancer 34:2056-2060, 1974.
4. Archer, V. E., J. K. Wagoner, and F. E. Lundin, Jr. Cancer mortality among uranium mill workers. J. Occup. Med. 15:11-14, 1973.
5. Archer, V. E., J. K. Wagoner, and F. E. Lundin. Lung cancer among uranium miners in the United States. Health Phys. 25: 351-371, 1973.
6. Archer, V. E., J. K. Wagoner, and F. E. Lundin, Jr. Uranium mining and cigarette smoking effects on man. J. Occup. Med. 15:204-211, 1973.
7. Auerbach, O., G. Saccomanno, M. Kuschner, R. D. Brown, and L. Garfinkel. Histologic findings in the tracheobronchial tree of uranium miners and non-miners with lung cancer. Cancer 42:483-489, 1978.
8. Axelson, O., C. Edling, and H. Kling. Lung cancer and residency--A care referent study on the possible impact of exposure to radon and its daughters in dwellings. Scand. J. Work Environ. Health 5:10-15, 1979.
9. Brandom, W. F., G. Saccomanno, V. E. Archer, P. G. Archer, and A. D. Bloom. Chromosome aberrations as a biological dose-response indicator of radiation exposure in uranium miners. Radiat. Res. 76:159-171, 1978.
10. Budnitz, R. J., J. V. Berk, C. D. Hollowell, W. W. Nazaroff, A. V. Nero, and A. H. Rosenfeld. Human Disease From Radon Exposures: The Impact of Energy Conservation in Residential Buildings. Lawrence Berkeley Laboratory Report LBL-7809, Revised EEB-Vent-78-5. Berkeley, Cal.: Lawrence Berkeley Laboratory, 1979.
11. Chameaud, J., R. Perraud, R. Masse, and J. Lafuma. Cancers induced by ^{222}Radon in the rat. Paper presented at the Specialist Meeting on Assessment of Radon and Daughter Exposure and Related Biological Effects, Rome, 1980.
12. Costa-Ribeiro, C., M. A. Barcinski, N. Figueiredo, E. Penna, F. Lobão, N. Lobão, and H. Krieger. Radiobiological aspects and radiation levels associated with the milling of monazite sand. Health Phys. 28:225-231, 1975.

13. Cross, F. T. Exposure standards for uranium mining. Health Phys. 37:765-772, 1979.
14. Cross, F. T., R. F. Palmer, R. H. Busch, and R. L. Buschbom. Influence of Radon Daughter Exposure Rate and Uranium Ore Dust Concentration on Occurrence of Lung Tumors. Paper presented for Battelle Pacific Northwest Laboratories at the Specialist Meeting on Assessment of Radon and Daughter Exposure and Related Biological Effects, Rome, 1980.
15. Cross, F. T., R. F. Palmer, R. E. Filipy, R. H. Busch, and B. O. Stuart. Study of the Combined Effects of Smoking and Inhalation of Uranium Ore Dust, Radon Daughters and Diesel Oil Exhuast Fumes in Hamsters and Dogs. Final Report. Battelle Pacific Northwest Laboratories Report No. PNL-2744-UC-48. Washington, D.C.: U.S. Department of Energy, 1978. 143 pp.
15a. Eadie, G. G., R. F. Kaufmann, D. J. Markley, and R. Williams. Report of Ambient Outdoor Radon and Indoor Radon Progeny Concentrations during November 1975 at Selected Locations in the Grants Mineral Belt, New Mexico. U.S. Environmental Protection Agency, Office of Radiation Programs Report No. ORP/LV-76-4. Las Vegas: U.S. Environmental Protection Agency, Las Vegas Facility, 1976. 53 pp.
16. Federal Radiation Council. Guidance for the Control of Radiation Hazards in Uranium Mining. Staff Report No. 8. Revised September 1967.
17. Fisher, D. R. Estimating Population Health Risk from Low-Level Environmental Radon. Paper presented for Battelle Pacific Northwest Laboratories at the Specialist Meeting on Assessment of Radon and Daughter Exposure and Related Biological Effects, Rome, 1980.
17a. Fitzgerald, J. E., Jr., R.-J. Guimond, and R. A. Shaw. A Preliminary Evaluation of the Control of Indoor Radon Daughter Levels in New Structures. U.S. Environmental Protection Agency Report No. EPA-520/4-76-018. Washington, D.C.: U.S. Environmental Protection Agency, Office of Radiation Programs, 1976. 88 pp.
18. George, A. C., and A. J. Breslin. The distribution of ambient radon and radon daughers in residential buildings in the New Jersey-New York area, pp. 1272-1292 (includes discussion). In T. F. Gesell and W. M. Lowder, Eds. Natural Radiation Environment III. Vol. 2. Proceedings of a Symposium Held at Houston, Texas, April 23-28, 1978. Oak Ridge, Tenn.: U.S. Department of Energy, Technical Information Center, 1980.
19. Guimond, R. J., Jr., W. H. Ellett, J. E. Fitzgerald, Jr., S. T. Windham, and P. A. Cuny. Indoor Radiation Exposure Due to Radium-226 in Florida Phosphate Lands. U.S. Environmental Protection Agency (Office of Radiation Programs) Report No. EPA 520/4-78-013. Washington, D.C.: U.S. Government Printing Office, [206] pp.
20. Hess, C. T., S. A. Norton, W. F. Brutsaert, R. E. Casparius, E. G. Coombs, and A. L. Hess. Radon-222 in potable water supplies in New England. J. N. Engl. Waterworks Assoc. 94:113-129, 1980.

21. Hofmann, W., S. Steinhäusler, and E. Pohl. Age-, sex-, and weight-dependent dose patterns due to inhaled natural radionuclides, pp. 1116-1144. In T. F. Gesell and W. M. Lowder, Eds. Natural Radiation Environment III. Vol. 2. Proceedings of a Symposium Held at Houston, Texas, April 23-28, 1978. Oak Ridge, Tenn.: U.S. Department of Energy, Technical Information Center, 1980.

22. Jacobi, W., and K. Eisfeld. Dose to tissues and effective dose equivalent of radon-222, radon-220 and their short-lived daughters. Institut für Strahlenschutz der Gesellschaft für Strahlen und Unweltforschung nbH, 1980.

23. Johnson, R. H., Jr., D. E. Bernhardt, N. S. Nelson, and H. W. Calley, Jr. Assessment of Potential Radiological Health Effects from Radon in Natural Gas. U.S. Environmental Protection Agency Report No. EPA-520/1-73-004. Washington, D.C.: U.S. Environmental Protection Agency, Office of Radiation Programs, 1973. 68 pp.

24. Kunz, E., J. Ševc, V. Plaček, and J. Horáček. Lung cancer in man in relation to different time distribution of radiation exposure. Health Phys. 36:699-706, 1979.

25. Letourneau, E. G., and D. T. Wigle. Mortality and Indoor Radon Daughter Concentrations in 13 Canadian Cities. Paper presented at the Specialist Meeting on Assessment of Radon and Daughter Exposure and Related Biological Effects, Rome, 1980.

26. Lundin, F. E., Jr., V. E. Archer, and J. K. Wagoner. An exposure-time-response model for lung cancer mortality in uranium miners: Effects of radiation exposure, age, and cigarette smoking, pp. 243-264. In Proceedings of Sims Conference on Energy and Health, 1978.

27. Lundin, F. E., Jr., J. K. Wagoner, and V. E. Archer. Radon Daughter Exposure and Respiratory Cancer: Quantitative and Temporal Aspects. National Institute for Occupational Safety and Health, and National Institute of Environmental Health Sciences Joint Monograph No. 1, 1971. 175 pp.

28. Masse, R. Histogenesis of lung tumors induced in rats by inhalation of alpha emitters, pp. 498-521. DOE Symposium Series 53, CONF 79 10002. Washington, D.C.: U.S. Department of Energy, 1979.

29. Momeni, M., J. B. Lindstrom, C. E. Dungey, and W. E. Kisieleski. Radon and Radon-Daughter Concentrations in Air in the Vicinity of the Anaconda Uranium Mill. U.S. Nuclear Regulatory Commission Report No. NUREG/CR-1133. Washington, D.C.: U.S. Nuclear Regulatory Commission, Office of Nuclear Regulatory Research, 1979. 97 pp.

30. National Research Council, Committee on the Biological Effects of Ionizing Radiations. The Effects on Populations of Exposure to Low Levels of Ionizing Radiation: 1980, pp. 308-331. Washington, D.C.: National Academy Press, 1980.

31. National Research Council, Advisory Committee on Biological Effects of Ionizing Radiations. The Effects on Populations of Exposure to Low Levels of Ionizing Radiation, pp. 145-157. Washington, D.C.: National Academy of Sciences, 1972.

32. Pohl-Rüling, J., and P. Fischer. An Epidemiological Study on Chromosome Aberrations in a Radon Spa. Paper presented at the Specialist Meeting on the Assessment of Radon and Daughter Exposure and Related Biological Effects, Rome, Italy, March 3-7, 1980.
33. Pohl-Rüling, J., and P. Fischer. The dose-effect relationship of chromosome aberrations to α and γ irradiation in a population subjected to an increased burden of natural radioactivity. Radiat. Res. 80:61-81, 1979.
34. Royal Commission on Health and Safety of Miners (Ontario). Report of the Royal Commission on Health and Safety of Workers in Mines. Toronto, Canada: Province of Ontario, Ministry of Attorney General, 1976. [349] pp.
35. Saccomanno, G., V. E. Archer, O. Auerbach, M. Kuschner, R. P. Saunders, and M. G. Klein. Histologic types of lung cancer among uranium miners. Cancer 27:515-523, 1971.
36. Saccomanno, G., V. E. Archer, O. Auerbach, R. P. Saunders, and L. M. Brennan. Development of carcinoma of the lung as reflected in exfoliated cells. Cancer 33:256-270, 1974.
37. Saccomanno, G., V. E. Archer, R. P. Saunders, O. Auerbach, and M. G. Klein. Early indices of cancer risk among uranium miners with reference to modifying factors. Ann. N.Y. Acad. Sci. 271:377-383, 1976.
38. Saccomanno, G., V. E. Archer, R. P. Saunders, L. A. James, and P. A. Beckler. Lung cancer of uranium miners on the Colorado Plateau. Health Phys. 10:1195-1201, 1964.
39. Saccomanno, G., R. P. Saunders, M. G. Klein, V. E. Archer, and L. Brennan. Cytology of the lung in reference to irritant, individual sensitivity and healing. Acta Cytol. 14:377-381, 1970.
40. Seltser, R. Lung cancer and uranium mining. Arch. Environ. Health 10:923-936, 1965.
41. Ševc, J., E. Kunz, and V. Plaček. Lung cancer in uranium miners and long-term exposure to radon daughter products. Health Phys. 30:430-437, 1976.
42. Snihs, J. O. The approach to radon problems in non-uranium mines in Sweden, pp. 900-911. In Proceedings of the Third International Congress of the International Radiation Protection Association, Washington, D.C., September 9-14, 1973. U.S. Atomic Energy Commission Report CONF-730907-P2. Washington, D.C.: U.S. Atomic Energy Commission, 1974.
43. Steinhäusler, F., E. Pohl, and W. Hofmann. On the Suitability of Epidemiological Studies of Population Groups Exposed to Elevated Levels of Radon and Daughters. Paper presented at the Specialist Meeting on the Assessment of Radon and Daughter Exposure and Related Biological Effects, Rome, Italy, March 3-7, 1980.
44. Turner, J. E., C. F. Holoway, and A. S. Loebl, Eds. Workshop on Dosimetry for Radon and Radon Daughters, Oak Ridge National Laboratory, April 12-13, 1977. Oak Ridge National Laboratory Report No. ORNL-4348. Oak Ridge, Tenn.: U.S. Department of Energy, Oak Ridge National Laboratory, 1978. 54 pp.

45. United Nations Scientific Committee on the Effects of Atomic Radiation. Sources and Effects of Ionizing Radiation, Report to the General Assembly, with Annexes, pp. 200-226, 394-399. New York: United Nations, 1977.
46. U.S. Environmental Protection Agency. Preliminary Findings: Radon Daughter Levels in Structures Constructed on Reclaimed Florida Phosphate Land. U.S. Environmental Protection Agency, Office of Radiation Programs Report No. ORP/CSD-75-4. Washington, D.C.: U.S. Environmental Protection Agency, 1975. 40 pp. Available from National Technical Information Service, Springfield, Va., as PB-257 679.
47. Wrenn, M. E., F. Steinhäusler, and G. Clemente, Eds. Proceedings of the Specialist Meeting on the Assessment of Radon Daughter Exposure and Related Biological Effects, Rome, Italy, March 3-7, 1980. (in press)

FORMALDEHYDE AND OTHER ORGANIC SUBSTANCES

This section reviews the potential effects of formaldehyde and some other indoor organic pollutants. The sources of such pollutants are discussed in Chapter IV. The concentrations of most of the organic substances in question are usually unknown, because they are identified mainly as peaks on chromatograms. Their health effects must be discussed as potential, rather than known, inasmuch as the characteristics of their presence indoors have not been studied. Committees of the National Research Council have reviewed the effects of particulate polycyclic organic matter[57] and vapor-phase organic pollutants.[58]

Adverse health effects due to formaldehyde may occur after exposure by inhalation, ingestion, or contact. Ascribing specific health effects to specific concentrations of formaldehyde is difficult, because people vary in their subjective responses and complaints. Moreover, hypersensitive persons, those with disease, and hyposensitive persons may not have been evaluated in epidemiologic studies. Thus, the threshold for response will not be constant among all segments of the population. Interpretation of the health effects of formaldehyde must also consider the duration of exposure of subjects. A short-term inhalation study cannot accurately predict the effects of formaldehyde exposures of residents of conventional or mobile homes who may be exposed continuously to low concentrations. Odor irritation and tolerance may develop after several hours of exposure and modify the response to formaldehyde.

EFFECTS OF FORMALDEHYDE IN ANIMALS

The reader is referred to the NRC report, Formaldehyde and Other Aldehydes,[56] for detailed discussions of the effects of formaldehyde exposure reported in experimental animal studies.

The carcinogenic effects of formaldehyde exposure in humans have not been assessed. However, studies in rats and mice done by the Chemical Industry Institute of Toxicology have found that formaldehyde induces nasopharyngeal carcinoma after several months of exposure at 15 and 6 ppm for 6 h/d, 5 d/wk. Dose-related histologic changes were observed in the nasal mucosa of rats exposed at 2 and 6 ppm for the same times. Formaldehyde has mutagenic activity in a variety of microorganisms and in some insects. More work is necessary to ascertain its mutagenic potential in germinal or somatic mammalian cells. Such information would be used to assess the potential hazard to persons exposed to formaldehyde. Formaldehyde has not been shown to be teratogenic in animals.

EFFECTS OF FORMALDEHYDE IN HUMANS*

The principal effect of low concentrations of formaldehyde observed in humans is irritation of the eyes and mucous membranes. A wide range of concentrations of airborn formaldehyde have been reported to cause specific human health effects. Table VII-3 shows the variability and overlap of responses among subjects. Some persons develop tolerance to olfactory, ocular, or upper respiratory tract irritation. Such factors as smoking habits, socioeconomic status, preexisting disease, various host factors, and interactions with other pollutants and aerosols are expected to modify these responses.

Eye

Human eyes are very sensitive to formaldehyde, responding to atmospheric concentrations of 0.01 ppm in some cases (when mixed with other pollutants) and producing a sensation of irritation at 0.05-0.5 ppm. Lacrimation is produced at 20 ppm, but damage is prevented by closing of the eyes in response to discomfort. See Table VII-4 for summaries of effects at various concentrations.

Olfactory System

The odor threshold of formaldehyde is usually around 1 ppm, but may be as low as 0.05 ppm in some people.[5 9 24 26 47 75 77 91 99] Olfactory fatigue--as determined on the basis of increased olfactory thresholds for rosemary, thymol, camphor, and tar--was reported among plywood and particleboard workers and is presumed to be associated with formaldehyde.[93 98] Olfactory fatigue can be important in the home owing to reduced sensitivity to odors; a person exposed to formaldehyde might not smell other substances, such as leaking gas or burning materials.

*Much of this discussion is derived from the National Research Council report, Formaldehyde and Other Aldehydes,[56] to which the reader is referred for additional details.

TABLE VII-3

Reported Health Effects of Formaldehyde at Various Concentrations[a]

Effects	Approximate Formaldehyde Concentration, ppm
None reported	0-0.05
Neurophysiologic effects	0.05-1.5
Odor threshold	0.05-1.0
Eye irritation	0.01-2.0[b]
Upper airway irritation	0.10-25
Lower airway and pulmonary effects	5-30
Pulmonary edema, inflammation, pneumonia	50-100
Death	100+

[a]Derived from National Research Council.[56]

[b]The low concentration (0.01 ppm) was observed in the presence of other pollutants that may have been acting synergistically.

TABLE VII-4

Eye Irritation Effects of Formaldehyde[a]

Formaldehyde Concentration, ppm	Duration of Exposure	Effects on Eyes
	Chamber--single:	
0.03-3.2	20-35 min; gradually increasing concentration	Increase in blink rate; irritation
13.8	30 min	Irritation (and nose irritation)
20	Less than 1 min	Discomfort and lacrimation
	Chamber--repeated:	
0.25	5 h/d for 4 d	19% "slight discomfort"
0.42	5 h/d for 4 d	31% "slight discomfort" and conjunctival irritation
0.83-1.6	5 h/d for 4 d	94% "slight discomfort" and conjunctival irritation
	Occupational:	
4-5	--	Irritation, lacrimation, and discomfort in 30 min
0.9-2.7	--	Tearing
0.3-2.7	--	Prickling and tearing
0.9-1.6	--	Intense irritation and itching
0.13-0.45	--	Stinging and burning
	Indoor residential:	
0.067-4.82	--	Tearing
0.02-4.15	--	Irritation
0.03-2.5	--	Irritation

[a]Reprinted from National Research Council.[56(p. 7-15)]

Respiratory Tract

The nose adjusts the temperature and water-vapor content of air and removes a large proportion of foreign gases and dusts,[69] and the nasal mucociliary system clears foreign material deposited on it. Nasal congestion from injury may lead to partial mouth-breathing; when nasal functions are impaired or the nose is otherwise bypassed for mouth-breathing, the burden of conditioning and cleaning the air falls on the oral airways and the lungs. If the nasal defense system is disturbed or if mouth-breathing occurs, greater concentrations of formaldehyde will reach the lungs, and other noxious materials ordinarily cleared from the airways may be retained.

Upper Airway Irritation. Symptoms of upper airway irritation include the feeling of a dry throat, tingling sensation of the nose, and sore throat, usually co-existent with tearing and pain in the eyes. Irritation occurs over a wide range of concentrations, usually beginning at approximately 0.1 ppm, but reported more frequently at 1-11 ppm[5,9,22,42,44,55,98,99] (see Table VII-3). Tolerance to eye and upper airway irritation may occur after 1-2 h of exposure.[5,42,75] However, even if tolerance develops, the irritation symptoms can return after a 1- to 2-h interruption of exposure.[2,5,39,42,53,75] As in the case of eye irritation, some persons seem to tolerate higher concentrations, 16-30 ppm; it is not known whether subjects develop tolerance.

Lower Airway and Pulmonary Effects. Lower airway irritation that is characterized clinically by cough, chest tightness, and wheezing is reported often in people exposed to formaldehyde at 5-30 ppm.[26,29,44,68,88,98,100] Chest x rays of persons apparently exposed to formaldehyde at high concentrations are usually normal, except for occasional reports of accentuated bronchovascular markings, but pulmonary-function test results may be abnormal.[100]

Pulmonary edema, pneumonitis, and death result from very high formaldehyde concentrations, 50-100 ppm.[6,68,100] It is not known what concentrations are lethal to humans, but concentrations exceeding 100 ppm would probably be extremely hazardous to most and might be fatal in sensitive persons.

Asthma. Formaldehyde has been shown to cause bronchial asthma in humans.[35,36,45,61,64,67,73,74,82,89] In some cases, asthmatic attacks are due specifically to formaldehyde sensitization or allergy; controlled inhalation studies with formaldehyde are positive in these instances.[35,36] More commonly, formaldehyde seems to act as a direct airway irritant in persons who have bronchial asthmatic attacks from other causes. Concentrations at which attacks occur are highly variable. Bronchial asthma is characterized by hyperreactivity of airways, and the airways respond to many nonspecific inhaled irritants, including formaldehyde.

The exact mechanism of the asthma syndrome related to formaldehyde exposure is not known. It has been suggested that an immunologic basis is sometimes operative. However, no studies have demonstrated the presence of specific circulating immunoglobulins (IgE or IgG) in affected persons. Although formaldehyde at low concentrations may cause asthmatic symptoms in some sensitized subjects, in irritant concentrations it produces bronchoconstriction in even normal persons. Inhalation of formaldehyde fumes may cause airway hyperreactivity, an important component of bronchial asthma.[19 46 65 79] Methacholine and histamine challenge tests have demonstrated this hyperreactivity with other environmental pollutants.[10 11 15 17 30]

Skin

Skin contact with formaldehyde has been reported to cause a variety of cutaneous problems in humans, including irritation, allergic contact dermatitis, and urticaria.[62 72 78] Allergic contact dermatitis from formaldehyde is relatively common, and formaldehyde is one of the more frequent causes of this condition both in the United States[25] and in other areas.[27] The North American Contact Dermatitis Group reported that formaldehyde is the tenth leading cause of skin reactions among dermatitis patients patch-tested for allergic contact dermatitis. Approximately 4% of 1,200 patients had positive skin reactions when tested with 2% formalin (0.8% formaldehyde) under an occlusive patch.[60] Minor epidemics of allergic contact dermatitis have been described in diverse situations, for example, among nurses who handled thermometers that had been immersed in a 10% solution of formaldehyde[71] and among those who were exposed to formaldehyde in hemodialysis units.[78]

In many cases, either the initiation or the elicitation of the allergy has been caused by contact with formaldehyde or formalin, but it may also result from formaldehyde-releasing agents used in cosmetics, medications, and germicides, from incompletely cured resins, and from the decomposition of formaldehyde-containing resins used in textiles.[49] People with cutaneous allergy to formaldehyde have particular problems because there are so many sources of formaldehyde exposure in ordinary daily life (for example, the FDA listed 846 cosmetic formulations containing formaldehyde[12]). The skin reaction rate from cosmetic formulations containing formaldehyde has not been excessive, because it is used mainly as a preservative in shampoos, whose contact time with skin is short. Formaldehyde-releasing cosmetic preservatives, such as Quaternium-15, have shown a greater reaction frequency than formaldehyde itself (unpublished data from Cosmetics Technology Division, Bureau of Foods, FDA).

Low concentrations of formaldehyde are associated with many sources, and repeated contact with them may be sufficient to provoke responses in people with allergic contact sensitization. These sources include components of plastics, glues, antifungal disinfectants, preservatives, paper, fabrics, leather, coal and wood smoke, fixatives for histology, and photographic materials.[28] Available data do not permit the determination of a degree of exposure to formaldehyde-containing products that would be safe once sensitization has occurred.

Most sensitized persons can tolerate topical axillary products containing formaldehyde at up to about 30 ppm.[41] With increasing concentration, one sees a higher frequency of responders,[51] probably because skin penetration by formaldehyde varies from one person to another and even from one site to another on the same person. Thus, different amounts of formaldehyde may reach different target sites. The dose needed to elicit a response depends on these factors and others, such as occlusion, temperature, contact time, and vehicle.

Allergic contact dermatitis is a manifestation of cell-mediated immunity. The standard diagnostic test for this condition is the epidermal patch test. Patch testing for skin sensitization to formaldehyde resin is performed with a 5-10% concentration of the resin in petrolatum.[1]

Although formaldehyde has been reported to cause contact urticaria, it is not yet clear whether this is immunologically mediated.[62] Formaldehyde is a potent sensitizer and irritant; repeated exposure to it may also result in dermatitis.

Central Nervous System

Central nervous system responses to formaldehyde have been tested in a variety of ways, including by determination of optical chronaxy,[6] electroencephalographically,[24] and by the sensitivity of the dark-adapted eyes to light.[52] Responses are reported in some persons at 0.05 ppm and are maximal at about 1.5 ppm. Formaldehyde at below 0.05 ppm probably has little or no objective adverse effect.[88] Fel'dman and Bonashevskaya reported that formaldehyde at 0.032 ppm produced no electroencephalographic changes and did not reach the odor threshold in five extremely sensitive subjects.[24] Melekhina demonstrated changes in the sensitivity of the dark-adapted eye to light at about 0.08 ppm.[52]

Alimentary Tract

Ingestion of formaldehyde has been reported to cause headache, upper gastrointestinal pain,[12 20 21 43 50 88] allergic reactions,[88] corrosive effects on gastrointestinal and respiratory tracts,[21 43 48] and systemic damage.[21 43 48] Accidental or suicidal poisoning with formaldehyde usually involves the ingestion of aqueous solutions; death occurs after the swallowing of as little as 30 ml of formalin.[6 43] Gastrointestinal tract damage is most marked in the stomach and lower esophagus, with the tongue, oral cavity, and pharynx generally not severely affected.[88] The small intestine may occasionally be involved; perforated appendix is a rare complication. When the chemical infiltrates around the epiglottis, injury to the larynx and trachea may occur.[6 43 88] After ingestion, there may be loss of consciousness, vascular collapse, pneumonia, hemorrhagic nephritis, and spontaneous abortion.[6 43] One autopsy report of a fatal ingestion described hardening of organs adjacent to the stomach (lung, liver, spleen, and pancreas), hyperemia and edema of the lungs, bilateral diffuse bronchopneumonia, fatty degeneration of the liver with subcapsular hemorrhage, renal tubular necrosis, and involvement of the brain.[6 43 70]

Consumer Complaints in Residential Environments

A number of studies have been undertaken to determine the magnitude and extent of formaldehyde exposure of persons in the residential environment.[2 13 31 37 39 80 85 94] Breysse reported a study of 325 persons living in 272 mobile homes, all of whom had eye and upper respiratory tract irritation.[13 80] Formaldehyde concentrations (measured in 138 instances) ranged from 0 to 2.5 ppm; approximately 90% were less than 1 ppm, and 9.4% were above 1.0 ppm.[13] Of 121 persons studied, 15% had no symptoms, and approximately 34% had three or more symptoms. Symptoms reported most often included eye irritation (about 30%), nose irritation (5%), respiratory tract involvement (24%), headache (21%), nausea (5%), and drowsiness.

In November 1977, the Connecticut Department of Health and Consumer Protection began receiving complaints from state residents who had urea-formaldehyde foam insulation installed in their homes.[31] By September 1978, 84 complaints had been received. The Department tested the 84 homes and found formaldehyde in the air in 75. The sensitivity of the testing system was reported to be less than 0.05 ppm. Health symptoms were reported by 224 residents of 74 homes, in which detectable concentrations of formaldehyde ranged between 0.5 and 10 ppm, with a mean of 1.8 ppm. The symptoms of the residents included eye, nose, and throat irritation; GI tract symptoms; headache; skin problems; and some miscellaneous complaints, such as fatigue, aches, and swollen glands. In 37%, however, symptoms occurred when formaldehyde was not detectable by the methods used. When formaldehyde was detectable (0.5-10 ppm), 49% of the occupants had eye irritation, 37% nose and throat irritation, 46% headache, and 22% GI tract symptoms; in homes with no detectable formaldehyde, 26% had eye symptoms, 41% nose and throat irritation, 26% headache, and 42% GI tract symptoms.

Occupational Standards for Formaldehyde

It is important to consider total exposure to formaldehyde. Therefore, it should be noted that some people are exposed to it at work. The present Occupational Safety and Health Administration (OSHA) standard for formaldehyde is 3 ppm, as a time-weighted average concentration over an 8-h workshift. In 1974, the American Conference of Governmental Industrial Hygienists (ACGIH) recommended a limit of 2 ppm, mainly because irritation might occur above this concentration. The National Institute for Occupational Safety and Health (NIOSH) has recommended a workplace ceiling limit of 1 ppm.[88]

Significance of Adverse Health Effects in Regard to Population at Risk

The total number of people who are exposed to formaldehyde and who manifest adverse health effects is difficult to determine. There is evidence that such responses may occur in a substantial proportion of the exposed population in the United States. The variability in response among exposed persons makes it particularly difficult to assess the problem.

Millions of persons live in mobile or conventional homes that contain either urea-formaldehyde (UF) foam insulation or particleboard made with UF resins. When measurements have been performed, a wide range of formaldehyde concentrations, from 0.01 ppm to 10.6 ppm, have been reported. In most indoor environments, 24-h average formaldehyde concentrations of 0.05-0.3 ppm are not uncommon today. Because people may spend over 70% of their time indoors, exposure to formaldehyde from gas cooking and smoking combined with that from UF foam, particleboard, and plywood could be substantial. In addition, people are exposed to formaldehyde from occupational sources, consumer products, and outdoor ambient air.

Formaldehyde concentrations measured in ambient air are lower than in residences. Concentrations vary, but atmospheric concentrations are usually less than 0.1 ppm and very often less than 0.05 ppm. The dose received by the 220 million people in the United States from outdoor exposure appears to be minimal, except for unusual circumstances of traffic, fuel use, or automobile density. Consumer exposures are mainly by direct contact, and contact dermatitis is an important consideration, as has been discussed.

Little is known about the magnitude of the population that is more susceptible to the effects of inhaling formaldehyde vapor. Asthmatics may constitute a segment of the general population that is more susceptible; inhalation even at low concentrations may precipitate acute symptoms. Airway hyperactivity may explain the susceptibility of asthmatics to formaldehyde at low concentrations. Using data gathered from over 1,500 methacholine challenge tests, one can estimate the prevalence of airway hyperreactivity in the population at large.[81] About 9 million people in the United States have bronchial asthma. Essentially all will react positively to methacholine challenge tests and thus be considered to have hyperreactive airways.[81] The degree of airway reactivity is variable and depends on a number of factors.[11] It has been estimated that 30% of atopic nonasthmatic people--perhaps 10 million--have positive methacholine tests.[81] Townley et al. reported that 5% of nonatopic persons--another 8.5 million--have positive methacholine tests.[81] Therefore, on the basis of calculations reported for positive methacholine challenge tests, it can be estimated that about 25 million persons in the United States, or 10-12% of the population, may be considered to have some degree of airway hyperreactivity. This population could potentially be more susceptible to formaldehyde.

Information on other assumed susceptible populations is limited. The U.S. Department of Health, Education, and Welfare, in a 1977 report on prevention, control, and elimination of respiratory disease, estimated that 10 million persons in the United States had chronic obstructive lung disease (excluding asthma).[87] An unknown percentage of them will have positive methacholine challenge tests. Britt _et al._[14] suggested that the presence of methacholine sensitivity and evidence of airway hyperreactivity are risk factors for the development of chronic obstructive pulmonary disease (COPD). Perhaps patients with COPD who manifest airway hyperreactivity constitute a susceptible population, inasmuch as they react more acutely to airborne irritants, including formaldehyde.

On the basis of sensitivity to methacholine, some atopic persons, some nonatopic subjects, and some COPD patients may constitute a potential formaldehyde-susceptible population. This population could also have greater eye and upper respiratory tract sensitivity. However, many apparently normal people also react to the irritant properties of formaldehyde; this makes it more difficult to determine the susceptible population.

In another attempt to estimate the susceptible population (particularly in relation to eye, nose, and throat sensitivity), information on a small number of healthy young adults exposed to formaldehyde at various concentrations for short periods was considered.[59] At 1.5-3.0 ppm, more than 30% of the subjects tested reported mild to moderate eye, nose, and throat irritation symptoms, and 10-20% had strong reactions. When test subjects were exposed at 0.5-1.5 ppm, slight or mild eye, nose, and throat irritation was noted in more than 30%, but 10-20% still had more marked reactions. Approximately 20% of the subjects had slight ear, nose, and throat irritation in response to formaldehyde at 0.25-0.5 ppm. Finally, at the lowest concentration tested, less than 0.25 ppm, some exposed subjects ("less than 20 percent") still reported minimal to slight eye, nose, and throat discomfort. These data might be interpreted as suggesting that there are subjects, perhaps 10-20% of those tested, who react to formaldehyde at any given concentration.

We may get further information from mobile-home surveys from which environmental and clinical data are available. Irritation symptoms were reported by 30-50% of subjects when formaldehyde concentration was greater than 0.5 ppm. When the concentration was less than 0.5 ppm, irritation symptoms were reported in fewer than 30% of subjects. Finally, in a more controlled study in which irritation symptoms were investigated, mild irritation responses (doubling of blinking rate) occurred in 11% of subjects tested at 0.5 ppm.

In summary, fewer than 20% but perhaps more than 10% of the general population may be susceptible to formaldehyde and may react acutely at any concentration, particularly if it is greater than 1.5 ppm. People report mild ENT discomfort and other symptoms at less than 0.5 ppm, with some noting symptoms at concentrations below 0.25 ppm. Low-concentration formaldehyde exposures may produce eye, nose, and throat symptoms and possibly lower-airway complaints. In some susceptible persons, an "allergic" reaction to formaldehyde may occur at very low concentrations, causing bronchoconstriction and asthmatic symptoms. This particular type of reaction to formaldehyde appears to be uncommon; its prevalence cannot now be estimated.

EFFECTS OF OTHER ORGANIC SUBSTANCES

Cardiac arrhythmia may occur through proteination of endogenous catecholamines produced by a number of chemicals present in the environment. Various environmental chemicals have structural similarities to other chemicals that may have similar effects on the myocardium; these chemicals have a lung tissue half-life that could represent a long-term hazard. Examples include the polyhalogenated hydrocarbons, which may cause sudden death. The polyhalogenated hydrocarbons also bind to estrogen receptors and have been shown to have estrogenic effects in animal systems. These effects may increase HDL cholesterol and triglyceride concentrations and thus increase coronary-heart-disease incidence or mortality risk.[86]

Disturbances of the nervous system may occur through exposure to such chemicals as polychlorinated biphenyls (PCBs), which may be stored in fatty tissue and result in a long-term body burden. PCBs inhibit growth in cell cultures and interfere with the activity of a variety of enzymes.[90]

Vapor-phase organic pollutants undergo biologic transformation sequences and metabolic reactions in the intestine, and the metabolites may be conjugated or excreted directly. Both forms may have a primary effect on the gastrointestinal tract.[58] The enzymatic activity of the microflora of the gastrointestinal tract may also lead to the conversion of ingested substances, such as nitrites to nitrates. Formation of nitrosamines by the reaction of secondary amines with nitrates may lead to cancer.

Vapor-phase organic pollutants are enzymatically converted in the kidney and in the liver to more polar compounds, which are then excreted.[58] These hydrocarbons may have nephrotoxic action. Solvents and chlorinated hydrocarbons may produce kidney and liver damage.[97]

Primary skin irritants include polycyclic organic matter and other vapor-phase organic pollutants. Various pathologic responses in man have been related to the use of polycyclic organic matter. Polycyclic aromatic hydrocarbons are reportedly associated with the same kinds of work exposures that have produced skin cancer. These materials include derivatives of fossil fuel, paraffin distillates, asphalt, and lubricating oils.[57] Polycyclic organic matter may produce changes in hair follicles and sebaceous glands.[57] Vapor-phase organic pollutants (like formaldehyde) may produce a variety of skin effects. They may produce eczematous contact dermatitis and dermal contact sensitivity.[83] They may be absorbed percutaneously because of solubility in the water-lipid system, they may produce skin paresthesis,[58] and they may produce eczematous reactions of an acute or chronic nature, including eruptions and exacerbations.[58, 76]

Highly water-soluble pollutants are most likely absorbed by the conjunctiva locally and systemically. The vapor-phase organic pollutants, for instance, will affect the conjunctival membranes, the cornea, and the nasal mucous membranes and cause mild to acute inflammation.[58]

REFERENCES

1. Adams, R. Occupational Contact Dermatitis. Philadelphia: J. B. Lippincott Co., 1969. 262 pp.
2. Ad Hoc Task Force--Epidemiology Study on Formaldehyde. Epidemiological Studies in the Context of Assessment of the Health Impact of Indoor Air Pollution. Summary and Recommendations. Bethesda, Md.: U.S. Consumer Product Safety Commission, May 10, 1979. 11 pp.
3. Andersen, I. Formaldehyde in the indoor environment--Health implications and the setting of standards, pp. 65-78 (includes discussion). In P. O. Fanger, and O. Valbjørn, Eds. Indoor Climate. Effects on Human Comfort, Performance, and Health in Residential, Commercial, and Light-Industry Buildings. Proceedings of the First International Indoor Climate Symposium, Copenhagen, August 30-September 1, 1978. Copenhagen: Danish Building Research Institute, 1979.
4. Barnes, E. C., and H. W. Speicher. The determination of formaldehyde in air. J. Ind. Hyg. Toxicol. 24:10-17, 1942.
5. Blejer, H. P., and B. H. Miller. Occupational Health Report of Formaldehyde Concentrations and Effects on Workers at the Bayly Manufacturing Company, Visalia. Study Report No. S-1806. Los Angeles: State of California Health and Welfare Agency, Department of Public Health, Bureau of Occupational Health, 1966. 6 pp.
6. Böhmer, K. Formalin poisoning. Dtsch. Z. Gesamte Gerichtl. Med. 23:7-18, 1934. (in German)
7. Boucher, R. C., P. D. Pare, and J. C. Hogg. Relationship between airway hyperreactivity and hyperpermeability in Ascaris-sensitive monkeys. J. Allergy Clin. Immunol. 64:197-201, 1979.
8. Bouhuys, A., and K. P. van de Woestijne. Respiratory mechanics and dust exposure in byssinosis. J. Clin. Invest. 49:106-118, 1970.
9. Bourne, H. G., Jr., and S. Seferian. Formaldehyde in wrinkle-proof apparel produces...tears for milady. Ind. Med. Surg. 28:232-233, 1959.
10. Boushey, H. A., D. W. Empey, and L. A. Laitinen. Meat wrapper's asthma: Effects of fumes of polyvinyl chloride on airways function. Physiologist 18:148, 1975.
11. Boushey, H. A., M. J. Holtzman, J. R. Sheller, and J. A. Nadel. Bronchial hyperreactivity. Am. Rev. Respir. Dis. 121:389-413, 1980.
12. Bower, A. J. Case of poisoning by formaldehyde. J. Am. Med. Assoc. 52: 1106, 1909.
13. Breysse, P. A. Formaldehyde exposure following urea-formaldehyde insulation. Environ. Health Safety News 26(1-12), 1978. 13 pp.
14. Britt, E. J., B. Cohen, H. Menkes, E. Bleecker, S. Permutt, R. Rosenthal, and P. Norman. Airways reactivity and functional deterioration in relatives of COPD patients. Chest 77(Suppl.):260-261, 1980.

15. Butcher, B. T., R. M. Karr, C. E. O'Neil, M. R. Wilson, V. Dharmarajan, J. E. Salvaggio, and H. Weill. Inhalation challenge and pharmacologic studies of toluene diisocyanate (TDI)-sensitive workers. J. Allergy Clin. Immunol. 64:146-152, 1979.
16. Butcher, B. T., J. E. Salvaggio, C. E. O'Neil, H. Weill, and O. Garg. Toluene diisocyanate pulmonary disease: Immunopharmacologic and mecholyl challenge studies. J. Allergy Clin. Immunol. 59:223-227, 1977.
17. Butcher, B. T., J. E. Salvaggio, H. Weill, and M. M. Ziskind. Toluene diisocyanate (TDI) pulmonary disease: Immunologic and inhalation challenge studies. J. Allergy Clin. Immunol. 58:89-100, 1976.
18. Crittenden, A. Built-in fumes plague homes. New York Times, Section 3. Business and Finance, Sunday, May 7, 1978.
19. Curry, J. J. Comparative action of acetyl-beta-methyl choline and histamine on the respiratory tract in normals, patients with hay fever, and subjects with bronchial asthma. J. Clin. Invest. 26:430-438, 1947.
20. Earp, S. E. The physiological and toxic actions of formaldehyde. With a report of three cases of poisoning by formalin. N.Y. Med. J. 104:391-392, 1916.
21. Ely, F. A. Formaldehyde poisoning. J. Am. Med. Assoc. 54:1140-1141, 1910.
22. Ettinger, I., and M. Jeremias. A study of the health hazards involved in working with flameproofed fabric. N.Y. State Dep. Labor, Div. Ind. Hyg. Monthly Rev. 34(7):25-27, 1955.
23. Fassett, D. W. Aldehydes and acetals, pp. 1959-1989. In F. A. Patty, Ed. Industrial Hygiene and Toxicology. 2nd rev. ed. D. F. Fassett and D. D. Irish, Eds. Vol. II. Toxicology. New York: John Wiley & Sons, Inc., 1963.
24. Fel'dman, Y. G., and T. I. Bonashevskaya. On the effects of low concentrations of formaldehyde. Hyg. Sanit. 36(5):174-180, 1971.
25. Fisher, A. A. Contact Dermatitis. 2nd ed. Philadelphia: Lea and Febiger, 1973. 448 pp.
26. Freeman, H. G., and W. C. Grendon. Formaldehyde detection and control in the wood industry. For. Prod. J. 21(9):54-57, 1971.
27. Fregert, S. Manual of Contract Dermatitis. Copenhagen: Munksgaard, 1974. 107 pp.
28. Fregert, S., and H. J. Bandmann. Patch Testing. New York: Springer-Verlag, 1975. 78 pp.
29. Gamble, J. F., A. J. McMichael, T. Williams, and M. Battigelli. Respiratory function and symptoms: An environmental-epidemiological study of rubber workers exposed to a phenol-formaldehyde type resin. Am. Ind. Hyg. Assoc. J. 37:499-513, 1976.
30. Golden, J. A., J. A. Nadel, and H. A. Boushey. Bronchial hyperirritability in healthy subjects after exposure to ozone. Am. Rev. Respir. Dis. 118:287-294, 1978.
31. Governor's Task Force on Insulation. Report on U-F Foam Insulation. Hartford, Ct: Connecticut Department of Consumer Protection, 1978. 106 pp.

32. Grossman, L. I. Paresthesia from N2 or N2 substitute. Report of a case. Oral Surg. Oral Med. Oral Pathol. 45:114-115, 1978.
33. Heling, B., Z. Ram, and I. Heling. The root treatment of teeth with Toxavit. Report of a case. Oral Surg. Oral Med. Oral Pathol. 43:306-309, 1977.
34. Helwig, H. Wie ungefährlich ist Formaldehyd? Dtsch. Med. Wochenschr. 102:1612-1613, 1977. (in German)
35. Hendrick, D. J., and D. J. Lane. Formalin asthma in hospital staff. Br. Med. J. 1:607-608, 1975.
36. Hendrick, D. J., and D. J. Lane. Occupational formalin asthma. Br. J. Ind. Med. 34:11-18, 1977.
37. Hilgemeier, M. W. Presentation on New Hampshire experiences with urea-formaldehyde foam, given at Ad Hoc Task Force Seminar on an Assessment of the Odor Problems from U-F Foam Insulations, Washington, D.C., December 1, 1978.
38. Hogg, J. C., P. D. Paré, and R. C. Boucher. Bronchial mucosal permeability. Fed. Proc. 38:197-201, 1979.
39. Hollowell, C. J. Presentation given at Ad Hoc Task Force Seminar on An Assessment of the Odor Problems from U-F Foam Insulations. Washington, D.C., December 1, 1978.
40. Humpstone, O. P., and W. Lintz. A case of formalin poisoning. J. Am. Med. Assoc. 52:380-381, 1909.
41. Jordan, W. P., Jr., W. T. Sherman, and S. E. King. Threshold responses in formaldehyde-sensitive subjects. J. Am. Acad. Dermatol. 1:44-48, 1979.
42. Kerfoot, E. J., and T. F. Mooney, Jr. Formaldehyde and paraformaldehyde study in funeral homes. Am. Ind. Hyg. Assoc. J. 36:533-537, 1975.
43. Kline, B. S. Formaldehyd [sic] poisoning. With report of a fatal case. Arch. Intern. Med. 36:220-228, 1925.
44. Kratochvil, I. The effect of formaldehyde on the health of workers employed the production of crease resistant ready made dresses. Pr. Lek. 23:374-375, 1971. (in Czech; English abstract)
45. Laffont, H., and J.-B. Noceto. A case of asthma due to sensitivity to formaldehyde. Algérie Méd. 65:777-781, 1961. (in French)
46. Lam, S., R. Wong, and M. Yeung. Nonspecific bronchial reactivity in occupational asthma. J. Allergy Clin. Immunol. 63:28-34, 1979.
47. Leonardos, G., D. Kendall, and N. Barnard. Odor threshold determinations of 53 odorant chemicals. J. Air Pollut. Control Assoc. 19:91-95, 1969.
48. Levison, L. A. A case of fatal formaldehyde poisoning. J. Am. Med. Assoc. 42:1492, 1904.
49. Logan, W. S., and H. O. Perry. Contact dermatitis to resin-containing casts. Clin. Orthop. Relat. Res. 90:150-152, 1973.
50. March, G. H. Formalin poisoning; recovery. Br. Med. J. 2:687, 1927.
51. Marzulli, F. N., and H. I. Maibach. The use of graded concentrations in studying skin sensitizers: Experimental contact sensitization in man. Food Cosmet. Toxicol. 12:219-227, 1974.

52. Melekhina, V. P. Hygienic evaluation of formaldehyde as an atmospheric air pollutant, pp. 9-18. In B. S. Levine (trans.) USSR Literature on Air Pollution and Related Occupational Diseases. Vol. 9. A Survey. Washington, D.C.: U.S. Public Health Service, 1963-1964. (available from National Technical Information Service, Springfield, Va., as TT64-11574)

53. Mills, J. CPSC warns about health hazard of foam home material. Washington Post, Real Estate Section, Saturday, August 11, 1979.

54. Montgomery, S. Paresthesia following endodontic treatment. J. Endodon. 2:345-347, 1976.

55. Morrill, E. E., Jr. Formaldehyde exposure from paper process solved by air sampling and current studies. Air Cond. Heat. Vent. 58(7):94-95, 1961.

56. National Research Council, Committee on Aldehydes. Formaldehyde and Other Aldehydes. Washington, D.C.: National Academy Press, 1981. [354] pp.

57. National Research Council, Committee on Biologic Effects of Atmospheric Pollutants. Particulate Polycyclic Organic Matter. Washington, D.C.: National Academy of Sciences, 1972. 361 pp.

58. National Research Council, Committee on Medical and Biologic Effects of Environmental Pollutants. Vapor-Phase Organic Pollutants. Volatile Hydrocarbons and Oxidation Products. Washington, D.C.: National Academy of Sciences, 1976. 411 pp.

59. National Research Council, Committee on Toxicology. Formaldehyde --An Assessment of Its Health Effects. Washington, D.C.: National Academy of Sciences, 1980. 38 pp.

60. North American Contact Dermatitis Group. Epidemiology of contact dermatitis in North America: 1972. Arch. Dermatol. 108:537-540, 1973.

61. Nova, H., and R. G. Touraine. Asthme au formol. Arch. Mal. Prof. 18:293-294, 1957. (in French)

62. Odom, R. B., and H. I. Maibach. Contact urticaria: A different contact dermatitis, pp. 441-453. Chapter 15 in F. N. Marzulli, and H. I. Maibach, Eds. Advances in Modern Toxicology. Vol. 4. Dermatotoxicology and Pharmacology. Washington, D.C.: Hemisphere Publishing Corporation, 1977.

63. Orringer, E. P., and W. D. Mattern. Formaldehyde-induced hemolysis during chronic hemodialysis. N. Engl. J. Med. 294:1416-1420, 1976.

64. Paliard, F., L. Roche, C. Exbrayat, and E. Sprunck. Chronic asthma due to formaldehyde. Arch. Mal. Prof. 10:528-530, 1949. (in French)

65. Parker, C. D., R. E. Bilbo, and C. E. Reed. Methacholine aerosol as test for bronchial asthma. Arch. Intern. Med. 115:452-458, 1965.

66. Pepys, J., C. A. C. Pickering, A. B. X. Breslin, and D. J. Terry. Asthma due to inhaled chemical agents--tolylene di-isocyanate. Clin. Allergy 2:225-236, 1972.

67. Popa, V., D. Teculescu, D. Stănescu, and N. Gavrilescu. Bronchial asthma and asthmatic bronchitis determined by simple chemicals. Dis. Chest 56:395-404, 1969.

68. Porter, J. A. H. Acute respiratory distress following formalin inhalation. Lancet 2:603-604, 1975.

69. Proctor, D. F. The upper airways. I. Nasal physiology and defense of the lungs. Am. Rev. Respir. Dis. 115:97-129, 1977.

70. Rathery, F., R. Piédelièvre, and J. Delarue. Death by absorption of formalin. Ann. Méd. Lég. Criminol. 20:201-206, 1940. (in French)

71. Rostenberg, A., Jr., B. Bairstow, and T. W. Luther. A study of eczematous sensitivity to formaldehyde. J. Invest. Dermatol. 19:459-462, 1952.

72. Roth, W. G. Tylosic palmar and plantar eczema caused by steaming clothes containing formalin. Berufsdermatosen 17:263-268, 1969.

73. Sakula, A. Formalin asthma in hospital laboratory staff. Lancet 2:816, 1975.

74. Schoenberg, J. B., and C. A. Mitchell. Airway disease caused by phenolic (phenol-formaldehyde) resin exposure. Arch. Environ. Health 30:574-577, 1975.

75. Shipkovitz, H. D. Formaldehyde Vapor Emissions in the Permanent-Press Fabrics Industry. Report No. TR-52. Cincinnati: U.S. Department of Health, Education, and Welfare, Public Health Service, Consumer Protection and Environmental Health Service, Environmental Control Administration, 1968. 18 pp.

76. Shy, C. M., J. R. Goldsmith, J. D. Hackney, M. D. Lebowitz, and D. B. Menzel. Health effects of air pollution. Paper presented at meeting of American Thoracic Society, Medical Section of American Lung Association, 1978.

77. Sim, V. M., and R. E. Pattle. Effect of possible smog irritants on human subjects. J. Am. Med. Assoc. 165:1908-1913, 1957.

78. Sneddon, I. B. Dermatitis in an intermittent haemodialysis unit. Br. Med. J. 1:183-184, 1968.

79. Spector, S. L., and R. S. Farr. A comparison of methacholine and histamine inhalations in asthmatics. J. Allergy Clin. Immunol. 56:308-316, 1975.

80. Tabershaw, I. R., H. N. Doyle, L. Gaudette, S. H. Lamm, and O. Wong. A Review of the Formaldehyde Problems in Mobile Homes. Report to National Particleboard Association. Rockville, Md.: Tabershaw Occupational Medicine Associates, P.A., 1979. 19 pp.

81. Townley, R. G., A. K. Bewtra, N. M. Nair, F. D. Brodkey, G. D. Watt, and K. M. Burke. Methacholine inhalation challenge studies. J. Allergy Clin. Immunol. 64:569-574, 1979.

82. Turiar, C. Asthma through sensitivity to formaldehyde. Soc. Franc. d'Allergie, Seance du 18 Nov. 1952.

83. Uehara, M. Follicular contact dermatitis due to formaldehyde. Dermatologica 156:48-54, 1978.

84. U.S. Consumer Product Safety Commission. News from CPSC. Wednesday, August 1, 1979.

85. U.S. Consumer Product Safety Commission, Directorate for Hazard Identification and Analysis--Epidemiology. Summaries of in-depth investigations, newspaper clippings, consumer complaints and state reports on urea-formaldehyde foam home insulation. Washington, D.C.: U.S. Consumer Product Safety Commission, July 1978.

86. U.S. Department of Health, Education, and Welfare, National Heart, Lung, and Blood Institute. Working Group on Heart Disease Epidemiology. DHEW (NIH) Publication No. 79-1667. Washington, D.C.: U.S. Department of Health, Education, and Welfare, 1979. 69 pp.

87. U.S. Department of Health, Education, and Welfare, National Heart, Lung and Blood Institute, Division of Lung Disease. Respiratory Diseases. Task Force Report on Prevention, Control, Education, pp. 84-91. DHEW Publication No. (NIH) 77-1248. Washington, D.C.: U.S. Government Printing Office, 1977.

88. U.S. Department of Health, Education, and Welfare, National Institute for Occupational Safety and Health. Criteria for a Recommended Standard... Occupational Exposure to Formaldehyde. DHEW (NIOSH) Publication No. 77-126. Washington, D.C.: U.S. Government Printing Office, 1976. 165 pp.

89. Vaughan, W. T. The Practice of Allergy, p. 677. St. Louis: The C. V. Mosby Company, 1939.

90. Waldbott, G. L. Health Effects of Environmental Pollutants. Saint Louis: The C. V. Mosby Company, 1973. 316 pp.

91. Walker, J. F. Formaldehyde, pp. 77-99. In A. Standen, Ed. Kirk-Othmer Encyclopedia of Chemical Technology. 2nd rev. ed. Vol. 10. New York: Interscience Publishers, 1966.

92. Wayne, L. G., R. J. Bryan, and K. Ziedman. Irritant Effects of Industrial Chemicals: Formaldehyde. DHEW (NIOSH) Publication No. 77-117. Washington, D.C.: U.S. Government Printing Office, 1976. [138] pp.

93. Weger, A. Thalamischer Symptomenkomplex bei Formalinintoxikation. Z. Ges. Neurol. Psych. 111:370-382, 1927. (in German)

94. Wisconsin Division of Health, Bureau of Prevention. Formaldehyde Case File Summary, October 23, 1978. Madison: Wisconsin Division of Health, 1978. 3 pp.

95. Wisconsin Division of Health, Bureau of Prevention. Statistics of particle board related formaldehyde cases through December 15, 1978. Madison: Wisconsin Division of Health, 1978. [4] pp.

96. Woodbury, J. W. Asthmatic syndrome following exposure to tolylene dissocyanate. Ind. Med. Surg. 25:540-543, 1956.

97. World Health Organization. Health Hazards of the Human Environment. Geneva: World Health Organization, 1972. 387 pp.

98. Yefremov, G. G. State of the upper respiratory tract in formaldehyde production workers. Zh. Ushn. Nos. Gorl. Bolezn. 30(5):11-15, 1970. (in Russian; English summary)

99. Zaeva, G. N., I. P. Ulanova, and L. A. Dueva. Materials for revision of the maximal permissible concentrations of formaldehyde in the inside atmosphere of industrial premises. Gig. Tr. Prof. Zabol. 12:16-20, 1968. (in Russian)

100. Zannini, D., and L. Russo. Long-standing lesions in the respiratory tract following acute poisoning with irritating gases. Lav. Um. 9:241-254, 1957. (in Italian; English summary)

FIBROUS BUILDING MATERIALS

Current knowledge of the adverse health effects of fibrous building materials is reasonably complete for only one such material: asbestos. Other materials, such as fibrous glass and rock wool, are becoming more widely used, but little is known about their health effects in humans. The continuing widespread use of fibrous building materials in the absence of an adequate understanding of their potential health impacts has led to public-health concern. The details of their production, use, emission, and control are reviewed in Chapter IV.

The nature of health problems varies with the organ system that may be affected. Fibrous materials can have a direct effect due to contact with the skin, can affect the lungs because of inhalation, and can affect the gastrointestinal tract because of inadvertent ingestion. Furthermore, every organ in the body may be affected through transport of fibers by the hematogenous or lymphatic systems. There does not appear to be any uptake of fibrous material through the skin.

Inhalation is the major route of entry of fibrous particulate matter. Deposition and retention depend on the usual factors of respiratory physiology and on fiber dimension. Most inhaled fibrous material is cleared by the mucociliary escalatory clearance mechanism, which results in inadvertent ingestion. Other inhaled particles are retained in the lung, where they accumulate. Some fibers, through uncertain routes, migrate to the pleura and stay there. Fiber translocation from the lung also occurs via the hematogenous and lymphatic systems, with eventual accumulation of fibers in virtually every organ of the body. The fractions that are disseminated by this mechanism and through the gastrointestinal system are not known.

That dispersion takes place through the gastrointestinal tract has been clearly shown in animals.[8] Gavage of asbestos or the presence of asbestos in drinking water or food results in transport across the intestinal wall into the peritoneal cavity and into the bloodstream, which leads to deposition throughout the body. After gavage, asbestos has been found in all organs examined, including kidney, liver, pancreas, and brain. It can also cross the placenta.[9]

Asbestos and other fibrous materials are not metabolized after entering the body, but there is leaching of chemical constituents, which varies with fiber type and size. Fibers can take up biologic residence in many organs, and most fibers remain uncoated. Some fibers become coated with an iron-protein matrix and form an "asbestos body" when asbestos is the core material. When the core is not identified, it is known as a "ferruginous body." The intracellular process that produces this coating has been well documented by Suzuki and Churg.[49] Some structural and compositional changes occur after fibers are taken up in tissue, particularly in the lung. In the case of chrysotile, magnesium leaches out of the fiber; this ultimately changes the structure of chrysotile and makes quantitative recovery from lung tissue less secure over time. The mechanism by which asbestos and other fibrous materials produce fibrosis appears to be different from that of silica, which acts by causing secondary lysosomal release of enzymes in macrophages that lead to a fibrotic tissue reaction. The interaction of asbestos and macrophages in vivo is less well understood, but some intact macrophages contain small fibers, and others surround parts of large fibers.

SPECIFIC HEALTH EFFECTS

Health effects resulting from exposure to fibrous materials can be divided into nonmalignant and malignant effects. A division into acute and chronic effects is of little value. Except for the acute cutaneous irritation produced by fibrous glass products, including those treated with resins and lubricants, the important effects of exposure to fibrous materials are chronic, and they often have very long latent periods. No initial short-term health effects of any major consequence are seen in either healthy adults or persons with pre-existing conditions, because of the long latent period of the effects.

All mineral fiber types have been shown in laboratory animals to be capable of producing a wide spectrum of disease when administered as long, thin fibers.

Nonmalignant Effects

Cutaneous Effects. Asbestos has been shown to produce granulomatous warts on the hands, and asbestos fibers have been found in these growths. Whether asbestos alone is the causative agent or viruses play a role has not been studied. It has been known for several decades that fibrous glass materials can produce severe irritation in those working with them. This may be a function of direct physical contact or of a chemical process related to resins and lubricants. It can be prevented by the wearing of long-sleeved clothing and gloves.

Findings in Sputum. Asbestos bodies were first described (although not so named until 1929) on the basis of autopsy findings in two deaths in 1906.[28] These fibers coated with iron-protein have been used as a marker for previous asbestos exposure, not as an indicator of disease or severity of disease. As a marker of exposure, their appearance works well. The coating of fibers in vivo varies with species.

Both uncoated fibers and asbestos bodies have been found in sputum and give some guidance as to extent of exposure, particularly if the exposure has been intense. Digestion and examination of lung tissue often reveal the presence of asbestos bodies and fibers; in approximately half of 3,000 consecutive autopsies in New York City, optical microscopy uncovered asbestos bodies in lung tissue.[24] Electron microscopic examination is preferred for more complete evaluation.

Pulmonary Disease. The most important nonmalignant health effect of asbestos exposure is the change in the pulmonary system. Asbestosis, lung scarring, can result from asbestos exposure and is a leading cause of death among some occupationally exposed groups. "Asbestosis" refers to the fibrotic process of either the lung parenchyma or the visceral pleura. If one is referring only to lung tissue, the term "parenchymal asbestosis" is preferable. "Asbestosis" is a clinical diagnosis that can be made in the absence of active symptoms or any apparent ill health. "Impairment" is a clinical judgment based on findings in the history and in physical and laboratory examinations. Not everyone with disease is impaired. The presence of disease without impairment should always suggest the possibility of future impairment or future additional disease. "Disability" is a legal concept related to the absence or presence of disease and impairment, often with regard to specific functions.

Fibrosis of the lung parenchyma after asbestos exposure is related to the degree of exposure and the period since the beginning of exposure. It has been documented[40] that increasing exposure leads to an increase in the incidence of asbestosis and an increase in its severity. The degree of asbestosis is usually measured in terms of the ILO-U/C classification for x-ray films (issued by the International Labour Office, 1971, Geneva), which evaluates changes on a 12-point scale and includes types of changes and associated findings.

Radiologically evident asbestosis generally develops only after considerable time has passed since first exposure, unless exposure has been intense. The appearance of extensive asbestosis in less than 20 yr is unusual. If asbestosis does occur within that period, it will probably not be far advanced. The exposure need not be continuous over this period; and relatively brief exposure may be sufficient to cause disease many years later.

In addition to parenchymal asbestosis, changes may develop in the pleura. Pleural fibrosis, with or without calcification, is a common finding after asbestos exposure. There is increasing evidence that the extent of exposure required to produce pleural changes may be less than that associated with parenchymal changes.[26] There is no reliable correlation between a finding of nonmalignant pulmonary changes and the predictability of development of neoplasia, which can develop without radiologic evidence of asbestosis.

Malignant Effects

A variety of malignant neoplasms are associated with exposure to asbestos.

Lung Cancer. Lung cancer is found in great excess among workers occupationally exposed to asbestos. This has been noted with all commercially important types of asbestos. Asbestos-related tumors tend to be in the lower lobes and peripheral, following the pattern of the parenchymal changes of asbestosis, although neoplasms are also increased in the upper lobes. The cell type distribution does not appear to be altered with asbestos-related cancers.[22] Cigarette-smoking acts synergistically to increase the risk of developing lung cancer.[41]

Mesothelioma. Both pleural and peritoneal mesotheliomas are seen after exposure to asbestos. Pleural mesotheliomas are generally more common; peritoneal mesotheliomas tend to take longer to develop and tend to occur in large numbers only among more heavily exposed populations. Their incidence is not related to cigarette-smoking.

Other Cancers. Other malignancies are found in excess after asbestos exposure. Of particular importance is the increase in gastrointestinal tract cancers, particularly of the colon and rectum, stomach, and esophagus.[42] Pancreatic cancer does not appear to be increased. Oropharyngeal cancers and laryngeal cancers are also increased.[42]

Among women, ovarian cancer has been reported after asbestos exposure,[15] but more data on this question are needed.

LABORATORY EVIDENCE OF HEALTH EFFECTS

An increasing number of reports have demonstrated adverse effects *in vivo* or *in vitro*, many in parallel with demonstrated human health effects.

In Vivo Effects

In vivo experimentation has documented all the major health effects caused by asbestos, and in some cases other materials have caused similar changes.

Inhalation. Inhalation studies with asbestos of several fiber types have documented the risk in animals of developing asbestosis and malignancy, including lung cancer and mesothelioma. Wagner *et al.*[55] and Davis *et al.*[10] have shown that all major fiber types produce both malignancies noted, and that the usual dose-response relationship holds. Wagner *et al.* and Bernstein *et al.*[4] are conducting studies to evaluate the hazards associated with inhalation of fibrous glass products and talc.

Injection. Intratracheal, intrapleural, and intraperitoneal injections into laboratory animals of asbestos, fibrous glass, and other fibrous material have been associated with disease production. Lung cancer has been produced in hamsters with intratracheal instillation.[46] Peritoneal mesothelioma has been produced by injections or other placement of asbestos into the pleural cavity. Stanton *et al.*[47] [48] produced neoplasms with long, thin fibers of various kinds, including asbestos, fibrous glass products, and such other fibrous materials as fibrous dawsonite, an aluminum carbonate. Peritoneal mesotheliomas have developed after intraperitoneal injections of asbestos.[44] Wright and Kuschner[58] instilled asbestos, glass, and other fibers into guinea pigs intratracheally; when the fibers were long and thin, all the materials produced lung fibrosis.

In Vitro Effects

In vitro techniques are increasingly being used to study the effects of asbestos and other fibrous materials. A comprehensive review on this subject was published by Harington *et al.*[17]

Macrophages. Asbestos and other fibrous materials may be toxic to macrophages. This was first shown by Harington *et al.*[17] in preparations of freshly prepared macrophages and more recently by Wade *et al.*[52] in continuously cultured malignant macrophage-like cells. Thus, the cytotoxic potential of a variety of fibers is clear. There is new evidence that the cytotoxic potential of fibrous materials may parallel their carcinogenic potential.[6,53] Miller has reviewed the effects of asbestos on macrophages as shown with electron microscopy.[30]

Fibroblasts. Addition of asbestos to cultures of fibroblasts has demonstrated alterations in both cellular biochemistry and morphologic appearance.[36,51] These chemical alterations have not yet been related to changes in human lung tissue.

Other Cell Systems. Other cell systems have been used to study the effects of asbestos. Schnitzer *et al.*[39] used erythrocytes as a test system for the evaluation of biologic effects of a variety of dusts, and work with mesothelial cell culture[21] has begun to increase the understanding of changes in this cell type brought about by exposure to such dusts as asbestos.

Organ Culture

Another approach to the understanding of the effects of asbestos and other dusts has been the use of organ culture systems. Mossman *et al.*[31] investigated the effects of crocidolite on hamster tracheal cultures, and Frank[12] studied the effects of amosite and chrysotile. Hyperplasia of basal cells was the most prominent morphologic change noted. Rajan *et al.*[35] studied the effects of asbestos on human pleura in organ culture and also noted hyperplasia after the addition of asbestos. Fibrous glass products are currently under study with organ culture techniques.

EPIDEMIOLOGY AND OCCUPATIONAL EXPOSURE

The first case of asbestosis reviewed for compensation purposes was observed in 1906 in England by Murray.[32] Additional cases were reported later, and the disease was better understood by 1930. In 1935, Lynch and Smith[27] reported the first case of lung cancer in a man who worked in an asbestos factory and suggested a causal relationship. The historical development of the understanding of asbestos-related disease was reviewed by Selikoff and Lee.[43]

Asbestosis

Cases of asbestosis were known earlier, but it was during 1930 that Merewether and Price[29] extensively reviewed the association of asbestos exposure and the development of fibrosis. More than 25% of 363 asbestos manufacturing workers had evidence of pulmonary fibrosis related to their exposure to asbestos. Similar findings were reported in the United States by Lanza et al.[25] For the development of asbestosis in relation to exposure and to the period from onset of exposure, the information reported by Selikoff and Lee[43] clearly demonstrated a dose-response relationship and showed that development and severity of asbestosis increase with time.

Other Chronic Lung Diseases

There have been few epidemiologic reports of other chronic lung disease after exposure to nonasbestos fibrous materials. Bayliss et al.[3] reported no increased lung-cancer mortality, but an increase in nonmalignant respiratory-disease deaths among fibrous-glass production workers. Boehlecke et al.[5] have reported on a group of workers exposed to wollastonite, a fibrous monocalcium silicate used as an asbestos substitute. They observed none of the clinical stigmata usually seen with asbestos exposure. Only 36% of workers had been exposed more than 15 yr before the study.

Cancer Epidemiology

The 1955 report by Doll[11] was a landmark in the establishment of a relationship between lung cancer and asbestos exposure. Other reports soon followed; several were represented at the New York Academy of Sciences in 1965.[57]

Before 1960, mesothelioma had been a rarely reported disease, although some cases had been seen in asbestos workers. In that year, Wagner et al.[56] reported 47 cases that had occurred during a 4-yr period in the asbestos-mining area of South Africa. Further reports appeared soon after, and the strong causal relationship with asbestos exposure has now been firmly established. Cochrane and Webster[7] found that 69 of 70 cases of mesothelioma at one South African hospital were associated with substantial asbestos exposure.

Principles

Through the epidemiologic investigations of occupationally exposed groups, several principles of exposure-disease relationship have become clear. The question of latency is now well understood: in general, 15-20 yr must pass before the signs of marked asbestos-related disease are detected, by either x-ray study, pulmonary-function results, or physical findings. The exposure may have been brief; if it was intense enough, disease may result, although in almost all cases there is a long latency period.

It has become clear that both the severity of asbestosis and the risk of developing lung cancer or mesothelioma depends on the total exposure to fibers. Because of the paucity of data on workplace or environmental concentrations, exposures are almost never known exactly, and only rough estimates of exposure can be agreed on by most authorities.

Exposure in the workplace can be either direct or indirect. In many cases, indirect, or "bystander," exposure has proved hazardous. This has been established in shipyard workers[18 19] and maintenance workers.[26] Thus, indirect exposure can be hazardous to people who work near those specifically assigned to handle asbestos.

Other Fibrous Mineral Materials and Dusts

No epidemiologic investigation has demonstrated substantial health hazards related to other fibrous materials and dusts that might contaminate the indoor environment. The asbestos substitutes used in construction are relatively new, at least with respect to their possibly producing human health effects, and little is known of their hazards. Those materials include the slag wools, rock wools, glass wools, and filaments. The subject of man-made mineral fibers has been reviewed by Hill[20] and Wagner *et al.*[54]

NONOCCUPATIONAL EXPOSURE

Much less is known about exposure to fibrous materials away from the workplace than about occupational exposure.

Neighborhood Exposure

There have been several studies of exposures of persons living near asbestos production facilities. Wagner's original report on cases of mesothelioma in South Africa included mainly persons who lived near asbestos mines or along the routes of transport. Newhouse and Thompson[33] showed that a substantial number of cases of mesothelioma at the London Hospital between 1910 and 1965 were in persons who lived within 0.5 mile of a production facility in East London. A more recent study showed no such relationship in persons who lived near an asbestos production facility in New Jersey, at least in less than 35 yr from onset of its operations.[16]

Of increasing recent interest is the situation in parts of Turkey, where it appears that naturally occurring fibrous zeolites in the environment give rise to numerous cases of mesothelioma among the general population.[2]

Family Exposure

Among the best recorded relationships between household contamination and disease development are those among families of asbestos workers. Particularly striking is the development of mesotheliomas among wives, children, and other family contacts of asbestos workers who have brought asbestos home on their persons and clothing. Contamination of the living environment resulted; 20 yr or more later, mesotheliomas appeared. In addition, roughly one-third of such persons had x-ray changes consistent with asbestosis. Studies on this subject were recently reported by Anderson et al.[1] and Tagnon et al.[50]

Exposure in Buildings

Only within the last several years has the scientific community become aware of the widespread use of asbestos in public buildings in ways that might be related to substantial risk. Asbestos is used in insulation and fireproofing materials, in ornamental decoration and soundproofing, and in large quantities on surfaces in public areas. Its use in school buildings has been the subject of recent reviews[34,38,45] that included discussion of potential health problems and suggestions of control measures. Public areas in other buildings can also become contaminated, especially during routine or other maintenance procedures, which may aerosolize friable asbestos coatings.

There have been few measurements of air concentration. Especially lacking are studies that put members of the general population or schoolchildren under prospective surveillance to see what (if any) adverse health effects occur. Refinement of risk estimates of effects in the general population awaits additional understanding and measurement of the potential effects of long-term low-concentration exposure. The relationship of total exposure to the age when first exposure occurs is not known.

Other Exposures

The general environment may become contaminated from ambient air or water pollution with asbestos or other fibrous material. This is of special concern for houses in communities with mining and processing facilities. Household exposures can also occur elsewhere, although few measurements have been made to demonstrate the extent of such exposures.

In Montgomery County, Maryland, widespread ambient-air contamination has resulted from the long use of crushed rock containing asbestos for the paving of roads, parking lots, and playgrounds. Air concentrations in the community were reported to be similar to those in some working environments,[37] and the particular asbestos that was contaminating the air was shown to have substantial biologic activity.[13]

Water contamination can also occur from the dumping of industrial wastes into lakes, as evidenced by the asbestos found in the water of some communities that take their municipal supplies from Lake Superior.[23]

Consumer products brought into the home may contain asbestos. The Consumer Product Safety Commission has banned artificial fireplace logs and hair-dryers containing asbestos. The subject of consumer-product asbestos hazards has been discussed by Franklin.[14]

It is clear that the household can become contaminated. Although scientific knowledge of specific health effects is sparse, the state of knowledge about the adverse consequences of asbestos exposure elsewhere justifies a cautious approach to any exposure.

REFERENCES

1. Anderson, H. A., R. Lilis, S. M. Daum, and I. J. Selikoff. Asbestosis among household contacts of asbestos factory workers. Ann. N.Y. Acad. Sci. 330:387-399, 1979.
2. Barış, Y. I., M. Artvınlı, and A. A. Şahin. Environmental mesothelioma in Turkey. Ann. N.Y. Acad. Sci. 330:423-432, 1979.
3. Bayliss, D. L., J. M. Dement, J. K. Wagoner, and H. P. Blejer. Mortality patterns among fibrous glass production workers. Ann. N.Y. Acad. Sci. 271:324-335, 1976.
4. Bernstein, D. M., R. T. Drew, and M. Kuschner. Experimental approaches for exposure to sized glass fibers. Environ. Health Perspect. 34:47-57, 1980. (includes comments)
5. Boehlecke, B. A., D. M. Shasby, M. R. Petersen, T. K. Hodous, and J. A. Merchant. Respiratory morbidity of wollastonite workers. Am. Rev. Respir. Dis. 117 (Suppl.):219, 1978. (abstract)
6. Chamberlain, M., R. C. Brown, R. Davies, and D. M. Griffiths. In vitro prediction of the pathogenicity of mineral dusts. Br. J. Exp. Pathol. 60:320-327, 1979.
7. Cochrane, J. C., and I. Webster. Mesothelioma in relation to asbestos fibre exposure. S. Afr. Med. J. 54:279-281, 1978.
8. Cunningham, H. M., and R. D. Pontefract. Asbestos fibers in beverages, drinking water, and tissues: Their passage through the intestinal wall and movement through the body. J. Assoc. Off. Anal. Chem. 56:976-981, 1973.
9. Cunningham, H. M., and R. D. Pontefract. Placental transfer of asbestos. Nature 249:177-178, 1974.
10. Davis, J. M. G., S. T. Beckett, R. E. Bolton, P. Collings, and A. P. Middleton. Mass and number of fibres in the pathogenesis of asbestos-related lung disease in rats. Br. J. Cancer 37:673-688, 1978.
11. Doll, R. Mortality from lung cancer in asbestos workers. Br. J. Ind. Med. 12:81-86, 1955.
12. Frank, A. L. Asbestos-induced changes in hamster trachea organ culture. In R. C. Brown, M. Chamberlain, R. Davies, and I. P. Gormley, Eds. The In Vitro Effects of Mineral Dusts. London: Academic Press Inc., London Ltd., 1980.

13. Frank, A. L., A. N. Rohl, M. J. Wade, and L. E. Lipkin. Biological activity in vitro of chrysotile compared to its quarried parent rock (platy serpentine). J. Environ. Pathol. Toxicol. 2:1041-1046, 1979.

14. Franklin, B. H. Public health control of environmental asbestos disease: Consumer products. Ann. N.Y. Acad. Sci. 330:497-501, 1979.

15. Graham, J., and R. Graham. Ovarian cancer and asbestos. Environ. Res. 1:115-128, 1967.

16. Hammond, E. C., L. Garfinkel, I. J. Selikoff, and W. J. Nicholson. Mortality experience of residents in the neighborhood of an asbestos factory. Ann. N.Y. Acad. Sci. 330:417-422, 1979.

17. Harington, J. S., A. C. Allison, and D. V. Badami. Mineral fibers: Chemical, physiochemical, and biological properties. Adv. Pharmacol. Chemother. 12:291-402, 1975.

18. Harries, P. G. Asbestos hazards in Naval Dockyards. Ann. Occup. Hyg. 11:135-145, 1968.

19. Harries, P. G. Experience with asbestos disease and its control in Great Britain's Naval Dockyards. Environ. Res. 11:261-267, 1976.

20. Hill, J. W. Health aspects of man-made mineral fibres. A review. Ann. Occup. Hyg. 20:161-173, 1977.

21. Jaurand, M.-C., H. Kaplan, J. Thiollet, M.-C. Pinchon, J.-F. Bernaudin, and J. Bignon. Phagocytosis of chrysotile fibers by pleural mesothelial cells in culture. Am. J. Pathol. 94:529-538, 1979.

22. Kannerstein, M., and J. Churg. Pathology of carcinoma of the lung associated with asbestos exposure. Cancer 30:14-21, 1972.

23. Langer, A. M., C. M. Maggiore, W. J. Nicholson, A. N. Rohl, I. B. Rubin, and I. J. Selikoff. The contamination of Lake Superior with amphibole gangue minerals. Ann. N.Y. Acad. Sci. 330:549-572, 1979.

24. Langer, A. M., I. J. Selikoff, and A. Sastre. Chrysotile asbestos in the lungs of persons in New York City. Arch. Environ. Health 22:348-361, 1971.

25. Lanza, A. J., W. J. McConnell, and J. W. Fehnel. Effects of the inhalation of asbestos dust on the lungs of asbestos workers. Public Health Rep. 50:1-12, 1935.

26. Lilis, R., S. Daum, H. Anderson, M. Sirota, G. Andrews, and I. J. Selikoff. Asbestos disease in maintenance workers of the chemical industry. Ann. N.Y. Acad. Sci. 330:127-135, 1979.

27. Lynch, K. M., and W. A. Smith. Pulmonary asbestosis. III. Carcinoma of lung in asbesto-silicosis. Am. J. Cancer 24:56-64, 1935.

28. Marchand, F. Über Eigentumliche Pigmentkristalle in den Lungen. Dtsch. Path. Ges. Verh. 17:223, 1906.

29. Merewether, E. R. A., and C. W. Price. Report on Effects on Asbestos Dust on the Lungs and Dust Suppression in the Asbestos Industry. London: His Majesty's Stationery Office, 1930. 34 pp.

30. Miller, K. The effects of asbestos on macrophages. CRC Crit. Rev. Toxicol. 5:319-354, 1978.

31. Mossman, B. T., J. B. Kessler, B. W. Ley, and J. E. Craighead. Interaction of crocidolite asbestos with hamster respiratory mucosa in organ culture. Lab. Invest. 36:131-139, 1977.

32. Murray, H. M. Report of the Departmental Committee on Compensation for Industrial Disease. London: His Majesty's Stationery Office, 1907.
33. Newhouse, M. L., and H. Thompson. Mesothelioma of pleura and peritoneum following exposure to asbestos in the London area. Br. J. Ind. Med. 22:261-269, 1965.
34. Nicholson, W. J., A. N. Rohl, R. N. Sawyer, E. J. Swoszowski, Jr., and J. D. Todaro. Control of sprayed asbestos surfaces in school buildings: A feasibility study. Report to the National Institute of Environmental Health Sciences. New York: City University of New York, Mount Sinai School of Medicine, Environmental Sciences Laboratory, 1978. [121] pp.
35. Rajan, K. T., J. C. Wagner, and P. H. Evans. The response of human pleura in organ culture to asbestos. Nature 238:346-347, 1972.
36. Richards, R. J., and F. Jacoby. Light microscope studies on the effects of chrysotile asbestos and fiber glass on the morphology and reticulin formation of cultured lung fibroblasts. Environ. Res. 11:112-121, 1976.
37. Rohl, A.N., A. M. Langer, and I. J. Selikoff. Environmental asbestos pollution related to use of quarried serpentine rock. Science. 196:1319-1322, 1977.
38. Sawyer, R. N., and E. J. Swoszowski, Jr. Asbestos abatement in schools: Observations and experiences. Ann. N.Y. Acad. Sci. 330:765-775, 1979.
39. Schnitzer, R. J., G. Bunescu, and V. Baden. Interactions of mineral fiber surfaces with cells in vitro. Ann. N.Y. Acad. Sci. 172:759-772, 1971.
40. Seidman, H., I. J. Selikoff, and E. C. Hammond. Short-term asbestos work exposure and long-term observation. Ann. N.Y. Acad. Sci. 330:61-89, 1979.
41. Selikoff, I. J., E. C. Hammond, and J. Churg. Asbestos exposure, smoking, and neoplasia. J. Am. Med. Assoc. 204:106-112, 1968.
42. Selikoff, I. J., E. C. Hammond, and H. Seidman. Mortality experience of insulation workers in the United States and Canada, 1943-1976. Ann. N.Y. Acad. Sci. 330:91-116, 1979.
43. Selikoff, I. J., and D. H. K. Lee. Asbestos and Disease. New York: Academic Press Inc., 1978. 549 pp.
44. Shin, M. L., and H. I. Firminger. Acute and chronic effects of intraperitoneal injection of two types of asbestos in rats with a study of the histopathogenesis and ultrastructure of resulting mesotheliomas. Am. J. Pathol. 70:291-313, 1973.
45. Silver, K. Z. Asbestos in school buildings: Results of a nation-wide survey. Ann. N.Y. Acad. Sci. 330:777-786, 1979.
46. Smith, W. E., L. Miller, R. E. Elsasser, and D. D. Hubert. Tests for carcinogenicity of asbestos. Ann. N.Y. Acad. Sci. 132:456-488, 1965.
47. Stanton, M. F., M. Layard, A. Tegeris, E. Miller, M. May, and E. Kent. Carcinogenicity of fibrous glass: Pleural response in the rat in relation to fiber dimension. J. Nat. Cancer Inst. 58:587-603, 1977.

48. Stanton, M., and C. Wrench. Mechanisms of mesothelioma induction with asbestos and fibrous glass. J. Nat. Cancer Inst. 48:797-821, 1972.
49. Suzuki, Y., and J. Churg. Structure and development of the asbestos body. Am. J. Pathol. 55:79-107, 1969.
50. Tagnon, I., W. J. Blot, R. B. Stroube, N. E. Day, L. E. Morris, D. B. Peace, and J. F. Fraumeni, Jr. Mesothelioma associated with the shipbuilding industry in coastal Virginia. J. Cancer Res. 40:3875-3879. 1980.
51. Wade, M. J., L. E. Lipkin, and A. L. Frank. Studies of in vitro asbestos-cell interaction. J. Environ. Pathol. Toxicol. 2:1029-1039, 1979.
52. Wade, M. J., L. E. Lipkin, R. W. Tucker, and A. L. Frank. Asbestos cytotoxicity in a long term macrophage-like cell culture. Nature 264:444-446, 1976.
53. Wade, M. J., L. E. Lipkin, M. F. Stanton, and A. L. Frank. $P388D_1$ in vitro cytotoxicity assay as applied to asbestos and other minerals: Its possible relevance to carcinogenicity. In R.C. Brown, M. Chamberlain, R. Davies, and I. P. Gormley, Eds. The In Vitro Effects of Mineral Dusts. London: Academic Press Inc., London Ltd., 1980.
54. Wagner, J. C., G. Berry, and F. D. Pooley. Carcinogenesis and mineral fibres. Br. Med. Bull. 36:53-56, 1980.
55. Wagner, J. C., G. Berry, J. W. Skidmore, and V. Timbrell. The effects of the inhalation of asbestos in rats. Br. J. Cancer 29:252-269, 1974.
56. Wagner, J. C., C. A. Sleggs, and P. Marchand. Diffuse pleural mesothelioma and asbestos exposure in the North Western Cape Province. Br. J. Ind. Med. 17:260-271, 1960.
57. Whipple, H. E., and P. E. van Reyen, Eds. Biological Effects of Asbestos. Ann. N.Y. Acad. Sci. 132:1-766, 1965.
58. Wright, G. W., and M. Kuschner. The influence of varying lengths of glass and asbestos fibres on tissue response in guinea pigs, pp. 455-472. In W. H. Walton Inhaled Particles IV. Proceedings of an International Symposium Organized by the British Occupational Hygiene Society, Edinburgh, 22-26 September 1975. Part 2. Oxford: Pergamon Press, 1977.

COMBUSTION PRODUCTS

This section concerns the effects of the exposure of people in buildings to the products of fossil-fuel combustion that takes place in those buildings. Such fuels are consumed in space- and water-heating, clothes-drying, cooking, and operating gas-powered refrigerators and propane torches. When unvented flames are used in maintenance, modification, and repairs or in hobby activities, some of the effluents are similar to those of flames used for cooking and space-heating--carbon monoxide and nitrogen dioxide. As discussed in Chapter IV, additional toxicants may also be released, depending on the composition of the objects heated and the temperatures attained.

The present discussion is limited to the effects of the products of fossil-fuel combustion. It excludes, for example, the materials vaporized by the application of a flame to a cooking pot, frying pan, or metal object involved in maintenance or hobby activities. With respect to cooking, this exclusion can be justified on the basis that a gas range and an electric range do not differ substantially in the composition and magnitude of pollutants released during cooking. The contribution of the cooking processes themselves to overall indoor pollution may be important, especially with respect to odor characteristics and the concentrations of suspended particles. But the effluents of cooking processes are highly variable, and their effects, if any, on the health of residents are generally not known. Cigarette combustion is also excluded, in that it is discussed in the next section of this chapter.

For the products of indoor combustion to constitute a health stressor, they must be able to cause toxic effects and they must be present in occupied spaces at sufficient concentrations and for sufficient durations to manifest their toxicity in a substantial part of the exposed population. The extent to which products of combustion contaminate indoor air depends on the composition of the fuel, the temperature of combustion, the efficiency of combustion, the efficiency of the venting of the combustion products to the outdoor air, and the isolation of discharged air from makeup air that enters the occupied space. The most important factor is usually the presence or absence of effective venting of the combustion products to the outdoor air. If the venting is effective, there should be relatively little buildup of combustion effluents indoors, even when liquid fuels (such as kerosene) or solid fuels (such as wood, charcoal, coke, and anthracite) are burned. However, the use of liquid and solid fuels makes it more difficult to achieve effective venting.

When venting of combustion effluents is incomplete, even the burning of the cleanest of fuels (natural gas) may liberate excessive amounts of toxic gaseous effluents, specifically carbon monoxide and nitrogen dioxide. There may also be measurable amounts of nitric oxide, unburned fuel (methane, ethane, propane, etc.), products of pyrosynthesis (e.g., aldehydes), and carbonaceous particles.

The products of indoor combustion that are most often of health concern are carbon monoxide and nitrogen dioxide. Airborne concentrations of these pollutants have been measured in a number of epidemiologic studies; but other air pollutants were also present, and their concentrations were usally not measured. At best, epidemiology can demonstrate an association, but it cannot establish causality.

CARBON MONOXIDE

Exposure to carbon monoxide (a product of incomplete combustion of any fossil fuel) constitutes a long-established and well known acute hazard. Exposure at over 500 ppm for more than 1 h can lead to approximately 20% of carboxyhemoglobin saturation. Exposure at 1,500 ppm for 1 h is dangerous to life.[1] Such high concentrations can result from improper combustion--e.g., without an adequate supply of combustion air. The issue of a threshold for adverse carbon monoxide effects was addressed in a 1977 National Research Council report:[32]

> Whether there is a threshold carboxyhemoglobin concentration for an adverse effect is still unknown. . . . The mechanism for adverse carbon monoxide effects is a fall in capillary oxygen partial pressure (pO_2) due to carbon monoxide binding to hemoglobin, and therefore a pertinent question is whether any fall in capillary pO_2, no matter how small, results in an adverse effect on tissues. It is known that many tissues, in order to keep intracellular pO_2 nearly constant, can adapt to acute falls in arterial pO_2 with resulting falls in capillary pO_2. The major adaptation mechanism in many tissues is probably recruitment of capillaries to give a decrease in oxygen diffusion distance between capillary blood and mitochondria. If such a mechanism occurs as carboxyhemoglobin increases, it is unlikely that adverse carbon monoxide effects occur at carboxyhemoglobin concentrations near zero and more probable that a threshold exists at a carboxyhemoglobin concentration where adaptation cannot compensate.
>
> . . . The tissues most sensitive to the adverse effect of carbon monoxide appear to be heart, brain, and exercising skeletal muscle. Evidence has been obtained that carboxyhemoglobin concentrations in the 3-5% saturation range may adversely affect the ability to detect small unpredictable environmental changes (vigilance). There is evidence that acute increases of carboxyhemoglobin to above 4-5% in patients with cardiovascular disease can exacerbate their symptoms when the carboxyhemoglobin is as low as 5%. . . . In the studies of the effect of carbon monoxide on vigilance and cardiovascular symptoms, there was no attempt either to determine the effect of lower carboxyhemoglobin concentrations or to look for a threshold. When aerobic metabolism of exercising skeletal muscle was studied, an apparent threshold was found. At a carboxyhemoglobin concentration below 5%, a measurable effect on oxygen uptake could not be demonstrated.

* * *

The current EPA standard for carbon monoxide is 9-ppm maximum for 8-hr average exposure, or 35-ppm maximum for 1-hr average exposure. Approximate calculated carbon monoxide uptakes for varying levels of activity after exposure to these concentrations are given below.

Exposure	Resting	Moderate Activity	Heavy Activity
9 ppm, 8 hr	1.3% sat	1.4% sat	1.4% sat
35 ppm, 1 hr	1.3%	2.2%	2.9%

. . . The current EPA standard is mainly justified on the basis of adverse carbon monoxide effects in patients with cardiac and peripheral vascular disease and effects of carbon monoxide on oxygenation of skeletal muscles in exercising normal human subjects. There appears to be an adequate safety factor between the lowest carboxyhemoglobin concentration that has been demonstrated to cause adverse effects and the maximal carboxyhemoglobin concentration that can occur at 9-ppm carbon monoxide for 8 hr or 35 ppm for 1 hr. (pp. 164-167)

The experimental studies on carbon monoxide health effects performed in recent years are summarized in Table VII-5. They have tended to confirm the judgments expressed in the 1977 NRC report.

Moderately severe exposures to carbon monoxide--e.g., at 50 ppm for up to about an hour--can occur in kitchens as a result of ordinary use of a gas range, especially when the cooking utensils divert or quench the flame. Higher exposures can be found indoors in public buildings, such as ice-skating rinks, where mean concentrations of carbon monoxide as high as 100 ppm have been measured (Spengler, personal communication). The health effects of indoor exposures to carbon monoxide are addressed in the next section of this chapter, which deals with involuntary smoking; carbon monoxide is, of course, a pollutant that is common to cigarette-smoke and fossil-fuel combustion effluents. The effects of carbon monoxide from indoor combustion cannot be adequately assessed without considering the influence of exposure to cigarette smoke. For smokers, mainstream smoke is the dominant source of carboxyhemoglobin (COHb) in the blood. For nonsmokers, the lower COHb values can be attributed to metabolism, to carbon monoxide in the outdoor air, to carbon monoxide in sidestream smoke, and to carbon monoxide from indoor combustion sources. The extent of COHb saturation associated with metabolism is 0.7%. Community air pollution or exposure to sidestream smoke can raise COHb in nonsmokers to 2-3%.[32] The influence of indoor combustion on COHb saturation depends on many factors. Carbon monoxide from indoor combustion may be dominant when there are no smokers in the occupied space and when outdoor carbon monoxide concentration is low. It can also be dominant even when cigarette use and outdoor concentration are high when the indoor combustion source is large (as in the case of unvented space-heaters) or when combustion efficiency is poor because of poor burner maintenance or blockage of the air supply. A summary of measurements of indoor carbon monoxide concentrations is provided in Chapter IV.

NITROGEN OXIDES

Indoor combustion can have an important effect on the indoor concentrations of nitric oxide and nitrogen dioxide. Nitric oxide binds to hemoglobin to produce methemoglobin. Many of the adverse effects reported in the past for carbon monoxide alone may be related to the combined action of COHb and methemoglobin, especially inasmuch

TABLE VII-5

Controlled Exposure to Carbon Monoxide

Species	Exposure	Health Effects Observed
Human (normal)[3]	100 ppm, 1 h	Mean exercise time until exhaustion significantly decreased
Human (n = 18)[28]	200 ppm, 3 h	No significant effect on scotopic sensitivity, reaction time, eye movements, visually evoked cortical potentials
Human[29]	15-20 mg/m^3, 13-18 ppm, 30 d	Increased albumin, β-globulins, total lipids, cholesterol, β-lipoproteins; decreased blood sugar
	10 mg/m^3, 9 ppm, 90 d	None
Human (n = 19)[37]	50 ppm, 4 h	No significant changes in lung function
Human (n = 20)[43]	150 ppm, 3.5 h	No effect on critical flicker-fusion frequency; in monotonous situation, relative "activation" of subjective feelings
Human[44]	3.2-4.7% COHb	Increased errors in auditory discrimination in open office[a]
	8% COHb	Less difficult task: no significant effects in isolation booth[a]
	4.92% COHb	Less difficult task: no significant effects[a]
Pigeon (normocholesterolemic and hypercholesterolemic)[2]	150 ppm, 52 and 84 wk	In hypercholesterolemic birds, atherosclerosis more severe
Rabbit (hypercholesterolemic)[10]	250 ppm, 10 wk	Coronary arterial atherosclerosis significantly higher
Dog (myocardial injury)[4]	100 ppm, 2 h	Decreased venticular fibrillation threshold
Dog (anesthetized, normal, open-chested)[5]	100 ppm, 2 h	Decreased venticular fibrillation threshold
Monkey[11]	100 ppm, 6 h, 9.3% (ave.) COHb	Ventricular fibrillation more easily induced
Rabbit[40]	180 ppm, 4 h	Focal intimal edema in aorta
Rat (n = 8-16)[6]	100 ppm, 200 ppm, 500 ppm; 4 h	Changes in blood glucose and lactic acid; no significant plasma corticosterone increase
Rat (n = 4)[7]	100-1,000 ppm, 1.5 h	Lever-pressing response rate decreased at increasing concentration[b]
Rat (exposed prenatally)[14]	150 ppm, continuous, 15% COHb	Reduced birthweight, decreased weight gain, lower behavioral activity, altered central catecholamine activity, less total brain protein at birth

[a]Environment and task difficulty may alter effects.

[b]Response decrease inconsistent at lower concentrations.

as sources that emit carbon monoxide often produce nitric oxide as well. If present data are indicative, nitric oxide at 3 ppm (3.75 mg/m^3) is physiologically comparable with carbon monoxide at 10-15 ppm (11-17 mg/m^3).[8] Thus, NO_x may increase cardiovascular stress due to hypoxia, although the NO_x does not shift the oxygen binding equilibrium for hemoglobin, and the effects of NO_x are not quantitatively identical. The work of Case et al.[8] has suggested that NO_x generated by household combustion appliances accounts for a substantial fraction of the total methemoglobin present in the blood of most humans. NO_x concentrations sufficient to generate 2% or more methemoglobin may be encountered often in the home (and in roadway tunnels).[18]

Nitrogen oxides may change heme by producing polycythemia with increased hematocrit and with decreased mean corpuscular volume. They may also produce leukocytosis and other hematologic abnormalities, as well as vascular membrane injury and leakage that lead to edema.[33] Nitrogen dioxide exposure affects the activity of several enzymes: it decreases erythrocyte membrane acetylcholinesterase, increases peroxidized erythrocyte lipids, and increases glucose-6-phosphate dehydrogenase. It also produces substantial decreases in hemoglobin and hematocrit values.[36]

Both nitric oxide and nitrogen dioxide are formed from atmospheric nitrogen and oxygen in the high-temperature part of a flame by the temperature-dependent process of nitrogen fixation. Acute toxicity is not to be expected from the nitrogen dioxide formed in unvented indoor combustion, because not enough nitrogen dioxide is generated. But nitrogen dioxide concentrations equal to or greater than the current ambient-air quality standard of 0.05 ppm are not unusual in kitchens where gas is used for cooking (see Chapter IV). At those concentrations, nitrogen dioxide may affect sensory perception, especially dark adaptation,[34] and produce eye irritation, especially with hydrocarbons.[17]

Nitrogen dioxide can produce transient and long-term damage to both small bronchial airways and alveolar tissue. In the bronchial airways, exposure of rats to nitrogen dioxide at as low as 2 ppm for 4 h stimulated the differentation of nonciliated cells into mature Clara cells and ciliated cells;[13] that effect raises the possibility that chronic exposure could lead to chronic bronchitis. The nitrogen dioxide also destroyed Type I epithelial cells and stimulated the proliferation of Type II cells. Thus, chronic exposure might contribute to the development of emphysema. In addition, in animals challenged with bacterial aerosols after exposure to nitrogen dixoide at 1.5 ppm for 2 h or at 0.5 ppm for 2 wk, there was significantly increased mortality, compared with that in animals challenged with the same bacterial aerosols without nitrogen dioxide.[16]

Further information on the acute and chronic toxicity of nitric oxide and nitrogen dioxide can be found in the 1977 NRC report on the nitrogen oxides[33] and in the 1981 EPA criteria document on NO_x.[42] Tables VII-6 and VII-7, from the NRC report, summarized observed effects of short-term exposures of humans to nitrogen dioxide at high and low concentrations. Table VII-8 summarizes some of the more recent experimental exposure studies.

TABLE VII-6

Human Effects of Acute Exposure to High Nitrogen Dioxide Concentrations[a]

Nitrogen Dioxide Concentration		Clinical Effect	Time between Exposure and Termination of Effect
mg/m^3	ppm		
940	500	Acute pulmonary edema--fatal	Within 48 h
564	300	Bronchopneumonia--fatal	2-10 d
282	150	Bronchiolitis fibrosa obliterans--fatal	3-5 wk
94	50	Bronchiolitis, focal pneumonitis--recovery	6-8 wk
47	25	Bronchitis, bronchopneumonia--recovery	6-8 wk

[a]Reprinted from National Research Council.[33(p. 269)]

TABLE VII-7

Summary of Human Responses to Short-Term Nitrogen Dioxide Exposures Alone[a]

Effect	Nitrogen Dioxide Concentration mg/m^3	ppm	Time to Effect
Odor threshold	0.23	0.12	Immediate
Threshold for dark adaptation	0.14	0.075	Not reported
	0.50	0.26	Not reported
Increased airway resistance	1.3-3.8	0.7-2.0	20 min[b]
	3.0-3.8	1.6-2.0	15 min
	2.8	1.5	45 min[c]
	3.8	2.0	45 min[d]
	5.6	3.0	45 min[e]
	7.5-9.4	4.0-5.0	40 min[f]
	9.4	5.0	15 min
	11.3-75.2	6.0-40.0	5 min
	13.2-31.8	7.0-17.0	10 min[g]
Decreased pulmonary diffusing capacity	7.5-9.4	4.0-5.0	15 min
Increased alveolar-arterial pO_2 difference	9.4	5.0	25 min[h]
No change in sputum histamine concentration	0.9-6.6	0.5-3.0	45 min

[a]Reprinted from National Research Council.[33]

[b]Exposure lasted 10 min. Effect on flow resistance was observed 10 min after termination of exposure.

[c]Effect was produced at this concentration when normal subjects and those with chronic respiratory disease exercised during exposure.

[d]Effect occurred at rest in subjects with chronic respiratory disease.

[e]Effect occurred at rest in normal subjects.

[f]Exposure lasted 10 min. Maximal effect on flow resistance was observed 30 min later.

[g]Also failed to find increased flow resistance over the range of nitrogen dioxide exposures from 5.1 to 30.1 mg/m^3 (2.7-16.0 ppm).

[h]Effect occurred 10 min after termination of 15-min exposure.

TABLE VII-8

Controlled Exposure to Nitrogen Oxides

Species	Exposure	Health Effects Observed
Mouse (6-8 wk old)[19]	NO_2, 10 ppm, 2 h/d, 5 d/wk, up to 30 wk	Lung damage, suppressed immune function with chronic exposure, enhanced immune reactivity with shorter exposures
	NO, 10 ppm	Paraseptal emphysema, suppressed immune function with chronic exposure, enhanced immune reactivity with shorter exposures
Mouse[27]	NO_2, 0.5-28 ppm, 6 mo to 1 yr	Mortality after *Streptococcus pyogenes*, mortality increased with increasing dose and exposure time
Guinea pig[25]	NO_x, 1 ppm, 6 mo	Disturbed glycolysis, enhanced catabolic processes in brain, inhibited respiration, decreased brain aminotransferase activity, morphologic alterations in blood vessels
Human (asthma, n = 13; bronchitis, n = 7)[23]	NO_2, 0.5 ppm, 2 h	Lightness in chest, burning of eyes, headache, or dyspnea; pulmonary-function changes; nasal discharge
Human (asthma, n = 20)[35]	NO_2, 0.1-0.2 ppm	Increased bronchoconstriction[a]
Cat[26]	NO_2, 80 ppm, 3 h	Diffuse alveolar damage
Guinea pig[41]	NO_2, 0.506 ppm; NO, 0.05 ppm; 122 d	In lungs: decreased phosphatidylethanolamine, sphingomyelin, phosphatidylserine, phosphatidic acid, phosphatidylglycerol 3-phosphate; increased lysophosphatidylethanolamine
Mouse[12]	NO_2, 1.5-5.0 ppm, 3 h	Mortality in mice challenged with *Streptococcus* aerosol significantly increased at 2.0 ppm and above
Mouse[38]	NO_2, 0.5 ppm; 10, 12, 14 d	Average protein content of lungs significantly higher
Hamster[24]	NO_2, 30 ppm, 3 wk	Loss of body weight, increased dry lung weight, decrease in lung elastin and collagen

[a]Carbachol provocation.

[b]Elastin and collagen later returned to normal.

The evidence of health effects after prolonged exposure at low concentrations is inconsistent. This should not be surprising, in that much of the evidence was obtained from epidemiologic studies in which the observed effects could have been due to the presence of other air pollutants or to their combined effects with nitrogen dioxide.

The 1977 NRC report summarized the effects of exposure to NO_x at low concentrations on respiratory function and disease as follows:[33]

> Two epidemiologic studies suggest that the combination of nitrogen dioxide at concentrations of 0.15 to 0.3 mg/m^3 (0.08 to 0.16 ppm) with other pollutants causes changes in ventilatory function. Two other studies in which lower levels of nitrogen dioxide were studied did not reveal these effects. Because of the disparity in populations and in pollutant conditions, conclusions cannot be reached regarding the effect, if any, of chronic exposure to nitrogen dioxide on ventilatory function.
>
> Some epidemiologic data support the idea that excess acute respiratory disease may occur in healthy populations following exposure to atmospheres containing nitrogen dioxide. Four studies have been reviewed in the search for an association between exposure to ambient concentrations of nitrogen dioxide from 0.10 to 0.58 mg/m^3 (0.053 to 0.309 ppm) and small excesses in respiratory illnesses. However, the variable pollutant exposures and conditions of study make it difficult to quantify the relationship of nitrogen dioxide by itself to the reported increases in respiratory disease. In each study air contaminants likely to enhance susceptibility to respiratory infection (sulfur dioxide, sulfuric acid, sulfates, nitrates, etc.) were also present.
>
> Evidence that nitrogen dioxide induces excess chronic respiratory disease is not convincing. Reports of excess chronic respiratory disease associated with low concentrations of ambient nitrogen dioxide (<0.10 mg/m^3 [0.053 ppm]) do not provide convincing evidence that other pollutants that were measured at relatively high concentrations were not the probable cause of the excess disease. In the presence of low concentrations of sulfur dioxide and particulates, three investigators failed to detect excess chronic respiratory disease in areas where nitrogen dioxide exposures were <0.10 mg/m^3 (0.053 ppm). (pp. 271-272)

SUMMARY OF RECENT EPIDEMIOLOGIC STUDIES OF INDOOR POLLUTION WITH SPECIAL REFERENCE TO NO_x EXPOSURE

The availability of inexpensive passive-diffusion-tube samplers for nitrogen dioxide has stimulated a series of studies of indoor nitrogen dioxide pollution effects in the United States and the United Kingdom.

Melia *et al.*[31] studied the relation between respiratory illness in primary-school children and the use of gas for cooking. In a 5-yr longitudinal study of schoolchildren in England and Scotland, 4,827 boys and girls aged 5-10 yr in 27 randomly selected areas were examined in 1977, the last year of the study. The authors reported that prevalence of one or more respiratory symptoms or diseases was higher in children from gas-cooking homes than in those from electric-cooking homes and that the association appeared to be independent of age, sex, social class, number of cigarette-smokers in the home, and latitude. However, it was found only in urban areas (for boys, $p < 0.005$; for girls, $p \approx 0.08$). In children aged 6-7.5 yr in 1973 who were followed until the last year of the study, there was some indication that the association between respiratory illness and gas cooking disappeared as the children grew older; this trend was not obvious in the children in the other age groups, who were follwed for 2-4 yr. The evidence from the 1977 study did show a relationship between gas cooking and respiratory illness that supported results of the 1973 study in the same group, although the results on cohorts showed some indication that the relationship may disappear as children grow older.

Florey et al.[15] examined the relation between lung function and respiratory illness in a population of 808 primary-school children aged 6-7 yr and the concentrations of nitrogen dioxide in the kitchens and bedrooms of their homes. Complete data were collected on about 66% of the population. The children lived in a defined 4-km^2 area in Middlesborough (United Kingdom). One-week average outdoor nitrogen dioxide concentrations varied little over the area: 25-43 $\mu g/m^3$ (14-24 ppb). The prevalence of respiratory illness was higher in children from gas-cooking than from electric-cooking homes ($p \approx 0.1$). Although prevalence was not related to kitchen nitrogen dioxide concentration (9-570 $\mu g/m^3$), it increased with increasing nitrogen dioxide in the children's bedrooms in gas-cooking homes (4-169 ppb; $p \approx 0.1$). Lung function was not related to nitrogen dioxide content in the kitchen or bedroom. Because of the very low nitrogen dioxide concentrations at which an association with illness was observed and the inconsistency between these results in the United Kingdom and those from several studies in the United States, the authors speculated that the nitrogen dioxide concentrations were a proxy for some other factor more directly related to respiratory disease, such as temperature or humidity.

A similar study by Keller et al.[21] in Columbus, Ohio, failed to establish any increase in respiratory disease or decrease in pulmonary function (FVC and $FEV_{0.75}$) associated with the use of gas for cooking. Their sample included 441 families, divided into two groups: those using gas and those using electricity in cooking. Family health and demographic data were obtained from the participants. Reports of acute respiratory illness were obtained through biweekly telephone calls to each household. Respondents were asked to report respiratory illness in any member of the household and to indicate the presence or absence of a set of signs and symptoms. Ambient air was analyzed indoors and outdoors in a sample of the households, and pulmonary-function tests were conducted on a subsample of the participants representing both types of households. The mean nitrogen dioxide concentrations were 0.05 ppm (90 $\mu g/m^3$) in the gas-cooking homes and 0.03 ppm (50 μg/m^3) in the electric-cooking homes. In an extension of this study, Keller et al.[22] selected 120 households with school-age children from the gas-cooking and electric-cooking cohorts. Reports of respiratory illness and symptoms were obtained by telephone interview every 2 wk for 13 mo by a nurse-epidemiologist. If the onset of respiratory illness occurred within 3 d of the call, a household visit was arranged to examine the person reported ill and to obtain a throat culture. In addition, two sets of "well" controls were examined. The results validated the reporting method and replicated earlier findings of no significant difference in incidence of acute respiratory illness between gas- and electric-cooking households.

The largest and most recently reported study of the effects of gas cooking on the health of children is that of Speizer et al.[39] As part of a long-range prospective study of the health effects of air pollution, they studied approximately 8,000 children aged 6-10 yr in six communities. Questionnaires were completed by their parents, and simple spirometry was performed in school. Comparisons were made between children living in homes with gas stoves and those living in homes with electric stoves. Children from households with gas stoves had a greater history of respiratory illness before age 2 (average difference, 32.5/1,000 children) and small but significantly lower FEV and FVC values corrected for height (average difference, 16 ml and 18 ml, respectively). These findings were not explained by differences in social class or in parental smoking habits. Measurements taken in the homes for 24-h periods showed that nitrogen dioxide concentrations were 4-7 times higher in homes with gas stoves than in homes with electric stoves. However, these 24-h measurements were generally well below the current federal 24-h outdoor standard of 100 μg/m^3. Short-term peak exposures, which were in excess of 1,100 μg/m^3, occurred regularly in kitchens. Further work will be required to determine the role of these short-term peaks in the effects noted.

REFERENCES

1. American Industrial Hygiene Association. Hygienic guide series. Carbon monoxide. Am. Ind. Hyg. Assoc. J. 26:431-434, 1965.
2. Armitage, A. K., R. F. Davies, and D. M. Turner. The effects of carbon monoxide on the development of atherosclerosis in the White Carneau pigeon. Atherosclerosis 23:333-344, 1976.
3. Aronow, W. S., and J. Cassidy. Effect of carbon monoxide on maximal treadmill exercise: A study in normal persons. Ann. Intern. Med. 83:496-499, 1975.
4. Aronow, W. S., E. A. Stemmer, B. Wood, S. Zweig, K. Tsao, and L. Raggio. Carbon monoxide and ventricular fibrillation threshold in dogs with acute myocardial injury. Am. Heart J. 95:754-756. 1978.
5. Aronow, W. S., E. A. Stemmer, and S. Zweig. Carbon monoxide and ventricular fibrillation threshold in normal dogs. Arch. Environ. Health 34:184-186, 1979.

6. Atland, P. D., and B. A. Rattner. Effects of nicotine and carbon monoxide on tissue and systemic changes in rats. Environ. Res. 19:202-212, 1979.
7. Ator, N. A., W. H. Merigan, Jr., and R. W. McIntire. The effects of brief exposures to carbon monoxide on temporally differentiated responding. Environ. Res. 12:81-91, 1976.
8. Case, G. D., J. S. Dixon, and J. C. Schooley. Interactions of blood metalloproteins with nitrogen oxides and oxidant air pollutions. Environ. Res. 20:43-65, 1979.
9. Case, G. D., J. C. Schooley, and S. D. Jonathan. Uptake and Metabolism of Nitrogen Oxides in Blood. Paper presented at the 20th Annual Meeting of the Biophysical Society, Seattle, Washington, February 24-27, 1976.
10. Davies, R. F., D. L. Topping, and D. M. Turner. The effect of intermittent carbon monoxide exposure on experimental atherosclerosis in the rabbit. Atherosclerosis 24:527-536, 1976.
11. DeBias, D. A., C. M. Banerjee, N. C. Birkhead, C. H. Greene, S. D. Scott, and W. V. Harrer. Effects of carbon monoxide inhalation on ventricular fibrillation. Arch. Environ. Health 31:42-46, 1976.
12. Ehrlich, R., J. C. Findlay, J. D. Fenters, and D. E. Gardner. Health effects of short-term inhalation of nitrogen dioxide and ozone mixtures. Environ. Res. 14:223-231, 1977.
13. Evans, M. J., and G. Freeman. Morphological and pathological effects of NO_2 on the rat lung, pp. 243-265. In S.D. Lee, Ed., Nitrogen Oxides and Their Effects on Health. Ann Arbor, Mich.: Ann Arbor Science Publishers, Inc., 1980.
14. Fechter, L. D., and Z. Annau. Toxicity of mild prenatal carbon monoxide exposure. Science 197:680-682, 1977.
15. Florey, C. du V., R. J. W. Melia, S. Chinn, B. D. Goldstein, A. G. F. Brooks, H. H. John, I. B. Craighead, and X. Webster. The relation between respiratory illness in primary schoolchildren and the use of gas for cooking. III. Nitrogen dioxide, respiratory illness and lung infection. Int. J. Epidemiol. 8:347-353, 1979.
16. Gardner, D. E., F. J. Miller, E. J. Blommer, and D. L. Coffin. Influence of exposure mode on the toxicity of NO_2. Environ. Health Perspect. 30:23-29, 1979.
17. Goldsmith, J. R. and L. T. Friberg. Effects of air pollution on human health, pp. 458-611. In A. C. Stern, Ed. Air Pollution. 3rd. ed. Vol. II. The Effects of Air Pollution. New York: Academic Press, Inc., 1977.
18. Hollowell, C. D., R. J. Budnitz, C. D. Case, and G. Traynor. Combustion Generated Indoor Air Pollution: Field Studies 8/75-10/75. Lawrence Berkeley Laboratory Publ. LBL-4416. Berkeley, Cal.: Lawrence Berkeley Laboratory, 1976.
19. Holt, P. G., L. M. Finlay-Jones, D. Keast, and J. M. Papadimitrou. Immunological function in mice chronically exposed to nitrogen oxides (NO_x). Environ. Res. 19:154-162, 1979.
20. Hugod, C., L. H. Hawkins, and P. Astrup. Exposure of passive smokers to tobacco smoke constituents. Int. Arch. Occup. Environ. Health 42:21-29, 1978.

21. Keller, M. D., R. R. Lanese, R. I. Mitchell, and R. W. Cote. Respiratory illness in households using gas and electricity for cooking. I. Survey of incidence. Environ. Res. 19:495-503, 1979.
22. Keller, M. D., R. R. Lanese, R. I. Mitchell, and R. W. Cote. Respiratory illness in households using gas and electricity for cooking. II. Symptoms and objective findings. Environ. Res. 19:504-515, 1979.
23. Kerr, H. D., T. J. Kulle, M. L. McIlhany, and P. Swidersky. Effects of nitrogen dioxide on pulmonary function in human subjects: An environmental chamber study. Environ. Res. 19:392-404, 1979.
24. Kleinerman, J., and M. P. C. Ip. Effects of nitrogen dioxide on elastin and collagen contents of lung. Arch. Environ. Health 34:228-232, 1979.
25. Kośmider, S., A. Misiewicz, and J. Pasiewicz. Effect of binding of nitrogen oxides with gaseous ammonia on the occurrence of changes in the central nervous system. Neuropatol Pol. 12:413-426, 1974. (in Polish; English summary)
26. Langloss, J. M., E. A. Hoover, and D. E. Kahn. Diffuse alveolar damage in cats induced by nitrogen dioxide or Feline Calicivirus. Am. J. Pathol. 89:637-648, 1977.
27. Larsen, R. I., D. E. Gardner, and D. L. Coffin. An air quality data analysis system for interrelating effects, standards, and needed source reductions: Part 5. NO_2 mortality in mice. J. Air Pollut. Control Assoc. 29:133-137, 1979.
28. Luria, S. M., and C. L. McKay. Effects of low levels of carbon monoxide on visions of smokers and nonsmokers. Arch. Environ. Health 34:38-44, 1979.
29. Markaryan, M. V., T. A. Smirnova, and O. S. Khokhlova. Effect of chronic exposure to carbon monoxide on the biochemical composition of human blood. Kosm. Biol. Aviakosm. Med. 11(4):46-50, 1977. (in Russian; English summary)
30. Melia, R. J. W., C. du V. Florey, D. G. Altman, and A. V. Swan. Association between gas cooking and respiratory disease in children. Br. Med. J. 2:149-152, 1977.
31. Melia, R. J. W., C. du V. Florey, and S. Chinn. The relation between respiratory illness in primary schoolchildren and the use of gas for cooking. I. Results from a national survey. Int. J. Epidemiol. 8:333-338, 1979.
32. National Research Council, Committee on Medical and Biologic Effects of Environmental Pollutants. Carbon Monoxide. Washington, D.C.: National Academy of Sciences, 1977. 239 pp.
33. National Research Council, Committee on Medical and Biologic Effects of Environmental Pollutants. Nitrogen Oxides. Washington, D.C.: National Academy of Sciences, 1977. 333 pp.
34. National Research Council, Committee on Medical and Biologic Effects of Environmental Pollutants. Vapor-Phase Organic Pollutants. Volatile Hydrocarbons and Oxidation Products. Washington, D.C.: National Academy of Sciences, 1976. 411 pp.
35. Orehek, J., J. P. Massari, P. Gayrard, C. Grimaud, and J. Charpin. Effect of short-term, low-level NO_2 exposure on bronchial sensitivity of asthmatic patients. J. Clin. Invest. 57:301-307, 1976.

36. Posin, C., K. Clark, M. P. Jones, J. V. Patterson, R. D. Buckley, and J. D. Hackney. Nitrogen dioxide inhalation and human blood biochemistry. Arch. Environ. Health 12:318-324, 1978.
37. Raven, P. B., J. A. Gliner, and J. C. Sutton. Dynamic lung function changes following long-term work in polluted environments. Environ. Res. 12:18-25, 1976.
38. Sherwin, R. P., and L. J. Layfield. Protein leakage in the lungs of mice exposed to 0.5 ppm nitrogen dioxide. A fluorescence assay for protein. Arch. Environ. Health 31:116-118, 1976.
39. Speizer, F. E., B. Ferris, Jr., Y. M. M. Bishop, and J. Spengler. Respiratory disease rates and pulmonary function in children associated with NO_2 exposure. Am. Rev. Respir. Dis. 121:3-10, 1980.
40. Thomsen, H. K., and K. Kjeldsen. Aortic intimal injury in rabbits: An evaluation of a threshold limit. Arch. Environ. Health 30:604-607, 1975.
41. Trzeciak, H. I., S. Kośmider, K. Kryk, and A. Kryk. The effects of nitrogen oxides and their neutralization products with ammonia on the lung phospholipids of guinea pigs. Environ. Res. 14:87-91, 1977.
42. U.S. Environmental Protection Agency. Air Quality Criteria for Oxides of Nitrogen. Research Triangle Park, N.C.: U.S. Environmental Protection Agency, Environmental Criteria and Assessment Office, 1981. (in press)
43. Weber, A., C. Jermini, and E. Grandjean. Effects of low carbon monoxide concentrations on flicker fusion frequency and on subjective feelings. Int. Arch. Occup. Environ. Health 36:87-103, 1975. (in German; English summary)
44. Wright, G. R., and R. J. Shephard. Carbon monoxide exposure and auditory duration discrimination. Arch. Environ. Health 33:226-235, 1978.

INVOLUNTARY SMOKING

The combustion of tobacco products is responsible for only a small fraction of the total atmospheric pollution,[64] and it is only in the enclosed indoor environment that smoking produces a major fraction of the airborne environmental contamination. The potential health effects of this contamination have recently become a subject of considerable concern and controversy.[46,55,61(pp.11-1--11-41),64] The health effects of smoking on smokers have been extensively studied.[61] But the health effects on nonsmokers have received far less study, and this section documents what is known about these effects.

The exposure of nonsmokers to environmental contamination by the combustion products of tobacco has been referred to as "passive smoking," "second-hand smoking," and "involuntary smoking." We use the term "involuntary smoking" for this kind of exposure; it provides exposure to many of the same constituents of tobacco smoke that voluntary smokers experience, and it is involuntary, in that the exposure occurs as an unavoidable consequence of breathing in a smoke-filled room.

The chemical constituents found in the atmosphere due to tobacco smoke are derived from two sources--mainstream and sidestream smoke. Mainstream smoke emerges from the tobacco product after being drawn through the tobacco during puffing. Sidestream smoke rises from the burning cone of tobacco. For several reasons, mainstream smoke and sidestream smoke contribute different concentrations of many substances to the atmosphere: different amounts of tobacco are consumed in the production of mainstream and sidestream smoke; the temperature of combustion for tobacco is different during puffing and during smoldering; and some substances are partially absorbed from the mainstream smoke by passage through the cigarette and the lungs of the smoker. The amount of a substance absorbed by the smoker depends on the characteristics of the substance and the depth of inhalation by the smoker. When the smoker does not inhale the smoke into his or her lungs, the smoke exhaled contains less than half its original amount of water-soluble volatile compounds, four-fifths of the original non-water-soluble compounds and particulate matter, and almost all the carbon monoxide.[21] When the smoker does inhale the mainstream smoke, that exhaled into the atmosphere contains less than one-seventh of the original amount of volatile and particulate substances and less than half the original concentration of exhaled carbon monoxide.[22] The differential impact of these factors on the extent of contamination is discussed elsewhere in this report.

The differences in chemical composition between sidestream and mainstream smoke and the differences between the low-dose, continuous exposure of the involuntary smoker and the high-dose, intermittent exposure of the voluntary smoker make the comparison of dosage in terms of "cigarette equivalents" highly speculative. The qualitative and quantitative differences between the two kinds of exposures prevent the extrapolation of the well-established health effects of cigarette-smoking to the involuntary smoker. We therefore try to identify health effects on the basis of actual exposures, rather than on the basis of effects on smokers.

ABSORPTION OF SMOKE CONSTITUENTS

There are no direct measurements of absorption of most of the constituents of tobacco smoke. However, Hugod et al.[28] found that the concentrations of carbon monoxide, nitric oxide, acrolein, hydrogen cyanide, and nitrogen dioxide in a sealed chamber decreased when nonsmokers were present, but not when the chamber was empty; hence, either absorption by the nonsmokers or adsorption onto their clothing occurs. A number of studies that have examined carbon monoxide absorption are summarized in Table VII-9. Carbon monoxide is often used as a measure of tobacco-smoke pollution and absorption, because it is readily measured and has been implicated in the pathogenesis of atherosclerosis. But there are several problems in the use of carbon monoxide as a measure of total smoke exposure. Smoking is only one source of carbon monoxide in the environment and great care must be taken to establish that the carbon monoxide measured is indeed from

TABLE VII-9

Absorption of Smoke Constituents from Environmental Exposure

Location and Dimensions	Ventilation[a]	Amount of Tobacco Burned	Concentrations of Constituents	Measure of Absorption
Room (80 m^3)[2]	6.4 ach	46 cigarettes and 3 pipefuls	CO, 4.5 ppm; nicotine, 377 mg/m^3	No change in COHb, 0.6%
Room (30.8 m^3)[3]	11.4 ach	15 cigarettes in 2 h	--	Nonsmokers: COHb, 1.25-1.77%
	None	15 cigarettes in 2 h	--	Nonsmokers: COHb, 1.30-2.28%
Isolated community[18]	--	--	CO, 1 ppm	Nonsmokers: COHb, 0.68%
Hospital outpatient department[18]	--	Smoking permitted	CO, 2-4 ppm	Nonsmokers: COHb, 0.97%
Office building[18]	--	Smoking permitted	CO, 3-8 ppm	Nonsmokers: COHb, 1.12%
Room (170 m^3)[24]	None	105 cigarettes	CO, 30 ppm	Smokers: COHb, 7.5% Nonsmokers: COHb, 2.1%
	1.2 ach	107 cigarettes	CO, 5 ppm	Smokers: COHb, 5.8% Nonsmokers: COHb, 1.3%
	2.3 ach	101 cigarettes	CO, 75 ppm	Smokers: COHb, 5.0% Nonsmokers: COHb, 1.6%

Room (68.1 m^3)[28]	one	20 cigarettes by machine; additional smoke added to keep smoke constant for 3 h	CO, 18-26 ppm	Nonsmokers: COHb, 0.73-1.63%
Chamber (14.6 m^3)[38]	None	4 cigarettes initially + 1 cigarette per 30 min by machine	CO, 24 ppm	Nonsmokers: COHb, 0.75-1%
Room (43 m^3)[44,45]	None	80 cigarettes and 2 cigars per hour	CO, 38 ppm	Smokers: COHb, 9.6%; urinary nicotine, 1,236 ng/ml Nonsmokers: COHb, 2.6%; urinary nicotine, 80 ng/ml
Restaurants and offices[51]	--	8-h workshift	CO, 2.5-15 ppm	No change in COHb, 2.1%
Room (37.5 m^3)[50]	None	126 cigarettes by smokers in 0.2 h	CO, 30 ppm	Smokers: COHb, 9.1% Nonsmokers: COHb, 2.2%
Car, engine off (2.09 m^3)[54]	None	10 cigarettes in 1 h	CO, 90 ppm	Smokers: COHb, 10% Nonsmokers: COHb, 5%
Office building[56]	--	--	CO, 2.7 ppm	Nonsmokers: COHb, 0.63-0.82%

[a]ach = air changes per hour.

cigarette smoke. Carbon monoxide is part of the gas phase of smoke and so does not settle out of the atmosphere passively, but is quite avidly absorbed from the atmosphere by breathing. As a result, the time course of carbon monoxide concentration differs from that of the particulate phase of the smoke, and the impact on carbon monoxide of filtration, ventilation, and number of persons in the room is also different from the impact on particulate constituents.

Because carbon monoxide is bound to hemoglobin with 210 times the affinity of oxygen, very low concentrations of carbon monoxide in the air can result in substantial carboxyhemoglobin (COHb) concentrations in the blood. A small amount of carbon monoxide is produced by the body, resulting in COHb content of approximately 0.7%. COHb values are about the same in rural communities[18] and increase to about 2.5-3% when there is marked smoke pollution. Two studies examined carbon monoxide absorption by nonsmoking workers under conditions where smoking was allowed as part of the normal work environment. Szadkowski et al.[56] found very low concentrations of carbon monoxide in an office setting and no change in COHb. Seppänen and Uusitalo[51] found higher carbon monoxide content in restaurants and offices (2.5-15 ppm) and still no change in COHb, but the workers began the day with COHb concentrations (2.1%) comparable with those that would be expected from such atmospheric carbon monoxide content and therefore would not be expected to change. A number of studies have documented increases in COHb secondary to smoke exposure under experimental conditions (see Table VII-9), and the greatest extent of smoke pollution that would normally be tolerated produces COHb of approximately 2.5-3% after 1-2 h. Srch[54] found COHb of 5% in nonsmokers, but there was more smoke than would be tolerated under normal conditions.

Russell et al.[44,45] also measured nicotine excreted by nonsmokers and found that they absorb measurable amounts of nicotine from the environment, but only 6.5% of that absorbed by smokers.

Repace and Lowrey[42] estimated the exposure of nonsmokers to respirable suspended particles from cigarette smoke. They predicted that a nonsmoker working in an office where smoking was allowed would inhale particles at a rate 3 times greater than without this exposure.

In summary, the literature suggests that nonsmokers would be expected to have slight increases in their COHb content (1-2%) from cigarette smoke in the normal working environment and more (2-3%) under conditions of heavy smoke pollution. Nonsmokers also absorb nicotine and an unknown quantity of other smoke constituents.

EFFECTS ON HEALTHY PERSONS

The effect of involuntary smoking on a person is determined not only by the qualitative and quantitative aspects of the smoke-filled environment, but also by the characteristics of the person. Reactions may vary with age and with the sensitivity of a person to the components of tobacco smoke.

Annoyance

In 1975, a national probability sample of U.S. telephone households[35] was asked to agree or disagree with the statement, "It is annoying to be near a person who is smoking cigarettes." Among "never smokers," 77% of the males and 80.5% of the females agreed with the statement; among current smokers, 35% of the males and 34.5% of the females also agreed with the statement.

Several federal agencies[62] cooperated to survey the symptoms experienced by travelers on military and commercial aircraft. They distributed a questionnaire to passengers on 20 military and eight commercial flights; 57% of the passengers on the military flights and 45% of the passengers on the commercial flights were smokers. The planes were well ventilated, and carbon monoxide content was always below 5 ppm, with low concentrations of other pollutants as well. In spite of the low measurable pollution, over 60% of the nonsmoking passengers and 15-22% of the smokers reported being annoyed by the other passengers' smoking. These feelings were even more prevalent among nonsmokers who had a history of respiratory disease. Seventy-three percent of the nonsmoking passengers on the commercial flights and 62% of the nonsmoking passengers on the military flights suggested that some remedial action be taken; 84% of those who suggested remedial action felt that segregating the smokers from nonsmokers would be a satisfactory solution. Such segregation is now required on commercial aircraft.

The annoyance reaction may be due to the odor, probably attributable to both the particulate and vapor phases; the odor threshold appears to be low.[46,47]

Irritation

Many of the substances in cigarette smoke are irritating; the major sites of irritation are the eyes and nasopharynx. Speer[53] assessed the nature of this irritation by interviewing 250 nonallergic patients about their reaction to cigarette smoke; 69.2% reported eye irritation, 31.6% headache, 29.2% nasal symptoms, and 25.2% cough. Barad[13] surveyed 21,366 employees of the Social Security Administration and found that nonsmoking workers reported high prevalences of conjunctival irritation (47.7%), nasal discomfort (34.7%), and cough, sore throat, or sneezing (30.3%) when exposed to cigarette smoke.

Weber *et al.*[65,66] exposed subjects to various concentrations of cigarette smoke in a sealed chamber and noted that the eyes were most sensitive to the irritants in the smoke, followed by the nose. Self-reported eye irritation was closely related to such objective signs as tear flow and eye closing or rubbing. Annoyance was the same for pollution caused by whole smoke and by only the gas phase; that indicates that it is the gas phase that is annoying. But whole smoke produced considerably more irritation as expressed by eye and nose symptoms, and that indicates that the particulate phase is responsible for irritation. Hugod *et al.*[28] confirmed that the eyes are the most sensitive site and found that acrolein at the concentrations found in smoke-filled environments did not cause significant irritation. Artho and Koch[9] have reported 11 unpleasant-smelling constituents in the volatile phase and 50 in the semivolatile phase of cigarette smoke.

The eye and nose irritation experienced by nonsmokers in a smoke-filled environment is influenced by the humidity of the air, as well as by the concentration of irritating substances. Johansson[29] and Johansson and Ronge[30] have shown that eye and nose irritation due to cigarette smoke is maximal in warm, dry air and decreases with a small rise in relative humidity.

Physiologic Responses to Smoke

At Rest. Harke and Bleichert[25] studied 18 adults (11 smokers and seven nonsmokers) in a 170-m^3 room in which 150 cigarettes were smoked or allowed to burn in ashtrays for 30 min. They noted that the subjects who smoked during the experiment had a significant lowering of skin temperature and a rise in blood pressure. Nonsmokers who were exposed to the same smoke-contaminated environment showed no change in either of these measures. Luquette *et al.*[34] performed a similar experiment with 40 children exposed alternately to smoke-contaminated and clean atmospheres, but otherwise under identical experimental conditions. Exposure to the smoke was associated with increases in heart rate (5 beats/min) and in systolic and diastolic blood pressure (4 and 5 mm Hg, respectively). The differences in results between these studies may be due, in part, to the age of the subjects: children may be more sensitive to the cardiovascular effects of involuntary smoking than adults. Or the increases in heart rate and blood pressure may be due to a difference between children and adults in the psychologic response to being in a smoke-filled atmosphere.

Pimm *et al.*[38] found a slight decline in heart rate in control subjects of both sexes (thought to be secondary to prolonged inactivity) and a similar decline in heart rate in males exposed to cigarette smoke. However, women exposed to smoke had a small but significant increase in resting heart rate. The authors suggested that this may be due to a difference in psychologic, rather than physiologic, response in the women.

Rummel *et al.*[43] examined this question with a group of 56 students exposed to cigarette smoke. There was a slight increase in the entire group in systolic blood pressure on exposure to smoke. When the group was divided into those who were indifferent to cigarette smoke and those who expressed a dislike for it, both groups again had a rise in systolic blood pressure on exposure to smoke, but the "dislike" group also had a significantly higher heart rate at the start and during the entire course of the study; that suggests that psychologic factors may play a role in the physiologic response to involuntary smoking.

Pimm *et al.*[38] examined the effect of exposure to machine-produced smoke on ventilatory function in healthy young adults. There were no significant changes in the subdivisions of lung volume, maximal expiratory flow volume, or single-breath nitrogen washout after exposure.

With Exercise. Several authors have found small decrements in maximal aerobic capacity at COHb contents corresponding to those associated with involuntary smoking;[4 23 41] for a given degree of exercise, there are reductions in exercise time to exhaustion and maximal oxygen consumption, and there is a higher heart rate. These effects were more pronounced in older than in younger subjects.

Gliner et al.[23] evaluated submaximal exercise and found no change with COHb at 3-6%. Pimm et al.[38] evaluated young adults after exposure to cigarette smoke for 2 h (COHb, 1%) and found no change with submaximal exercise. Shephard et al.[52] studied 23 healthy young adults after 2 h of passive smoke exposure with intermittent bicycle ergometer work sufficient to increase respiratory minute volumes by a factor of 2.5. Carbon monoxide equivalents of 20 ppm and 31 ppm did not change static lung volumes and produced small changes (3-4%) in FVC, FEV, V_{max} 50%, and V_{max} 75%, equivalent to cigarette consumption of less than 0.5 cigarette in 2 h.

Psychomotor Function

There has been some concern over the effects of relatively low concentrations of carbon monoxide on psychomotor functions (which involve perception of and reaction to stimuli), especially those related to driving an automobile. There is an extensive and sometimes contradictory literature on this subject; but it is beyond the scope of this report, and the reader is referred to several recent reviews.[31 36 69] Most of the documented effects occur at COHb concentrations well above those produced by involuntary smoking; however, slight changes in acoustic and visual vigilance have been reported at COHb as low as 2%. The impact of these changes on complex functions such as driving, and their interactions with fatigue and alcohol have not been evaluated for COHb in the range of 2-3%.

Long-Term Effects

The question of long-term effects on the nonsmoker of exposure to cigarette smoke has only recently been raised. The difficulty of measuring the exposure, the complex interaction of cigarette-smoking with behavioral and socioeconomic factors, and the problem of controlling for past smoking history, air pollution, and industrial exposure make it very difficult to isolate the effect of cigarette smoke on the nonsmoker. Recent population studies that accounted for these confounding factors indicated that passive smoke exposures are associated with increased incidences of respiratory mechanical function abnormalities.[27 60 68]

White and Froeb[68] examined the relationship of exposure to cigarette smoke in the workplace and tests of lung function. They found that nonsmokers who had worked where smoking was allowed had unadjusted midexpiratory (FEF 25-75%) and end-expiratory (FEF 75-85%) flow rates lower than those of workers in workplaces where smoking was restricted, but not significantly when adjusted for sex, age, and height. They suggested that the differences represent small-airway dysfunction produced by smoke exposure, and small-airway dysfunction is thought to be an early precursor of clinically significant chronic obstructive lung disease. They controlled for occupational and air-pollution exposure and for smoking in the home. It is difficult to prove an association from a single study, especially in a subject as complex as involuntary smoking; however, their data do suggest that exposure to cigarette smoke may have a deleterious effect on the health and function of the healthy nonsmoker in the work environment.

Hirayama,[27] in a study of mortality records in 29 health-center districts in Japan, followed 91,540 nonsmoking wives, aged 40 and above, for 14 yr (1966-1979) and assessed the standardized mortality rates for lung cancer according to the smoking habits of their husbands. Wives of heavy smokers (greater than 20 cigarettes/d) were found to have a relative risk of developing lung cancer of 2.1, whereas wives of ex-smokers and of smokers of fewer than 20 cigarettes/d had a relative risk of 1.6. The relation between a husband's smoking and a wife's risk of developing lung cancer showed a similar pattern when analyzed by age and occupation of the husband. The husband's smoking habits did not affect his wife's risk of dying from other diseases, such as stomach cancer, cervical cancer, and ischemic heart disease. The risk of developing emphysema and asthma seemed to be higher in nonsmoking wives of heavy smokers, but the effect was not statistically significant. The husbands' drinking habits seemed to have no effect on any cause of death in their wives, including lung cancer.

Trichopoulos et al.[60] interviewed 51 women with lung cancer and 163 other hospital patients in Greece regarding their smoking habits and their husbands' smoking habits. Forty of the lung-cancer patients and 149 of the other patients were nonsmokers. Among the nonsmoking women, there was a statistically significant difference between the cancer patients and the other patients with respect to their husbands' smoking habits. Estimates of relative risk of lung cancer associated with having a husband who smokes were 2.4 for smokers of less than one pack per day and 3.4 for smokers of more than one pack per day.

In two studies indicating a similar effect, it appears that chronic passive smoking significantly increases the incidence of lung cancer.

EFFECTS ON SPECIAL POPULATIONS

The studies mentioned examined the effects of involuntary smoking on relatively healthy populations. An exposure that is harmless for someone who is healthy may have a very different effect on someone with heart or lung disease or hypersensitivity to substances found in smoke. Effects may differ in children, owing to their greater ventilation per unit of body weight. This section reviews the evidence on the effects of involuntary smoking on each of these special populations.

Cardiovascular Disease

Carbon monoxide impairs oxygen transport in two ways. First, it competes with oxygen for hemoglobin binding sites. Second, it increases the affinity of oxygen for the remaining hemoglobin, thereby requiring a larger gradient in pO_2 between the blood and tissue to deliver a given amount of oxygen. Carbon monoxide also binds to other heme-containing pigments, most notably myoglobin, for which it has an even greater affinity than for hemoglobin at low pO_2. The significance of this binding is unclear, but it may be important in some tissues (such as heart muscle) that have both high oxygen requirements and large amounts of myoglobin.

In healthy people, the COHb content due to involuntary smoking is probably functionally insignificant, with small changes demonstrable only under extreme exertion. In those with a limited cardiovascular reserve, however, any reduction in the oxygen-carrying capacity of the blood may be of greater importance.

Ayres et al.[10,11] exposed a group of patients to various concentrations of carbon monoxide (COHb, 9%) and found that they had lower arterial, mixed venous, and coronary sinus pO_2 and decreased lactate extraction.

Aronow and Isbell[8] and Anderson et al.[1] have shown a decrease in the mean duration of exercise before onset of pain in patients with angina pectoris exposed to carbon monoxide at low concentrations (50 and 100 ppm). Carboxyhemoglobin was significantly increased (2.9% after 50 ppm; 4.5% after 100 ppm), and the systolic blood pressure, heart rate, and product of systolic blood pressure and heart rate (a measure of cardiac work) were all significantly lower at the onset of angina pectoris.

In a continuation of this work, Aronow et al.[5,7] studied eight patients with angiographically demonstrated coronary arterial disease (>75% obstruction of at least one coronary artery) during two cardiac catheterizations. During the first, each patient smoked three cigarettes; during the second, each patient inhaled carbon monoxide until the maximal coronary sinus COHb content equaled that produced by smoking during the first catheterization. Smoking increased the systolic and diastolic blood pressure, heart rate, left ventricular end-diastolic pressure (LVEDP), and coronary sinus, arterial, and venous COHb; no changes were noted in left ventricular contractility (dp/dt), aortic systolic ejection period, or cardiac index; and there were decreases in stroke index and coronary sinus, arterial, and venous pO_2. When carbon monoxide was inhaled, increased LVEDP and coronary sinus, arterial, and venous COHb were noted; there were no changes in systolic and diastolic blood pressure, heart rate, or systolic ejection period; and there were decreases in left ventricular dp/dt, stroke index, cardiac index, and coronary sinus, arterial, and venous pO_2. These data suggest that carbon monoxide has a negative ionotropic effect on myocardial tissue, which results in the decreased dp/dt and stroke index. When the positive effect of nicotine on contractility and heart rate is added by smoking, the net effect is increased cardiac work for the same cardiac output.

Aronow[3] also examined the effect of involuntary smoking on patients with angina pectoris. Ten patients (two smokers and eight nonsmokers) exercised after a control exposure to uncontaminated air, after exposure to 15 cigarettes smoked over 2 h in a well-ventilated 30.8-m^3 room, and after exposure to 15 cigarettes smoked over 2 h in an unventilated 30.8-m^3 room. Carboxyhemoglobin rose from 1.25% in the control situation to 1.77% after exposure in the ventilated room, and to 2.28% after exposure in the unventilated room. The mean time of exercise until onset of angina decreased by 22% after exposure in the ventilated room and by 38% after exposure in the unventilated room. The patients also had onset of angina at a lower heart rate and systolic blood pressure, and they had increases in heart rate and systolic and diastolic blood pressures. Aronow attributed this to the possible absorption of nicotine (nicotine was not measured). The relatively low nicotine absorption documented under these conditions (see the previous section) makes it unlikely that nicotine would be responsible for these physiologic changes. Another possible explanation is that anxiety or aggravation associated with being in the smoke-filled room resulted in a stress response.[63] The combination of increased blood pressure and pulse at the start of exercise and the increase in carboxyhemoglobin resulted in a greater decline in exercise time until angina for the measured carboxyhemoglobin than had been shown for carbon monoxide exposure alone.

In summary, there is evidence that increases in carboxyhemoglobin capable of being produced by involuntary smoking can reduce the exercise duration required to induce angina in some patients with coronary arterial disease.

Chronic Obstructive Lung Disease

Patients with chronic lung disease constitute a second group who are limited in their ability to exercise and who might be particularly susceptible to involuntary smoking. Aronow *et al.*[6] had 10 patients with hypoxic chronic lung disease ($pO_2 < 70$ torr) exercise before and after a 1-h exposure to carbon monoxide at 100 ppm (COHb increased from 1.43% to 4.08%). There was a significant reduction in mean exercise time until marked dyspnea, from 218.5 s to 146.6 s. There was no difference in exercise mean systolic or diastolic blood pressure, heart rate, product of systolic blood pressure and heart rate, or arterial pO_2, pCO_2, or pH before or after carbon monoxide exposure. The mechanism for this earlier induction of dyspnea remains unclear, because decreased oxygen transport to the exercising tissues should have been reflected in a shift to anaerobic metabolism and the development of acidosis.

Persons with Allergies

The existence of a true tobacco allergy remains unclear. There is no proof that specific sensitization to cigarette smoke occurs.[59]

However, it is clear that allergic patients are more sensitive to a variety of environmental irritants, including tobacco smoke.[59] The manifestations of this irritation may often mimic the allergic symptoms experienced by these patients when they come into contact with well-established allergens. It has also been demonstrated that cigarette-smoking by parents is a significant exacerbating factor in childhood asthma.[37]

Infants and Children

Children have a higher incidence of acute respiratory illness than adults and may be more susceptible to air pollutants than adults, owing to their greater minute ventilation per unit of body weight. Several researchers have investigated the effects of parental smoking on the health of children.

Colley[19] found a relationship between parental smoking habits and the prevalence of respiratory illness in children. However, an even stronger relationship was found between cough and phlegm production in parents and respiratory infections in children. They postulated that the latter relationship resulted from the greater infectivity of these parents due to their cough and phlegm production. The relationship between parental cigarette-smoking and respiratory infection in their children would then occur because cigarette-smoking caused the parents to cough and produce phlegm and would not be indicative of a direct effect of smoke-filled air on the children. Bland et al.[14] reported similar relationships.

There have been several other research reports of associations between passive smoking in the home and symptoms or illnesses in children.[16,48] A telephone survey[16] confirmed an earlier survey,[15] in that children and adult nonsmokers subjected to household tobacco smoke had had a higher prevalence of acute respiratory illness in the preceding week than children and adult nonsmokers not so exposed. Reporting biases of the telephone respondents were not examined or controlled. Said and Zalokar[48] questioned Parisian high-school students (aged 9-19) about parental smoking and about their history of adenoidectomies and tonsillectomies. They found increases in the latter related to amount of tobacco smoked by either or both parents. This relationship was not always consistent (e.g., the% of such operations did not increase with increases in maternal smoking if paternal smoking was high). The prevalence of appendectomies was related to maternal smoking as well (and appendectomies were significantly correlated with the other operations). Social status was not controlled; and the effect of actual (voluntary) smoking, a critical factor in this age group,[12] was not evaluated in this study, thereby limiting its usefulness.

Harlap and Davies[26] studied infant admissions to Hadassah Hospital in West Jerusalem and found a relationship between admissions for bronchitis and pneumonia in the first year of life and maternal smoking habits during pregnancy. Data on postnatal maternal smoking habits were not obtained, but it can be assumed that most of the mothers who smoked during pregnancy continued to smoke during the following year. A relationship between infant admission and maternal smoking habits was demonstrable only between the sixth and ninth months of infant life and was more pronounced during the winter months. Mothers who smoke during pregnancy are known to have infants with a lower average birthweight than nonsmoking mothers. The relationship between maternal smoking and infant admission to the hospital found in this study was greater for low-birthweight infants, but the same relationship was found for normal-birthweight infants. Harlap and Davies[26] demonstrated a dose-response relationship between maternal smoking and infant admission for bronchitis and pneumonia; however, they also found a relationship between maternal smoking and infant admission for poisoning and injuries. This may indicate a bias in the study due to relationships that may exist between smoking and such factors as parental neglect and socioeconomic class. In addition, hospital admission rates may not be an accurate index of infant morbidity.

Colley _et al._[20] and Leeder _et al._[33] studied the incidence of pneumonia and bronchitis in 2,205 children over the first 5 yr of life in relation to the smoking habits of both parents. They found that a relationship between parental smoking habits and respiratory infection in children occurred only during the first year of life. They also showed a relationship between infant infection and parental cough and phlegm production that was independent of the effect of parental smoking habits. The relationship between parental smoking and infant infection was greater when both parents smoked and increased with the number of cigarettes smoked per day. The relationship persisted when social class and birthweight were controlled for.

Rantakallio[39,40] has also found an increased incidence of pneumonia in children under the age of 5. She studied 12,000 children born in northern Finland in 1966 and matched smoking mothers with nonsmoking mothers for various factors, including marital status, maternal age, and socioeconomic status. Children of smoking mothers had significantly higher morbidity ($p < 0.001$) and were more likely to be hospitalized ($p < 0.001$), and their hospitalizations were longer. Most of this excess morbidity was due to respiratory illness and was present in the first 5 yr of life, with the most pronounced effect occurring in the first year of life.

Cederlöf and Colley[17] stated that, "when parents' respiratory symptoms were taken into account, exposure of the child to cigarette smoke generated by the parents' smoking had little if any effect upon the child's respiratory symptoms." Lebowitz and Burrows[32] and Schilling _et al._[49] reported nonsignificant relationships between parental smoking and children's symptoms when parental symptoms were taken into account. They concluded that parents had a "bias" toward reporting symptoms in their children when they themselves had such symptoms.

Tager _et al._[57,58] examined the relationship of parental smoking habits and expiratory flow rates in children. They found a dose-dependent decline in the FEF (25-75%) in the children, with a greater decline occurring if both parents smoked than if one parent smoked, and with the decline increasing with number of cigarettes smoked. This effect was independent of the smoking habits of the children. Pulmonary infection early in life has been shown to affect pulmonary function in children and adults adversely, and the decline in flow rates reported by Tager et al. may be secondary to the excess risk of pneumonia in infants whose parents smoke. They attempted to examine this by retrospectively asking the parents about childhood illness, but did not show an association between parental smoking and childhood infection, in contrast with the results of Rantakallio and Harlap. It is not clear whether this represents a true difference in the populations.

In a further study of 5- to 9-yr-old children in the same population, Weiss et al.[67] reported that parental cigarette-smoking was linearly related to the occurrence of persistent wheeze ($p = 0.012$) and lower degrees of mean forced midexpiratory flow. Current persistent wheeze occurred in one of 57 children (1.8%) from households where both parents had never smoked; in 10 of 146 children (6.8%) with one parent currently smoking; and in 20 of 169 children (12%) with both parents currently smoking. When the analysis was repeated with the exclusion of mothers with wheeze, the results were similar--0, 1.8, and 7.7% wheeze in children with no smoking parents, one smoking parent, and two smoking parents, respectively. Exclusion of fathers with wheeze gave 0, 6.7, and 14% wheeze, respectively.

In summary, children of smoking parents have an increased incidence of persistent wheeze and may be at excess risk of repiratory infection at least for the first year of life. They may also have reduced pulmonary function as adults. The exact interplay among the effects of maternal smoking during pregnancy, involuntary smoking by children, and actual occurrence of infection has not been established.

CONCLUSIONS

- Tobacco smoke is a major source of pollution in the indoor environment.
- The nonsmoker absorbs measurable amounts of carbon monoxide and nicotine and may absorb small amounts of other constituents, owing to involuntary smoking.
- The amount of carbon monoxide absorbed owing to exposure to tobacco smoke in the environment varies from negligible amounts in well-ventilated office buildings to enough to raise carboxyhemoglobin contents by 2-3% in a 1- to 2-h exposure.
- The carboxyhemoglobin produced by the most severe involuntary smoking exposures likely to occur in everyday living can reduce the maximal exercise capacity in normal, healthy adults, but does not effect submaximal exercise to any measurable degree.
- Involuntary smoking has not been shown to produce acute change in lung volumes, expiratory flow rates, closing volumes, or the slope of phase III of the single-breath nitrogen washout in normal, healthy adults; but long-term exposure to cigarette smoke is related to small-airway dysfunction and an increased incidence of lung cancer in healthy nonsmoking adults.

• Small changes in visual and auditory vigilance have been demonstrated at carboxyhemoglobin contents capable of being produced by involuntary smoking, but no change in tests of complex function has been demonstrated. The interaction of fatigue, alcohol, and carbon monoxide exposure on complex functions, such as automobile driving, has not been investigated for COHb contents capable of being produced under normal conditions of involuntary smoking.

• Patients with angina pectoris have reduced exercise tolerance after involuntary smoking that may be a combination of psychologic stress and a reduction in oxygen delivery to the myocardium induced by carbon monoxide. Carbon monoxide clearly reduces the amount of exercise possible until the onset of angina in patients with angina pectoris at COHb contents that may be reached as a result of involuntary smoking.

• Carbon monoxide has been shown in one study to reduce the amount of exercise that patients with hypoxic chronic obstructive lung disease can perform until the onset of dyspnea.

• Most nonsmokers find it annoying to be exposed to cigarette smoke. This annoyance is probably due to substances in the gas phase of the smoke.

• Cigarette-smoke exposure results in eye, nose, throat, and respiratory irritation. The eyes are most sensitive, followed by the nose and throat. The particulate phase of cigarette smoke seems to be predominantly responsible for this irritation.

• Persons with allergies are more sensitive to the irritant effects of cigarette smoke. However, there is no proof of tobacco allergy.

• Children whose parents smoke may be more likely to have respiratory symptoms, bronchitis, and pneumonia as infants and may have poorer pulmonary function as adults, compared with children of nonsmoking parents. This relationship is not independent of parental symptoms, socioeconomic class, and the smoking habits of the children; and it is associated with the number of cigarettes smoked per day by the parents.

REFERENCES

1. Anderson, E. W., R. J. Andelman, J. M. Strauch, N. J. Fortuin, and J. J. Knelson. Effect of low-level carbon monoxide exposure on onset and duration of angina pectoris. A study in ten patients with ischemic heart disease. Ann. Intern. Med. 79:46-50, 1973.
2. Anderson, G., and T. Dalhamn. The risks to health of passive smoking. Läkartidningen 70:2833-2836, 1973.
3. Aronow, W. S. Effects of passive smoking on angina pectoris. N. Engl. J. Med. 299:21-24, 1978.
4. Aronow, W. S., and J. Cassidy. Effect of carbon monoxide on maximal treadmill exercise. A study in normal persons. Ann. Intern. Med. 83:496-499, 1975.
5. Aronow, W. S., J. Cassidy, J. S. Vangrow, H. March, J. C. Kern, J. R. Goldsmith, M. Khemka, J. Pagano, and M. Vawter. Effect of

cigarette smoking and breathing carbon monoxide on cardiovascular hemodynamics in anginal patients. Circulation 50(2):340-347, 1974.

6. Aronow, W. S., J. Ferlinz, and F. Glauser. Effect of carbon monoxide on exercise performance in chronic obstructive pulmonary disease. Am. J. Med. 63:904-908, 1977.
7. Aronow, W. S., J. R. Goldsmith, J. C. Kern, and L. L. Johnson. Effects of smoking cigarettes on cardiovascular hemodynamics. Arch. Environ. Health 28:330-332, 1974.
8. Aronow, W. S., and M. W. Isbell. Carbon monoxide effect on exercise-induced angina pectoris. Ann. Intern. Med. 79:392-395, 1973.
9. Artho, A., and R. Koch. Caracterisation olfactive des composes de la fumée de cigarettes (Characterization of the olfactory properties of cigarette smoke components). Annales du Tabac (Section 1-11):37-45, 1973.
10. Ayres, S. M., S. Giannelli, Jr., and H. Mueller. Myocardial and systemic responses to carboxyhemoglobin. Ann. N.Y. Acad. Sci. 174:268-293, 1970.
11. Ayres, S. M., H. S. Mueller, J. J. Gregory, S. Giannelli, Jr., and J. L. Penny. Systemic and myocardial hemodynamic responses to relatively small concentrations of carboxyhemoglobin (COHB). Arch. Environ. Health 18:699-709, 1969.
12. Banks, M. H., B. R. Bewley, J. M. Bland, J. R. Dean, and V. Pollard. Long term study of smoking by secondary school-children. Arch. Dis. Child. 53:12-19, 1978.
13. Barad, C. B. Smoking on the job: The controversy heats up. Occup. Health Saf. Jan-Feb. 1979, p. 21.
14. Bland, M., B. R. Bewley, V. Pollard, and M. H. Banks. Effects of children's and parent's smoking on respiratory symptoms. Arch. Dis. Childhood 51:100-105, 1978.
15. Cameron, P., J. S. Kostin, J. M. Zaks, J. H. Wolfe, G. Tighe, B. Oselett, R. Stocker, and J. Winton. The health of smokers' and nonsmokers' children. J. Allergy 43:336-341, 1969.
16. Cameron, P., and D. Robertson. Effect of home environmental tobacco smoke on family health. J. Appl. Physiol. 57:142-147, 1973.
17. Cederlöf, R., and J. Colley. Epidemiological investigations on environmental tobacco smoke. Scand. J. Respir. Dis. (Suppl. 91):47-49, 1974.
18. Cole, P. V. Comparative effects of atmospheric pollution and cigarette smoking on carboxyhemoglobin levels in man. Nature 255:699-701, 1975.
19. Colley, J. R. T. Respiratory symptoms in children and parental smoking and phlegm production. Br. Med. J. 2:201-204, 1974.
20. Colley, J. R. T., W. W. Holland, and R. T. Corkhill. Influence of passive smoking and parental phlegm on pneumonia and bronchitis in early childhood. Lancet 2:1031-1034, 1974.
21. Dalhamn, T., M.-L. Edfors, and R. Rylander. Mouth absorption of various compounds in cigarette smoke. Arch. Environ. Health 16:831-835, 1968.
22. Dalhamn, T., M.-L. Edfors, and R. Rylander. Retention of cigarette smoke components in human lungs. Arch. Environ. Health 17:746-748, 1968.

23. Gliner, J. A., P. B. Raven, S. M. Horvath, B. L. Drinkwater, and J. C. Sutton. Man's physiologic response to long-term work during thermal pollutant stress. J. Appl. Physiol. 39:628-632, 1975.
24. Harke, H.-P. The problem of "passive smoking." Münch. Med. Wochenschr. 112:2328-2334, 1970. (in German; English summary)
25. Harke, H.-P., and A. Bleichert. On the problem of passive smoking. Int. Arch. Arbeitsmed. 29:312-322, 1972. (in German; English summary)
26. Harlap, S., and A. M. Davies. Infant admissions to hospital and maternal smoking. Lancet 1:529-532, 1974.
27. Hirayama, T. Non-smoking wives of heavy smokers have a higher risk of lung cancer: A study from Japan. Br. Med. J. 282:183-185, 1981.
28. Hugod, C., L. H. Hawkins, and P. Astrup. Exposure of passive smokers to tobacco smoke constituents. Int. Arch. Occup. Environ. Health 42:21-29, 1978.
29. Johansson, C. R. Tobacco smoke in room air--an experimental investigation of odour perception and irritating effects. Build. Services Eng. 43:254-262, 1976.
30. Johansson, C. R., and H. Ronge. Akuta irritationseffekter av tobaksrök i rumsluft. (Acute irritation effects of tobacco smoke in the room atmosphere). Nord. Hyg. Tidskr. 46:45-50, 1965.
31. Laties, V. G., and W. H. Merigan. Behavioral effects of carbon monoxide on animals and man. Ann. Rev. Pharmacol. Toxicol. 19:357-392, 1979.
32. Lebowitz, M. D., and B. Burrows. Respiratory symptoms related to smoking habits of family adults. Chest 69:48-50, 1976.
33. Leeder, S. R., R. Corkhill, L. M. Irwig, W. W. Holland, and J. R. T. Colley. Influence of family factors on the incidence of lower respiratory illness during the first year of life. Br. J. Prevent. Social Med. 30:203-212, 1976.
34. Luquette, A. J., C. W. Landiss, and D. J. Merki. Some immediate effects of a smoking environment on children of elementary school age. J. School Health 40:533-536, 1970.
35. National Clearinghouse for Smoking and Health. Adult Use of Tobacco, 1975. U.S. Department of Health, Education, and Welfare, National Clearinghouse for Smoking and Health, June 1976. 23 pp.
36. National Research Council, Committee on Medical and Biologic Effects of Environmental Pollutants. Carbon Monoxide. Washington, D.C.: National Academy of Sciences, 1977. 239 pp.
37. O'Connell, E. J., and G. B. Logan. Parental smoking in childhood asthma. Ann. Allergy 32:142-145, 1974.
38. Pimm, P. E., F. Silverman, and R. J. Shephard. Physiological effects of acute passive exposure to cigarette smoke. Arch. Environ. Health 33:201-213, 1978.
39. Rantakallio, P. Relationship of maternal smoking to morbidity and mortality of the child up to the age of five. Acta Paediatr. Scand. 67:621-631, 1978.
40. Rantakallio, P. The effect of maternal smoking on birth weight and the subsequent health of the child. Early Human Dev. 2:371-382, 1978.

41. Raven, P. B., B. L. Drinkwater, S. M. Horvath, R. O. Ruhling, J. A. Gliner, J. C. Sutton, and N. W. Bolduan. Age, smoking habits, heat stress, and their interactive effects with carbon monoxide and peroxyacetylnitrate on man's aerobic power. Int. J. Biometeorol. 18:222-232, 1974.
42. Repace, J. L., and A. H. Lowrey. Indoor air pollution, tobacco smoke, and public health. Science 208:464-472, 1980.
43. Rummel, R. M., M. Crawford, and P. Bruce. The physiological effects of inhaling exhaled cigarette smoke in relation to attitude of the nonsmoker. J. School Health 45:524-529, 1975.
44. Russell, M. A. H., P. V. Cole, and E. Brown. Absorption by non-smokers of carbon monoxide from room air polluted by tobacco smoke. Lancet 1:576-579, 1973.
45. Russell, M. A. H., and C. Feyerabend. Blood and urinary nicotine in non-smokers. Lancet 1:179-181, 1975.
46. Rylander, R., Ed. Environmental Tobacco Smoke Effects on the Nonsmoker. Report from a Workshop. Scand. J. Respir. Dis. (Suppl. 91):1-90, 1974.
47. Rylander, R. Perspectives on environmental tobacco smoke effects. Scand. J. Respir. Dis. (Suppl. 91):79-87, 1974.
48. Said, G., and J. Zalokar. Incidence of upper respiratory tract disorders in children of smokers. Ann. d'Oto-laryngol. Chir. Cervico-Fac. 95:236-240, 1978.
49. Schilling, R. S. F., A. D. Letai, S. L. Hui, G. J. Beck, J. B. Schoenberg, and A. Bouhuys. Lung function, respiratory disease, and smoking in families. Am. J. Epidemiol. 106:274-283, 1977.
50. Seppänen, A. Smoking in closed space and its effect on carboxyhaemoglobin saturation of smoking and nonsmoking subjects. Ann. Clin. Res. 9:281-283, 1977.
51. Seppänen, A., and A. J. Uusitalo. Carboxyhaemoglobin saturation in relation to smoking and various occupational conditions. Ann. Clin. Res. 9:261-268, 1977.
52. Shephard, R. J., R. Collins, and F. Silverman. Responses of exercising subjects to acute "passive" cigarette smoke exposure. Environ. Res. 19:279-291, 1979.
53. Speer, F. Tobacco and the nonsmoker. A study of subjective symptoms. Arch. Environ. Health 16:443-446, 1968.
54. Srch, M. Über die Bedeutung des Kohlenoxyds beim Zigarettenrauchen im Personenkraftwageninneren. Dtsch. Z. Gesamte Gerichtl. Med. 60(3):80-89, 1967. (in German)
55. Sterling, T. D., and D. M. Kobayashi. Exposure to pollutants in enclosed "living spaces." Environ. Res. 13:1-35, 1977.
56. Szadkowski, D., H.-P. Harke, and J. Angerer. Body burden of carbon monoxide from passive smoking in offices. Innere Med. 3:310-313, 1976.
57. Tager, I. B., B. Rosner, P. V. Tishler, F. E. Speizer, and E. H. Kass. Household aggregation of pulmonary function and chronic bronchitis. Am. Rev. Respir. Dis. 114:485-492, 1976.
58. Tager, I. B., S. T. Weiss, B. Rosner, and F. E. Speizer. Effect of parental cigarette smoking on the pulmonary function of children. Am. J. Epidemiol. 110:15-26, 1979.

59. Taylor, G. Tobacco smoke allergy--Does it exist? Scand. J. Respir. Dis. (Suppl. 91):50-55, 1974.
60. Trichopoulos, D., A. Kalandidi, L. Sparros, and B. MacMahon. Lung cancer and passive smoking. Int. J. Cancer 27:1-4, 1981.
61. U.S. Department of Health, Education, and Welfare, Public Health Service. Smoking and Health. A Report of the Surgeon General. DHEW Publication No. (PHS) 79-50066. Washington, D.C.: U.S. Government Printing Office, 1979. [1250] pp.
62. U.S. Department of Transportation, Federal Aviation Administration, and U.S. Department of Health, Education, and Welfare, National Institute for Occupational Safety and Health. Report on Health Aspects of Smoking in Transport Aircraft. Washington, D.C.: U.S. Department of Health, Education, and Welfare, National Institute for Occupational Safety and Health, Division of Technical Services, 1971. 85 pp.
63. Waite, C. L. Letter to the editor. N. Engl. J. Med. 299:897, 1978.
64. Wakeham, H. R. R. Environmental carbon monoxide from cigarette smoking--A critique. Prev. Med. 6:526-534, 1977.
65. Weber, A., T. Fischer, and E. Grandjean. Passive smoking: Irritating effects of the total smoke and the gas phase. Int. Arch. Occup. Environ. Health 43:183-193, 1979.
66. Weber, A., C. Jermini, and E. Grandjean. Irritating effects on man of air pollution due to cigarette smoke. Am. J. Public Health 66:672-676, 1976.
67. Weiss, S. T., I. B. Tager, F. E. Speizer, and B. Rosner. Persistent wheeze. Its relation to respiratory illness, cigarette smoking, and level of pulmonary function in a population sample of children. Am. Rev. Respir. Dis. 122:697-707, 1980.
68. White, J. R., and H. F. Froeb. Small-airways dysfunction in nonsmokers chronically exposed to tobacco smoke. N. Engl. J. Med. 302:720-723, 1980.
69. Yabroff, I., E. Meyers, V. Fend, N. David, M. Robertson, R. Wright, and R. Braun. The role of atmospheric carbon monoxide in vehicle accidents. Menlo Park, Cal.: Stanford Research Institute, February 1974.

INDOOR AIRBORNE CONTAGION

Among the pollutants of indoor air are biologic aerosols produced by people when they cough, sneeze, sing, spit, blow their noses, or even talk. Discussion of airborne infection is as old as recorded history, but refined concepts of contagion, expressible in quantitative terms, are surprisingly recent. Less then 50 yr ago, William F. Wells synthesized a coherent theory that has now been tested and amplified.[91] Even though the ideas are not yet imbedded in medical thinking and teaching, they pertain to a very important medical and public-health problem. Airborne contagion is the mechanism of transmission of most acute respiratory infections, and these are the greatest of all causes of morbidity. Primary pulmonary tuberculosis is also transmitted in this way. Airborne contagion from person to person is mostly an indoor phenomenon.

ASSESSING INDOOR BIOGENIC POLLUTANTS

In approaching biogenic pollutants, the dearth of data on indoor prevalence and the lack of satisfactory study methods constitute a single complex problem. Although most indoor biologic agents are distinctive microscopically, many categories of biogenic particles are not. For those which also fail to grow recognizably in culture, no practical enumeration is yet possible.

The penetration of biologic particles into buildings has been little studied, but seems to depend most on the extent of mass flow through windows and doors. Additional factors in ventilation[64,84] include incident-wind speed and direction, negative pressurization by exhaust fans, and stack effects (which may be minor in warm periods). Air leaks between structural members ("crackage") foster ventilation; window and door frames contribute less.[64] Tracer gases often have suggested brisk infiltration of air into structures; however, the capacity of windborne particles to negotiate minute cracks, certainly less, remains to be estimated. Sampling for biologic agents both indoors and outdoors is fundamental to studies of their sources. Furthermore, because particles may remain indoors for a long time after infiltration from free air, analyses of indoor-outdoor relationships must be sensitive to the resulting lag effects.[84]

Biogenic pollutants bear complex and varied organic structures, which defy automated chemical assay. Culture or direct microscopic enumeration offers a workable, although tedious, alternative for some particles; for others, immunofluorescence and multiphasic microscopy[11] have demonstrated potential. Concentrations of airborne biogenic dusts that lack morphologically defined units might be estimated by subjecting extracts of bulk aerometric samples to immunoassay; methods suitable for amorphous components of house dust have been discussed elsewhere.[87]

Because many biologic pollutants are relatively large aerodynamically, whereas others are quite small or undefined, precautions to minimize size-related collection bias are essential. Rates of circulation indoors are generally lower and often more nearly constant than those outdoors. However, the velocities generated by fans, human activity, and pronounced convection are important; they readily bias recoveries based on fallout and may affect collection with suction traps.[89] As a result, differences in particle recovery between indoor and outdoor sites may reflect prevailing flow conditions more than real transmural differences in aerosol prevalence.

Despite their longstanding popularity, "gravitational" methods involve particularly marked, size-related bias in collections.[82] And the small samples obtained and the lack of volumentric capability further limit the usefulness of data obtained in this way. Suction devices have been used sucessfully for indoor studies, and miniature impactors are adaptable for this purpose. In all applications, the siting of samplers vis-a-vis probable pollutant sources should be considered and points of low flow avoided.

In contrast with chemically simpler pollutants, biogenic agents exhibit limited direct toxicity, more often provoking infection.

Airborne transmission of infectious agents is facilitated indoors by the prompt dispersion of particles.[12] In general, indoor bacteria appear to have indoor sources; data on outdoor airborne bacteria are severely limited. One study has reported bacteria in outdoor air at up to 4,000/m^3.[7] Another study presented indoor-to-outdoor ratios of 0.76-14.25,[78] with much site-to-site variation. Outdoor concentrations may depend on high wind velocity and temperature[7] and are apparently highest in summer.[7 78]

Bacteria from outdoors do contaminate interiors, but to an unknown degree. Clostridium perfringens, primarily a soil bacterium, has been found in room air and house dust.[79]

The primary source of bacteria in most indoor places is the human body.[53] Although the major source is the respiratory tract,[52 67] there are other sources. According to Clark and Cox,[14] 7 million skin scales are shed per minute per person, with an average of four viable bacteria per scale. Abrasion is the primary factor in the rate of loss[14 56] and showering increases the rate of loss of bacteria.[14 17 83]

EVIDENCE OF INDOOR AIRBORNE INFECTION

Droplet nuclei are the dried residues of the smallest respiratory droplets. They range in size from 1 to 3 µm, disperse rapidly throughout the air of a room, and are carried wherever the air goes. Settling velocity is negligible in comparison with the velocity of air movement in occupied rooms. The concentration of viable organisms attached to droplet nuclei may be reduced by natural die-away, air filtration, or exchange with outdoor air. Standard filters used in ventilating systems remove a small fraction. There is no reservoir of infectious droplet nuclei other than the respiratory tracts of people carrying the organisms. Wells[90] believed that aerial transmission from person to person occurs indoors where droplet nuclei are in sufficient concentration to be a hazard. Infectious contact (contagion) requires proximity in time and space between host and victim, but can be extended to the confines of the enclosed atmosphere and to a shared ventilating system if the air within the system is recirculated;[68] the recirculating system then becomes a common enclosed atmosphere.

Tuberculosis

In the 1950s, a study was carried out at the Veterans' Administration Hospital in Baltimore in which guinea pigs breathed air vented from a tuberculosis ward.[70] Tubercle bacilli from the patients, which had gone through the ventilating ducts and through the upper respiratory tracts of the guinea pigs, were positively identified in the lungs of infected animals.[69] On the basis of this and other evidence, it is generally agreed that the initial infection of the lungs with tuberculosis is airborne.

Measles

An early demonstration of airborne infection involved the indirect approach--the control of epidemic spread by disinfecting the air. The study was carried out in the 1940s in schools in Germantown and Swarthmore, Pennsylvania, by Wells et al.[92] Shortly after ultraviolet (UV) air-disinfection fixtures had been installed in the test schools, a major measles epidemic struck. In both communities, the nonirradiated schools had sharp outbreaks of measles, and the UV-irradiated schools did not. The reduced spread of infection in the irradiated schools could be attributed to the single factor that was different, namely, the concentration of viable airborne measles virus.

During the spring of 1974, a sharp outbreak of measles occurred in an elementary school near Rochester, New York.[68] Measles was introduced into the school by a girl in the second grade. Twenty-eight secondary cases followed after an incubation period of about 10 d; these were distributed among 14 classrooms served by the same ventilating system. The wide distribution of the 28 secondary cases among children who had never even occupied the same room as the child with the index case and the fact that about 70% of the air was recirculated, and hence shared by all children served by the ventilating system, led to the conclusion that measles reached the various classrooms via the ventilating system.

Asian Influenza

During the 1957-1958 pandemic of Asian influenza, the main building of the Veterans' Hospital in Livermore, California, had UV air disinfection installed throughout, and the patients in the building constituted a test group.[57] Patients in neighboring nonirradiated buildings served as controls. No mixing of test and control groups was allowed. Hospital personnel and visitors who mingled in the outside community were relied on to introduce infection into the two patient populations. The incidence of serologically diagnosed influenza among the 209 patients living in the UV-irradiated building was 2%; among the 396 patients living in nonirradiated buildings, it was 19%. Like the measles study, this demonstration provided evidence consistent with transmission of the infectious virus by air.

Schulman demonstrated that natural transmission of influenza from mouse to mouse is airborne,[76] and the studies of various viral infections by Knight and collaborators are compatible with aerial transmission in humans.[49]

Smallpox

In 1970, in a West German hospital, a patient with smallpox infected 19 other persons, despite rigid isolation procedures. Investigators from the World Health Organization and West Germany demonstrated that smallpox, like influenza, could be transmitted by air currents, but smallpox has been eradicated and need not be considered further.[43]

These selected epidemiologic studies show the kind of evidence supporting the droplet-nucleus concept of indoor airborne contagion. There have, of course, been many other studies, including unsuccessful attempts at confirmation of some of those mentioned. In retrospect, the failures appear to be attributable to inadequacies of experimental design. The droplet-nucleus mechanism of Wells is emerging as the successor, for most respiratory infections, to the direct-contact mechanism of Chapin.

Other Types of Airborne Infection

Infections in hospitals have not been shown to be primarily airborne, and such organisms as staphylococci, streptoccci, and gram-negative bacilli are not characteristically transmitted by air. Nevertheless, hospital-acquired infections of the lower respiratory tract are presumptively airborne, inasmuch as inspired air is the most likely vehicle for carrying organisms to the lungs. Hospital patients are often hypersusceptible to infection, and transmission may occur in ways not often seen in the general population.

A major epidemic of Legionnaire's disease occurred in a hospital into which outside air contaminated with Legionella pneumophila leaked during adjacent construction.[86] This organism is unusual among bacterial pathogens, in that it apparently exists in outdoor natural reservoirs (soils) and infection is possible through inhalation of contaminated outdoor air.[18,31] The most common mode of spread of Legionnaire's disease involves air-cooling equipment that becomes contaminated and produces concentrated bacterial aerosols.[18,37,58]

Air-conditioning and -humidifying equipment can be a source of intramural bacterial aerosols. Cool-mist vaporizers and nebulizers that can produce heavily contaminated aerosols are of special concern.[1,4,10,19,23,40,44,48,52,63,77,81] Apparently, bacterial contamination of such units approaches 100%.[20,71] Pseudomonas appears to be the most commonly isolated bacterial genus.[10,23,40,48,63] Smith[81] reported several cases of Acinetobacter infection resulting from contaminated cool-mist vaporizers. Evaporative humidifiers, although often contaminated with bacteria, are less likely to produce bacteria-laden aerosols.[4,9,19] Disinfection of any humidifying unit is effective only temporarily,[9,19,40] and Rosenzweig[71] has recommended banning cool-mist humidifiers for home or hospital use.

Other appliances reported to be potential sources of indoor bacterial aerosols are flush toilets.[22] Ice machines are also potential foci for bacterial contamination.[5] Rylander et al.[73] discussed carpeting as a focus for bacterial contamination, but concluded that carpeting can, in fact, reduce airborne bacterial concentrations by trapping bacteria-laden particles in the pile.

There are specific sites at which bacteria may become airborne at high concentrations. Factories that process organic materials may contain dense bacterial aerosols.[13,41,51,61,94]

In some interior situations, even low bacterial concentrations are of concern. A submarine constitutes a closed system in which human-source bacteria could accumulate to an undesirable extent. Morris and Fallon[59 60] and Wright et al.[93] discussed this subject and concluded that modern air-cleaning in submarines creates an environment unusually low in bacteria.

Also of concern are bacterial concentrations and patterns of bacterial distribution in aircraft, especially those used to transport infectious patients. Clayton et al.[15] reported that, whereas in the Boeing 707 artificially aerosolized indicator bacteria are confined to the rear portions of the aircraft, in the C130E Hercules such bacteria become rapidly disseminated throughout the vehicle.

Bacteria (both surface and airborne) in the hospital environment warrant attention. Bacterial content in a hospital environment depends primarily on the presence of humans and on the degree and types of their activity.[5 34 46 52 62 74]

Bacterial products may contaminate indoor air in the absence of bacterial cells. Fine dust in a detergent factory was found to contain Bacillus subtilis enzymes.[35] Workers became ill when exposed to sewage-sludge dust; the active factor was presumed to be airborne endotoxin.[55] Finally, laboratory illness has occurred as the result of inhalation of tuberculin aerosols during operation of a high-speed centrifuge.[66]

Several fungi--Blastomyces, Cryptococcus, Coccidioides, and Histoplasma, all known primarily as human pathogens--exist in natural reservoirs, usually associated with bird or animal emanations.[2 6 21 47 54 85] The extent of contamination of interior situations by these fungi is unknown. However, all are known to enter the body by the respiratory route,[2 21 25 26 32 36 65] and Coccidioides and Histoplasma are known to be highly infective.[65] Thus, natural reservoirs near human habitation will surely result in some interior contamination leading to a possible risk of infection.[2 3 6 30 33 42 85] For example, 5×10^7 viable Cryptococcus spores have been found per gram of dry pigeon fecal material,[2] and the spores were present in more than half the pigeon droppings examined[85]--droppings that are often abundant in areas of dense human population.

Candida albicans and dermatophytes have been recovered from air and dust samples in clinic rooms especially during and after examination of infected patients.[16 39 44 75] However, Friedrich[38] reported only 3% of air samples and 14% of dust samples positive in examining rooms during periods with no patients.

IMPORTANCE OF AIRBORNE CONTAGION

In a 9.5-yr study of 85 families in Cleveland, Ohio, Dingle found that 63% of all illnesses were respiratory.[24] According to the National Health Survey, respiratory conditions (predominantly upper respiratory disease and "influenza") account for more than half of all acute conditions, including illnesses and injuries.[88] The incidence of respiratory conditions is just under one per person per year, and, on an average, each person's activity is restricted for 4.5 d. If one grants that the respiratory conditions referred to are mostly in the category of indoor airborne contagion, the problem is seen to be enormous. Loss of time from work or from school exceeds that from any other cause.

PREVENTION OF INDOOR AIRBORNE CONTAGION

Less-crowded living conditions, isolation, and vaccination have helped to reduce airborne contagion. We consider here a measure that, although neglected in the past, is assuming increasing importance: air disinfection in buildings.

Control of epidemic spread of airborne contagion requires that each infectious case beget, on the average, no more than one new case. The concentration of infectious droplet nuclei must be reduced to the point where susceptible people stand but a small chance of inhaling an infectious particle. In relatively airtight buildings where the capacity of the ventilating system, the fraction of fresh-air makeup, and the efficiency of the filters are known, where the number of infections in each generation of an epidemic is available from records, and where the pulmonary ventilation and duration of exposure of the occupants can be estimated, the essential characteristics of airborne contagion can be dealt with quantitatively. In the 1974 measles epidemic in a school near Rochester, New York, this was done.[68] During the first generation, the number of infectious particles (quanta of infection) produced per minute in the index case was 93--an amount that produced a concentration in recirculated air of 1 per 5.17 m^3. Twenty-six susceptible children breathing this sparsely infected air acquired measles and appeared as cases in the second generation. Such calculations provide architects and engineers with an appreciation of the particulate nature and the quantitive aspects of a characteristic airborne infection.

Thus, the routes of transmission are airborne through infiltration and ventilation, from person to person, and via fomites. The effects of ventilation rates are unknown.[31 53] There are interactions between microorganisms and pollutants, as between indoor combustion and smoking, in producing respiratory illness, especially in children and the infirm.[13 76 78]

REFERENCES

1. Airoldi, T., and W. Litsky. Factors contributing to the microbial contamination of cold-water humidifiers. Am. J. Med. Technol. 38:491-495, 1972.
2. Ajello, L. Comparative ecology of respiratory mycotic disease agents. Bacteriol. Rev. 31:6-24, 1967.
3. Ajello, L., K. Maddy, G. Crecelius, P. G. Hugenholtz, and L. B. Hall. Recovery of Coccidioides immitis from the air. Sabouraudia 4:92-95, June, 1965.

4. Bamert, P., and F. Roth. Bacterial transmission caused by air humidifiers. Schweiz. Med. Wochenschr. 104(50):1856-1859, 1974. (in German; English summary)
5. Blevins, A., D. Armstrong, T. E. Kiehn, and L. Borch. The coordinating role of the microbiologist in hospital epidemiology. (Abstract) Abstracts of the Annual Meeting Am. Soc. Microbiol. 79:316, 1979.
6. Botard, R. W., and D. C. Kelley. A survey to determine the occurrence of Histoplasma capsulatum and Cryptococcus neoformans in air-conditioners. Mycopathol. Mycol. Appl. 37(4):372-376, 1969.
7. Bovallius, A., B. Bucht, R. Roffey, and P. Ånäs. Three-year investigation of the natural airborne bacterial flora at four localities in Sweden. Appl. Environ. Microbiol. 35(5):847-852, May, 1978.
8. Buhles, W.C., Jr. Airborne staphylococcic contamination in experimental procedures on laboratory animals. Lab. Anim. Care 19:465-469, 1969.
9. Burge, H. A., W. R. Solomon, and J. R. Boise. Microbial prevalence in domestic humidifiers. Appl. Environ. Microbiol. 39(4):840-844, 1980.
10. Cartwright, R. Y., and P. R. Hargrave. Pseudomonas in ventilators. Lancet 1:40, 1970.
11. Case, S. K., S. P. Almeida, W. J. Dallas, J. M. Fournier, K. Pritz, J. Cairns Jr., K. L. Dickson, and P. A. Pryfogle. Coherent microscopy and matched spatial filtering for real-time recognition of diatom species. Environ. Sci. Technol. 12:940-946, 1978.
12. Chatigny, M. A., and R. L. Dimmick. Transport of aerosols in the intramural environment, pp. 95-110. In R. L. Edmonds, Ed. Aerobiology. The Ecological Systems Approach. Stroudsburg, Pa.: Dowden, Hutchinson and Ross, Inc., 1979.
13. Cinkotai, F. F., M. G. Lockwood, and R. Rylander. Airborne micro-organisms and prevalence of byssinotic symptoms in cotton mills. Am. Ind. Hyg. Assoc. J. 38:554-559, 1977.
14. Clark, R. P., and R. N. Cox. The generation of aerosols from the human body, pp. 413-426. In J. F. P. Hers and K. C. Winkler, Eds. Airborne Transmission and Airborne Infection. New York: John Wiley & Sons, Inc., 1973.
15. Clayton, A. J., D. C. O'Connell, R. A. Gaunt, and R. E. Clarke. Study of the microbiological environment within long- and medium-range Canadian forces aircraft. Aviat. Space Environ. Med. 47:471-482, 1976.
16. Clayton, Y. M., and G. Midgley. Estimation of dermatophytes (ringworm fungi) and Candida spores in the environment. J. Med. Microbiol. 4(2):Piii-Piv, 1971. (abstract)
17. Cleton, F. J., Y. S. van der Mark, and M. J. van Toorn. Effect of shower-bathing on dispersal of recently acquired transient skin flora. Lancet 1:865, 1968.
18. Cordes, L. G., D. W. Fraser, P. Skaliy, C. A. Perlino, W. R. Elsea, G. F. Mallison, and P. S. Hayes. Legionnaires' disease outbreak at an Atlanta, Georgia, country club: Evidence for spread from an evaporative condenser. Am. J. Epidemiol. 111:425-431, 1980.

19. Covelli, H. D., J. Kleeman, J. E. Martin, W. L. Landau, and R. L. Hughes. Bacterial emission from both vapor and aerosol humidifiers. Am. Rev. Respir. Dis. 108:698-701, 1973.
20. Crowley, T. P. Contaminated humidifiers. J. Am. Med. Assoc. 240:348, 1978.
21. D'Alessio, D. J., R. H. Heeren, S. L. Hendricks, P. Ogilvie, and M. L. Furcolow. A starling roost as the source of urban epidemic histoplasmosis in an area of low incidence. Am. Rev. Respir. Dis. 92:725-731, 1965.
22. Darlow, H. M., and W. R. Bale. Infective hazards of water-closets. Lancet 1:1196-1200, 1959.
23. Dickgiesser, N. Examinations about the behavior of grampositive and gramnegative bacteria in dry and moist atmosphere. Zentralbl. Bakteriol. Parasitenkd. Infektionskr. Hyg. Abt. 1 Orig. Reihe B 167:48-62, 1978. (in German; English abstract)
24. Dingle, J. H. An epidemiological study of illness in families. Harvey Lectures 53:1-24, 1957.
25. Doto, I. L., F. E. Tosh, S. F. Farnsworth, and M. L. Furcolow. Coccidioidin, histoplasmin, and tuberculin sensitivity among school children in Maricopa County, Arizona. Am. J. Epidemiol. 95:464-474, 1972.
26. Drutz, D. J. Urban coccidioidomycosis and histoplasmosis. N. Engl. J. Med. 301:381-382, 1979.
27. Dutkiewicz, J. Exposure to dust-borne bacteria in agriculture. I. Environmental studies. Arch. Environ. Health 33:250-259, 1978.
28. Dutkiewicz, J. Exposure to dust-borne bacteria in agriculture. II. Immunological survey. Arch. Environ. Health 33:260-270, 1978.
29. Dutkiewicz, J., and A. Molocznik. Correlation between dust concentration and microorganism count in the air of grain mills and grain silos. Arch. Hyg. Bakteriol. 154:371-377, 1970. (in German; English summary)
30. Eckmann, B. H., G. L. Schaefer, and M. Huppert. Bedside interhuman transmission of coccidioidomycosis via growth on fomites. An epidemic involving six persons. Am. Rev. Respir. Dis. 89:175-185, 1964.
31. Eickhoff, T. C. Epidemiology of Legionnaires' disease. Ann. Intern. Med. 90:499-502, 1979.
32. Emmons, C. W. The natural occurrence of pathogenic fungi, pp. 22-30. In E. W. Chick, A. Balows, and M. L. Furcolow, Eds. Opportunistic Fungal Infections. Proceedings of the Second International Conference. Springfield, Ill.: Charles C Thomas, Publisher, 1975.
33. Fass, R. J., and S. Saslaw. Earth Day histoplasmosis. A new type of urban pollution. Arch. Intern. Med. 128:588-590, 1971.
34. Fitzgerald, R. H., Jr. Microbiologic environment of the conventional operating room. Arch. Surg. 114:772-775, 1979.
35. Flindt, M. L. H. Pulmonary disease due to inhalation of derivatives of Bacillus subtilis containing proteolytic enzyme. Lancet 1:1177-1181, 1969.
36. Flynn, N. M., P. D. Hoeprich, M. M. Kawachi, K. K. Lee, R. M. Lawrence, E. Goldstein, G. W. Jordan, R. S. Kundargi, and G. A.

Wong. An unusual outbreak of windborne coccidioidomycosis. N. Engl. J. Med. 301:358-361, 1979.

37. Fraser, D. W., D. C. Deubner, D. L. Hill, and D. K. Gilliam. Nonpneumonic, short-incubation-period legionellosis (Pontiac fever) in men who cleaned a steam turbine condenser. Science 205:690-691, 1979.

38. Friedrich, E., and R. Blaschke-Hellmessen. Candida in the rooms of a dermatological clinical center. Mykosen 18:97-105, 1975. (in German; English summary)

39. Gip, L. Investigation of the Occurrence of Dermatophytes on the Floor and in the Air of Indoor Environments. Acta Derm. Venereol. (Stockholm) 46(Suppl. 58):1-54, 1966.

40. Grieble, H. G., F. R. Colton, T. J. Bird, A. Toigo, and L. G. Griffith. Fine-particle humidifiers. Source of Pseudomonas aeruginosa infections in a respiratory-disease unit. N. Engl. J. Med. 282:531-535, 1970.

41. Grunnet, K., and J. C. Hansen. Risk of infection from heavily contaminated air. Scand. J. Work Environ. Health 4(4):336-338, 1978.

42. Hasenclever, H. F. Impact of airborne pathogens in outdoor systems: histoplasmosis, pp. 199-208. In R. L. Edmonds, Ed. Aerobiology: The Ecological Systems Approach. Stroudsburg, Pa.: Dowden, Hutchinson and Ross, Inc., 1979.

43. Henderson, D. A. The eradication of smallpox. Sci. Am. 235(4):25-33, 1976.

44. Hodges, G. R., J. N. Fink, and D. P. Schlueter. Hypersensitivity pneumonitis caused by a contaminated cool-mist vaporizer. Ann. Intern. Med. 80:501-504, 1974.

45. Hojovec, J., and A. Fišer. The microflora of the atmosphere in chicken houses for broilers. Dtsch. Tierärztl. Wochenschr. 75:483-486, 1968. (in German; English summary)

46. Jopke, W. H., and D. R. Hass. Contamination of dishwashing facilities. Hospitals 44(6):124-127, March 16, 1970.

47. Khan, Z. U., M. Pal, H. S. Randhawa, and R. S. Sandhu. Carriage of Cryptococcus neoformans in the crops of pigeons. J. Med. Microbiol. 11:215-218, 1978.

48. Klein, H.-J., and M. Kunze. Experimental investigations on the spread of Pseudomonas aeruginosa by a cold aerosol apparatus for moistening of the room atmosphere. Zentralbl. Bakteriol. Parasitenkd. Infektionskr. Hyg. Abt. 1 Orig. 216(2):199-209, 1971. (in German; Engish abstract)

49. Knight, V. Airborne transmission and pulmonary deposition of respiratory viruses, pp. 1-9. In V. Knight, Ed. Viral and Mycoplasmal Infections of the Respiratory Tract. Philadelphia: Lea and Febiger, 1973.

50. Kösters, J., and W. Müller. Exposure of personnel to bacteria in mass poultry husbandry. Zentralbl. Veterinärmed. Reihe B 17:154-158, 1970. (in German; English summary)

51. Lacey, J. Microorganisms in air of cotton mills. Lancet 2:455-456, 1977.

52. Leedom, J. M., and C. G. Loosli. Airborne pathogens in the indoor environment with special reference to nosocomial (hospital)

infections, pp. 208-237. In R. L. Edmonds, Ed. Aerobiology. The Ecological Systems Approach. Stroudsburg, Pa.: Dowden, Hutchinson and Ross, Inc., 1979.

53. Lewis, H. E., A. R. Foster, B. J. Mullan, R. N. Cox, and R. P. Clark. Aerodynamics of the human microenvironment. Lancet 1:1273-1277, 1969.

54. Mantovani, A. The role of animals in the epidemiology of the mycoses. Mycopathologia 65:61-66, 1978.

55. Mattsby, I., and R. Rylander. Clinical and immunological findings in workers exposed to sewage dust. J. Occup. Med. 20:690-692, 1978.

56. May, K. R., and N. P. Pomeroy. Bacterial dispersion from the body surface, pp. 426-432. In J. F. Hers and K. C. Winkler, Eds. Airborne Transmission and Airborne Infection. New York: John Wiley and Sons, Inc., 1973.

57. McLean, R. L. The effect of ultraviolet radiation upon the transmission of epidemic influenza in long-term hospital patients. Am. Rev. Respir. Dis. (Suppl. 83):36, 1961.

58. McNall, P. E., Jr. Practical methods of reducing airborne contaminants in interior spaces. Arch. Environ. Health 30:552-556, 1975.

59. Morris, J. E. W. Microbiology of the submarine environment. Proc. R. Soc. Med. 65:799-800, 1972.

60. Morris, J. E. W., and R. J. Fallon. Studies on the microbial flora in the air of submarines and the nasopharyngeal flora of the crew. J. Hyg. 71:761-770, 1973.

61. Mundt, J. O., E. J. Anandam, and I. E. McCarty. Streptococceae in the atmosphere of plants processing vegetables for freezing. Health Lab. Sci. 3:207-213, 1966.

62. Nelson, C. L. Environmental bacteriology in the unidirectional (horizontal) operating room. Arch. Surg. 114:778-782, 1979.

63. Pennington, J. H., J. Lumley, and F. O'Grady. The growth of Pseudomonas pyocyanea in Garthur condenser humidifiers. An experimental study. Anaesthesia 21:211-215, 1966.

64. Peterson, J. E. Estimating air filtration into houses: An analytical approach. ASHRAE J. 21(1):60-63, 1979.

65. Pike, R. M. Laboratory-associated infections: Incidence, fatalities, causes, and prevention. Ann. Rev. Microbiol. 33:41-66, 1979.

66. Radonic, M. Systemic allergic reactions due to occupational inhalation of tuberculin aerosols. Ind. Med. Surg. 35:24-26, 1966.

67. Reznikov, M., J. H. Leggo, and D. J. Dawson. Investigation by seroagglutination of strains of the Mycobacterium intracellulare-M. scrofulaceum group from house dusts and sputum in southeastern Queensland. Am. Rev. Respir. Dis. 104:951-953, 1971.

68. Riley, E. C., G. Murphy, and R. L. Riley. Airborne spread of measles in a suburban elementary school. Am. J. Epidemiol. 107:421-432, 1978.

69. Riley, R. L., C. C. Mills, F. O'Grady, L. U. Sultan, F. Wittstadt, and D. N. Shivpuri. Infectiousness of air from a tuberculosis ward: Ultraviolet irradiation of infected air: Comparative infectiousness of different patients. Am. Rev. Respir. Dis. 85:511-525, 1962.

70. Riley, R. L., W. F. Wells, C. C. Mills, W. Nyka, and R. L. McLean. Air hygiene in tuberculosis: Quantitative studies of infectivity and control in a pilot ward. Am. Rev. Tuberc. Pulm. Dis. 75:420-431, 1957.
71. Rosenzweig, A. L. Contaminated humidifiers. N. Engl. J. Med. 283:1056, 1970.
72. Rutter, D. A., and C. G. T. Evans. Aerosol hazards from some clinical laboratory apparatus. Br. Med. J. 1:594-597, 1972.
73. Rylander, R., K.-E. Myrbäck, B. Verner-Carlson, and M. Öhrström. Bacteriological investigation of wall-to-wall carpeting. Am. J. Public Health 62(2):163-168, 1974.
74. Sayer, W. J., N. M. MacKnight, and H. W. Wilson. Hospital airborne bacteria as estimated by the Andersen sampler versus the gravity settling culture plate. Am. J. Clin. Pathol. 58:558-566, 1972.
75. Schönborn, C., and F. Winden. Occurrence of fungi in the air and dust of clinic rooms. Mykosen 16:385-391, 1973. (in German; English summary)
76. Schulman, J. L. The use of an animal model to study transmission of influenza virus infection. Am. J. Public Health 58:2092-2096, 1968.
77. Scott, C. C., and I. Jacobson. Pseudomonas in ventilators. Lancet 1:239, 1970.
78. Seisaburo, S., K. Kiyoko, and N. Tatsuko. Free dust particles and airborne microflora. Bull. Dept. Home Econ., Osaka City University (Osaka) 4:31-37, 1959.
79. Sidorenko, G. I. Data on the distribution of Clostridium perfringens in the environment of man. Communication 1. J. Hyg. Epidemiol. Microbiol. Immunol. (Prague) 11:171-177, 1967.
80. Singh, R. P. Incidence of Salmonella spp. in poultry farms and hatcheries and their pathogenicity. Indian Vet. J. 44:833-837, 1967.
81. Smith, P. W. Room humidifiers as the source of Acinetobacter infections. J. Am. Med. Assoc. 237:795-797, 1977.
82. Solomon, W. R. Assessing fungus prevalence in domestic interiors. J. Allergy Clin. Immunol. 56:235-242, 1975.
83. Speers, R., Jr., F. W. O'Grady, R. A. Shooter, H. R. Bernard, and W. R. Cole. Increased dispersal of skin bacteria into the air after shower baths: The effect of hexachlorophene. Lancet 1:1298-1299, 1966.
84. Spendlove, J. C. Penetration of structures by microbial aerosols. Dev. Ind. Microbiol. 16:427-436, 1975.
85. Swinne-Desgain, D. Cryptococcus neoformans of saprophytic origin. Sabouraudia 13:303-308, 1975.
86. Thacker, S. B., J. V. Bennett, T. F. Tsai, D. W. Fraser, J. E. McDade, C. C. Shepard, K. H. Williams, Jr., W. H. Stuart, H. B. Dull, and T. C. Eickhoff. An outbreak in 1965 of severe respiratory illness caused by the Legionnaires' disease bacterium. J. Infect. Dis. 138:512-519, 1978.
87. Tovey, E. R., and R. A. Vandenberg. Effect of reagins and allergen extracts on radioallergosorbent assays for mite allergen. Clin. Allergy 8:329-339, 1978.

88. U.S. Department of Health, Education, and Welfare, National Center for Health Statistics. Acute Conditions. Incidence and Associated Disability, United States, July 1973-June 1974. Data from the National Health Survey, Vital and Health Statistics Series 10, No. 102. DHEW Publication No. (HRA)76-1529. Rockville, Md.: U.S. Department of Health, Education, and Welfare, National Center for Health Statistics, 1975.
89. Watson, H.H. Errors due to anisokinetic sampling of aerosols. Am. Ind. Hyg. Assoc. Q. 15:21-25, 1954.
90. Wells, W. F. Apparatus for study of the bacterial behavior of air. Am. J. Public Health 23:58-59, 1933.
91. Wells, W. F. On air-borne infection. Study II. Droplets and droplet nuclei. Am. J. Hyg. 20:611-618, 1934.
92. Wells, W. F., M. W. Wells, and T. S. Wilder. The environmental control of epidemic contagion. I. An epidemiologic study of radiant disinfection of air in day schools. Am. J. Hyg. 35:97-121, 1942.
93. Wright, D. N., E. M. K. Vaichulis, and M. A. Chatigny. Biohazard determination of crowded living-working spaces: Airborne bacteria aboard two naval vessels. Am. Ind. Hyg. Assoc. J. 29:574-581, 1968.
94. Young, L. S., J. C. Feeley, and P. S. Brachman. Vaporized formaldehyde treatment of a textile mill contaminated with *Bacillus anthracis*. Arch. Environ. Health 20:400-403, 1970.

ALLERGIC REACTIONS IN THE INDOOR ENVIRONMENT

Only a few airborne allergens are found in enclosed spaces. Although human exposure to them is recurrent and of variable duration, the health effects of exposure to them alone are difficult to estimate. Despite this uncertainty, the impact of some agents is clearly appreciable. House dust and pollen, for example, are acknowledged as two of the most important factors in provoking symptoms of allergic rhinitis and asthma in many locales. Clinically evident allergy to animal danders is due to both the popularity of house pets and the strong sensitizing capacities they exhibit. However, the contribution of other indoor exposures to the overall toll exacted by allergic diseases remains entirely speculative.

A broad array of pollens, fungi, algae, actinomycetes, arthropod fragments, dusts, and pumices have been confirmed as airborne antigen sources that evoke human responses; evidence similarly implicating airborne bacteria, protozoa, and other groups is still emerging. Analyses of health impact are further complicated by the varied tissue processes that may be evoked, separately or in combination, by antigen challenge.

Particles recovered from indoor air often are assumed to have arisen within the enclosure studied. However, a large proportion of indoor particles reflect natural sources, especially when local growing conditions are favorable.[3,138] Inward flux is especially evident for pollen, but also affects interior loads of fungi, insects, and algae.[140]

Allergic reactions can occur on the skin and in the nose, airways, and alveoli. Although increasingly recognized as important causes of allergic lung diseases, occupational agents are considerably less prevalent as causes of these diseases than such allergens as pollen, moulds, mites, and animal dander and excreta, to which exposure occurs in the home environment.

ALLERGIC REACTIONS ON THE SKIN

Both primary irritants and allergic sensitizers may produce inflammation of the skin or an eczematous process. Primary irritation causes contact dermatitis. The effect is through direct action on the skin. Irritants act by removing lipid films, producing denaturation of keratin, or interfering with the barrier layer. Through the production of dehydration, the effects may occur by protein precipitation or oxidation. Sensitizers produce cutaneous changes after previous contacts, either immediately on recontact or shortly thereafter. Almost any chemical may be a sensitizer. A sensitizer stimulates the immune mechanisms by producing an antigen, usually by combining with a protein. Immediate hyperreactivity is produced by a binding with IgE on basal cells. Delayed hyperreactivity may be produced by IgG mediation.[2,3]

Photoallergic reactions may be produced by ultraviolet (UV) light, which leads to an inflammatory response. The antigenic agent may be a UV-mediated degradation product or a visible-light photosensitizer that produces an immediate hypersensitivity response.

Secondary effects may occur after the cutaneous defenses have been removed. Some fatty acids on the surface lipid film may act as antimicrobials.[1-3] Bacteria may grow on an oozing or fissured surface. Other toxins may also enter the system at that point. Lesions may be highly variable, with a range from slight inflammation to tumor. Acute contact eczematous dermatitis can be due to a primary irritant or a sensitizer and is characterized by inflammation changes, crusts, and sloughing.

ALLERGIC REACTIONS IN THE RESPIRATORY TRACT

Allergic reactions in the respiratory tract can be distinguished by the site and the nature of the reaction and by the underlying immunologic mechanisms.

Site and Nature of Reactions

Allergic responses to inhaled materials cause a local inflammatory reaction that affects predominantly the nose (allergic rhinitis), the airways (allergic asthma), the airways and adjacent alveolar spaces (allergic asthma with pulmonary eosinophilia--allergic bronchopulmonary aspergillosis), or alveoli and peripheral bronchioles (hypersensitivity pneumonitis or extrinsic allergic bronchioloalveolitis). Similar reactions can occur in each of these separate sites, often producing similar pathologic changes, in the absence of any recognized extrinsic cause.

Allergic rhinitis is characterized by vasodilatation and edema of the nasal mucosa with mucus hypersecretion, which causes nasal discharge and obstruction.

Asthma is most usefully defined in functional terms as partial narrowing of the airways that is reversible over short periods, either spontaneously or as a result of treatment. The defining criterion of asthma is therefore variable airway narrowing. This airway narrowing may be due to contractions of airway smooth muscle, edema of the bronchial mucosa, accumulation of bronchial mucosal secretions, or any combination of those. If the cause of the airway narrowing can be identified as an allergic reaction to an extrinsic agent, the term "allergic asthma" can be used.

Pulmonary eosinophilia (pulmonary infiltration with eosinophilia, or PIE) is defined as transient shadowing on a chest roentgenogram accompanied by an increased blood eosinophil count. Histologically, the alveolar spaces in the affected parts of the lung are consolidated with eosinophils. Pulmonary eosinophilia may be caused by a reaction to drugs or to helminths migrating through the lung. *Aspergillus fumigatus* is the only important inhaled allergen identified as a cause of the syndrome; when it is the cause, the syndrome is called "allergic bronchopulmonary aspergillosis," or ABPA. ABPA episodes are often accompanied by asthmatic attacks.

Hypersensitivity pneumonitis is characterized, at least during its early stages, by the infiltration of alveolar walls and peripheral bronchioles with mononuclear cells. The condition is often associated with the formation of epithelioid and giant-cell granulomata. The disease may be accompanied by progressive fibrosis, which makes it difficult to distinguish the changes in the lung from other causes of alveolar wall fibrosis.

Mechanisms of Reactions

Allergic reactions are conventionally distinguished from reactions of protective immunity by the extent of tissue damage. The immunologic mechanisms underlying allergic reactions differ little from those involved in immune reactions. They are distinguished by outcome: whereas little or no tissue damage occurs in immune reactions, allergic reactions are characterized by the disproportionate damage caused in most tissues. The different types of immunologic reaction that may cause tissue damage have been classified by Gell *et al.*[51] Of their four types of allergic reactions, three are of particular importance in relation to allergic lung disease: immediate (Type I) reactions, Arthus or local immune complex (Type III) reactions, and cell-mediated delayed-hypersensitivity (Type IV) reactions. It is important to appreciate that, although these types are considered separately, more than one type is involved in most, if not all, cases of allergic lung disease.

Immediate (Type I) Reactions. In this type, antigen reacts with antibody, primarily IgE antibody on the surface of circulating basophils and mast cells, which are present in the submucosa in the nose and airways (as well as in the skin and gastrointestinal tract). IgE antibody is attached to surface receptors on the cell through its Fc portion. Bridging of two adjacent IgE molecules by reaction of antigen with the Fab portion of the molecules stimulates intracellular metabolic changes that lead to the release of pharmacologically active mediators from the cytoplasmic granules of the cells. The mediators released include histamine and slow-reacting substance of anaphylaxis (SRS-A), both of which increase the permeability of small blood vessels to intravascular protein molecules, as well as stimulating smooth-muscle contraction. In addition, eosinophil chemotactic factor of anaphylaxis (ECF-A) is released. The mediators released from these cells during the reaction have pharmacologic properties that could mediate many of the changes observed in the nose and airways in allergic rhinitis and asthma.

Arthus Local Immune Complex (Type III) Reactions. Immune complexes formed in tissue spaces of antigen and IgG antibody in relative antigen excess can cause local tissue damage. Such complexes fix complement, and several of the complement cleavage products released have pharmacologic activity. C_{3a} and C_{5a} are anaphylotoxins that stimulate histamine release from mast cells. C_{567} is chemotactic both for neutrophils and eosinophils. Neutrophils that have surface receptors for C_{3b} and the Fc portion of IgG phagocytose immune complexes formed in antigen excess when ingested and provoke the release of proteolytic lysosomal enzymes from neutrophils. This is referred to as "regurgitation during feeding" and causes local tissue damage. This mechanism is probably important in the tissue damage that occurs in the airways in ABPA and in the inflammatory reaction in alveolar walls in hypersensitivity pneumonitis.

Cell-Mediated Delayed-Hypersensitivity (Type IV) Reactions. In these reactions, antigen reacts not with antibody, but with specifically sensitized T-lymphocytes. The reaction stimulates the release from the lymphocyte of a number of biologically active soluble substances known as "lymphokines," which are capable of mediating a local inflammatory reaction. Among these biologic activities, lymphokines have been shown to be chemotactic for macrophages, to induce activation of macrophages, and to inhibit their migration. The interaction of antigen with specifically sensitized T-lymphocytes can therefore stimulate local recruitment and activation of macrophages, in addition to maintaining them at the site of reaction. This type of reaction is well recognized in the delayed hypersensitivity to tuberculin and has also been shown to participate in the inflammatory reaction in alveolar walls in hypersensitivity pneumonitis. (Type IV reactions occur with some bacteria, such as *M. tuberculosis*, and with fungi, as discussed in the previous section.)

These immunologic reactions participate in the different allergic reactions to inhaled materials in the respiratory tree, as shown in Table VII-10. Other immunologic mechanisms, not yet clearly identified, may also be involved in these diseases.

TABLE VII-10

Immunologic Mechanisms in Allergic Lung Diseases

Reaction Site	Disease	Immunologic Mechanism
Nose	Allergic rhinitis	IgE
Airways	Allergic rhinitis	IgE
Airways and alveolar spaces	Asthma with pulmonary eosinophilia, allergic bronchopulmonary aspergillosis	IgE, IgG, immune complexes
Alveolar walls and peripheral bronchioles	Hypersensitivity, pneumonitis (extrinsic allergic bronchiolo-alveolitis	IgG, immune complexes, sensitized T-lymphocytes

FACTORS THAT DETERMINE ALLERGIC REACTIONS IN THE RESPIRATORY TRACT

Allergic reactions to inhaled materials in the respiratory tract are determined by a number of identifiable factors, which include the physical and chemical properties of the inhaled particles, the immunologic reactivity of the host, and the pattern of exposure.

Nature of the Inhaled Material

Most inhaled materials that stimulate an allergic response in the respiratory tract are particulate and approximately spherical. To cause this kind of response, particles must remain suspended in atmospheric air long enough to be inhaled; for an immunologic reaction to occur, they must penetrate to a reaction site in the lungs. Inhaled particles are deposited on the surfaces of the nose, airways, and alveoli. Three separate mechanisms contribute to this deposition: gravitational sedimentation, inertial impaction, and diffusion. Any particles suspended in air will fall under the force of gravity and reach a terminal velocity that is determined by diameter and density. In addition, a particle suspended in an airstream that is changing in direction, as occurs in the nasopharyngeal or branching bronchial airways, will continue for some distance in the original direction of airflow, owing to inertia. The distance traveled by a particle in the original direction of airflow is determined by its density and diameter, the velocity of airflow, and the angle of change of direction. Impaction of particles due to this mechanism is of most importance with large particles (aerodynamic diameter, greater than 3 µm) in the nose and proximal airways, the portions of the respiratory tract in which the velocity of airflow and angles of change of direction are greatest. Gravitational sedimentation is of greater importance in determining deposition of smaller particles (aerodynamic diameter, 0.5-3 µm) in the smaller, more peripheral airways and in the alveoli. Very small particles (aerodynamic diameter, less than 1 µm) may be deposited by diffusion, owing to their Brownian movement resulting from the impact of surrounding gaseous molecules.

Most particles whose aerodynamic diameter is 20 µm or more and 50% of particles down to 5 µm are deposited in the nose. Almost complete deposition of particles of 5 µm or more occurs in the tracheobronchial tree. Alveolar deposition (the so-called respirable fraction) is maximal for particles whose aerodynamic diameter is between 2 and 4 µm. Once deposited, a particle is not resuspended in respired air. Particles not deposited are expelled in the exhaled air.

Particles deposited in the airways are trapped in the mucus blanket on the mucosal surface and are moved centrally by the coordinated beating of the cilia, which extend distally as far as the terminal bronchioles (the "mucociliary escalator"). Some particles deposited in the airways are usually cleared within 24 h. Particles deposited in the alveoli are phagocytosed and cleared by alveolar macrophages; they migrate from the alveoli with their engulfed particles either proximally into the airways on to the mucociliary escalator or out into the draining lymphatics. Clearance of particles deposited in the alveoli takes place in several phases, with characteristic times measured in hours, days, weeks, and years.

Of the potentially allergenic particles, the moulds and organic dusts (including animal excreta) have aerodynamic diameters that allow their penetration into the airways and alveoli. The majority of pollens, however, have diameters greater than 12 μm, with aerodynamic diameters somewhat different because of their density. However, pollen grains have been recovered from the airways by endoscopy and from resected lungs.[100]

The chemical properties of a molecule that determine its antigenicity are poorly understood. In general, "complete" antigens that are able to stimulate antibody production are proteins or polysaccharides of high molecular weight. Both organic and inorganic molecules of low molecular weight (less than 1,000) can act as haptenes, stimulating antibody production when coupled to high-molecular-weight carrier molecules.

Immunologic Reactivity of the Host--Atopy

Coca and Cooke[28] introduced the term "atopy" to describe persons who were readily sensitized to proteins in their environment. These reactions have since been shown to be mediated by IgE antibody.[67] Pepys[113] has redefined "atopy" as "that form of immunological reactivity of the subject in which reaginic antibody, now identifiable as IgE antibody, is readily produced in response to the common allergens of the subject's environment." The presence of IgE antibody specific to a particular allergen can be demonstrated by its ability to elicit an immediate "weal and flare" skin prick test reaction. The atopic status of a person--but not clinical reactivity--can therefore be defined by reactions to skin tests with allergens appropriate to the particular environment.

Several studies[30,129] have shown atopy to be familial. Its genetic basis, however, remains unclear. Atopy (defined as one or more positive skin test reactions to common inhalant allergens) in asthmatics is strongly related to the frequency in first-degree relatives of eczema and hay fever and less to the frequency of asthma.

Circumstances of Exposure to Inhaled Allergens

Evidence from the investigations of allergic reactions to inhaled particles in industry has suggested a relationship between exposure and disease. Juniper *et al.*[70] showed, in a study of IgE-mediated allergic reactions to *B. subtilis* enzymes, particularly alcalase, in the enzyme-detergent industry that the incidence of skin test reactions to alcalase was increased in groups with heavier exposures, but that the incidence in atopics exceeded that in nonatopics at each exposure. The effects of differences in intensity and duration of exposure in determining sensitization have not been studied.

Exposure to allergens during the early months of life may be important in determining the tendency to produce IgE antibody to inhalant allergens. Taylor et al.[152] demonstrated an association between transient IgA deficiency in infancy and later development of atopy. They suggested that inhalant and ingested (particularly milk) allergens traverse mucosal barriers in IgA-deficient persons during this "vulnerable" period and stimulate the production of IgE antibody. If this hypothesis is correct, it presents an opportunity for the prevention of atopy: exclusion of potential allergens from the environment during this period may prevent the stimulus to IgE production.

ALLERGIC LUNG DISEASES AND THEIR CAUSAL ALLERGENS

Allergic Rhinitis and Asthma

IgE-mediated rhinitis and asthma may be caused by a wide variety of allergens common in the environment, which include the house dust mite, pollens, moulds, and animal dander and excreta. The particular allergens that cause these diseases have great geographic variation: exposure to different moulds and pollens and to the house dust mite varies greatly according to climatic conditions, which determine their prevalence in a particular environment.

Pollen. Among biogenic particles, pollen grains can perhaps be most confidently ascribed to sources in nature, and their presence in enclosed spaces generally reflects incursions from outdoor air. Wind-pollinated plants typically have drab, scentless, individually minute flowers, unlike the large showy blooms of many popular house plants. However, with sizable indoor planting, pollens of cyclamen and impatiens have been found to reach concentrations of hundreds of grains per cubic meter; their possible effects have not been studied (H. Burge and W. Solomon, unpublished data).

The few studies of the flux of pollens into homes have emphasized the importance of free ventilation and indicated that even partial closure of windows can substantially exclude these particles.[3] Undisturbed, closed rooms were recognized as pollen "refuges" many years ago. However, pollen grains are known to enter fully closed structures through faults between structural members ("crackage")--an effect positively related to outdoor wind speed[32] and probably heavily dependent on gust-induced effects. The opening of doors as occupants enter and leave further enhances the entrance of pollen and limits the barrier function of any building. After penetrating indoors, pollen may remain airborne or be deposited, with secondary reflotation possibly due to scouring air currents and disturbance of substrates.

Many different pollens from grass, trees, and flowers have been shown to cause allergic rhinitis and asthma. Pollen counts vary greatly during the year and from year to year, and symptoms in sensitized people are closely related to the prevalence of the particular pollen causing disease. Grass and ragweed pollen, which are particularly common causes of allergic rhinitis and asthma, are most abundant in the summer months. In some areas, the pollens of mulberry and other trees are also important.

Although soluble extracts of pollen have been shown to provoke asthmatic reactions in symptomatic persons, pollen grains are generally too large to penetrate into the respiratory tract beyond the trachea. Various suggestions have been made to explain how in these circumstances pollen can penetrate into the airways and react with IgE on the surface of submucosal mast cells to provoke an asthmatic reaction. Pollen grains and fragments of pollen grains (or plant bract) may be small enough to penetrate into the airways and have been so found in an autopsy study.[100] Kimura *et al.*[75] have identified basophils and mast cells in the lumen of the airways. Inhaled pollen grains may therefore stimulate an IgE-dependent mast cell and basophil degranulation that occurs initially in the bronchial lumen.

House Dust Mite. Interest in acarids has been stimulated largely by some pyroglyphid mites that contribute sensitizing materials to domestic dusts. House dust is a poorly characterized and variable substance that is universally recognized to produce allergic rhinitis and asthma.[77,95,159] In many areas, two mite species, *Dermatophagoides pteronyssinus* and *D. farinae*, are abundant indoors and may be recovered from air in domestic interiors.[31,78,107,159] Parallel allergic reactivity to these agents and to crude house dust has been described widely,[103,104,159] although exceptions are well documented.[73] A possible role for mite allergy has been proposed in urticaria (hives)[33] and in other skin disorders.[62]

House dust mites are most abundantly associated with mattresses, bedclothes, and heavily used upholstered furniture.[77,78,159] A temperature of around 25°C appears most favorable, and a relative humidity of at least 45% is essential to prevent death from desiccation. Populations vary with atmospheric moisture, often being highest in autumn, but also in summer or winter, especially in damp houses.[78,107,159,164] Pyroglyphid mites have varied diets, but relish especially human epidermal scales; specific fungi may also be favored,[136] but nutritive factors may not be important for concentration.[77] Indoor temperature does not appear to be important.[77,107] Mite numbers may be reduced by decreasing indoor humidity, avoiding fibrous floor and furniture coverings, and encasing pillows and mattresses in plastic. Safe and effective chemical miticides for the home are still needed.[164]

Extensive surveys of indoor acarids have disclosed additional commonly recovered genera (*Glycyphagus*, *Hirstia*, *Tyrophagus*, and *Euroglyphus*) that inhabit dust. In addition, interiors used to process or store agricultural products often yield large and specialized mite populations.[78,159,166] Although the whole house dust mite is about 300 μm long, the allergens are probably present in its debris, particularly the excreta, whose particles are of an inhalable size.

Moulds. The moulds that may cause allergic rhinitis and asthma are predominantly of the genera *Alternaria*, *Cladosporium*, and *Aspergillus*, as well as *Merulius lachrymans*, the cause of dry rot. Atmospheric spore counts of these moulds are usually highest in the late summer and autumn, although spore counts of *Aspergillus fumigatus* are maximal in the autumn and winter.

Cladosporium was by far the most frequent taxon recovered both outdoors and in domestic and other "clean" interiors during summer in the United States,[19 64 69 79 91 92 121 125 138 167] in Europe,[1 54 116 122] and in Asia,[72] but was always more abundant outdoors than indoors.[3 10 19] *Penicillium* usually dominated wintertime U.S. collections[19 85 121 134 138] and some European sites[47 68 156 168] and is often considered an "indoor" fungus, being frequently more abundant indoors than outdoors.[1 10 47 68 123] *Penicillium* concentrations also increase substantially with housecleaning and repair.[63 122] *Alternaria* was the most frequent indoor fungus in the summertime in two southwestern U.S. studies,[34 85] although the contribution of outdoor air was uncertain. *Aspergillus* was dominant in only two studies, one in China[25] and one in Great Britain,[86] but is considered one of the most common groups of indoor fungi.[1 29] *Aspergillus* species were usually other *glaucus* groups (e.g., *A. amestelodami* and *A. repens*) or *A. versicolor*, with relatively few *A. fumigatus* or *A. flavus* recoveries. *Mucor* was consistently more frequent indoors[10] than outdoors.

Cases of contamination of domestic interiors usually involved outdoor fungi that increased indoors on specific substrates. Floodwater disasters often produce abundant mold growth with attendant increase in airborne spore counts.[19 79] Any organic material may support mold growth when wet. Damp walls may acquire abundant *Cladosporium cladosporioides* and *Aureobasidum*,[87 91] and damp leather, cotton, and paper often are covered with *Penicillium* or *Aspergillus* spores. Fireproofing materials,[4] furniture stuffing (e.g., kapok, feathers, and hair[29 155 160]), carpets,[64] and stored organic material[87 88] all have been implicated as foci of mold contamination. House dust contained 10,000-3,000,000 spores/g in one study,[149] and dust-raising activities clearly increased spore counts in libraries[19] and domestic interiors.[97 121] Repair work increased counts up to 20-fold, presumably because of dust dispersion; and contamination was not restricted to the actual repair site, but spread throughout the home.[98] Fungal taxa in dust may occur in proportions different from those in air;[53] in fact, some fungi may grow in dust.[124 155] Several additional reports have implicated house dust[53 125 161] as a source of airborne fungi. House plants have been implicated as sources of increased *A. fumigatus* concentrations in homes and hospital rooms;[146 147] however, Burge and Solomon (unpublished data) did not find evidence of this ubiquitous soil fungus in domestic air associated with house plants. Pets have also been blamed for increased *A. fumigatus* counts,[64] but Burge *et al.*[21] did not find evidence of direct animal contributions to the *A. fumigatus* in a series of animal-care rooms. Poor landscaping practices, including accumulation of organic debris and high shade,[79] and such appliances as evaporative humidifiers[20 115 151] and air-conditioners[142 165] have been named as potential sources of airborne fungal contamination. The wood-rotting fungus Merulius lacrymans, contaminating damp timbers in old and damaged English homes, may produce potentially sensitizing spores indoors at up to 360,000/m^3 of air.[49 56] Spores released in (or allowed to enter) one area of a home spread rapidly to all areas with open doors.[26]

The situation differs markedly in interiors used specifically for processing or handling of biologic materials.[6 9 17 24 27 29 38 42 52 57 58 65 66 81-84 88 110 117 118 120 123 128 144 154]

Aspergillus fumigatus is a ubiquitous soil fungus and can be present wherever organic material provides a suitable substrate.[89 119] In outdoor urban air, A. fumigatus rarely exceeds 150 spores/m^3 and is not necessarily frequent.[19 140 141] In the presence of compost, counts can rise into the millions per cubic meter,[102 157] with the possibility of attendant indoor pollution. In relatively clean interiors, A. fumigatus counts generally are low: 0-200/m^3.[5 19 61 86 111 141] However, in interiors in which organic material is stored or handled, they can exceed 2 x 10^{10} spores/m^3.[6 57 58 65]

The incidence of invasive aspergillosis in the normal population appears quite low, even in persons exposed to spores at high counts,[58 65] but the risk of hypersensitivity disease in overexposed people is substantial.[4-6 50 61 86 94 133 135 145 146 165]

A wide variety of fungi that are normally saprophytic (as is A. fumigatus) may opportunistically become invasive human pathogens.[39] Some data on indoor contamination by these organisms are available from general surveys of indoor mould concentrations,[2 6 9 19 21 55 57 79 83 139] but none has been studied with respect to possible risk factors that foster human infection.

Algal Particles. Algal cells, including viable units, are regular components of outdoor aerosol and have been regarded as potential human allergens for decades.[60 99] Algae reside in soil and on natural surfaces, as well as in aquatic habitats, from which they may become airborne in the bursting of bubbles[126] or the fragmentation of foams.[127] These processes, as well as dry dispersion from soil, all increase with rising wind speed and with physical (e.g., agricultural) disturbance of substrates.[16]

Although concentrations of algae in indoor air have not been systematically studied, they are often present in water reservoirs and in house dust and are undoubtedly dispersed periodically. Soil particles and aerosols from water reservoirs are the most probable sources of algae recovered indoors. Sensitization to dustborne green algae (Chlorella spp.) has been amply documented; however, the impact of indoor exposure to these or related organisms remains uncertain.[11]

Animal Dander and Excreta. Domestic animals, particularly cats, are important causes of allergic rhinitis and asthma. Other animals that can cause these diseases include dogs, rabbits, guinea pigs, and horses. It has been generally accepted that the source of allergens is the animal dander. Recent studies of laboratory-animal workers with rhinitis and asthma due to IgE-mediated reactions to rats and mice have identified the animals' urine as an important source of allergenic protein. The urine of the other animals has not yet been investigated as a possible source of allergens.

Sensitivities to pets are discrete, showing species specificity and often breed specificity.[106] Washed hair is generally antigen-poor,[41] but epidermal scales and body fluids[105] appear to carry potent activity.

Highly sensitive persons may develop urticaria or flares of allergic eczema from direct physical contact with the implicated species. Whether airborne antigen alone can cause these skin problems is not clear. However, respiratory symptoms often continue for many weeks after an implicated pet has left the home, and that suggests that minute inhaled doses of dander can sustain a clinical response. Because the airborne antigen has not been measured, more explicit dose-response relationships also remain unknown.

Although more commonly associated with hypersensitivity pneumonitis,[162] serum proteins in the feces of birds, particularly of parakeets (budgerigars) and pigeons, can stimulate IgE antibody production, especially in atopic persons,[40] and cause asthma.

Feathers, present largely in bedding and clothing, are familiar sensitizers that seem to acquire increasing antigenic potency with age,[95] although this change could reflect progressive invasion by mites (or microbial contamination); differences between purified antigens of feathers and house dust have been claimed.[14] The offending materials seem to be derived primarily from serum and intestinal secretory components,[36] although feather and egg proteins also may contribute antigens.

The cause of rhinitis and asthma associated with an IgE-mediated reaction can usually be readily identified from the history and from skin prick test reactions. A history of symptoms that are temporarily related to a particular exposure (i.e., seasonal variation in symptoms or symptoms occurring after exposure to domestic animals or bedmakings) strongly suggests the relationship. The presence of IgE antibody specific to the particular allergen will be revealed through skin prick testing with the putative allergens.

Fragments of epidermis and skin appendages (danders) are shed by all vertebrates and undoubtedly are dispersed indoors. With components of sweat, saliva, and waste discharges, these particles may contribute substantial quantities of species-specific materials to the indoor environment. Although nontoxic, the dispersed materials are strong sensitizers and commonly elicit allergic rhinitis and asthma. In addition, emanations of birds and of at least one small mammal, the gerbil,[137] have been implicated as agents of allergic alveolitis.

Insects. Insect emanations are strong sensitizers, can elicit respiratory allergic reactions, and are often evident in outdoor air.[95] Symptoms have occurred as a result of local swarming of caddis flies,[112] mushroom flies,[74] may flies,[43] and box elder beetles;[108] and antibody responses to other types are easily demonstrable. In addition, species that establish themselves indoors--including roaches, houseflies, bedbugs, and carpet beetles--have been implicated as human sensitizers.[114] Roaches have received special attention because of their widespread abundance and the capacity of aqueous extracts of common species to elicit respiratory responses in challenged sensitive subjects.[71] Roach fecal pellets appear to share antigens with the bodies of the insects that produce them and, on dissolution, may contribute importantly to air contamination.[13] Sensitization to roach antigens is particularly frequent in persons of low income; this apparently reflects an association with poor sanitary conditions.[12] In addition to pests, a variety of insects exploited commercially for research purposes can produce indoor contamination and respiratory allergy in exposed persons.[48 96 148]

Human dander. Epidermal scales of human origin are often the most abundant and microscopically distinctive component of indoor dusts.[15 55] These structures readily serve as "rafts" for aerial transport of bacteria and other microorganisms. However, the role of human dander as an antigen for allergic subjects is highly controversial. Antibody responses to this material have been reported,[15 158] but confirmatory data and evidence of clinical impact are still awaited.

Asthma and Pulmonary Eosinophilia Due to Allergic Bronchopulmonary Aspergillosis

ABPA is caused by an allergic reaction in the airways to inhaled *Aspergillus fumigatus*. The disease is characterized by recurrent episodes of pulmonary eosinophilia, usually associated with attacks of asthma. In addition to asthma and pulmonary eosinophilia, as the disease progresses, bronchiectasis (typically proximal, but sometimes widespread), airway narrowing that becomes increasingly less reversible, and pulmonary fibrosis that particularly involves the upper lobes may also develop.

Aspergillus fumigatus is widely distributed. Its spores are about 3 μm in diameter, but tend (unlike such other members of the genus as *A. clavatus*, the cause of malt-worker's lung) to form spore chains up to 10 μm in length, which when inhaled deposit in proximal airways. Unlike the other moulds that may cause asthma, such as the genera *Alternaria* and *Cladosporium*, *A. fumigatus* grows at body temperature, and its septate hyphae may be found in sputum. Spore counts are usually highest during the winter, when exacerbations of the disease most frequently occur.

In common with such allergens as *Dermatophlagoides pteronyssinus* and grass pollen, *A. fumigatus* inhaled into the respiratory tree can stimulate production of specific IgE antibody. Its ability to persist and to grow in the airways can also stimulate IgG antibody production. *A. fumigatus* inhaled into the airways of a person with IgE and IgG antibody not only provokes release of mediators from mast cells, but can form immune complexes, which in antigen excess cause a local tissue-damaging inflammatory reaction. This is thought to be the cause of the proximal bronchiectasis that is characteristic of the disease.

Eosinophil chemotactic factors released from mast cells and from complement activation may be responsible for the associated eosinophilic consolidation.

Allergic bronchopulmonary aspergillosis is an uncommon disease, although in the United Kingdom it is the commonest cause of pulmonary eosinophilia. In one series, it was responsible for 116 of 143 cases of pulmonary eosinophilia. It appears to be less prevalent in the United States than in the United Kingdom. That is probably due to climatic differences: conditions in the United States are generally less favorable to the growth of *A. fumigatus*, although the difference may be due in part to differences in diagnostic convention.

The disease should be suspected in those with episodes of pulmonary eosinophilia, usually with asthma, which occur particularly during periods of the year when *A. fumigatus* spores are most prevalent. Those with asthma whose chest roentgenograms show changes of bronchiectasis or upper lobe fibrosis are also likely to have ABPA.

Hypersensitivity Pneumonitis (Extrinsic Allergic Alveolitis)

Hypersensitivity pneumonitis is an inflammatory reaction in alveolar walls and peripheral bronchioles due to an allergic reaction between inhaled organic particles and circulating antibodies and sensitized lymphocytes. In the acute stages of the disease, alveolar walls and peripheral bronchioles are infiltrated with mononuclear cells, which form noncascading giant and epithelioid cell granulomata. With progression of the disease (due to repeated or continuous allergen exposure), pulmonary fibrosis, particularly involving the upper lobes, may develop. The disease is thought to be the result both of the formation of immune complexes in antigen excess in alveolar walls and bronchioles and of a reaction between inhaled allergen and sensitized lymphocytes.

An increasing number of organic materials have been identified as capable of causing hypersensitivity pneumonitis. Exposure to most of these, such as *Micropolyspora faeni* (the cause of farmer's lung), is occupational. Two important causes of the disease are nonoccupational: bird fancier's lung and ventilation pneumonitis. Bird fancier's lung is caused by the inhalation of serum proteins in the feces of pigeons and parakeets. The disease may develop in pigeon fanciers or in those who keep parakeets in their homes. Ventilation pneumonitis is due to a reaction to thermophilic actinomycetes growing in ventilation systems; they have been shown to cause hypersensitivity pneumonitis both in offices and in homes with ventilation systems contaminated by these organisms. Thermophilic actinomycetes are not common in outdoor air, but may be extremely abundant in interiors where organic material is handled[18,23,83,90,109,131] and are apparently common in domestic interiors.[7,20] Concentrations in barns and cotton mills can exceed 30,000/m^3 of air,[18,22,23,59,81,84,90,109,131] whereas recoveries in domestic interiors rarely exceed 3,000/m^3.[20]

Domestic sources of actinomycetes are less clearly identified. Thermophilic actinomycetes have been recovered from humidifier fluid,[18 20 44-46 80 130 132 150 153] air-conditioners,[8 44 80 163] and an evaporative cooler.[93] However, their presence in a humidifier does not imply dispersion in the air,[7 20 101] and airborne taxa may differ from those commonly recovered from humidifier fluid.[20]

The pattern of symptoms in those affected is related primarily to the circumstances of exposure to the causal allergen. Those with intermittent exposure to high concentrations of allergens, as occurs in pigeon fanciers, develop recurrent episodes of breathlessness accompanied by flulike symptoms of malaise, headache, myalgia, and fever. Measurements of lung function during such an acute episode show a restrictive ventilatory defect with a decrease in gas transfer. In the absence of further exposure to the causal allergen, symptoms resolve over a period of 7-10 d, with improvement in lung-function measurements and chest-roentgenogram abnormalities over a month. With further exposures, lung-function tests and radiographic abnormalities can persist, and pulmonary fibrosis develop.

Those who have more continuous exposure to low concentrations of allergen, such as those exposed to parakeet excreta in their homes, often do not develop constitutional symptoms, but later, less reversible stages of the disease with increasing exertional dyspnea. The abnormalities of lung function are similar to those found in acute disease: a restrictive ventilatory defect with impairment of gas transfer. There may also be loss of volume of the upper lobes with linear shadows and cystic change due to fibrosis.

A disease that is probably due to an allergic reaction in the alveolar wall to contaminants of humidification systems, but which has several important features that distinguish it from typical hypersensitivity pneumonitis, has recently been described and called "humidifier fever."[35 37] The particular contaminants responsible are unknown, but may be amebae growing in the water. Those affected have recurrent episodes of flulike symptoms and fever that are often severe enough to overshadow the associated breathlessness. Symptoms develop 4-6 h after the onset of exposure and resolve spontaneously, whether or not exposure continues; and they recur only on reexposure after an absence of several days from exposure. Lung-function measurements during an attack show a restrictive ventilatory defect with impairment of gas transfer that improves over a period of days with the resolution of symptoms, despite continuing exposure. Unlike hypersensitivity pneumonitis, it is not accompanied by abnormalities on the chest roentgenogram during the acute attack, and pulmonary fibrosis does not occur, even in those who have had recurrent episodes of the disease for several years. Precipitins to an extract of the humidifier water or of the "jelly" growing in the humidifier are found in the serum of those affected, but may also be found in the serum of other exposed persons who do not get the disease. Immunofluorescent antibodies to various species of amebae, particularly *Negleria gruberi* and *Acanthamoebae*, have been found in the serum of those with precipitins to the humidifier water, but the relationship of these antibodies to disease remains unclear.

Concentrations of protozoa in interior air have not been sytematically reported; however, their occurrence indoors from both external and intramural sources may be anticipated. Protozoa have been recovered, in culture, from free air by several investigators, as summarized by Schlichting,[127] although the indicated concentrations have been well below those of pollens, algae, and fungal spores. Wind scouring of dry soil has been favored as a source of airborne isolates, although foams and such factors as sewage-processing may contribute locally.[127]

Indoor fluid collections--such as aquariums, humidifier reservoirs, and physiotherapy pools--are among the sites of potential colonization by protozoa. Recoveries of an ameba (Hartmannella castellanii) from air in a pediatric respiratory-care facility may implicate similar sources; however, many strains of the same species, as well as Naegleria and Schizopyrenus, also were taken from outside air.[76] Suspicion has also been cast on protozoa as agents responsible for "humidifier fever" in office and factory workers.[35,37] Some species (i.e., Naegleria fowler and Acanthamoeba spp.) are known to cause dangerous neurologic infections, although aerial transmission has not been demonstrated.[35]

REFERENCES

1. Ackermann, H.-W., B. Schmidt, and V. Lenk. Mycological studies of the outdoor and indoor air in Berlin. Mykosen 12:309-320, 1969.
2. Acosta, F., Jr., and G. W. Robertstad. Chrysosporium species as fungal air pollutants. Ann. Allergy 42:11-13, 1979.
3. Adams, K. F., and H. A. Hyde. Pollen grains and fungal spores indoors and out at Cardiff. J. Palynol. (Lucknow) 1:67-69, 1965.
4. Aisner, J., S. C. Schimpff, J. E. Bennett, V. M. Young, and P. H. Wiernik. Aspergillus infections in cancer patients. Association with fireproofing materials in a new hospital. J. Am. Med. Assoc. 235:411-412, 1976.
5. Arnow, P. M., R. L. Anderson, P. D. Mainous, and E. J. Smith. Pulmonary aspergillosis during hospital renovation. Am. Rev. Respir. Dis. 118:49-53, 1978.
6. Austwick, P. K. C. Ecology of Aspergillus fumigatus and the pathogenic phycomycetes, pp. 644-651. In N. E. Gibbons, Ed. Recent Progress in Microbiology. Toronto: University of Toronto Press, 1963.
7. Banaszak, E. F., J. Barboriak, J. Fink, G. Scanlon, D. P. Schlueter, A. Sosman, W. Thiede, and G. Unger. Epidemiologic studies relating thermophilic fungi and hypersensitivity lung syndromes. Am. Rev. Respir. Dis. 110:585-591, 1974.
8. Banaszak, E. F., W. H. Thiede, and J. N. Fink. Hypersensitivity pneumonitis due to contamination of an air conditioner. N. Engl. J. Med. 283:271-276, 1970.
9. Baruah, H. K. The air spora of a cowshed. J. Gen. Microbiol. 25:483-491, 1961.
10. Benson, F. B., J. J. Henderson, and D. E. Caldwell. Indoor-Outdoor Air Pollution Relationships: A Literature Review. U.S. Environmental Protection Agency, National Environmental Research

Center Publication No. AP-112. Washington, D.C.: U.S. Government Printing Office, 1972. 73 pp.

11. Bernstein, I. L., and R. S. Safferman. Sensitivity of skin and bronchial mucosa to green algae. J. Allergy 38:166-173, 1966.
12. Bernton, H. S., and H. Brown. Cockroach allergy. I. The relation of infestation to sensitization. South. Med. J. 60:852-855, 1967.
13. Bernton, H. S., and H. Brown. Insect allergy: The allergenicity of the excrement of the cockroach. Ann. Allergy 28:543-547, 1970.
14. Berrens, L. On the composition of feather extracts used in allergy practice. Int. Arch. Allergy Appl. Immunol. 34:81-94, 1968.
15. Berrens, L., J. H. Morris, and R. Versie. The complexity of house dust, with special reference to the presence of human dandruff allergen. Int. Arch. Allergy. Appl. Immunol. 27:129-144, 1965.
16. Brown, R. M., Jr., D. A. Larson, and H. C. Bold. Airborne algae: Their abundance and heterogeneity. Science 143:583-585, 1964.
17. Bruun, E., and M. Schwartz. Svampeallergi. Nord. Med. 26:1219-1225, 1945.
18. Buechner, H. A., A. L. Prevatt, J. Thompson, and O. Blitz. Bagassosis. A review with further historical data, studies of pulmonary function, and results of adrenal steroid therapy. Am. J. Med. 25:234-247, 1958.
19. Burge, H. A., J. R. Boise, W. R. Solomon, and E. Bandera. Fungi in libraries: An aerometric survey. Mycopathologia 64:67-72, 1978.
20. Burge, H. A., W. R. Solomon, and J. R. Boise. Microbial prevalence in domestic humidifiers. Appl. Environ. Microbiol. 39(4):840-844, 1980.
21. Burge, H. A., W. R. Solomon, and P. Williams. Aerometric study of viable fungus spores in an animal care facility. Lab. Anim. 13:333-338, 1979.
22. Burke, G. W., C. B. Carrington, R. Strauss, J. N. Fink, and E. A. Gaensler. Allergic alveolitis caused by home humidifiers. Unusual clinical features and electron microscopic findings. J. Am. Med. Assoc. 238:2705-2708, 1977.
23. Burrell, R., and M. J. McCullough. Production of thermophilic actinomycete-hay aerosols for use in experimental hypersensitivity pneumonitis. Appl. Environ. Microbiol. 34:715-719, 1977.
24. Charpin, J., M. Lauriol-Mallea, M. Renard, and H. Charpin. Study of fungal pollution in bake shops. Bull. Acad. Nat. Méd. Paris 155:52-55, 1971. (in French)
25. Chen, C. Y., and C-Y. Chuang. Fungi isolated from asthmatic homes in the Taipei area. Chin. J. Microbiol. 8(4):253-258, 1975.
26. Christensen, C. M. Intramural dissemination of spores of Hormodendrum resinae. J. Allergy 21:409-413, 1950.
27. Cobe, H. M. Asthma due to a mold. Hypersensitivity due to Cladosporium fulvum, Cooke. A case report. J. Allergy 3: 389-391, 1931.
28. Coca, A. F., and R. A. Cooke. On the classification of the phenomenon of hypersensitiveness. J. Immunol. 8:162-183, 1923.

29. Conant, N. F., H. C. Wagner, and F. M. Rackemann. Fungi found in pillows, mattresses, and furniture. J. Allergy 7:234-237, 1936.
30. Cooke and VanderVeer, A., Jr. Human sensitisation. J. Immunol. 1:201-305, 1916.
31. Cunnington, A. M., and P. H. Gregory. Mites in bedroom air. Nature 217: 1271-1272, 1968.
32. Dingle, A. N. Meteorological considerations in ragweed hay fever research. Fed. Proc. 16(2):615-627, 1957.
33. Dixit, I. P. Dust-mite urticaria. Practitioner 210:664, 1973.
34. Dworin, M. A study of atmospheric mold spores in Tucson, Arizona. Ann. Allergy 24:31-36, 1966.
35. Edwards, J. H. Humidifier fever. Thorax 32:653-663, 1977.
36. Edwards, J. H., J. J. Barboriak, and J. N. Fink. Antigens in pigeon breeders' disease. Immunology 19:729-734, 1970.
37. Edwards, J. H., A. J. Griffiths, and J. Mullins. Protozoa as sources of antigen in humidifier fever. Nature 264:438-439, 1976.
38. Emanuel, D. A., B. R. Lawton, and F. J. Wenzel. Maple-bark disease. Pneumonitis due to Coniosporium corticale. N. Engl. J. Med. 266:333-337, 1962.
39. Emmons, C. W. Natural occurrence of opportunistic fungi. Lab. Invest. 11:1026-1034, 1962.
40. Faux, J. A., L. Wide, F. E. Hargreave, J. L. Longbottom, and J. Pepys. Immunological aspects of respiratory allergy in budgerigar (Conelopsittacus undulatus) fanciers. Clin. Allergy 1:149, 1971.
41. Feinberg, S. Environmental factors and host responses in asthma. Acta Allergol. 29 (Suppl. 11):7-14, 1974.
42. Fergus, C. L. Thermophilic and thermotolerant molds and actinomycetes of mushroom compost during peak heating. Mycologia 56:267-284, 1964.
43. Figley, K. D. Asthma due to the Mayfly. Am. J. Med. Sci. 178:338-345, 1929.
44. Fink, J. N., E. F. Banaszak, J. J. Barboriak, G. T. Hensley, V. P. Kurup, G. T. Scanlon, D. P. Schlueter, A. J. Sosman, W. H. Thiede, and G. F. Unger. Insterstitial lung disease due to contamination of forced air systems. Ann. Intern. Med. 84:406-413, 1976.
45. Fink, J. N., E. F. Banaszak, W. H. Thiede, and J. J. Barboriak. Interstitial pneumonitis due to hypersensitivity to an organism contaminating a heating system. Ann. Intern. Med. 74:80-83, 1971.
46. Fink, J. N., A. J. Resnick, and J. Salvaggio. Presence of thermophilic actinomycetes in residential heating systems. Appl. Microbiol. 22:730-731, 1971.
47. Flensborg, E. W., and T. Samsøe-Jensen. Studies in mold allergy: 3. Mold spore counts in Copenhagen. Acta Allergol. 3:49-65, 1950.
48. Frankland, A. W. Locust sensitivity. Ann. Allergy 11:445-453, 1953.
49. Frankland, A. W., and M. J. Hay. Dry rot as a cause of allergic complaints. Acta Allergol. 4:186-200, 1951.

50. Gage, A. A., D. C. Dean, G. Schimert, and N. Minsley. Aspergillus infection after cardiac surgery. Arch. Surg. 101:384-387, 1970.

51. Gell, P. G. H., R. R. A. Coombs, and P. J. Lachmann. Clinical Aspects of Immunology. 3rd ed. Oxford: Blackwell Scientific Publications, 1975. 1356 pp.

52. Gemeinhardt, H., and I. Bergmann. Moulds in bakery dusts. Zentralbl. Bakteriol. Abt. 2. Naturwiss. 132:44-54, 1977. (in German; English summary)

53. Gravesen, S. Identification and prevalence of culturable mesophilic microfungi in house dust from 100 Danish homes. Comparison between airborne and dust-bound fungi. Allergy 33:268-272, 1978.

54. Gravesen, S. Identification and quantitation of indoor airborne microfungi during 12 months from 44 Danish homes. Acta Allergol. 27:337-354, 1972.

55. Gregory, P. H. The Microbiology of the Atmosphere, pp. 57-70. New York: John Wiley & Sons, Inc., 1973.

56. Gregory, P. H., J. M. Hirst, and F. T. Last. Concentrations of basidiospores of the dry rot fungus (Merulius lacrymans) in the air of buildings. Acta Allergol. 6:168-174, 1953.

57. Gregory, P. H., and M. E. Lacey. Mycological examination of dust from mouldy hay associated with farmer's lung disease. J. Gen. Microbiol. 30:75-88, 1963.

58. Halweg, H., P. Krakówka, O. Podsiadło, J. Owczarek, A. Ponahajba, and L. Pawlicka. Studies on air pollution by fungal spores at selected working posts in a paper factory. Pneumonol. Polska 46:577-585, 1978.

59. Hearn, C. E. D.: Bagassosis: An epidemiological, environmental and clinical survey. Br. J. Ind. Med. 25:267-282, 1968.

60. Heise, H. A. Symptoms of hay fever caused by algae. J. Allergy 20:383-385, 1949.

61. Herman, L. G. Aspergillus in patient care areas. Ann. N.Y. Acad. Sci. 353:140-146, 1980.

62. Hewitt, M., G. I. Barrow, D. C. Miller, F. Turk, and S. Turk. Mites in the personal environment and their role in skin disorders. Br. J. Dermatol. 89:401-409, 1973.

63. Hill, J., A. Howell, and R. Blowers. Effect of clothing on dispersal of Staphylococcus aureus by males and females. Lancet 2:1131-1133, 1974.

64. Hirsch, S. R., and J. A. Sosman. A one-year survey of mold growth inside twelve homes. Ann. Allergy 36:30-38, 1976.

65. Hořejší, M., J. Šach, A. Tomšiková, and A. Mecl. A syndrome resembling farmer's lung in workers inhaling spores of aspergillus and penicillia moulds. Thorax 15: 212-217, 1960.

66. Hughes, W. T., and J. W. Crosier. Thermophilic fungi in the mycoflora of man and environmental air. Mycopathol. Mycol. Appl. (The Hague) 49:147-152, 1973.

67. Ishizaka, K., T. Ishizaka, and N. M. Hornbrook. Physico-chemical properties of human reaginic antibody. IV. Presence of a unique immunoglobulin as a carrier of reaginic antibody activity. J. Immunol. 97:75-85, 1966.

68. Jimenez-Diaz, C., J. M. Alés, F. Ortiz, F. Lahoz, L. M. Garcia Puente, and G. Canto. The aetiologic role of molds in bronchial asthma. Acta Allergol. Suppl. 7:139-149, 1960.

69. Jopke, W. H., and D. R. Hass. Contamination of dishwashing facilities. Hospitals 44(6):124-127, March 16, 1970.

70. Juniper, C. P., M. J. How, B. F. J. Goodwin, and A. K. Kimsholt. Bacillus subtilis enzymes: A 7-year clinical epidemiological and immunological study of an industrial allergen. J. Soc. Occup. Med. 23:3, 1977.

71. Kang, B. Study on cockroach antigen as a probable causative agent in bronchial asthma. J. Allergy Clin. Immunol. 58:357-365, 1976.

72. Kanno, S. Indoor contamination by fungi. Japan. J. Bacteriol. 30:458-460, 1975. (in Japanese)

73. Kawai, T., D. G. Marsh, L. M. Lichtenstein, and P. S. Norman. The allergens responsible for house dust allergy. 1. Comparison of Dermatophagoides pteronyssinus and house dust extracts by assay of histamine release from allergic human leukocytes. J. Allergy Clin. Immunol. 50:117-127, 1972.

74. Kern, R. A. Asthma due to sensitization to a mushroom fly (Aphiochaeta agarici). J. Allergy 9:604-606, 1938.

75. Kimura, I., Y. Moritari, and Y. Tanizaki. Basophils in bronchial asthma with reference to reagin type allergy. Clin. Allergy 3:195, 1973.

76. Kingston, D., and D. C. Warhurst. Isolation of amoebae from the air. J. Med. Microbiol. 2:27-36, 1969.

77. Korsgaard, J. House-dust mites and allergy to house-dust. Ugeskr. Laeg. 141:892-897, 1979. (in Danish; English summary)

78. Korsgaard, J. House-dust mites (Pyroglyphidae, acari) in Danish homes. Ugeskr. Laeg. 141:888-892, 1979. (in Danish; English summary)

79. Kozak, P. P., Jr., J. Gallup, L. H. Cummins, and S. A. Gillman. Factors of importance in determining the prevalence of indoor molds. J. Allergy Clin. Immunol. 61:185, 1978. (Abstract No. 189)

80. Kurup, V. P., J. N. Fink, and D. M. Bauman. Thermophilic actinomycetes from the environment. Mycologia 68:662-666, 1976.

81. Lacey, J. Microorganisms in air of cotton mills. Lancet 2:455-456, 1977.

82. Lacey, J. Potential hazards to animals and man from microorganisms in fodder and grain. Br. Mycol. Soc. Trans. 65:171-184, 1975.

83. Lacey, J. The air spora of a Portuguese cork factory. Ann. Occup. Hyg. 16:223-230, 1973.

84. Lacey, J., J. Pepys, and T. Cross. Actinomycete and fungus spores in air as respiratory allergens, pp. 151-184. In D. A. Shapton and R. G. Board, Eds. Safety in Microbiology. New York: Academic Press, Inc., 1972.

85. Levetin, E., and D. Hurewitz. A one-year survey of the airborne molds of Tulsa, Oklahoma. II. Indoor survey. Ann. Allergy 41:25-27, 1978.

86. Lidwell, O. M., and W. C. Noble. Fungi and clostridia in hospital air: The effect of air-conditioning. J. Appl. Bacteriol. 39:251-261 1975.

87. Liebeskind, A. Diagnostic value of culture procedures and provocation tests in suspected mold allergies. Acta Allergol. 26:106-116, 1971.

88. Liebeskind, A. Mold allergy in factories. Allerg. Asthma (Leipzig) 11:62-65, 1965. (in German; English summary)

89. Llamas, R., D. R. Hart, and N. S. Schneider. Allergic bronchopulmonary aspergillosis associated with smoking moldy marihuana. Chest 73:871-872, 1978.

90. Lockwood, M. G., and R. W. Attwell. Thermophilic actinomycetes in air of cotton mills. Lancet 2:45-46, 1977.

91. Lumpkins, E. D., Sr., and S. Corbit. Airborne fungi survey. II. Culture plate survey of the home environment. Ann. Allergy 36:40-44, 1976.

92. Lumpkins, E. D., Sr., S. L. Corbit, and G. M. Tiedeman. Airborne fungi survey. 1. Culture-plate survey of the home environment. Ann. Allergy 31:361-370, 1973.

93. Marinkovich, V. A., and A. Hill. Hypersensitivity alveolitis. J. Am. Med. Assoc. 231:944-947, 1975.

94. Marsh, P. B., P. D. Millner, and J. M. Kla. A guide to the recent literature on aspergillosis as caused by A. fumigatus. USDA Manual ARM-NE-5. Washington, D.C.: U.S. Department of Agriculture, 1979.

95. Mathews, K. P. Other inhalant allergens, pp. 945-956. In E. Middleton, Jr., C. E. Reed, and E. F. Ellis, Eds. Allergy: Principles and Practice. Vol. 2. Saint Louis: The C. V. Mosby Company, 1978.

96. Matsumura, T., K. Tateno, S. Yugami, and T. Kimura. Four cases of asthma caused by silk inhalation. J. Asthma Res. 4:205-208, 1967.

97. Maunsell, K. Air-borne fungal spores before and after raising dust. (Sampling by sedimentation.) Int. Arch. Allergy Appl. Immunol. 3:93-102, 1952.

98. Maunsell, K. Concentration of airborne spores in dwellings under normal conditions and under repair. Int. Arch. Allergy Appl. Immunol. 5:373-376, 1954.

99. McGovern, J. P., T. R. McElhenney, and R. M. Brown. Airborne algae and their allergenicity. Part I. Air sampling and delineation of the problem. Ann. Allergy 23:47-50, 1965.

100. Michel, B., J. P. Marty, L. Quet, and P. Cour. Penetration of inhaled pollen into the respiratory tract. Am. Rev. Respir. Dis. 115:609-616, 1977.

101. Miller, M. M., R. Patterson, J. N. Fink, and M. Roberts. Chronic hypersensitivity lung disease with recurrent episodes of hypersensitivity pneumonitis due to a contaminated central humidifier. Clin. Allergy 6:451-462, 1976.

102. Millner, P. D., P. B. Marsh, R. B. Snowden, and J. F. Parr. Occurrence of Aspergillus fumigatus during composting of sewage sludge. Appl. Environ. Microbiol. 34:765-772, 1977.

103. Mitchell, W. F., G. W. Wharton, D. G. Larson, and R. Modic. House dust, mites and insects. Ann. Allergy 27:93-99, 1969.

104. Miyamoto, T., S. Oshima, and T. Ishizaki. Antigenic relation between house dust and a dust mite, Dermatophagoides farinae

Hughes, 1961, by a fractionation method. J. Allergy 44:282-291, 1969.
105. Moore, B. S., and J. S. Hyde. Characterization of breed-specific dog dander and serum allergens. J. Allergy Clin. Immunol. 63:206, 1979. (Abstract No. 247)
106. Moore, B. S., J. S. Hyde, and L. M. Manaligod. A comparative study of allergens of canine origin. Ann. Allergy 39:240-245, 1977.
107. Murray, A. B., and P. Zuk. The seasonal variation in a population of house dust mites in a North American city. J. Allergy Clin. Immunol. 64:266-269, 1979.
108. Murray, F. J., H. Brown, and H. S. Bernton. A case of asthma caused by the box elder beetle. J. Allergy 45:103, 1970. (Abstract No. 12)
109. Nicholson, D. P. Bagasse worker's lung. Am. Rev. Respir. Dis. 97:546-560, 1968.
110. Nilsby, I. Allergy to moulds in Sweden. A botanical and clinical study. Acta Allergol. 2:57-90, 1949.
111. Noble, W. C., and Y. M. Clayton. Fungi in the air of hospital wards. J. Gen. Microbiol. 32:397-402, 1963.
112. Parlato, S. J. The sand fly (caddis fly) as an exciting cause of allergic coryza and asthma. II. Its relevant frequency. J. Allergy 1:307-312, 1930.
113. Pepys, J. Atopy, p. 877. In G. H. Gell, R. R. A. Coombs, and P. J. Lachmann, Eds. Clinical Aspects of Immunology. Oxford: Blackwell, 1975.
114. Perlman, F. Insects as inhalant allergens. Consideration of aerobiology, biochemistry, preparation of material, and clinical observations. J. Allergy 29:302-328, 1958.
115. Peterson, J. E. Estimating air filtration into houses: An analytical approach. ASHRAE J. 21(1):60-63, 1979.
116. Popescu, I. G., and E. Capetti. Study of mold spores in houses of asthmatics. Rev. Roum. Méd. Interne (Bucharest) 8:357-361, 1971.
117. Popescu, I. G., E. Capetti, C. Galalaie, and I. Spiegler. Study of atmospheric fungi in a big cereal silo over a period of one year. Rev. Roum. Méd. Médecine Interne (Bucharest) 13:221-226, 1975.
118. Prince, H. E., M. B. Morrow, and G. H. Meyer. Molds in occupational environments as causative factors in inhalant allergic diseases. A report of two cases. Ann. Allergy 22:688-692, 1964.
119. Raper, K. B., and D. I. Fennell. The genus _Aspergillus_. Baltimore: The Williams & Wilkins Company, 1965. 686 pp.
120. Refai, M., and A. Loot. Studies of mould contaminations of meat in slaughter houses, butcher's shops and in cold stores. Mykosen 12:621-624, 1969.
121. Richards, M. Atmospheric mold spores in and out of doors. J. Allergy 25:429-439, 1954.
122. Ripe, E. Mould allergy. I. An investigation of the airborne fungal spores in Stockholm, Sweden. Acta Allergol. 17:130-159, 1962.

123. Samsøe-Jensen, T. Mould allergy. Sensitization by special exposure illustrated by two cases of allergy to Cladosporium fulvum. Acta Allergol. 9:38-44, 1955.
124. Samson, R. A., and B. van der Lustgraaf. Aspergillus penicilloides and Eurotium halophilicum in association with house-dust mites. Mycopathologia 64:13-16, 1978.
125. Schaffer, N., E. E. Seidmon, and S. Bruskin. The clinical evaluation of air-borne and house dust fungi in New Jersey. J. Allergy 24:348-354, 1953.
126. Schlichting, H. E., Jr. Ejection of microalgae into the air via bursting bubbles. J. Allergy Clin. Immunol. 53:185-188, 1974.
127. Schlichting, H. E., Jr. The importance of airborne algae and protozoa. J. Air Pollut. Control Assoc. 19:946-951, 1969.
128. Schlueter, D. P., J. N. Fink, and G. T. Hensley. Wood-pulp workers' disease: A hypersensitivity pneumonitis caused by Alternaria. Ann. Intern. Med. 77:907-914, 1972.
129. Schwartz, M. Heredity in bronchial asthma. Acta Allergol. 5 (Suppl. II), 1952.
130. Seabury, J., B. Becker, and J. Salvaggio. Home humidifier thermophilic actinomycete isolates. J. Allergy Clin. Immunol. 57:174-176, 1976.
131. Seabury, J., J. Salvaggio, H. Buechner, and V. G. Kundur. Bagassosis. III. Isolation of thermophilic and mesophilic actinomycetes and fungi from moldy bagasse. Proc. Soc. Exp. Biol. Med. 129:351-360, 1968.
132. Seabury, J., J. Salvaggio, J. Domer, J. Fink, and T. Kawai. Characterization of thermophilic actinomycetes isolated from residential heating and humidification systems. J. Allergy Clin. Immunol. 51:161-173, 1973.
133. Segretain, G. Infection by fungi that ordinarily are saprophytes. Pulmonary aspergillosis. Lab. Invest. 11:1046-1052, 1962.
134. Sherman, H., and D. Merksamer. Skin test reactions in mold-sensitive patients in relation to presence of molds in their homes. N.Y. State J. Med. 64:2533-2535, 1964.
135. Sidransky, H. Experimental studies with aspergillosis, pp. 165-176. In E. W. Chick, A. Balows, and M. Furcolow, Eds. Opportunistic Fungal Infections. Proceedings of the Second International Conference. Springfield, Ill.: Charles C Thomas Publishers, 1975.
136. Sinha, R. N., J. E. M. H. van Bronswijk, and H. A. H. Wallace. House dust allergy, mites and their fungal associations. Can. Med. Assoc. J. 103:300-301, 1970.
137. Slavin, R. G., and P. Winzenburger. Epidemiologic aspects of allergic aspergillosis. Ann. Allergy 38:215-218, 1977.
138. Solomon, W. R. Assessing fungus prevalence in domestic interiors. J. Allergy Clin. Immunol. 56:235-242, 1975.
139. Solomon, W. R. Fungus aerosols arising from cold-mist vaporizers. J. Allergy Clin. Immunol. 54:222-228, 1974.
140. Solomon, W. R., and H. P. Burge. Aspergillus fumigatus levels in- and out-of-doors in urban air. J. Allergy Clin. Immunol. 55:90-91, 1975.

141. Solomon, W. R., H. P. Burge, and J. R. Boise. Airborne Aspergillus fumigatus levels outside and within a large clinical center. J. Allergy Clin. Immunol. 62:56-60, 1978.
142. Solomon, W. R., H. P. Burge, and J. R. Boise. Exclusion of particulate allergens by window air conditioners. J. Allergy Clin. Immunol. 63:215, 1979. (Abstract No. 274)
143. Spendlove, J. C. Penetration of structures by microbial aerosols. Dev. Ind. Microbiol. 16:427-436, 1975.
144. Sreeramulu, T. Concentrations of fungus spores in the air inside a cattle shed. Acta Allergol. 16:337-346, 1961.
145. Staib, F., T. Abel, S. K. Mishra, G. Grosse, M. Focking, and A. Blisse. Occurrence of Aspergillus fumigatus in West Berlin--Contribution to the epidemiology of aspergillosis. Zentralbl. Bakteriol. Parasitenkd. Infektionskr. Hyg. Abt. 1 Orig. Reihe A 241:337-357, 1978. (in German; English abstract)
146. Staib, F., U. Folkens, B. Tompak, T. Abel, and D. Thiel. A comparative study of antigens of Aspergillus fumigatus isolates from patients and soil of ornamental plants in the immunodiffusion test. Zentralbl. Bakteriol. Parasitenkd. Infektionskr. Hyg. Abt. 1 Orig. Reihe A 242:93-99, 1978.
147. Staib, F., B. Tompak, D. Thiel, and A. Blisse. Aspergillus fumigatus and Aspergillus niger in two potted ornamental plants, cactus (Epiphyllum truncatum) and clivia (Clivia miniata). Biological and epidemiological aspects. Mycopathologia 66:27-30, 1978.
148. Stevenson, D. D., and K. P. Mathews. Occupational asthma following inhalation of moth particles. J. Allergy 39:274-283, 1967.
149. Swaebly, M. A., and C. M. Christensen. Molds in house dust, furniture stuffing and in the air within homes. J. Allergy 23:370-374, 1952.
150. Sweet, L. C., J. A. Anderson, Q. C. Callies, and E. O. Coates, Jr. Hypersensitivity pneumonitis related to a home furnace humidifier. J. Allergy Clin. Immunol. 48:171-178, 1971.
151. Taylor, A. N., C. A. C. Pickering, J. Pepys, and M. Turner-Warwick. Respiratory allergy to a factory humidifier contaminant. Clin. Allergy 6:411-412, 1976. (abstract)
152. Taylor, B., A. P. Norman, H. A. Orgel, C. R. Stokes, M. W. Turner, and J. F. Soothill. Transient IgA deficiency and pathogenesis of infantile atopy. Lancet 2:111, 1973.
153. Tourville, D. R., W. I. Weiss, P. T. Wertlake, and G. M. Leudermann. Hypersensitivity pneumonitis due to contamination of home humidifier. J. Allergy Clin. Immunol. 49:245-251, 1972.
154. Towey, J. W., H. C. Sweany, and W. H. Huron. Severe bronchial asthma apparently due to fungus spores found in maple bark. J. Am. Med. Assoc. 99:453-459, 1932.
155. van der Lustgraaf, B. Xerophilic fungi in mattress dust. Mycosen 20:101-106, 1977. (in English)
156. van der Werff, P. J. Mould Fungi and Bronchial Asthma. A Mycological and Clinical Study. Springfield, Ill.: Charles C Thomas, 1958. 174 pp.

157. von Klopotek, A. Über das Vorkommen und Verhalten von Schimmelpilzen bei der Kompostierung städtischer Abfallstoffe. Antonie van Leeuwenhoek J. Microbiol. Serol. 28:141-160, 1962. (in German)
158. Voorhorst, R. The human dander atopy. I. The prototype of auto-atopy. Ann. Allergy 39:205-212, 1977.
159. Voorhorst, R., F. T. M. Spieksman, H. Varekamp, M. J. Leupen, and A. W. Lyklema. The house-dust mite (Dermatophagoides pteronyssinus) and the allergens it produces. Identity with the house-dust allergen. J. Allergy 39:325-329, 1967.
160. Wagner, H. C., and F. M. Rackemann. Kapok and molds: An important combination. Ann. Intern. Med. 11:505-513, 1937.
161. Wallace, M. E., R. H. Weaver, and M. Scherago. A weekly mold survey of air and dust in Lexington, Kentucky. Ann. Allergy 8:202-211, and 228, 1950.
162. Warren, W. P. Hypersensitivity pneumonitis due to exposure to budgerigars. Chest 62:170-174, 1972.
163. Weiss, N. S., and Y. Soleymani. Hypersensitivity lung disease caused by contamination of an air-conditioning system. Ann. Allergy 29:154-156, 1971.
164. Wharton, G. W. House dust mites. J. Med. Entomol. 12:577-621, 1976.
165. Wolf, F. T. Observations on an outbreak of pulmonary aspergillosis. Mycopathol. Mycol. Appl. 38:359-361, 1969.
166. Wraith, D. G., A. M. Cunnington, and W. M. Seymour. The role and allergenic importance of storage mites in house dust and other environments. Clin. Allergy 9:545-561, 1979.
167. Wray, B. B. Mycotoxin-producing fungi from house associated with leukemia. Arch. Environ. Health 30:571-573, 1975.
168. Yulu, G. N., and S. K. S. Timur. Indoor and outdoor fungal flora of Anakara. Mikrobiyol. Bül. 11:355-364, 1977.

VIII

EFFECTS OF INDOOR POLLUTION ON HUMAN WELFARE

"Indoors" is the place of escape from extremes of temperature, humidity, and environmental conditions and from exposure to some pollutants found in the "outdoors." It is the place where rest, relaxation, and the general welfare afforded by bodily comforts are sought. It is known that attempts to reduce energy consumption in buildings can affect the quality of indoor environments. This chapter discusses some of the effects on human welfare--e.g., discomfort, decreased productivity, soiling, corrosion, and maintenance and housekeeping needs--caused by alterations in environmental control systems.

Discomfort is the result of undesirable sensory stimuli, such as noise, malodors, glare, and extremes of humidity and temperature. These often invoke a human response, identified as "discomfort," that is straightforward and physical and that may sometimes be relieved by attenuation of the stimulus. However, mere attenuation of the sensory stimulus sometimes does <u>not</u> suffice. Discomfort is a sensitive indicator of the need for adjustments in environmental quality control.

The relationships between indoor pollution and productivity can be evaluated only after one carefully defines productivity and determines how it is to be assessed. Originaly, productivity was conceived simply as quantity of output; but it has come to be addressed in terms of economy--the cost per unit of production. This chapter discusses some attempts to measure the effects of environmental quality, with productivity as a tool.

Indoor air pollution is a source of soiling and contributes to the deterioration and corrosion of equipment, furnishings, and appliances. Soiling increases needs for maintenance and housekeeping and for some equipment in the ventilation system.

RELATIONSHIPS BETWEEN SOCIOECONOMIC STATUS AND INDOOR POLLUTION

The relationships between housing characteristics and the health of the occupants among the various socioeconomic groups are not well known. The available information, although limited, is important if we are to understand and identify the problems involved and if we are to learn the relationships between housing types, housing quality, indoor environmental quality, and pollutant types, on the one hand, and the health and welfare of the people in the several socioeconomic groups, especially those in the lower groups, on the other hand.

A comprehensive treatment of socioeconomic status (SES) and indoor air pollution may be important to the formulation of control strategies (local, state, or federal) in matters that influence indoor pollution, such as energy-conservation assistance programs and low-income and rent-subsidy programs.

Housing characteristics are related to social status or income level.[35 59 67] Status and income often have been shown to be related to health and probably constitute an intervening variable in the relationship between selected housing characteristics and health.[36 51] The role of housing itself in determining health is still unclear.[33 35 59 67 69] Crowding indoors is thought to be an important contributor to the spread of infectious diseases and a potential source of physiologic stress.[37 51 67 69] A substantially higher proportion of persons in low-income groups have chronic health conditions that limit their activities[44] and keep them indoors.

Some characteristics of housing constitute definite risks to health--e.g., carbon monoxide poisoning from faulty venting of space-heating systems[72] and lead poisoning from paints.[40] The two mentioned are also related to low-income houses, which often have greater rates of air infiltration and, because they are close to sources of pollution, transport pollution more freely from outdoors to indoors.[35 54] Spivey and Radford[59] found that a high proportion of gas stoves and gas space-heaters (60% in lower-SES homes in east Baltimore) had higher indoor than outdoor concentrations of carbon monoxide (8-8.9 ppm versus 5.5-6.1 ppm). In two sets of homes studied, the amount of passive smoking did not appear to be related to any differences observed in indoor carbon monoxide concentrations. In over 70% of these homes, the lead content in dust and paint samples exceeded currently recommended standards. Blood lead contents are lower in persons who live in SES-equivalent houses with air-conditioning than without.[20] Binder et al.[12] found that indoor respirable-particle concentrations were higher in homes with higher ratios of persons to room volume.

The following tentative conclusions can be drawn: Homes with controlled ventilation systems, air filtration, good maintenance, and properly working appliances have lower concentrations of indoor pollutants. That implies that the middle and upper socioeconomic groups are at lower risk. However, there are sources of pollution other than those mentioned in upper-income houses, specifically, newer and more carpets, curtains, and furniture. Low-income housing is more likely to have improper ventilation, poor maintenance, defective appliances (such as improperly operating stoves and space-heaters), and lead-based paint--all of which contribute to higher indoor concentrations of pollutants.[20 51 56 60] Furthermore, persons in the low-income groups are more likely to live in mobile homes or apartments,[37] which frequently are crowded (high ratio of persons to volume).[51] Mobile homes generally are very airtight, and crowding can result in high concentrations of indoor pollutants.[56 60] Recreational vans and trailers have many of the physical characteristics of mobile homes and can have similar pollution problems. Those who can afford to "tighten" their conventional homes for energy conservation may also have higher indoor concentrations of some pollutants, although one would expect an eventual balance between "tightening" and proper ventilation in those homes.

HUMAN DISCOMFORT

The incentive to control the indoor environment is derived as much from consideration of human comfort as from consideration of health. Discomfort provides an immediate incentive to control the quality of the indoor environment. Undesirable sensory signals (e.g., noise, glare, and cold) register as discomfort. These signals have straightforward physical correlates (e.g., sound pressure, contrast ratio, and temperature) with the need for controls, such as the installation of sound-absorbing tiles. A person annoyed initially by the loud conversation of co-workers may eventually become annoyed even by whispered exhanges; thus, mere attentuation of noise may not suffice. A person annoyed frequently by sidestream cigarette smoke from the person at the next desk may eventually become angered by the slightest trace of tobacco-smoke odor. Such time-dependent changes in sensitivity show a cognitive contribution to discomfort. Some persons can become annoyed merely by the information carried by a stimulus, and this reaction can be as important as a reaction to the stimulus itself.

Whether discomfort is caused by the intensity of stimulation or by the conditioned response resulting from sensitization, the questions arise: Will avoidance or elimination of discomfort ensure a reasonably healthful indoor climate? Does endurance of discomfort take a psychologic or physiologic toll?

Our senses are remarkably adaptive. Therefore, they do not provide infallible sensory signals about the safety of the environment, owing to their inability to register some types of energy or potential stimuli. For instance, a person may view a solar eclipse without knowledge that the ultraviolet rays, unregistered by the photoreceptors, may damage the eye. A person may bask in the warmth of the summer sun without awareness that ultraviolet rays, poorly registered in this case by cutaneous receptors, may cause serious, even lethal burns. Similarly, a person may eat a bacteria-laden, although delicious, meal without any sensory warning of the ptomaine toxins present. The sense of smell also fails to register some harmful stimuli, such as carbon monoxide. With only a limited number of notable exceptions, however, the absence of annoying stimuli indoors may be misleading, but generally does signify safe conditions of occupancy.

Regarding the endurance of discomfort, possible long-term effects include irritability, depression, inability to concentrate, anxiety, indigestion, headaches, back pain, and insomnia.[57] Short-term effects of discomfort are often rather specific to a particular modality. Hence, malodors may cause symptoms of digestive upset, poor lighting may cause headaches, and cold drafts may cause muscle stiffness. Objective verification of direct causes of these various symptoms is difficult. For this reason, the symptoms, even when severe, fail to qualify as adverse health effects.

This section briefly discusses some of the indoor-polution aspects of discomfort.

MALODORS

The olfactory senses signal the presence of some harmful airborne stimuli, but sometimes they fail to do so, and there are frequent "false alarms." As mentioned in Chapter IV, people have historically avoided bad-smelling air for fear that it signaled illness-causing conditions. In the nineteenth century, the criteria for ventilation commonly arose from the notion that odorous air contained harmful ingredients known variously as crowd poison, morbific matter, and anthropotoxin.[18] For instance, Russell stated in The Atmosphere in Relation to Human Life and Health:[52]

> Organic matter is given off from the lungs and skin, of which neither the exact amount nor the composition has been hitherto ascertained. Their quantity is very small, but of its importance there can be no doubt. . . . Since this organic matter has been proved to be highly poisonous, even apart from carbon dioxide and vapor, we may safely infer that much of the mischief resulting from the inspiration of rebreathed air is due to the special poisons exhaled by the body.

In the absence of instrumentation to detect the presence of small amounts of odorous organic vapors, the nose remains a sensitive indicator. Surprisingly, even today there are no good rules for laymen or scientists to relate perceived odor quality to toxicity. Some odorous signals are used to warn about toxic hazards (e.g., mercaptans are used in natural gas to signal leaks). We may know from experience that some foul-smelling living spaces pose no overt danger, but people will still avoid such places. We may argue that this avoidance is derived from mere discomfort, but occupants may fail to see the situation in such benign terms.

In the early twentieth century, the New York State Commission on Ventilation performed a set of experiments regarding the effects of occupancy odor on human comfort and performance.[46] In a popular synopsis of this 8-yr effort, Winslow[70] stated:

> We may summarize our discussion of the physiology of ventilation as follows: The chemical vitiation of the air of an occupied room (unless poisons or dusts from industrial processes or defective heating appliances are involved) is of relatively slight importance. The organic substances present,

> manifest as body odors, may exert a depressing effect upon inclination to work and upon appetite; therefore, occupied rooms should be free from body odors which are obvious to anyone entering from without. (Such odors are never perceived by those who have been continuously in the room while they have been accumulating.) Objectionable effects of this sort have only been demonstrated, however, with a carbon dioxide content of over .2 per cent, which would correspond to an air change of less than 6 cubic feet per person per minute.

During the 1930s, Winslow and Herrington[71] demonstrated that "dust odor" similar to that from a heating system could also depress appetite.

Winslow implied that the olfactory sense generally adapts to prevailing odorous stimulation in such a way as to reduce discomfort. Similarly, Cain reported that a temporary reduction in olfactory sensitivity, perhaps in conjunction with affective habituation, presumably explains why workers in some malodorous industries eventually find the odorous atmosphere unobjectionable.[17] In contrast, people who live near malodorous sources of pollution seem to experience adverse olfactory reactions of constant or even increasing severity. For example, residents exposed frequently to malodorous emission of factories complained of chronic headaches, nausea, coughing, disturbance of sleep, and loss of appetite.[68] Those adverse reactions seem to arise as a consequence of industrial odors that are more or less unremitting and are beyond the residents' control. But when the source is in the occupied space, some control (or avoidance) may well be possible. (Tobacco smoke, traditionally the most bothersome odorant, is a common exception.)

Complaints about irritation of the eyes, throat, and nose are common and increasing among people in newly constructed or newly renovated offices.[4] These complaints may arise from a confluence of low, energy-conserving rates of ventilation and emission of odorous or irritating substances, such as formaldehyde, from new furnishings. Tobacco smoke may exacerbate the problem. The course of the reaction of the common chemical sense of those exposed may vary considerably.[16] One person may notice irritation immediately; another may notice it only after occupying a space for a few hours, but continue to experience it long after leaving the space and possibly fail to associate the irritation with its source. As a further complication, it has long been suspected that formaldehyde acts as an olfactory anesthetic.[58]

NOISE

Discomfort due to noise has received more attention than that related to any other type of sensory stimulation. Noise-induced discomfort occurs in a great variety of situations, ranging from disturbance of sleep to difficulty in hearing in the workplace. Noise standards and regulations abound: for outdoor noises, for sound insulation in buildings, for controlling the risk of occupation-related deafness, and for guarding against hearing difficulty and annoyance in offices, schools, and hospitals. The context can have a strong bearing on the degree of annoyance. Nemecek and Grandjean[45] surveyed a large office and found that most of the employees were disturbed by noise that was considered well within professional design standards. The "noise" came from conversations, and it was content, rather than intensity, that was the disturbing attribute.

Experiments in both human beings and animals have shown that stressful effects from nondeafening noise arise without respect to the "meaning" of the auditory stimulation.[62] Physical attributes that seem particularly relevant to annoyance include intensity, concentration of energy within high frequencies, temporal and spectral complexity, duration, and the suddenness of sounds.[34] Table VIII-1 shows results of a survey made to determine the importance of various physical and perceived attributes of annoying sounds.[23] The respondents judged loudness the most important attribute, with suddenness next in line. The next three most important attributes comprised cognitive features (sound is man-made, sound cannot be turned off, sound is unnecessary). The preeminence of loudness in the determination of annoyance has led to recommendations, such as those in Table VIII-2, for tolerable maximal loudness in various types of rooms.[34] The loudness values listed here refer to continuous noise in the period between 7 a.m. and 10 p.m.

Both human and animal laboratory experiments have shown hormonal effects of noxious, although nondeafening, noise exposure. Even exposures of about 70 dB can increase the output of adrenal corticosteroids.[6,26] Sound intensity this low can also cause constriction of peripheral blood vessels.[38] Such changes, and other physiologic manifestations, usually fail to outlast the stimulus, but do cause concern that noise might eventually lead to more chronic symptoms of stress or affect sleep. Frequent interruption of sleep or alteration in the normal progression of sleep patterns may be thought to jeopardize physical or mental health eventually. Fortunately, adaptive alterations in the pattern of sleep seem to minimize most short-term consequences of disruption by noise.[32]

In addition to physiologic manifestations, noise exposure produces adverse behavioral manifestations. Experimental exposure to noise diminished the quality of interpersonal contact,[15] increased aggressiveness,[27] and impaired willingness to help persons in need.[41] Loud noise, particularly intermittent noise, may alter productivity. The effect may be facilitative, rather than inhibitory; that has led to the speculation that noise may interact with other environmental factors and with personal factors to achieve a degree of arousal desirable for work.[29]

TEMPERATURE

There is little scientific information on the connection between thermal conditions and productivity.[42] In laboratory experiments at 65-85°F (18-29°C), productivity often reached a peak at nonpreferred temperatures.[73] In an apparel factory, productivity (i.e.,

TABLE VIII-1

Contributions of Various Characteristics of Sound to Annoyance[a]

Sound Characteristic	Relative Annoyance (Scale Value)
Steady high-pitched sounds	3.94
Steady low-pitched sounds	3.66
Intermittent high-pitched sounds	4.54
Intermittent low-pitched sounds	3.81
Loudness of sounds	6.46
Suddenness of sounds	5.80
Feeling that a sound cannot be turned off	5.55
Feeling that a sound is unnecessary	5.38
Feeling that a sound comes from a source of little benefit	4.81
Sounds that clash (unharmonious)	4.43
Sounds that catch one's attention at a distance and then get louder and louder	5.23
Sound is man-made	5.65

[a]Data from Dunn.[23]

TABLE VIII-2

Suggested Maximal Tolerable Intensities in Various Indoor Locations for More or Less Continuous Noise between 7 a.m. and 10 p.m.[a]

Type of Space	Intensity, dB(A)
Broadcast studio	28
Concert hall	28
Legitimate theater (500 seats, no amplification)	33
Music room	35
Schoolroom (no amplification)	35
Apartments, hotel	38
Assembly hall	38
Home	40
Motion-picture theater	40
Hospital	40
Church	40
Courtroom	40
Library	40
Office	
Executive	35
Secretarial (mostly typing)	50
Drafting	45
Meeting room (sound amplification)	45
Retail store	47
Restaurant	55

[a]Data from Kryter.[34]

piecework) varied little, if at all, with thermal conditions (note, however, that workers were paid by the piece).[48] When given the opportunity to express an opinion, people will be consistent in their preference regarding environmental conditions. The "comfort vote" has literal meaning in research on thermal acceptability. It refers to a subjective rating on a seven-point scale of comfort, on which the midpoint signifies thermal neutrality. A large body of research has made it possible to determine, by means of "comfort equations," the combinations of several factors--notably air temperature, humidity, radiant temperature, air velocity, degree of activity, and type of clothing--that will minimize discomfort. The range of acceptable combinations of environmental conditions is known as the "comfort zone."

Figure VIII-1 depicts summer and winter comfort zones adopted in 1981 by The American Society of Heating, Refrigerating, and Air-Conditioning Engineers (ASHRAE).[2,10] The comfort zones show the relationship of comfort to temperature and humidity during "light" activity. At least 80% of occupants should feel comfortable--no more than slightly warm or slightly cool--in these zones. The comfort zone is different between summer and winter, because people wear more clothing during the winter. The thermal resistance of a clothing ensemble can be measured precisely in "clo" units. Table VIII-3 offers an example of how a change in clothing will be reflected quantitatively in clo values and optimal operative temperatures. Operative temperature is determined on the basis of air temperature and average radiant temperature. In an interior zone with only a slight radiant component, the operative temperature approximately equals dry-bulb temperature.

Insulation from clothing and degree of activity interact in determining acceptable temperature. The ASHRAE standard therefore offers an equation to convert acceptable operative temperature (°C) for sedentary occupancy (1.2 mets) to that for more active occupancy (e.g., housework at 2 mets, garage work at 3 mets): $t_{o(active)} = t_{o(sedentary)} - 3(1 + clo)(met - 1.1)$, where t_o represents operative temperature. In addition to steady-state features of the thermal environment, the standard considers temporal nonuniformities (e.g., temperature cycling) and spatial nonuniformities (e.g., vertical temperature differences). Some limited nonuniformities, such as monotonic temperature drifts, may prove both economical and acceptable.[11]

Conditions for thermal comfort seem to vary little, if at all, with such factors as geographic location, sex, body build, ethnic background, and even age.[24] The effects of aging seem to merit some special consideration. Basal metabolic rate decreases progressively with age, but, according to Fanger,[24] evaporative heat loss does, also. The two changes seem to offset each other, although the elderly spend much more time than the young in sedentary activities. Furthermore, with the lower temperatures now common indoors during winter, the elderly seem to have a narrower temperature range over which they can increase their thermal resistance.[53] Because of sensory adaptation, a sedentary old person may fail to notice the symptoms of impending hypothermia until it becomes severe. Adequate

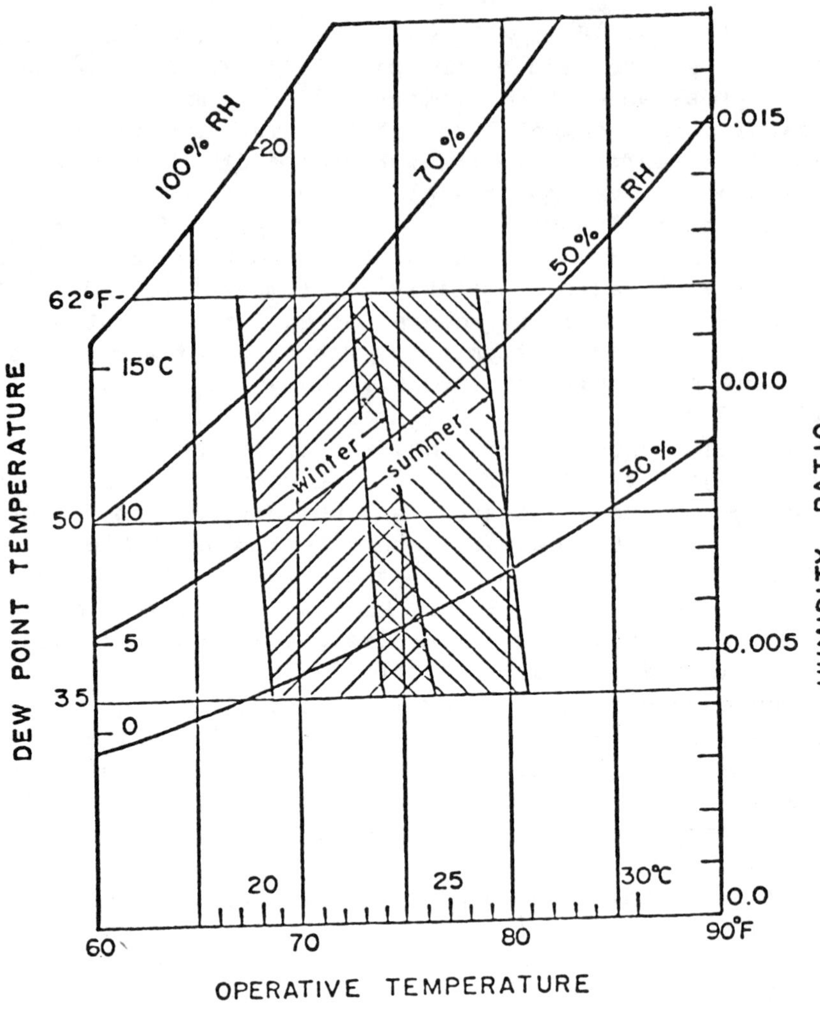

FIGURE VIII-1 Acceptable ranges of operative temperature and humidity for persons wearing typical summer clothing and typical winter clothing. These "comfort zones" assume that occupants are engaged in only light activity. Reprinted with permission from American Society of Heating, Refrigerating, and Air-Conditioning Engineers.[2]

TABLE VIII-3

Temperatures for Thermal Acceptability (Comfort) of Sedentary or Slightly Active Persons ($\leq$1.2 mets) at 50% Relative Humidity[a]

Season	Typical Clothing	clo[d]	Optimal Operative Temperature[b] °C	Optimal Operative Temperature[b] °F	Operative Temperature for 80% Thermal Acceptability[c] °C	Operative Temperature for 80% Thermal Acceptability[c] °F
Winter	Heavy slacks, long-sleeved shirt, sweater	0.9	21.7	71	20-23.6	68-74.5
Summer	Light slacks, short-sleeved shirt	0.5	24.4	76	22.8-26.1	73-79
	Minimal	0.05	27.2	81	26-29	79-84

[a]Data from American Society of Heating, Refrigerating, and Air-Conditioning Engineers.[2] Other than clothing, there are no seasonal or sex variations for the temperatures listed. For infants, some elderly persons, and physically disabled persons, the lower temperature limits should be avoided. met = measure of energy production per unit of surface area of a seated person at rest. 1 met = 58.2 W/m^2 or 18.4 Btu/h•ft^2. Surface area of average man = 1.8 m^2 (19 ft^2).

[b]Indoor operative temperature is a weighted mean of average air and average radiant temperatures.

[c]Slow air movement ($\leq$0.15 m/s = 30 ft/min).

[d]clo = measure of thermal resistance of a clothing ensemble: 1 clo = 0.155 m^2•K/W or 0.88 ft^2•h•°F/Btu.

clothing is the best precaution against cold distress. In the United States, people were not energy-conscous until rather recently, and both young and elderly seem to need more education regarding the way to match clothing to the thermal load of the environment.

INTERRELATIONSHIPS OF ENVIRONMENTAL FACTORS

Other prominent factors in the indoor environment include lighting, furnishings, and the size and configuration of the space. Control of type and quality of illumination is often an aspect of design. Professional and aesthetic preferences can govern the choice of intensity, placement of sources, hue, and degree of contrast of lighting. These matters often receive much attention in the workplace. The question of whether light, temperature, and sound are optimal should be viewed in terms of such needs as productivity and accident prevention. In the home, the considerations are different from those of the workplace--questions of efficiency and productivity place few constraints on the physical environment at home, and few persons obtain professional advice regarding ways to maximize comfort and minimize hazards in the home.

Proshansky and colleagues[50] stated that "behavior in relation to a physical setting is dynamically organized: a change in any component of the setting has varying degress of effects on all other components in that setting, thereby changing the characteristic behavior pattern of the setting as a whole." That conclusion may seem obvious; however, the need to consider it arises in experiments where a source of discomfort is expected to decrease productivity, but increases it instead. Such studies may often fail to give precise answers regarding the importance of one or another environmental factor, but they can help to heighten our awareness. Awareness is a powerful tool in recognizing and dealing with the complex interplay of safety, health, comfort, and productivity indoors.

SUMMARY

A person's perception of discomfort can provide a useful indicator of possible adverse effects of environmental agents. Discomfort gives immediate incentive to avoid or to correct environmental deficiencies. There is little information regarding whether long-term exposure to sources of discomfort will eventually cause adverse health effects. This question has no global answer. The discomfort caused by a thermally variable environment may lead to physiologically useful acclimatization and to behavioral strategies that diminish the impact of environmental challenges. In contrast, exposure to moderately intense noise--e.g., 80 dB(A)--leads to no such physiologic accommodation in the auditory system and may eventually cause hearing loss in susceptible persons. As an added complication, low-intensity sound may cause considerable discomfort and even intense autonomic reactions in persons sensitized to the "meaning" of the sound. The long-term deleterious effects of continuous activation of the autonomic nervous system are not known, and efforts to measure such symptoms as nausea, headaches, and dizziness and learn their clinical significance should be encouraged.

Some airborne chemical contaminants cause discomfort via stimulation of the olfactory sense or the common chemical sense. This probably serves a useful purpose, inasmuch as people will often avoid bad-smelling atmospheres, regardless of any known toxic properties. The discomfort can also lead to closer investigation of the source of malodors.

A generic relationship between discomfort and productivity has eluded specification. It seems possible at best to state only that the point of maximal productivity may not coincide with the point of minimal discomfort, but will hardly fall at the point of maximal discomfort. Comfort is derived from the harmonious interactions of many things, including physical factors, context, motivation, social factors, attitudes, and skill at the task at hand. Therefore, it is related to all aspects of a person's behavior and may prove just as difficult to predict. Nevertheless, appropriate attention to the maintenance of proper lighting, air, and thermal conditions increases the diversity of activities and the numbers of people that can be accommodated in comfort.

RECOMMENDATIONS

- Subtle forms of discomfort often arise from the use of manufactured products, building materials, and consumer products. Therefore, identification of the products leading to irritation and of its duration can be used by manufacturers in the design of safer products.
- The relationships of subjective complaints of discomfort to associated symptoms--such as headaches, nausea, and other health effects--should be studied in persons of different ages and in different categories of other kinds, such as socioeconomic status. Objective data from these studies could be useful in targeting the design characteristics of buildings.
- For each type of discomfort (e.g., noise-induced discomfort and odor-induced discomfort), there is a need for research on how to relate stimulation to discomfort.

DECREASED PRODUCTIVITY

Direct relationships between indoor pollution and decreased productivity can be evaluated only if one carefully defines "productivity" and how it is to be assessed. Productivity was originally conceived of as simply the quantity of output; Sumerian documents dating to 5000 B.C. have been identified as organizational records of productivity. Although employee counseling appears in Egyptian records of around 4000 B.C., productivity at the beginning of the twentieth century was still viewed as essentially the output of robot-like workmen; the production line of the 1920s was simply a way of organizing the work to increase output. As the numbers of workers rose and hourly wages and investment in equipment increased, productivity began to be measured in economic terms, such as cost per unit of production. More attention was later given to devising work methods for decreasing costs per unit of production. This focus on increasing productivity while reducing costs led, almost inevitably, to a degradation of product quality. Therefore, quality became a consideration in measuring productivity, and again the definition of "productivity" had to be revised.

DEFINITION OF "PRODUCTIVITY"

There is national recognition that our resources are not infinite, and this recognition has led to reexamination of the earlier definitions of "productivity." The demands for increasing productivity have had a serious impact on the physical and mental well-being of both the workforce and the consumers of its production, to say nothing of the impact on "quality of life" in our society. Thus, the measure of productivity was expanded to consider the "efficiency of the output." Productivity is currently addressed in terms of cost effectiveness--"Does it work?"--with considerations of timeliness, effects obtained or results achieved, and such humanistic elements as manner of performance and methods of achieving the results. The periods of redefinition of "productivity," at the national level, can be dated by Presidencies. The simple concept that productivity equaled output was displaced, during the Franklin D. Roosevelt era, by consideration of cost per unit of production. This idea was displaced, during and after the Kennedy administration, by consideration of the effectiveness of policies to improve productivity. Over roughly the last 50 yr, the definition of "productivity" has used a complex of interacting entities and characteristics, including quantity and quality of product, monetary cost, timeliness, and human costs. Human costs include those engendered by the manner of performance, the method of achieving the results, and the actual benefits, as compared with the social costs. As long as people are involved in the definition, productivity can be adversely affected by pollution (defined as the presence of any unwanted or unnecessary element in the environment).

PRODUCTIVITY IN INDUSTRIAL ENVIRONMENTS

Pollution will affect productivity at two distinct levels: its physical effects on the means of production, or on the product itself, which are directly related to the quality of the product; and the effects on the health of the worker. Air contaminants can be categorized into particulate and gaseous, organic and inorganic, visible and invisible, submicroscopic and microscopic, or toxic and harmless.

With even "clean," country air containing particles bigger than 0.3 μm at over $10^6/ft^3$ and atmospheric dust loads of 20-200 $tons/mi^2$ per month in cities, the physical effects of pollution can be substantial. Solid particulate contaminants accumulate on surfaces, contaminate food products, and discolor walls, ceilings, floors, and furnishings; nonparticulate contaminants (vapors and gases) also affect food products, discolor surfaces and furnishings, and cause deterioration of fabrics and finishes. When one considers that unnecessary cleaning, repairs, and painting and untimely replacement are nonproductive, one can see that the physical effects of pollution are a drain on productivity. The loss of light to dirt on windows, the role of dirty car windows in causing accidents, the inefficiencies of dirty cooling coils or heating elements, the erosion of building structures, and the diversion of resources (both money and time) because of these problems all contribute to a lowering of productivity. Indeed, tobacco smoke and cooking and body odors form the primary requirement for ventilation in nonindustrial occupied spaces. About 17% of the national energy use is devoted to moving, heating, or cooling air for ventilation; such pollutants can be considered a major factor in limiting the energy available for productivity.

Even though many contaminant effects are related to specific products or processes, their effects on the health of workers are not, and they can be dealt with generically. As opposed to the physical classification of contaminants mentioned above, a spectrum of health effects of pollutants can be suggested: lethal-disabling-sickening-irritating-annoying-distracting-discomforting.

Although the lethal end of this spectrum has captured more attention (e.g., consider lead poisoning and asbestos exposure), in the present social climate, where productivity depends more on what people will produce than on what they can produce, the greatest effects on productivity will probably be incurred toward the discomfort end of the spectrum. Obviously, premature death or chronic disability removes the individual producer. But even a slight increase in illness or malaise can reduce productivity by absenteeism or by "taking it easy" for a few days. Indeed, even momentary distraction, discomfort, annoyance, or physical irritation will reduce the quantity or quality of production.

In a 1979 NIOSH pilot study in industrial plants in Oregon and Washington, workers were examined for occupational diseases and other conditions. Hearing loss (noise pollution?) was the most frequent (28%), and skin conditions were next (18%), followed by lower respiratory conditions (14%), toxic and low-grade toxic effects (14%), upper respiratory conditions (11%), and eye conditions (9%).[61] No such data exist for nonindustrial environments, but, if one considers today's nonindustrial and social environments (discos, lounges, rock concerts, radios, and record-players and their sound intensities), it seems probable that decrements in hearing due to "noise pollution" represent one of the leading correlates of productivity losses. Although the incidences of the other health effects mentioned above may differ in a nonindustrial setting, they are all likely to occur.

The impact on productivity from pollutants that are simply annoying, distracting, or discomforting (temperature, odor, and soiling) has been largely ignored until recently. However, insight into their anticipated effects can be gained by examination of the tables of "relaxation allowances" established to develop production standards for jobs or of "environmental differential-pay plans" developed to provide extra compensation for putting up with a variety of undesirable conditions. Allowances of formal rest breaks and pay differentials, based primarily on physical strain, began to be common around 1950. These allowances have expanded; they were based increasingly on psychologic factors in the 1960s and on environmental factors in the 1970s. Some rest allowance seems appropriate to lessen the abuse of the worker, but growing public concern over environmental factors may have led to increases in rest allowances and pay differentials. These provide documentation of the costs of the adverse effects of environmental pollution better than anything else available.

The relaxation allowances consider four elements. A standard 10% time break, 18 min every 3 h, is considered adequate for personal needs, such as a trip to the rest room or a coffee break, although in practice it tends to be more generous in most industrial settings. A second set of relaxation allowances are based on such physiologic factors as energy demands, postures, body motions, and restrictive protective clothing; a third is based on psychologic factors associated with timing, monotony, and the required concentration (diligence). A fourth deals with environmental factors, such as thermal quality, humidity, other air pollution, noise, dirt, and vibration. Williams[65] suggested that relaxation allowances (i.e., percent of productive time lost) be determined as a function of environmental conditions, as follows:

1. Thermal and atmospheric conditions:
Consider whether, despite or in the absence of protective clothing or equipment, and extractors or air-conditioning equipment, the air conditions in terms of temperature and purity are such that additional demands are made when performing the work; air conditions are defined as:

A. Adequate ventilation and circulation with normal climatic humidity.
B. Inadequate ventilation and circulation with non-standard climatic conditions causing some discomfort.
C. Very poor ventilation and circulation. Fumes, dust, steam, causing irritation to eyes, skin, nose, throat.

Temperature		Air Condition		
°F	°C	A	B	C
		"relaxation allowance"		
55° to 75°	13° to 24°	0%	0 to 5%	5 to 10%
76° to 100°	24° to 38°	1 to 10%	5 to 15%	10 to 20%
55° to 30°	13° to -1°	1 to 10%	5 to 15%	10 to 20%
Below 30°	-1°C	10 to 20%	20 to 25%	20 to 30%

2. Physical conditions, including noise.
Consider the general physical conditions of the environment in relation to the work being performed and the degree of discomfort caused by dirt, oil, grease or water and other liquids, ice, chemicals, etc. Consider also whether noise is irritating by irregularity, or uncomfortable pitch or volume.

A.	Clean, bright, dry surroundings. Normal "machine" and human noise.	0%
B.	Dirty, wet, greasy and contaminated surroundings	0 to 3%
C.	Uncomfortable noise	0 to 4%
D.	Combination of several factors	0 to 8%

In an effort to check the allowances, some 16 establishments and 145 different jobs, including about 6% female workers, were examined. In general, the findings supported the relaxation-allowance approach. The allowances for these environmental factors are obviously only suggestions. Therefore, it is doubtful whether additional research would provide any reliable refinement of the productivity losses due to environmental factors, because such psychologic factors as motivation, leadership, expectation, and need (and their interactions) are as important as the environmental factors in determining productivity.

In the environmental pay-differential approach, Federal Personnel Manual letter 532-17 established specific pay differentials for exposure after November, 1970, to "various degrees of hazards, physical hardships and working conditions of an unusual nature," as follows:

<u>Dirty Work</u>: Performing work which subjects the employee to soil of body or clothing: a) beyond that normally expected in performing the duties of the classification; and b) where not adequately alleviated by mechanical equipment or protective devices . . .; or c) when their use results in an unusual degree of discomfort. 4%

<u>Cold or Hot Work</u>: At or below 32°F or above 110°F 4%
Working with or near:

A. Poisons (Toxic Chemicals)

High hazard	8%
Low hazard	4%

B. Micro-Organisms - High hazard

High hazard	8%
Low hazard	4%

Although these pay differentials are not directly relatable to productivity decrements, the increases in direct costs of protection are explicit, and productivity decrements are therefore also explicit; hence productivity losses can be inferred. However, the basis for such pay differentials is at least as much political as factual; additional research along these lines is not likely to be very informative.

Determining productivity losses caused by pollution is extremely complex. Even with careful definitions and measurement, it appears unlikely that any simple cause-effect correlations can be established that would not be destroyed by alterations in motivation, leadership, expectation, and need.

PRODUCTIVITY IN NONINDUSTRIAL ENVIRONMENTS

These very limited considerations of interaction between air quality and productivity can be defined in terms of units of production, percentage of rejects, or costs per unit of salable product. Models have been developed to show the influence of heat exposure on productivity,[8] but, as with the comfort models, there has not been much work on validating them. Thus, the models provide only a theoretical prediction of reductions in work capacity. Few studies have been carried out on the causes of productivity decreases in industry, and even fewer in institutional settings. The American Society of Heating, Refrigerating and Air-conditioning Engineers (ASHRAE) has supported studies of the potential benefits of air-conditioning in schools.[49] Air-conditioned classrooms and libraries were heavily preferred (by about 95%), but it could be inferred that air-conditioned schools attract better teachers and that better teachers get better results. Similarly ambiguous results have attended most of the numerous ASHRAE studies on air-conditioning criteria.[1] The difficulties rest in part with the variability of actual environmental conditions, as distinct from those supposedly maintained by the control systems, and in part with the difficulty (suggested by Wyon[74]) of defining the criteria for such environmental qualities as "comfort" and air quality, as distinct from the criteria for performance.

There has been growing recognition of the difficulties in demonstrating linkages between environmental quality and productivity, and the pace of research in this subject appears to have slackened in the last few years. Concerns about productivity have been focused more and more on workplace layout and worker motivation; that is probably appropriate, because they have direct and tangible impacts on productivity. The most tangible effects of air quality on productivity and quality of life are the adverse effects on health and longevity; even so, experimental confounding easily blurs any direct linkages. E.g., when the U.S. Army introduced its "MUST" field hospital, which used air-conditioning, in Vietnam, patient survival and hospitalization time were clearly improved, but argument arose as to whether the improvements were caused by air-conditioning or by staffing.

SOILING, CORROSION, MAINTENANCE, AND HOUSEKEEPING

Indoor air pollution is a source of soiling and contributes to the deterioration and corrosion of equipment, furnishings, and appliances. Changes in ventilation, such as a decrease in the amount of outside air used in ventilation to save energy or an increase to accomplish the same end by making greater use of natural ventilation during mild weather, can affect the rates of soiling and deterioration. Even if indoor pollutants do not adversely affect occupants or the rate of soiling, deterioration, or corrosion, they increase requirements for housekeeping and associated environmental control systems to maintain the value of materials and property.

PARTICLE DEPOSITION

Deposition of dust particles on walls and other surfaces is the most common cause of soiling. The number and surface and mass relationships of particles are important in soiling. A 5-μm-diameter spherical particle has 1,000 times the mass of a 0.5-μm particle of the same material, but only 100 times the surface area. Thus, it is the submicrometer particles that have greater soiling potential, although the relationship between particle size, optical characteristics, and soiling is complex. However, larger particles contribute more to abrasion, and lint can foul equipment. Mechanical heating, cooling, and ventilating systems commonly include air filters to remove lint and larger particles. In some manufacturing operations, such as production of microelectronic circuits,[7] it is essential to have very-high-efficiency filtration for removal of submicrometer particles. The average home or place of business does not approach these high standards of air cleanliness, although an increasing number of residences are using electronic air-cleaners capable of removing submicrometer particles.

Larger particles settle faster than smaller ones. Gravity sedimentation is an important mode of deposition, but it may be comparatively unimportant in deposition of very small airborne particles. Figure VIII-2 is a plot of Stokes diameter of a particle as a function of time required to settle 1 m in air. Settling times are plotted for particles with densities of 1 and 2 g/cm^3. Water and oil droplets have densities of about 1 g/cm^3. Figure VIII-2 is a somewhat idealized representation, but it permits a visual estimate of the relative sedimentation rates of large and small particles. Particles larger than 5 μm in Stokes diameter settle in a

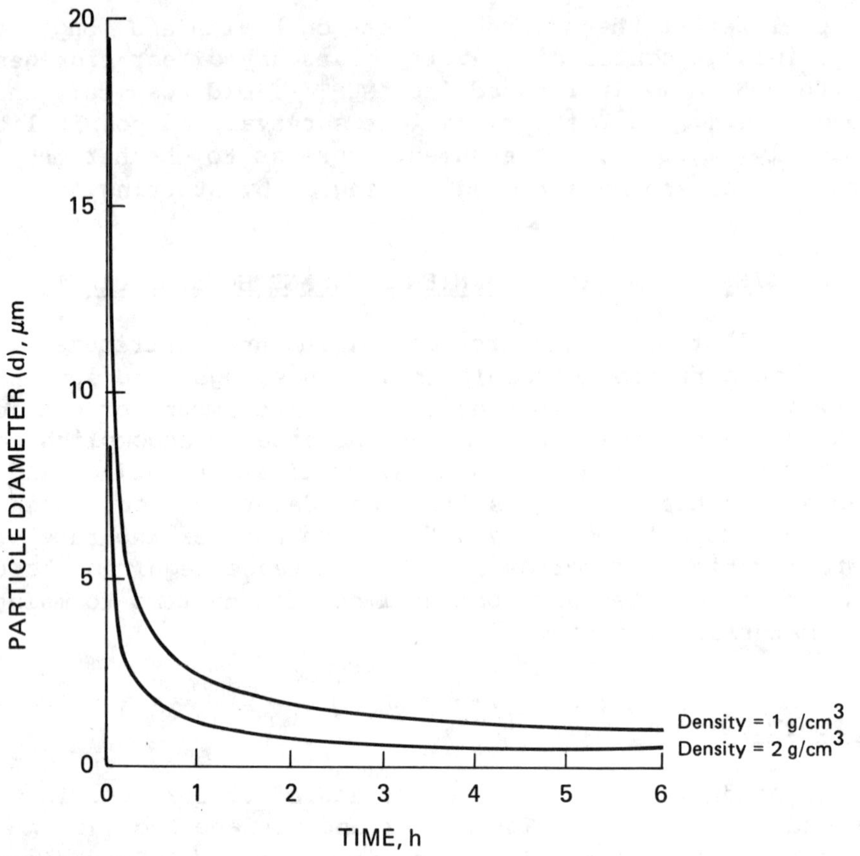

FIGURE VIII-2 Particle diameter (d) and density (ρ_p) as a function of time required to settle 1 m in air, according to Stokes's law.

comparatively short time; particles smaller than 1 μm may remain suspended for hours, unless they become attached to other particles, walls, or surfaces. Davies has reviewed deposition from moving aerosols.[22]

Electrostatic and thermal precipitation are two important mechanisms by which particles are deposited. Penney and Ziesse[47] have measured the mobilities of airborne dust particles under the influence of thermal and electrostatic gradients and have estimated an average effective thermal mobility of 2.4×10^{-8} $m^2/°C \cdot s$ and an effective electric mobility of about 11×10^{-8} $m^2/V \cdot s$. These values can vary widely for different dust particles, but they are useful approximations for the design of dust-collecting equipment. Penney and Ziesse also noted that an electrostatic precipitator that does not capture all particles causes more soiling than an air-cleaner of the same efficiency that does not charge particles. Apparently, the particles become electrically charged, and that causes them to attach to surfaces more readily. Thus, it is important that the precipitator be designed for maximal capture.

The force of attraction between two molecules (London-Van der Waals force) varies as the inverse of the 7th power of the distance between them[39] and plays a role in interparticle adhesion or adhesion to surfaces. The electrostatic attraction of particles to surfaces is very strong at distances of a few angstroms, but diminishes rapidly with increasing distance. From the standpoint of soiling, London-Van der Waals forces are probably important in particle retention after a particle contacts a surface. Corn[19] calculated the electrostatic attraction between a charged particle 1 μm in diameter and an adhering particle or surface in which it induces an equal and opposite charge. Assuming a particle charge of 15 electrostatic units (e.s.u.)--i.e., 15×10^{-9} coulombs--and a separation of 1 nm (10 Å), he estimated a force of 5.2×10^{-3} dynes, which is about 10^7 times the gravitational force, assuming a density of 1 g/cm^3. However, this is only one one-thousandth of the estimated Van der Waals force.

Capillary attraction is a mechanism of particle retention due to adsorbed liquid films. Capillary attraction is probably more important in fouling (where air comes into contact with damp coils or pipes) or in particle filtration (where adhesive liquids are applied to the filter) than in most everyday soiling of walls and surfaces. When the radius of the liquid film at the point of contact is small, compared with the radius of the particle, the force of attraction between a sphere and a plane surface, with a film of liquid interposed, may be expressed by the relationship $F = 4\Pi\gamma r$, where F is the capillary force, γ is the surface tension of the liquid, and r is the particle radius.[25] Corn[19] has suggested that that equation is approached only at relative humidities near 100%, where water is in the liquid phase. At lower vapor pressures, the force is less.

The surface-to-volume relationship of particles increases dramatically as particles become very small, and this relationship is important in soiling. Surface forces have a much greater role in determining soiling properties of small particles than of larger particles. Very fine particles cling to a glass slide when the slide is inverted. Walker and Fish[64] demonstrated that removing small particles by either liquids, airstreams, brushing, or gravity was more difficult than removing large particles.

Human activities can cause agitation that resuspends deposited particles. Primarily, it is the larger particles that are more readily redispersed by this means. Hunt,[31] in experiments using a light-scattering-particle counter, showed that vacuum-cleaning a rug or operating an electric fan caused a severalfold increase in the number of particles larger than 3 μm, but only a slight perturbation in the number of smaller particles. But other activities--such as smoking, heating, or cooking--produced primarily submicrometer particles. Also, aerosols in this size range are probably produced by condensation from the vapor phase, rather than by dispersing preexisting particles from surfaces or from a powder.

MOISTURE AND FUNGAL GROWTH

Fungal growth is another cause of soiling and deterioration that generally occurs in areas with high humidity and low ventilation. Microbial slimes in air-cooling and -humidifying units, plumbing fixtures, condensation trays, and drains cause serious and often costly mechanical problems. These and other airborne organisms can discolor paint, weaken fabrics, and degrade foodstuffs. Microorganisms can also lead to odors, such as the musty smell of a damp basement. Schaffer[55] has reviewed many of the effects of moisture in buildings, including the promotion of fungal growth. Moisture can be generated internally from combustion during heating and cooking, drying clothes, bathing, and even breathing, and it can come from the outside during periods of high humidity. Moisture generated indoors can result in high humidities when there is no dehumidification, when ventilation rates are low, or when a structure has tight vapor barriers in walls and partitions. Fungal growth in ducts or on walls and surfaces has been observed after the use of large amounts of outside air for ventilation during damp periods.

Water vapor is not ordinarily regarded as a pollutant. Not only is it essential to support the growth of microorganisms, but, if it is present in excessive amounts, it can cause more visible effects, such as peeling of paint and wallpaper. It also has an effect on comfort (as discussed earlier), and it can enhance the effect of other pollutants. Hermance et al.,[30] for example, have noted this in studying damage to telephone contacts by airborne nitrates.

GASEOUS POLLUTANTS

The important gaseous pollutants--such as ozone, sulfur dioxide, oxides of nitrogen, and carbon monoxide--affect the corrosion and deterioration of materials. Ozone can cause cracking of rubber and some other elastomers. The amount or rate of cracking of stretched rubber bands has been used as a method for determining low concentrations of ozone.[14 63] Not only does ozone occur in the outdoor air, but trace amounts can be produced indoors by arcing of electric motors in tools and appliances and by corona discharges of electrostatic air-cleaners. Sulfur dioxide and oxides of nitrogen may also contribute to corrosion and deterioration, but they are more often considered as potential health hazards. Carbon monoxide is comparatively inert and does not react on surfaces; although it is a hazard to health and safety, it does not normally cause soiling or deterioration.

EFFECTS OF TIGHT CONSTRUCTION

Reduction of infiltration resulting from tighter construction decreases the amounts of pollutants coming from outside, but can cause increases in the concentrations of those generated indoors, unless there is a change in ventilation rate. To achieve the full benefit of tight construction without increasing soiling, corrosion, and deterioration, provision must be made to abate or eliminate indoor-generated moisture and the indoor pollutants at their source. Particles and moisture are probably the most important agents that affect the rates of soiling, corrosion, and deterioration. Particle counts are usually lower indoors,[9] but not always. Cooking, cleaning, and other indoor activities intermittently distribute particles, as well as moisture. Sources of many other pollutants are discussed in Chapter IV.

As mentioned earlier, increased tightness of buildings can result in increased moisture indoors. Previously, moisture generated indoors has leaked out through the building structure, but, as these paths of elimination are reduced, it may be necessary to use dehumidifiers.

EFFECTS ON MAINTENANCE FOR CORROSION AND DETERIORATION

Andrews[5] estimated that the cost of corrosion in the United States exceeds $25 billion per year. This expense is reported to be due to additional fuel, maintenance, or replacement costs. Although the fraction of these costs caused by indoor pollution was not reported, it can be assumed that even a small percentage could represent a great financial impact over the lifetime of a building. Four types of corrosion, which must be controlled in building environmental control systems, are shown in Table VIII-4, with some methods of prevention.

If the quality of the indoor air is degraded, the increased concentration of contaminants can aggravate scaling of heat-exchanger surfaces.[5] For example, the air in a space with relatively high moisture content often is recirculated across a cooling coil for dehumidification. Increased carbon dioxide and sulfur dioxide of the indoor air may react with the condensed water and accelerate corrosion on the cooling coil.

TABLE VIII-4

Types of Corrosion and Methods of Environmental Control in Buildings[a]

Type of Corrosion	Result	Maintenance Action
Uniform	Direct chemical attack	Apply protective coatings
Pitting	Local deposits of particles on metal surfaces	Inspect and remove solid deposits Remove solids in suspension
Galvanic	Electrochemical reaction between dissimilar metals (less noble metal is corroded)	Apply such coatings as plastics, paints, and asphaltum (protect both metals with same material) Apply appropriate chemical inhibitors
Stress	Corrosion attacks stress-weakened metal	Replace

[a]Data from Andrews.[5]

Reports of increased maintenance of heat-exchangers or rotating equipment necessitated by degradation of indoor air quality were not found in the literature, but the appropriate conditions for increased corrosion have been reported.[5,19,25,31,47,55] For example, Hermance et al.[30] reported that telephone switching equipment required increased maintenance because of nitrates.

Inasmuch as nitrogen oxides and sulfur oxides can be present in indoor environments, either from indoor sources or from outdoors, the potential exists for corrosion of electric components in most indoor environments.

EFFECTS ON HOUSEKEEPING

Cleaning and care of materials and properties in institutional spaces represent approximately 15-20% of the total annual operating costs of these facilities (W. W. Whitman, personal communication). In turn, annual operating costs can be approximately 50-75% of the annualized initial investment of buildings.* Thus, any degradation of the indoor air quality that causes an increase in housekeeping can seriously affect the life-cycle cost of a building.

As buildings have become more energy-efficient, the moisture content has been generally reported to have increased, owing to decreased infiltration.[55] Additionally, the concentrations of smoke particles and other contaminants from smoking and other indoor activities have increased (see Chapter IV). Thus, the rates of soiling and deterioration of exposed surfaces may be accelerated, as a result of degradation of indoor air quality.

Windows are a primary site for accelerated soiling, especially during the heating season. Because resistance to heat transfer through windows is usually one-tenth to one-third that of adjacent walls, the inside surface temperatures of the windows will be much lower than those of the walls. If the inside surface temperatures of the windows are lower than the dewpoint temperature of the occupied space, condensation will occur at these surfaces. Particles and gaseous contaminants in equilibrium with the water vapor will be deposited on the window surfaces with the condensate. As the condensate leaves the windows by evaporation or draining, the other contaminants will be left on the surfaces as residue, thus increasing the required frequency of cleaning. Boyce[13] reported that, when windows are not thoroughly cleaned periodically, a cloudy film builds up that can be removed only with muriatic acid. To combat pollution in Los Angeles, Boyce stated, aluminum mullions and transoms on the CNA Park Plaza Building must be cleaned annually with mild steel wool and oil must then be applied to protect the metal. If outdoor pollutants are transported indoors, or if similar pollutants are generated indoors, the interior surfaces of windows might require similar treatment.

---

*The annualized initial investment is based on a present cost of $70/ft^2 amortized over 50 yr at an inflation rate of 9%. Current annual operating costs are approximately $3/ft^2.

Indoor lighting efficiency is also affected by indoor air quality. Williams[66] reported that dirt accumulations on lamps and fixtures can reduce light output by 10-50% over the rated "end-of-life" of the lamps. Thus, as dirt and film accumulate on fixtures and lamps, cleaning and relamping frequencies must be increased to maintain proper illumination.

Another major category of housekeeping expense is related to the care of floors and carpeting. Darling[21] reported that, on a national average, 40-60% of the working hours of cleaning crews is required for floors and carpeting and that carpeting soils more quickly in industrial centers than in suburban areas, where air pollution is less severe.

Furniture, paintings, sculptures, and musical instruments are also affected by indoor air quality. The special requirements for environmental control in museums, art galleries, and auditoriums are indicative of the care that is required to protect these properties.[3]

METHOD OF TREATMENT

There are ways to reduce the indoor pollution that causes soiling and deterioration. For example, air filtration reduces the amount of airborne dust. Most central heating and air-conditioning systems contain air filters. Although these are usually not of high efficiency, they do reduce dust. An electronic air-cleaner designed for a specific system can remove still more particles.

The visible effects of undesirable thermal precipitation of dust on walls near grilles and radiators may be reduced by shields that direct air away from walls.

Dehumidifiers remove excessive moisture. However, during the heating season, humidity is often low indoors, and it may be necessary to add moisture to the air, to prevent stress cracking in furniture and other wood products due to excessive drying. (The relationship between human comfort and humidity and temperature is discussed earlier in this chapter and in Chapter IV.) Tobacco-smoking places an added burden on air-cleaning and ventilation systems. In public buildings, smoking is often prohibited or restricted to specified areas.

Particles and other airborne materials generated in cooking may be largely removed by exhaust systems near the point of generation.

Activated carbon and other adsorbent air-cleaners are sometimes used in buildings in high-pollution areas to remove gaseous pollutants. However, these are not in general use, and they present some special problems. For example, it is harder to determine when an adsorbent filter needs to be changed than a particulate filter (see also Chapter IX).

RECOMMENDATIONS

Some of the commonly recognized agents that produce soiling and deterioration have been discussed in the foregoing paragraphs, but additional questions need investigation. With regard to removal of indoor particles, where is the point of diminishing returns in improving the efficiency of particulate filters? Likewise, where is the point of diminishing returns reached in increasing the rate at which air is removed from an occupied space and filtered? Dust composition may also be important. There have been a few analyses of indoor dust,[28 43] but much less work that has tried to relate soiling, corrosion, or other deleterious effects to dust composition and particle size. Thus, the effectiveness of dust removal technology and the specific nature of the dust, as they relate to soiling and deterioration, need further investigation.

Information on the role of gaseous pollutants in soiling or corrosion is lacking.

REFERENCES

1. American Society of Heating, Refrigerating and Air-Conditioning Engineers. Symposium Bulletin. Air Conditioning Criteria for Man's Living Environment, Louisville, Kentucky, June 24-28, 1973. New York: American Society of Heating, Refrigerating and Air-Conditioning Engineers, Inc., 1973. 33 pp.
2. American Society of Heating, Refrigerating and Air-Conditioning Engineers. ASHRAE Draft Standard 55-74R. Thermal Environmental Conditions for Human Occupancy. New York: American Society of Heating, Refrigerating and Air-Conditioning Engineers, Inc., April 1980.
3. American Society of Heating, Refrigerating and Air-Conditioning Engineers. Commercial and public buildings, pp. 3.1-3.16. In ASHRAE Handbook and Product Directory. 1978 Applications. New York: American Society of Heating, Refrigerating, and Air Conditioning Engineers, Inc., 1978.
4. Andersen, I. Formaldehyde in the indoor environment--Health implications and the setting of standards, pp. 65-77, and discussion, pp. 77-87. In P. O. Fanger and O. Vølbjorn, Eds. Indoor Climate. Effects on Human Comfort, Performance, and Health in Residential, Commercial, and Light-Industry Buildings. Proceedings of the First International Indoor Climate Symposium, Copenhagen, August 30-September 1, 1978. Copenhagen: Danish Building Research Institute, 1979.
5. Andrews, F. T. Building Mechanical Systems, pp. 117-124. New York: McGraw-Hill Book Company, 1977.
6. Arguelles, A. E., D. Ibeas, J. P. Ottone, and M. Chekherdemian. Pituitary-adrenal stimulation by sound of different frequencies. J. Clin. Endocrinol. Metab. 22:846-852, 1962.
7. Austin, P. R., and S. W. Timmerman. Design and Operation of Clean Rooms 1965, pp. 96-135. Birmingham, Mich.: Business News Publishing Company, 1965.
8. Axelsen, O. Influence of heat exposure on productivity. Work Environ. Health 11:94-99, 1974.

9. Benson, F. B., J. J. Henderson, and D. E. Caldwell. Indoor-Outdoor Pollutant Relationships: A Literature Review. U.S. Environmental Protection Agency (National Environmental Research Center) Publication No. AP-112. Washington, D.C.: U.S. Government Printing Office, 1972. 73 pp.
10. Berglund, L. G. New horizons for 55-74: Implications for energy conservation and comfort. ASHRAE Trans. 86 (Pt. 1):507-515, 1980.
11. Berglund, L. G., and R. R. Gonzalez. Application of acceptable temperature drifts to built environments as a mode of energy conservation. ASHRAE Trans. 84 (Pt. 1):110-121, 1978.
12. Binder, R. E., C. A. Mitchell, H. R. Hosein, and A. Bouhuys. Importance of the indoor environment in air pollution exposure. Arch. Environ. Health 31:277-279, 1976.
13. Boyce, S. Reflections on a clean glass building, pp. 36-37. In Maintenance Guide for Commercial Buildings. Cedar Rapids: Stamats Publishing Co., 1975.
14. Bradley, E. C., and A. J. Haagen-Smit. The application of rubber in the quantitative determination of ozone. Rubber Chem. Technol. 24:750-755, 1951.
15. Broadbent, D. E. Noise in relation to annoyance, performance, and mental health. J. Acoustical Soc. America 68:15-17, 1980.
16. Cain, W. S. Contribution of the trigeminal nerve to perceived odor magnitude. Ann. N.Y. Acad. Sci. 237:28-34, 1974.
17. Cain, W. S. Lability of odor pleasantness, pp. 303-315. In J. H. A. Kroeze, Ed. Preference Behaviour and Chemoreception. London: Information Retrieval Ltd., 1979.
18. Cain, W. S., L. G. Berglund, R. A. Duffee, and A. Turk. Ventilation and odor control: Prospects for energy efficiency. Lawrence Berkeley Laboratory Report LBL-9578. Berkeley, Cal.: Lawrence Berkeley Laboratory, Energy and Environment Division, 1979. 61 pp.
19. Corn, M. Adhesion of particles, pp. 359-392. In C. N. Davies, Ed. Aerosol Science. New York: Academic Press, Inc., 1966.
20. Daines, R. H., D. W. Smith, A. Feliciano, and J. R. Trout. Air levels of lead inside and outside of homes. Ind. Med. 41(10):26-28, 1972.
21. Darling, W. E. A lot more of what you're looking for on carpet care, pp. 22-25. In Maintenance Guide for Commercial Buildings. Cedar Rapids, Iowa: Stamats Publishing Company, 1975.
22. Davies, C. N. Deposition from moving aerosols, pp. 393-446. In C. N. Davies, Ed. Aerosol Science. New York: Academic Press, Inc., 1966.
23. Dunn, B. E. The noise environment of man, pp. 193-257. In H. W. Jones, Ed. Noise in the Human Environment. Vol. 2. Edmonton, Alberta: Environment Council of Alberta, 1979.
24. Fanger, P. O. Thermal Comfort. Analysis and Applications in Environmental Engineering. Copenhagen: Danish Technical Press, 1972. 244 pp.
25. Fuchs, N. A. The Mechanics of Aerosols, p. 362. New York: Pergamon Press, 1964.
26. Geber, W. F., T. A. Anderson, and B. Van Dyne. Physiologic responses of the albino rat to chronic noise stress. Arch. Environ. Health 12:751-754, 1966.

27. Geen, R. G., and E. C. O'Neal. Activation of cue-elicited aggression by general arousal. J. Personality Soc. Psychol. 11:289-292, 1969.
28. Gieseke, J. A., E. R. Blosser, and R. B. Reif. Collection and characterization of airborne particulate matter in buildings. ASHRAE Trans. 84(Pt. 1):572-589, 1978.
29. Glass, D. C., and J. E. Singer. Urban Stress. Experiments on Noise and Social Stressors. New York: Academic Press, Inc., 1972. 182 pp.
30. Hermance, H. W., C. A. Russell, E. J. Bauer, T. F. Egan, and H. V. Wadlow. Relation of airborne nitrate to telephone equipment damage. Environ. Sci. Technol. 5:781-785, 1971.
31. Hunt, C. M. Simple Observations of Some Common Indoor Activities as Producers of Airborne Particulates. Paper presented at ASHRAE Symposium on Cleaner Indoor Air--Progress and Problems Cl-72-1, Cincinnati, Ohio, October 19-22, 1972.
32. Jovanović, U. J. Normal Sleep in Man. An Experimental Contribution to Our Knowledge of the Phenomenology of Sleep. Stuttgart: Hippokrates Verlag Gmblt., 1971. 327 pp.
33. Kasl, S. V. The effects of the residential environment on health and behavior: A review, pp. 65-127. In L. E. Hinkle, Jr., and W. C. Loring, Eds. The Effect of the Man-Made Environment on Health and Behavior. DHEW Publication No. (CDC) 77-8318. U.S. Department of Health, Education, and Welfare, Center for Disease Control. Washington, D.C.: U.S. Government Printing Office, 1977.
34. Kryter, K. D. The Effects of Noise on Man. New York: Academic Press, Inc., 1970. 633 pp.
35. Lebowitz, M. D. A critical examination of factorial ecology and social area analysis for epidemiological research. Ariz. Acad. Sci. 12(2):86-90, 1977.
36. Lebowitz, [M.] D. Social environment and health. Public Health Rev. 4:327-351, 1975.
37. Lebowitz, M. D. The relationship of socio-environmental factors to the prevalence of obstructive lung diseases and other chronic conditions. J. Chron. Dis. 30:599-611, 1977.
38. Lehmann, G., and J. Tamm. Über Veränderungen der Kreislaufdynamik des ruhenden Menschen unter Einwirkung von Geräuschen. Int. Z. Angew. Physiol. einschl. Arbeitsphysiol. 16:217-227, 1956. (in German)
39. Lennard-Jones, J. E. Cohesion. Proc. Physical Soc. (London) 43:461-482, 1931.
40. Lin-Fu, J. S. Vulnerability of children to lead exposure and toxicity (First of two parts). N. Engl. J. Med. 289:1229-1233, 1973.
41. Mathews, K. E., Jr., and L. K. Canon. Environmental noise level as a determinant of helping behavior. J. Personality Soc. Psychol. 32:571-577, 1975.
42. McNall, P. E., Jr. The relation of thermal comfort to learning and performance: A state-of-the-art report. ASHRAE Trans. 85 (Pt. 1), 759-767, 1979.
43. Moschandreas, D. J., J. W. Winchester, J. W. Nelson, and R. M. Burton. Fine particle residential indoor air pollution. Atmos. Environ. 13:1413-1418, 1979.

44. National Center for Health Statistics. Medical Care, Health Status and Family Income. Series 10, No. 9. Washington, D.C.: U.S. Government Printing Office, 1964.
45. Nemecek, J., and E. Grandjean. Results of an ergonomic investigation of large-space offices. Human Factors 15:111-124, 1973.
46. New York State Commission on Ventilation. Ventilation. New York: Dutton, 1923.
47. Penney, G. W., and N. G. Ziesse. Soiling of surfaces by fine particles. ASHRAE Trans. 74(Pt. 1):VI.3.1-VI.3.13, 1968.
48. Pepler, R. D. A study of productivity and absenteeism in an apparel factory with and without air conditioning. ASHRAE Trans. 79(Pt. 2):81-86, 1973.
49. Pepler, R. D., and R. E. Wamer. Temperature and learning: An experimental study. ASHRAE Trans. 74(Pt. 2):211, 1968.
50. Proshansky, H. M., W. H. Ittelson, and L. G. Rivlin. The influence of the physical environment on behavior: Some basic assumptions, pp. 27-37. In H. M. Proshansky, W. H. Ittelson, and L. G. Rivlin, Eds. Environmental Psychology: Man and His Physical Setting. New York: Holt, Rinehart and Winston, Inc., 1970.
51. Radford, E. P. Health aspects of housing. J. Occup. Med. 18:105-108, 1976.
52. Russell, F. A. R. The Atmosphere in Relation to Human Life and Health. Publication No. 1072. Washington, D.C.: Smithsonian Institution, 1896. Compiled in Smithsonian Misc. Collections 39:Article III, 1899. 148 pp.
53. Sacher, G. A. Energy metabolism and thermoregulation in old age. ASHRAE Trans. 85(Pt. 1):775-783, 1979.
54. Schaefer, V. J., V. A. Mohnen, and V. R. Veirs. Air quality of American homes. Science 175:173-175, 1972.
55. Schaffer, E. L. A survey of some moisture and other problems influenced by building tightness. ASTM-DOE Symposium on Air Infiltration and Air Change Rate Measurement, Washington, D.C., March 16, 1978 (in press).
56. Schaplowsky, A. F., L. G. Polk, F. B. Oglesbay, J. H. Morrison, R. E. Gallagher, and W. Berman. Carbon monoxide contamination of the living environment: A national survey of home air specimens and children's blood samples. Presented at American Public Health Association Meeting, November 7, 1973. U.S. Department of Health, Education, and Welfare, Center for Disease Control.
57. Selye, H. The Stress of Life. rev. ed. New York: McGraw-Hill Book Company, Inc., 1976. 516 pp.
58. Spealman, C. R. Odors, odorants, and deodorants in aviation. Ann. N.Y. Acad. Sci. 58:40-43, 1954.
59. Spivey, G. H., and E. P. Radford. Inner-city housing and respiratory disease in children: A pilot study. Arch. Environ. Health 34:23-29, 1979.
60. Sterling, T. D., and D. M. Kobayashi. Exposure to pollutants in enclosed "living spaces." Environ. Res. 13:1-35, 1977.

61. U.S. Department of Health, Education, and Welfare, National Institute for Occupational Safety and Health. National Occupational Hazard Survey. Pilot Study. DHEW (NIOSH) Publication No. 75-162. Washington, D.C.: U.S. Department of Health, Education, and Welfare, May 1975.

62. U.S. Environmental Protection Agency, Office of Noise Abatement and Control. Information on Levels of Environmental Noise Requisite to Protect Public Health and Welfare with an Adequate Margin of Safety. U.S. Environmental Protection Agency Report No. 550/9-74-004. Washington, D.C.: U.S. Environmental Protection Agency, 1974. 46 pp. + appendices.

63. Vega, T., and C. J. Seymour. A simplified method for determining ozone levels in community air pollution surveys. J. Air Pollut. Control Assoc. 11:28-33,44, 1961.

64. Walker, R. L., and B. R. Fish. Adhesion of Particles to Surfaces in Liquid and Gaseous Environments. Paper presented at 4th Annual Meeting of the American Association for Contamination Control, Miami, Fla., May 25-28, 1965.

65. Williams, H. Developing a table of relaxation allowances. Ind. Eng. 5(12):18-22, 1973.

66. Williams, H. G. More light with less manpower, pp. 56-68. In Maintenance Guide for Commercial Buildings. Cedar Rapids: Stamats Publishing Co., 1975.

67. Wilner, R., R. Walkey, T. Pinkerton, and M. Tayback. The Housing Environment and Family Life. Baltimore: The Johns Hopkins Press, 1962. 338 pp.

68. Winneke, G., and J. Kastka. Odor pollution and odor annoyance reactions in industrial areas of the Rhine-Ruhr region, pp. 471-479. In J. Le Magnen and P. MacLeod, Eds. Proceedings of the Sixth International Symposium on Olfaction and Taste. London: Information Retrieval Ltd., 1977.

69. Winnick, L. American Housing and Its Use: The Demand for Shelter Space. Census Monograph Series. New York: John Wiley & Sons, Inc., 1957. 143 pp.

70. Winslow, C.-E. A. Fresh Air and Ventilation. New York: E. P. Dutton & Company, 1926. 182 pp.

71. Winslow, C.-E. A., and L. P. Herrington. The influence of odor upon appetite. Am. J. Hyg. 23:143-156, 1936.

72. World Health Organization. Health Hazards of the Human Environment. Geneva: World Health Organization, 1972. 387 pp.

73. Wyon, D. P. Human productivity in thermal environments between 65F and 85F (18-30C), pp. 192-216. In J. A. J. Stolwijk, Ed. Energy Conservation Strategies in Buildings. New Haven: John B. Pierce Foundation of Connecticut, Inc., 1978.

74. Wyon, D. P. The role of the environment in buildings today: Thermal aspects. Factors affecting the choice of a suitable room temperature. Build Int. 6:39-54, 1973.

IX

CONTROL OF INDOOR POLLUTION

The quality of the environment in a building is inherently dependent on the design and operation of the building's environmental control system. Several factors that affect the design and operation of control systems are identified in Chapter V, including human activities and geographic and building characteristics. Optimally, control systems are designed to maximize human comfort, and it is essential to know the acceptable ranges for environmental characteristics (comfort and air-quality factors). Some constraints that must be imposed on control systems are related to cost and energy consumption. As a result of the application of these constraints, the goal of maximal comfort is usually compromised. The ranges of conditions within which control systems operate are usually based on codes and standards that have been developed and promulgated to protect the health and welfare of occupants. This chapter begins with a review of codes and standards that pertain to indoor pollution.

Codes and standards have been developed as prescriptive guidelines based on consensus, but, as energy conservation and operating cost become more important, the need for evaluation of control-system performance increases. Criteria of system acceptability are also changing--codes and standards are becoming oriented more toward performance, and life-cycle costs are receiving more attention.

Changes in the attitude toward environmental control present several difficulties. Feedback control for acceptable indoor air quality is recognized and needed, but the availability of reliable and inexpensive controllers is seriously limited. Performance-oriented standards have not been widely accepted by contractors and enforcement officials, because of barriers in technology transfer and increased costs of implementation and liability. And economic decisions based on life-cycle costing have not been accepted by contractors and building developers, who have resisted because of a lack of incentives, such as amortization periods and allowance of pass-through of operating costs, and because of the high cost of capital. Appendix B considers energy, environmental, and economic factors and presents a method for providing acceptable control of indoor air quality at acceptable costs of money and energy.

VENTILATION CODES AND STANDARDS

Control of indoor environments in residential and commercial buildings to achieve what is termed "comfortable" or an "acceptable" thermal quality requires approximately one-third of the total annual energy consumption in the United States.[65] An additional 10% may be required to maintain conditions that are acceptable for occupants in industrial facilities. Ventilation systems have been reported to require as much as 50% or 60% of the total energy consumed in buildings.[33,59,80]

For energy conservation, rather arbitrary changes in building codes and standards are being proposed.[64,65] Reduction of ventilation in residential, commercial, and industrial buildings could jeopardize the health, safety, or welfare of those who occupy them. Reduction of energy consumption is a necessary but insufficient step in the development of acceptable building energy management programs. Also required is the maintenance of environmental conditions that are not deleterious to the occupants or harmful to property. These conditions include spatial, thermal, illumination, and acoustic qualities of the environment, as well as the gaseous and particulate qualities of the air. Ventilation is the historically and currently practical means of providing acceptable indoor air quality.

To protect the health, safety, and welfare of the general public, building codes have been adopted and enforced by local, state, and federal government agencies. These codes generally specify minimal acceptable ventilation criteria to be maintained in the buildings. Note that "ventilation air," as used here and elsewhere in this document, refers to outdoor air or recirculated, treated air. Compliance with building codes is usually the responsibility of licensed professional engineers and architects during design. Responsibility for compliance during operation often is vague, if specified at all. After a building has been designed and constructed, the owner or manager usually assumes responsibility for maintaining the quality of the indoor environment, and there is normally no official enforcement.

State and local building codes are normally based, directly or with modification, on one of three model building codes published in the United States: The BOCA Basic Building Code[15,16] of the Building Officials and Code Administrators International (BOCA); the Uniform Building Code[29] of the International Conference of Building Officials (ICBO); and the Southern Building Code[55] of the Southern Building Code Congress International, Inc. (SBCCI).

Building codes are usually derived from standards that have been promulgated by authoritative bodies, such as the American National Standards Institute (ANSI), the National Fire Protection Association (NFPA), and the American Society for Testing and Materials (ASTM). Other organizations that publish standards for the building industry are the American Society of Heating, Refrigerating, and Air-Conditioning Engineers (ASHRAE), the American Society of Mechanical Engineers (ASME), the Illuminating Engineering Society (IES), the

American Concrete Institute (ACI), the Air Conditioning and Refrigeration Institute (ARI), and the Sheet Metal Contractors Association (SMACNA). (In preparing proposed procedures for listing voluntary standards bodies for federal agency support and participation, the Department of Commerce held discussions with some 37 voluntary standards bodies.[62])

Standards published by these organizations are usually developed by a consensus method and are known as "voluntary standards" or "consensus standards."[1 62 73] Voluntary standards are usually adopted, after periods of open review, as guidelines of recommended practice or minimal performance criteria by which an organization may govern itself. However, a voluntary standard may become mandatory if it is adopted within legal documents, such as government standards or building codes.

Standards also are developed in response to state or federal laws. These are known as "mandatory standards"[62 73] and are promulgated in the form of state or federal regulations after they have been subjected to public hearings. Agencies responsible for the promulgation and enforcement of mandatory standards relevant to the building industry include the Department of Housing and Urban Development (HUD), the Department of Health and Human Services (DHHS, formerly the Department of Health, Education, and Welfare, or DHEW), and the Department of Energy (DOE).

BACKGROUND

By selecting the site, size, shape, and orientation of housing, man has nearly always taken advantage of natural ventilation for thermal and air-quality control. Ventilation requirements in buildings have been specified since the eighteenth century. The early history of the development of ventilation codes and standards has been reviewed by Nevins,[43] Klauss _et al._,[32] and Arnold and O'Sheridan, Inc.[11]

As shown in Figure IX-1, ventilation rates increased from 4 cfm/person in 1824 to 30 cfm/person in 1895. A minimal requirement of 30 cfm/person dominated design of ventilation systems during the first quarter of the twentieth century, as evidenced by the fact that in 1925 the codes of 22 states required a minimal ventilation rate of 30 cfm of outdoor air per person.[32]

A major change in ventilation standards resulted from experimental work reported by Yaglou _et al._[86] in the 1930s. These studies recognized the importance of controlling indoor air quality, as well as ventilation-air quantity, and reported ventilation rates in cubic feet per minute per person required to provide "odorfree" environments as functions of available air space per person. It should be noted that these ventilation rates were based on the assumption that outdoor air ("fresh air") was odorfree.

The Yaglou studies, conducted under controlled experimental conditions, have served as the primary reference in codes and standards for the last 40 yr. However, because of the difficulty in accurately estimating occupancy and the lack of feedback control methods for

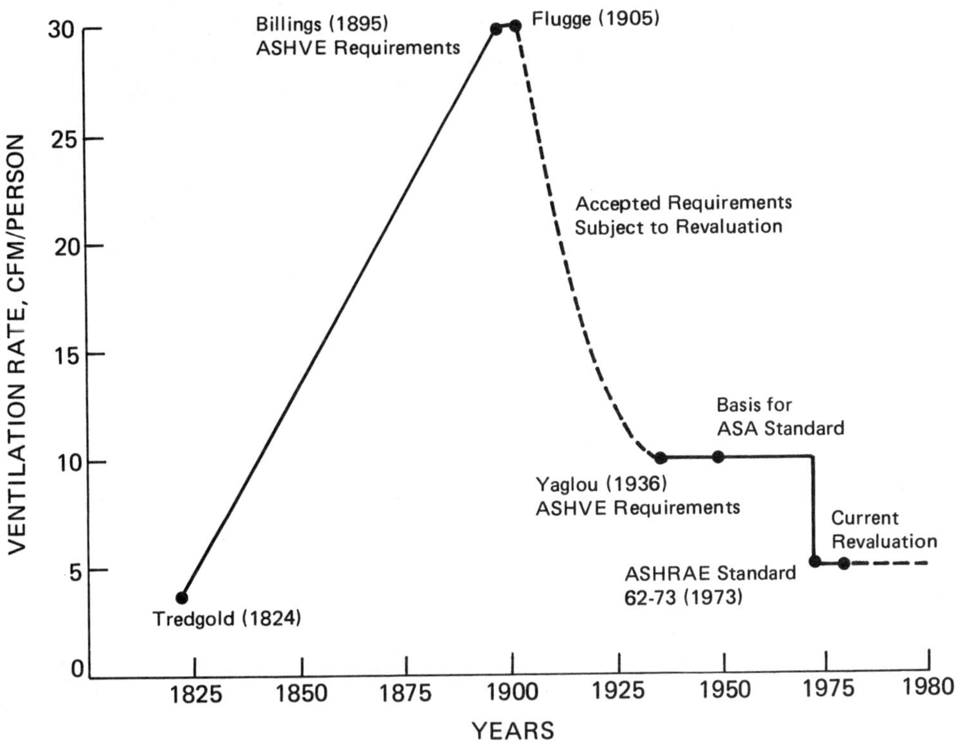

FIGURE IX-1 Historical development of ASHRAE Standard 62-73. After Klauss *et al.*[32]

ventilation, many codes and standards, including several now in effect,[40-42 68-71] have specified ventilation requirements as room-air changes per hour, rather than exchange rate per person. Theoretically, these criteria should be synonymous, but they are not. When ventilation rates are specified as room-air changes per hour, sensitivities to spatial dimensions and occupancy are lost. For example, 5 air changes per hour (ach) in a theater with a 20-ft (6.1-m) ceiling height and a sparse occupancy of 100 ft^2 (9.3 m^2) of floor area per person would result in 167 cfm (79 L/s) per person, whereas the same room-air exchange rate and occupancy in a classroom with an 8-ft (2.4-m) ceiling would mean 67 cfm (32 L/s) per person. However, at full-load occupancies of 10 ft^2 (0.9 m^2) per person in the theater and 20 ft^2 (1.9 m^2) per person in the classroom, 5 ach would result in 17 cfm (8 L/s) per person in the theater and 13 cfm (6 L/s) per person in the classroom. Thus, at less than full-load occupancies, the ventilation rates per person would exceed the values shown in Figure IX-2, whereas at full loads, the ventilation would be insufficient to provide "odorfree" air.

The inherent problems associated with specifying air changes per hour have been recognized in some standards for several years. In 1946, the American Standard Building Requirements for Light and Ventilation, A53.1, was published by the American Standards Association (ASA) with primary criteria in cubic feet per minute per square foot of floor area.[10] A revision and update of A53.1 was published in 1973 by ASHRAE with primary criteria in cubic feet per minute per person.[6] The latter standard was adopted by the ANSI (formerly ASA) in 1977 and has been designated ANSI Standard B194.1. For the first time in a ventilation standard, Standard 62-73 provided a quantitative definition of "acceptable outdoor air" and specified conditions under which recirculated air could be used. Both minimal and recommended ventilation rates were specified in the ASHRAE standard to accommodate fuel economy (minimal values) or comfort in odorfree environments (recommended values). Energy savings at design summer and winter conditions resulting from minimal ventilation rates specified in Standard 62-73 have been estimated to range from 27 to 81% for various occupied spaces, compared with rates in Standard A53.1.[80]

In response to demands for energy-efficient buildings, ASHRAE developed a new standard, which was published in 1975: Standard 90-75, Energy Conservation in New Building Design.[7] Through a contract with DOE, the National Conference of States on Building Codes and Standards, Inc. (NCSBCS), undertook, with the three model-code groups recognized in the United States, to write a model Code for Energy Conservation in New Building Construction.[17] This model code was based on ASHRAE Standard 90-75 and is generally considered to be its codified counterpart. By 1980, legislation either had been passed or was being considered by 45 states for energy-conservation regulations based on these two documents.[87]

ASHRAE Standard 90-75 was expected to reduce energy requirements in new buildings by 15-60%,[12] but efforts to promulgate the standard resulted in a conflict with Standard 62-73. Standard 90-75 stated that the "minimum" column in Standard 62-73 for each type of occupancy

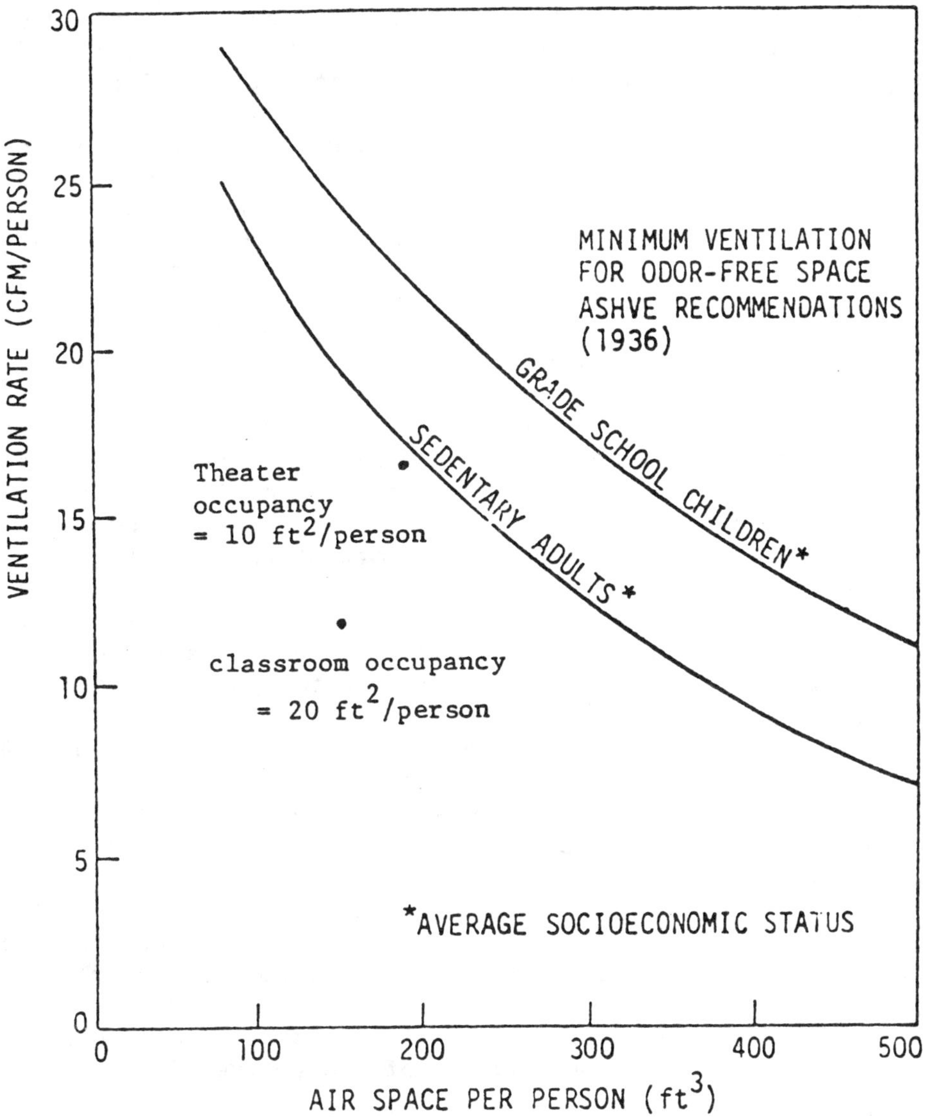

FIGURE IX-2 Ventilation rates resulting from the Yaglou studies.

"shall" be used for design purposes. This statement in Standard 90-75 effectively deleted the "recommended" column in Standard 62-73 and caused serious concern regarding the possibility of insufficient ventilation in new buildings. For example, when smoking was allowed in a room ventilated at the minimal rate of 5 cfm (2.4 L/s) per person, the carbon monoxide concentrations approached the limits specified by the EPA primary ambient-air quality standards, and particle concentrations exceeded the proposed limits by a factor of 30-60.[12,60] There is still controversy about what are acceptable concentrations of pollutants and ventilation rates.

In January 1981, ASHRAE adopted Standard 62-1981[9] in an effort to resolve some of the problems with Standard 90-75 and to reflect newer design requirements, equipment, systems, and instruments. A comparison of the newly revised Standard 62-1981, Standard 62-73, and the obsolete Standard A53.1 is shown in Table IX-1. Several major revisions have been made in an effort to resolve the apparent conflict between operating ventilation control systems for energy savings and operating them for protection of the health and comfort of the occupants.

• The quality of outdoor air to be used for dilution and control of indoor air pollution has been defined, not only in terms of the EPA primary standards, but also in terms of other recognized guidelines and professional judgment.

• Values for minimal and recommended ventilation rates have been replaced with required values for smoking and nonsmoking areas. Nonsmoking areas have proposed values similar to the existing minimal values, and those for smoking areas are similar to or greater than the values currently recommended.

• A method has been specified that will determine the amount of recirculation air required to compensate for allowable reductions in outdoor air. The amount is determined as a function of air-cleaner efficiency.

• The operation of mechanical ventilation systems during periods of occupancy is specified as a function of the source of indoor pollutants.

• An alternative method specifies both objective and subjective criteria for indoor air-quality, but the method of achieving control is left to the discretion of the operator.

With the advent of performance criteria for indoor pollutant control, conflicts between various codes and standards could become more intensive.

IMPLEMENTATION OF CODES AND STANDARDS

Ventilation codes and standards have been published by several agencies and organizations. As a result, the designer or operator of a system has the responsibility of reviewing the relevant documents and then deciding which of them apply. In many cases, the values in these codes and standards will not be consistent. Thus, it can present a

TABLE IX-1

Comparison of Ventilation Requirements in ASHRAE Standard 62-73 (1973), ASA Standard A 53.1 (1946), and ANSI/ASHRAE Standard 62-1981 (1981)

Subject	ASA 53.1[10]	ASHRAE 62-73[6]	ANSI/ASHRAE 62-1981[9]
Outdoor-air quality	Accepted ambient conditions	Acceptable criteria specified	Acceptable criteria more precisely specified
Ventilation air	Same as for outdoor air	Treated recirculated air allowable	Treatment more precisely specified
Mechanical ventilation	Requirements same as for natural ventilation	Reduced rates allowable	Methods for variable ventilation described
Primary ventilation criteria	cfm/ft^2, minimal values	cfm/person, minimal and recommended values	Required values of cfm/person specified for smoking and nonsmoking areas
Indoor-air quality	Not specified	Not specified	Alternative performance method specified

challenge to the building designer and operator to select a ventilation rate that will meet the requirements of all relevant codes and standards. Under these circumstances, the usual procedure has been to select the largest value that would satisfy the requirements of all the codes and standards.

Because of recent concerns regarding energy consumption and costs, some regulations have been promulgated or proposed that are in direct conflict with those promulgated to protect the health or comfort of occupants; an example is the 1977 Assembly Bill 983 of Wisconsin, Ventilation Requirements for Public Buildings and Places of Employment. Bill 983 would have eliminated mandatory minimal ventilation requirements specified in the state building code (i.e., 5 cfm per person) during the period October 1 to April 1 of each year. Building owners would have been allowed to close or otherwise regulate outside-air intakes to conserve energy during these periods. Bill 983 was passed by the 1977 General Assembly and vetoed by the governor; the veto was overridden by the Senate and sustained by the House. This legislation was reintroduced as a rider to an appropriations bill in the 1979 General Assembly. It was later amended to allow reduced ventilation only through adminstrative action; in that form, it passed. The state Department of Industry, Labor and Human Relations, previously responsible for ventilation requirements, will administer the law.

A summary of the most commonly cited ventilation codes and standards is shown in Table IX-2. Several model codes and ASHRAE Standard 62-73 may be applied to each of the nine functional catagories of buildings listed in Table IX-2.[50] Other voluntary and mandatory standards are shown as they apply to particular functions. It should also be noted that the NCSBCS model Code for Energy Conservation in New Building Construction was developed with the three model-code groups and applies to all functional catagories.[17] This model code specifies ventilation rates for energy calculations as the minimal values in Standard 62-73. The ASHRAE standard, in turn, defers to other standards or codes when they have precedence and require higher ventilation rates.

Domiciles

As indicated in Table IX-2, the two primary sources for ventilation requirements are ASHRAE standards[6,9] and the HUD Minimum Property Standards (MPS).[69-71] Both sets of standards are considered voluntary, but may become mandatory under specific conditions--Standard 62-73 when adopted as part of a state energy code, and the MPS if housing is financed through the Federal Housing Administration (FHA).

Ventilation rates for various spaces throughout private dwelling places are specified in Standard 62-73 as 5-20 cfm/person (minimum) and 7-50 cfm/person (recommended). The higher rates are for bathrooms and kitchens and are for intermittent operation. The MPS also set intermittent exhaust rates in kitchens and baths at 15 and 8 ach, respectively. The 1979 revisions of the MPS allow ventilation by

TABLE IX-2

Sources of Ventilation Codes and Standards for Occupied Spaces

Building-Function Category	Voluntary Standards	Mandatory Standards	Model Building Codes
Domicile: place of residence, such as a single-family dwelling, multifamily dwelling, public housing, rowhouse, apartment, or condominium	ASHRAE[6,9] MPS 4900[69] MPS 4910[70]	--	BOCA[15,16] UBC[29] SBCCI[55] NCSBCS[17]
Educational: building used for classrooms or instruction	ASHRAE[6,9]	--	BOCA[15,16] UBC[29] SBCCI[55] NCSBCS[17]
Laboratory: building used predominantly for research and diagnostic work, and not necessarily for instruction	ILAR guide and standards[39-42] NIH guidelines[66] ASHRAE[6,9]	9 CFR 1.1, 1979[74] 29 CFR 1910, 1979[72]	BOCA[15,16] UBC[29] SBCCI[55] NCSBCS[17]
Medical: building used for health-care facilities, such as hospital, clinic, medical center, sanitarium, day nursery, infirmary, orphanage, nursing home, or mental-health institution	ASHRAE[6,9] MPS 4920[71]	HRA 79-14500[68]	BOCA[15,16] UBC[29] SBCCI[55] NCSBCS[17]
Office: such buildings as used for offices, civil administration, or radio or television station	ASHRAE[6,9]	--	BOCA[15,16] UBC[29] SBCCI[55] NCSBCS[17]
Public assembly: building where groups can meet for such functions as theater, restaurant, cafeteria, retail store, art gallery, museum, bank, post office, courthouse, assembly hall, church, dance hall, coliseum, passenger terminal, or library	ASHRAE[6,9]	--	BOCA[15,16] UBC[29] SBCCI[55] NCSBCS[17]

Table IX-2 (contd)

Building-Function Category	Voluntary Standards	Mandatory Standards	Model Building Codes
Rehabilitation: non-health-care building used for instruction, but not of the regimented classroom type; pertains more to readjustment, such as jail, prison, reformatory, or half-way houses	ASHRAE[6,9]	--	BOCA[15,16] UBC[29] SBCCI[55] NCSBCS[17]
Warehouse: building used for storage of materials and supplies, such as storage facility, maintenance facility, garage, airplane hangar, or bus barn	ASHRAE[6,9]	OSHA[72]	BOCA[15,16] UBC[29] SBCCI[55] NCSBCS[17]
Industrial: such buildings as factories, assembly plants, foundries, mills, power plants, telephone-exchange facilities, water and waste-water treatment plants, solid-refuse plants, zoos, greenhouses, aviaries, arboretums, or others requiring environmental control for process control	ASHRAE[6,9]	OSHA[72]	BOCA[15,16] UBC[29] SBCCI[55] NCSBCS[17]

infiltration rates of 0.5 ach and natural ventilation through operable windows, which must have a total area of at least one-twentieth of the floor area of the room. ANSI/ASHRAE Standard 62-1981 specifies 10 cfm (5 L/s) per room for spaces other than bathrooms and kitchens, for which values are set at 50 and 100 cfm (24 and 47 L/s) per room, respectively. Although the ventilation rates are specified differently in these voluntary standards, the results are intended to be equivalent. Moreover, the 1979 revisions to the MPS and the values in Standard 62-1981 are in close agreement with values recommended internationally.[23 85]

Educational Facilities

The mechanical-ventilation rate for classrooms is specified in Standard 62-73 and in the model codes as a minimum of 5 cfm/person for a full-load occupancy of 20 ft^2/person. However, the minimal supply-air rate (i.e., ventilation plus recirculation) is specified as 10 cfm/person in Standard 62-73 and 15 cfm/person in the model codes. Natural ventilation is specified in the model codes as that obtainable through operable windows with areas one-twentieth of the floor areas; Standard 62-73 specifies minimal and recommended natural-ventilation rates of 10 and 10-15 cfm/person, respectively. Standard 62-1981 specifies required ventilation rates of 5 cfm (2.5 L/s) and 25 cfm (12.5 L/s) per person for nonsmoking and smoking areas, respectively, in classrooms.

Laboratories

Specific controls for ventilation in laboratory spaces are required for protection of the health and comfort of laboratory personnel and for the preservation of specimens and critical experimentation conducted in the facilities. A differential in air pressure may be required between laboratory areas and public spaces, such as meeting rooms and reception areas, to protect the general public. Thus, the nature of ventilation control is more complex in these facilities than in most other indoor environments.

Toxic and hazardous materials used in the laboratory must be controlled to within the limits prescribed by OSHA.[72] Control may be by isolation or enclosure of the pollutants, dilution, or air-cleaning, but OSHA does not mandate a particular control method. This type of standard has become known as a "performance standard."

Indoor areas in which substances suspected of being carcinogenic are used or where recombinant-DNA research is conducted must be kept under negative static pressure relative to the surrounding areas.[66 72] Local exhaust and clean makeup air may be used for pressure control, but the exhaust must be decontaminated before discharge. Also, "experiments," procedures, and equipment that could produce aerosols must be confined to laboratory hoods or glove boxes.[72]

When laboratory animals are used in experiments, their care and well-being must also be maintained. The Animal Welfare Act[74] specifies many procedures for the care and handling of the animals, but is vague and nonspecific about environmental control in the laboratory or the cage.[84] The standards published by the Institute of Laboratory Animal Resources (ILAR) of the National Research Council are somewhat more specific in "recommending" ventilation rates.[39] However, these standards often require 10-20 ach with 100% outside air, which is energy-intensive, and the use of 100% outside air may have little or no impact in the cage microenvironment.[82]

ASHRAE Standard 62-73 specifies 15 cfm/person as minimal and 20-25 cfm/person as recommended ventilation rates for spaces without animals. With animals, the minimal rate is 40 cfm/person and the recommended rate is 45-50 cfm/person. These outdoor-air requirements may be reduced by two-thirds for mechanical ventilation systems with adequate particle filtration.

Standard 62-1981 specifies a required ventilation rate of 10 cfm (5 L/s) per person for nonsmoking areas and recognizes that other standards may override this rate.[9]

Medical Facilities

Ventilation and control of biologic contamination in medical facilities, especially in some hospital treatment areas, have been the subject of much research since the middle of the nineteenth century.[13,36] The Health Resource Administration (HRA) publishes requirements[68] that must be maintained if federal funds (i.e., Hill-Burton funds) are used for new construction or major modifications.

Since 1969, these regulations have allowed recirculation in sensitive areas, such as operating rooms. However, changes have occurred in the specified number of air changes per hour of supply air and the percentage of outside air.[13] Currently, HRA allows recirculation of air in all areas of hospitals, with the following restrictions:[68]

- In sensitive areas, such as operating rooms, two air filters are required--a prefilter and a final filter, rated at 25% and 90% efficiency, respectively, according to ASHRAE Standard 52-76.[4]
- Each space in which inhalation anesthetic agents are administered must be supplied with a separate scavenging system for exhausting waste anesthetic gases.
- Appropriate air-pressure relationships must be maintained with respect to adjacent areas.

Changes specified in the minimal requirements of construction and equipment for hospital and medical facilities[68] allow reductions of up to 25% when specific rooms are unoccupied, provided that the specified pressure relationships are maintained when they are occupied. When this feature is used, positive provisions, such as an electric interconnect between the ventilation system and room lights, must be included, to ensure that the specified ventilation rates are automatically resumed when the rooms are reoccupied.

Standard 62-73 specifies ventilation rates for hospitals and nursing and convalescent homes in terms of minimal and recommended cubic feet per minute per person and allows reductions in the use of outdoor air to one-third of the specified values when mechanical ventilation is used.

Standard 62-1981 specifies required ventilation rates for patient rooms as 35 cfm (17.5 L/s) and 7 cfm (3.5 L/s) per bed for smoking and nonsmoking spaces, respectively. In other hospital areas, Standard 62-1981 values are per person for nonsmoking spaces and are similar to the minimal values previously specified.

As shown in Table IX-3, the ventilation and total air-supply rates specified in HRA 79-14500 are generally greater than those specified in ASHRAE Standard 62-73 or 62-1981.[79]

The ventilation rates specified in MPS 4920[71] are primarily in terms of allowable infiltration rates and exhaust rates for kitchens and patient-room lavatories.

Other Nonindustrial Spaces

The ASHRAE standards[6,9] are primary sources for ventilation rates for offices, public-assembly buildings, and rehabilitation facilities. Currently, no other standards are generally used in the United States. In Standard 62-73, ventilation rates are specified as minimal and recommended rates per person with reductions in outdoor air of one-third of the specified values allowed for mechanical ventilation, if adequate filtration is provided. Standard 62-1981 specifies ventilation rates as required for smoking and nonsmoking areas with reductions in outdoor air allowed for mechanical ventilation as a function of filter efficiency.

SUMMARY

State and local building codes usually are based on one of the three model-code documents. These become legal documents when adopted by appropriate government agencies. The ventilation rates specified in the building codes are usually derived from standards, such as those published by ASHRAE.

Traditionally, other mandatory standards have taken precedence over a building code when the values in standards exceeded those in the building code. However, model energy-conservation codes have been promulgated by the model-code groups, and there can now be conflicts in required ventilation rates between codes and mandatory standards. With the advent of indoor-pollutant criteria, the conflicts could become more extensive, because methods of pollution control that do not require the traditional ventilation rates may be used.

TABLE IX-3

Comparison of Hospital Ventilation Standards[a]

Area	ASHRAE Standards: Equivalent Air Changes per Hour[b]				Hill-Burton Standards							
	ASHRAE 62-73		ANSI/ASHRAE 62-1981		Outdoor-Air Changes per Hour, Minimum				Total-Air Changes per Hour, Minimum			
	Minimum	Recommended	Smoking	Nonsmoking	Pre-1969	1969	1974	1979	Pre-1969	1969	1974	1979
Operating room	5	--	--	1.2	12	5	5	5	12	12	25	25
Recovery room	4[c]	--	--	3[d]	6	2	2	2	6	6	6	6
Patient room	1	1.5-2	2.5	0.5	2	2	2	2	2	2	2	2
Ward	1.5	2-2.5	2.5	0.5	2	2	2	2	2	2	2	2
Medical procedure (treatment)	--	--	2.5	0.5	6	2	2	2	6	6	6	6
Physical therapy	2	2.5-3	--	2.5	6	2	2	2	6	6	6	6
Autopsy[e]	2	2.5-3	--	7	6	2	2	2	15	12	12	12

[a]From Woods.[79]

[b]Ceiling height assumed to be 10 ft (3 m).

[c]Special requirements or codes may determine ventilation rates.

[d]Activities generating contaminants may require higher rates.

[e]Air shall not be recirculated.

RECOMMENDATIONS

The general public is not aware of the distinction between ventilation control and indoor air-quality control. The techniques and terminology used in air-quality control and ventilation design, operation, and codes should be described in clear and consistent language.

We recommend that professional and government organizations coordinate to establish a model code for indoor air quality that would meet health, energy, and economic criteria.

Responsibility for enforcement of acceptable control of indoor air quality should be defined for various building categories. Enforcement procedures should be considered with respect to building construction and building operation.

AIR DIFFUSION CONTROL

Air is supplied to ventilate an enclosed space (i.e., a room or group of rooms in a building) for two main reasons:

- To maintain acceptable oxygen concentration and to dilute (and remove) carbon dioxide and other contaminants for safety of the occupants (and sometimes to provide a differential in air pressure as required by building codes or standards). Ventilation air flow rates are specified in codes and standards.[6] It should be noted that supplying the specified or mandated rates for ventilation does not guarantee adequate dilution or removal of contaminants if the air is not uniformly diffused throughout the occupied space.
- To provide a thermally controlled environment that is acceptable to the occupants. An acceptable thermal environment has been defined as one in which at least 80% of the occupants clothed normally and engaged in sedentary or near-sedentary activities would express thermal comfort, which is defined as "that condition of mind which expresses satisfaction with the thermal environment."[5] Depending on the activity and typical clothing of the occupants, the combination of air temperature, mean radiant temperature, relative humidity, and air velocity must be appropriate for the occupants to feel comfortable (see Chapters IV and VII).[24 25 52]

Conventionally, air diffusion control has been designed and installed to meet the criteria for thermal comfort, with the assumption that the air-quality criteria will be met simultaneously.

AIR DIFFUSION EQUIPMENT

The supply air for ventilation is usually treated at a central location (i.e., filtered and conditioned for an appropriate dry-bulb and dew-point temperature) and then distributed by a duct system to the intended space. The amount of air supplied to each space is controlled by the terminal units of the duct system.

Four types of terminal units are commonly available:[3] grilles, slot diffusers, ceiling diffusers, and perforated ceilings.

Grilles, which can have different configurations (e.g., adjustable bar grilles, fixed bar grilles, stamped grilles, and variable-area grilles), are usually in a high sidewall position, in a perimeter installation, or in the ceiling. The air from a high sidewall position is thrown across the ceiling and drops toward the floor as it traverses the room. From a floor or sill grille, the air is directed vertically upward along the perimeter walls to which the airstream adheres, owing to the Coanda effect. (The Coanda effect can be defined as the ability of a jet to cling to a curved or deflected surface while increasing its mass flow rate along the flow path.[20]) When grilles are installed in the ceiling, curved vanes deflect the air along the ceiling so that the Coanda effect causes the airstream to follow a horizontal distribution.

A slot diffuser is usually installed in long continuous lengths in several different locations similar to those described for grilles.

Ceiling diffusers usually are series of rings or louvers (not necessarily circular) that direct the airstream across the ceiling.

In perforated ceilings, the air is contained in a supply plenum above the ceiling and delivered through holes or slots in the ceiling material.

AIR DIFFUSION CRITERIA

The velocity of the air is important--if the appropriate velocity is exceeded, conditions can become drafty and thus uncomfortable.[38, 53] The force of the air supplied must be such that it stirs the air already in the space so that mixing is accomplished, to reduce the variance of air properties, both thermal and chemical, throughout the space. However, complete mixing of the air in the space is seldom achieved.

In some cases, especially when there are high ceilings (i.e., commercial or institutional spaces), various zones can be identified as occupied and unoccupied spaces. There is little thermal and respiratory exchange between people and the air above head level, and the space between head level and the ceiling is called "unoccupied space." Uniform mixing of the air is necessary for the comfort of those in the occupied space, but is not needed in unoccupied space.[8] Because there may be incomplete mixing of the air in the unoccupied space, the chemical and thermal composition may be noticeably different from that of the occupied space.

Nonuniform mixing may be caused by the type and location of the terminal units selected for the space or by such deficiencies as:

- Direct air flow from the terminal supply unit to the exhaust or return air grilles that bypasses a part (or most) of the occupied space.[30]
- An air circulation pattern that causes secondary air currents where the supplied air does not have sufficient force to cause complete mixing, thus leading to air stratification within the enclosure.[43]

Mathematical models to determine the effective mixing rate that occurs in a space have been proposed, but extensive research is still needed to obtain reliable methods to quantify mixing.[33 49 82 83]

When mixing in an occupied space is nonuniform, comfortable conditions cannot be ensured for the occupants in the stagnant (secondary flow) zones. To minimize nonuniform mixing, the location, type, and size of the terminal units must be selected correctly. There are very few definite criteria to make this selection for a particular application, but the concept of air-diffusion performance index (ADPI)[44] is commonly used to characterize a terminal unit.

The ADPI is based on subjective responses to drafts. The effective draft temperature (θ) is determined from the local velocity (V_x), in feet per minute, and the difference in dry-bulb temperature, in degrees Fahrenheit, between the local point (t_x) and the control temperature (t_c): $\theta = (t_x - t_c) - 0.07(V_x - 30)$. The ADPI is the percentage of the total number of measured points that have effective draft temperatures of -3.0 to +2.0°F and local velocities of 70 ft/min or less (see Figure IX-3). ADPI values have been experimentally calculated for typical applications of terminal units as a function of the airflow characteristics from the units and the thermal loads of the spaces (see Tables IX-4, IX-5, and IX-6 and Figure IX-4). From the values listed in Table IX-4, types and sizes of terminal units can be selected to provide acceptable mixing in the occupied space.

The ADPI, although practical, may not yield the best selection in all cases. Several points should be considered:

• The location of the exhaust outlet influences air movement only in a small zone near the outlet itself. Thus, the ADPI does not depend on the location of the exhaust outlet.[43] However, there are studies[45 46] that have shown that different locations of the supply and exhaust units cause different patterns of airflow, some of which may be unstable and some unacceptable for thermal comfort.

• Airflow patterns are different during heating and cooling cycles. Commonly, the same terminal units are used for both situations. Thus, the cool air from a ceiling diffuser would drop into the occupied space, but hot air supplied by the same terminal mixes to provide thermal comfort of the occupants. However, the ceiling location may not be appropriate for a heating situation, inasmuch as the hot air supplied by the ceiling diffuser would tend to stay near the ceiling, owing to buoyancy; this results in air stratification near the floor. A similar situation may occur when the terminal units are placed low in the occupied space; the hot-air supply tends to rise and affect the whole room, whereas the cold-air supply tends to stay low, thus possibly causing development of a stagnant layer near the ceiling. Therefore, for spaces where both heating and cooling are needed, care should be taken to ensure that the terminal unit will have an appropriate ADPI in both situations.

• Calculation of an ADPI assumes that steady conditions exist and that the room has no airflow obstructions. Some attempts have been made to study the effects of obstructions in the occupied space,[81]

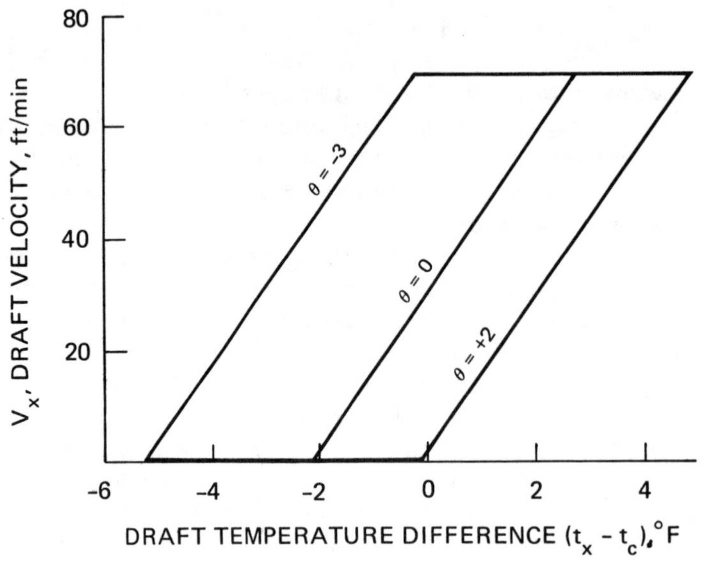

FIGURE IX-3 Comfort criteria used to evaluate the air-diffusion performance index (ADPI).

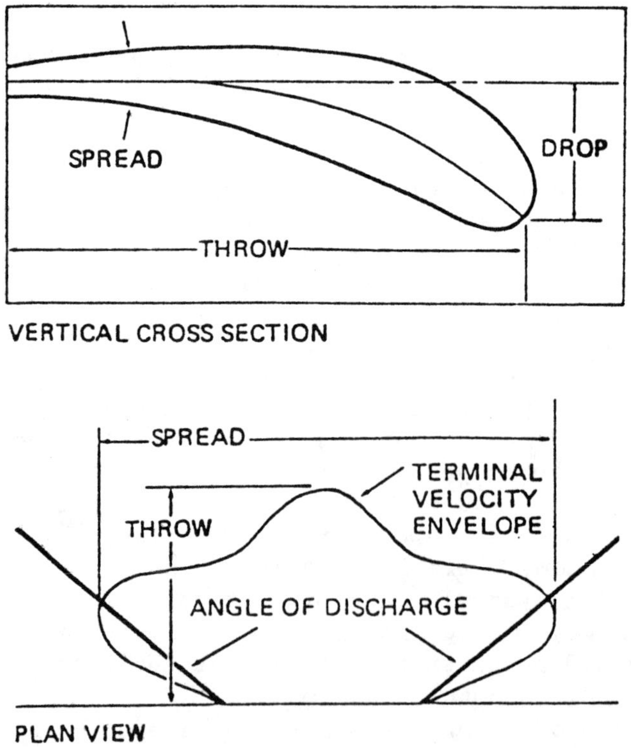

FIGURE IX-4 Airstream characteristics. Reprinted with permission from Nevins.[43]

TABLE IX-4

Air Diffusion Performance Index for Common Diffuser Applications[a]

Terminal Device	Room Load, $Btuh/ft^2$	T_{50}/L for Max. ADPI	Maximal ADPI	For ADPI Greater Than	Range of T_{50}/L
High-sidewall grilles	80	1.8	68	--	--
	60	1.8	72	70	1.5-2.2
	40	1.6	78	70	1.2-2.3
	20	1.5	85	80	1.0-1.9
Circular ceiling diffusers	30	0.8	76	70	0.7-1.3
	60	0.8	83	80	0.7-1.2
	40	0.8	88	80	0.5-1.5
	20	0.8	93	90	0.7-1.3
Sill grille straight vanes	80	1.7	61	60	1.5-1.7
	60	1.7	72	70	1.4-1.7
	40	1.3	86	80	1.2-1.8
	20	0.9	95	90	0.8-1.3
Sill grille spread vanes	80	0.7	94	90	0.8-1.5
	60	0.7	94	80	0.6-1.7
	40	0.7	94	--	--
	20	0.7	94	--	--
Ceiling slot diffusers	80	0.3[b]	85	80	0.3-0.7
	60	0.3[b]	88	80	0.3-0.8
	40	0.3[b]	91	80	0.3-1.1
	20	0.3[b]	92	80	0.3-1.5
Light troffer diffusers	60	2.5	86	80	<3.8
	40	1.0	92	90	<3.0
	20	1.0	95	90	<4.5
Perforated and louvered ceiling diffusers	11-51	2.0	96	90	1.4-2.7
				80	1.0-3.4

[a]Reprinted with permission from Nevins.[43] T_{50} = throw of isothermal airstream to a terminal velocity of 50 ft/min (see Figure IX-4 and Table IX-5). L = characteristic dimension of the space (see Table IX-6).

[b]T_{100}/L.

TABLE IX-5

Definitions of Airstream Characteristics

Throw:	Horizontal distance measured from plane of supply diffuser to farthest point of airstream center line at which airstream velocity equals selected velocity, i.e., terminal velocity (T_{50} if terminal velocity is 50 ppm or T_{100} if terminal velocity is 100 fpm).[43]
Spread:	Divergence of airstream in horizntal or vertical plane.
Drop:	Vertical distance between center line of terminal unit and point to which throw is measured.

TABLE IX-6

Characteristic Dimensions for Different Air Diffusers[a]

Diffuser Type	Characteristic Length
High-sidewall grille	Distance to wall perpendicular to jet
Circular ceiling diffuser	Distance to closest wall or intersecting air jet
Sill grille	Length of room in direction of jet flow
Ceiling slot diffuser	Distance to wall or midplane between outlets
Light troffer diffusers	Distance to midplane between outlets plus distance from ceiling to top of occupied zone
Perforated, louvered ceiling diffusers	Distance to wall or midplane between outlets

[a]Reprinted with permission from Nevins.[43]

but more research is needed to analyze the effects better. Moreover, actual conditions in the occupied space (i.e., normal working conditions) may be very different from those predicted by the ADPI method, because the obstructions, people, and appliances in the space may cause a different air pattern.

• ADPI relates comfort to local air temperature and local air velocity as they deviate from a setpoint suitable for providing thermally comfortable conditions. This setpoint must be established by other methods, such as percentage of people dissatisfied (PPD),[24] KSU thermal sensation index,[52] and standard effective temperature (SET).[25] All these methods determine the proper combination of ambient air temperature, relative humidity, mean radiant temperature, and air velocity that must exist in the room as a function of the occupants' activity and clothing.

CONCLUSIONS

Currently, air-diffusion systems are designed for two main purposes: to supply ventilation air according to type of room and intended use, as required by codes; and to locate supply-air terminal units and define setpoints on the basis of occupant comfort, which depends on thermal factors.

RECOMMENDATIONS

Interactions between thermal factors and mass factors (i.e., concentrations of water vapor, odors, and other gaseous contaminants and suspended particles) that influence the comfort or health of occupants must be studied, and the results must be incorporated into the design procedure. The measurement of air quality in a space is still quite difficult.[26] It is recommended that research be conducted to provide design guidelines for the selection and placement of air-supply and -return units that will ensure both the mass air quality and the thermal air quality required in a space under conditions of use.

AIR CLEANING EQUIPMENT

The principles that govern the process of cleaning air to improve its quality for use indoors are similar to those for industrial processes to remove effluent gases before discharging exhaust air to the atmosphere. However, these processes and the equipment involved to process air for ventilation are radically different from their industrial counterparts.

LOCATION OF INDOOR-AIR CLEANERS

As shown schematically in Figure IX-5, the location of air-cleaning equipment in a ventilation system will vary with the type of system and its application. (Strategies for control of indoor pollutants are discussed in the final section of this chapter.) First, if the concentrations of contaminants in outdoor air are unacceptable,[2 6] the outdoor air must be cleaned. Second, recirculated air from occupied spaces must be cleaned to achieve the same quality as specified for the outdoor air used for ventilation. Third, special ventilation systems, such as fume hoods, may use air-cleaners in the supply and exhaust airstreams.

TYPES OF AIR-CLEANERS

An air-cleaner capable of controlling particulate, vaporous, and gaseous contaminants does not exist. Filters and electronic cleaners are used to remove airborne particles from ventilation air; and commercial and institutional facilities may use wet collectors. Viable biologic particles are usually removed by special filters, electronic air-cleaners, or wet collectors; in some cases, ultraviolet (UV) lamps may be used to inactivate the viable contaminants. It should be noted that UV radiation is used to kill bacteria, but not to remove them from the airstream.

Filters and electronic air-cleaners are not effective for removing gases or vapors. Sorption devices are usually selected to remove these contaminants from the airstream. If both particles and gases or vapors must be removed, air-cleaners first remove the particles, then the gases or vapors. In some critical applications, such as hospital operating rooms, a "final" filter may be required to remove residues or particles sloughed from the gas-removal devices.[67]

Devices for Particle Removal

Airborne particles are commonly removed by mechanical filter units that use one or a combination of the following mechanism:[22]

- Inertial impingement (impaction): An abrupt change in direction of the airstream causes airborne particles to collide with the filter fiber. This method of collection is most effective with larger particles.
- Interception: This method of collection is a special case of impingement in which a particle collides with a fiber, independently of inertia. This method may be more effective than impingement at low velocities.
- Straining: Airborne particles are captured as they attempt to pass between two adjacent fibers.
- Diffusion: Very small airborne particles are driven to the filter fiber by random molecular bombardment by air molecules. This method is most effective for the smallest particles.

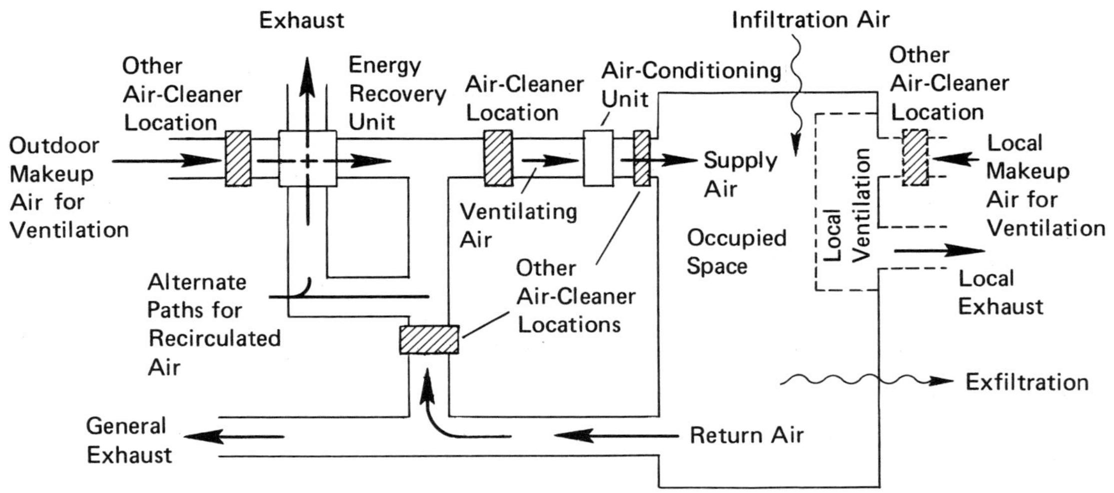

FIGURE IX-5 Schematic of a ventilation system. Adapted from ANSI/ASHRAE 62-1981.[2]

The relationship between filter efficiency and particle size is shown in Figure IX-6. The effectiveness of diffusion is greater for smaller particles, whereas the effectiveness of impingement, interception, and straining is greater for larger particles. Thus, a characteristic performance curve, similar to the upper curve in Figure IX-6, shows a minimal effectiveness of a filter to remove a given particle size.[21]

The performance of a mechanical filter is usually expressed in terms of its particle removal efficiency, its loading capacity, and its resistance to air flow.[28] Removal efficiency may be expressed as a function of the mass, physical or aerodynamic size (e.g., Stokes diameter), or number of particles removed.[28,56] A relationship among these efficiencies can be expressed mathematically,[56] but the filters must be tested to express the appropriate efficiencies numerically. Although many test methods for evaluating filter efficiencies have been published,[19,34,54] those generally accepted for indoor environments in the United States are ASHRAE Standard 52-76 for mass and size efficiencies[4] and MIL Standard 282 for number efficiency.[63] A method for rating the loading (i.e., dust-holding) capacity is also specified in ASHRAE Standard 52-76. Both standards result in single value ratings. MIL Standard 282 specifies a means for rating the efficiency of DOP (dioctylphthalate) produced by a special generator in removing 0.3-μm particles. ASHRAE Standard 52-76 specifies a means for rating the size removal efficiency of a filter challenged with a standardized "atmospheric dust." A "weight arrestance" (i.e., mass removal efficiency) procedure specified in ASHRAE Standard 52-76 results in a single value rating for a filter challenged with a "synthetic dust." This dust is used to rate the dust-holding capacity of a filter (i.e., the amount of dust a filter can retain before a specified pressue drop is reached). A major shortcoming of these standards is the lack of defined procedures to rate mass, size, and particle removal efficiencies as functions of particle size.

The other characteristics necessary to rate the performance of mechanical filters are the air flow rate at which the efficiencies are determined and the air pressure drops imposed by the filter when it is clean and when it is fully loaded. Both standards specify procedures for determining these characteristics. Characteristics for several types of mechanical filters are summarized in Table IX-7.

Mechanical filters are used for three kinds of applications in which the three types of removal efficiencies are required:

- Filters used to remove the largest and heaviest particles from an airstream are usually rated by weight efficiency (see Table IX-7). They are often described as low-efficiency filters and are used as upstream prefilters to remove some of the load before the final filters[3] or to protect such mechanical devices as fans and heat-exchangers. Probably the most common use for this type of filter is in residential furnaces and central air-conditioning systems.
- Medium-efficiency filters, usually rated by size or dust-spot efficiency,[4] are used when smaller particles must be removed from the air. They are more expensive than low-efficiency filters. They are

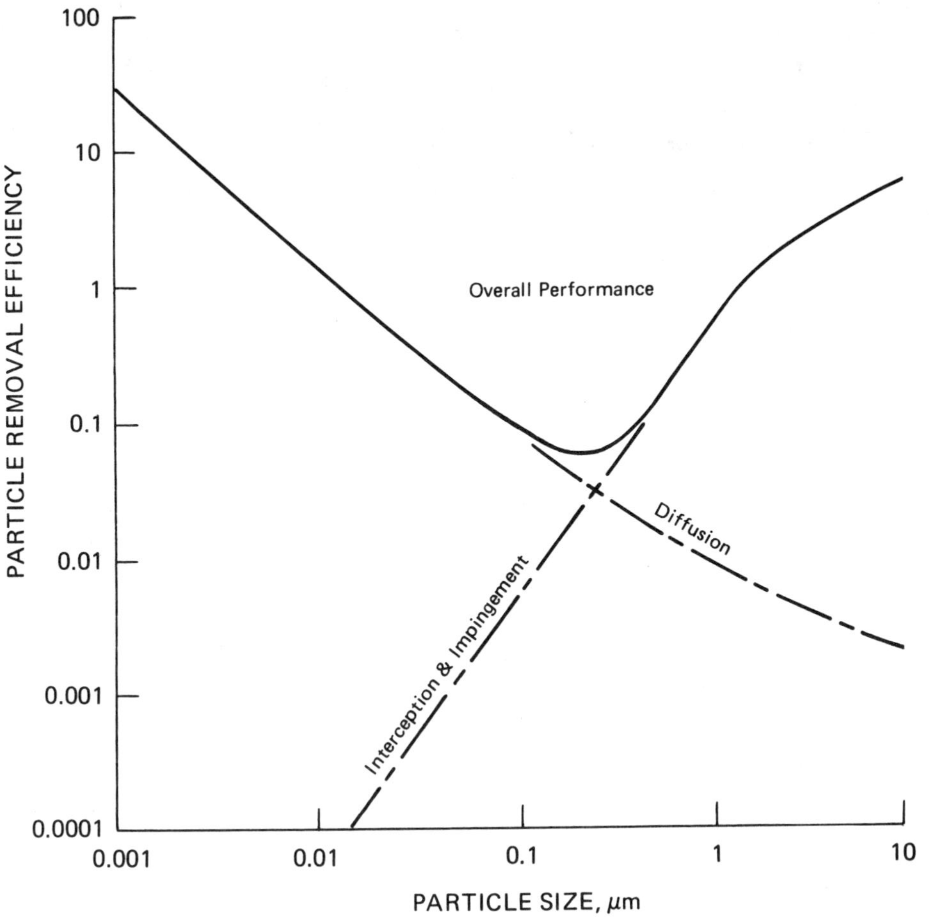

FIGURE IX-6 Filter efficiency as a function of particle size for a typical impingement filter. Adapted from Crawford.[21]

TABLE IX-7

Typical Characteristics of Mechanical Filters[a]

Type of Air Cleaner	Control Mechanism	Pressure Drop,[b] Pa	Efficiency, % Weight[c]	Size[d]	Number[e]	Dust-Holding Capacity,[f] $g/(1,700\ m^3/h)$
Viscous-impingement filter	Impingement	<125	20-50	5-10	--	70-140
			50-75	5-15	--	120-560
			60-80	5-20	--	180-540
			70-85	10-25	--	240-760
Dry-type extended surface:	Impingement and interception					
Open-cell foams		12-250	70-80	15-30	--	180-425
Cellulose glass-fiber mats			80-90	20-35	--	90-180
Multi-ply glass-fiber mats			85-90	25-40	5-10	90-180
5- to 10-μm fibers, 6-12 mm thick			90-95	40-60	15-20	270-540
3- to 5-μm fibers, 6-20 mm thick			>95	60-80	35-40	180-450
1- to 4-μm fibers and asbestos			>95	80-90	50-55	180-360
0.5- to 2-μm glass fibers			--	90-98	75-90	90-270
0.1- to 1-μm fibers (HEPA filters)			--	--	99.97-99.999	500-1,000

[a]Adapted from ASHRAE.[3]

[b]Pressure drop at air velocity of 1-4 m/s through media; filter-face velocity usually higher.

[c]Also described as dust arrestance in ASHRAE Standard 52-76.[4]

[d]Also described as atmospheric dust spot efficiency in ASHRAE Standard 52-76.[4]

[e]Particle-removal efficiency of 0.3 μm (i.e., DOP method), as described in Mil Standard 282.[63]

[f]As described in ASHRAE Standard 52-76.[4]

often used with prefilters to extend their useful lives. They are also specified for sensitive areas in hospitals[67] and are used for removal of tobacco smoke or for protection of materials from soiling.

• High-efficiency particulate air (HEPA) filters are rated by number efficiency[63] and are used when "absolute" filtration is required. These filters were originally developed for industrial applications and are generally used in nuclear reactor facilities and for cleanroom applications. HEPA filters are expensive and are therefore normally protected by prefilters. Medical facilities use them in isolation wards, pharmacies, and surgical suites.[36(pp. 51-73)] In these areas, it is often necessary to test the filters after installation. Test procedures developed for some of these applications are available in the literature.[36(pp. 193-213)] Small HEPA filters have recently been used to create "clean-air zones," especially about the heads of allergic persons during sleep. Air is drawn through the filters and distributed from headboard emission ports as a discrete laminar-flow field.[88] Claims of removal of various inhalant allergic substances and reduction in associated problems have been made; however, the effect on removal of biogenic particles requires more study.

It should be noted that mechanical filters are effective only when the particles remain airborne in the ventilation system, and particles from an occupied space must be transported to a filter in the system (Figure IX-5). Particles that have settled are a residual source of contamination if re-entrained. In removal or cleaning of filters, care must be exercised to minimize re-entrainment and exposure of maintenance personnel.

Electrostatic precipitators are also commonly used to remove airborne particles. The precipitation process consists of providing an electric charge on the particle, establishing an electric field, and removing the particle from the precipitator.[47(p. 192)]

Electrostatic precipitators, used for outdoor pollution control, typically are of single-stage design and use a high direct-current (d-c) voltage (20-100 kV) to produce a negative corona (see Figure IX-7). The corona generated provides the necessary charge to particles for the electric field to cause them to drift to the collecting electrodes.

Electrostatic precipitators, which are used only for cleaning ventilation air, are designated as "electronic air-cleaners."[3] Three types of electronic air-cleaners are commonly used for control of particulate matter in residential and commercial environments:

• <u>Ionizing-plate type</u>: These devices are typically of the two-stage design, as shown in Figure IX-8. A high d-c voltage (e.g., 12 kV) produces a positive charge on the airborne dust, which is then precipitated on the collection plates. The positively charged corona is less effective as a particle-collector than the negatively charged, but produces much less ozone.

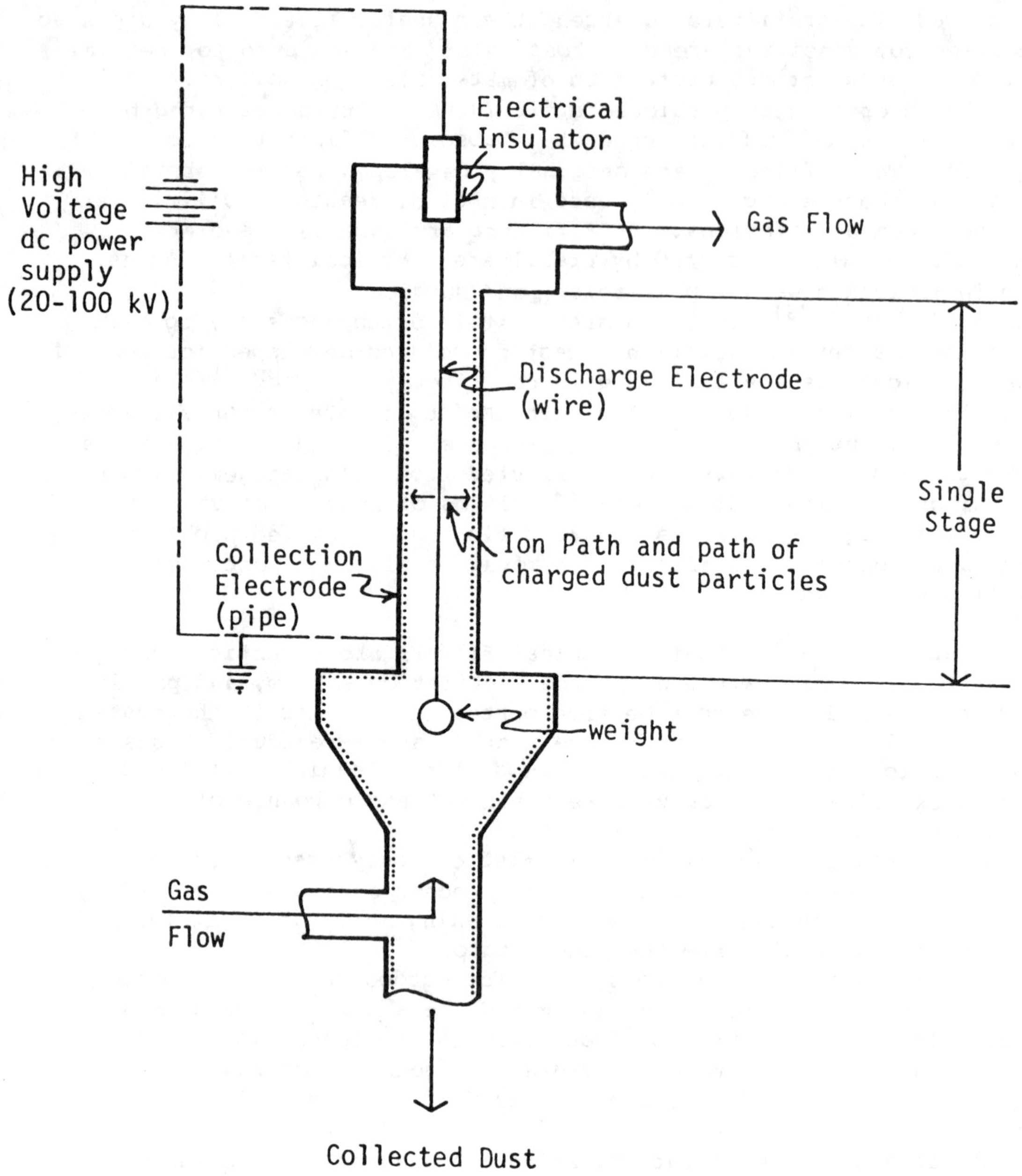

FIGURE IX-7 Schematic of a single-stage (wire and pipe) electrostatic precipitator. After Oglesby and Nichols.[47]

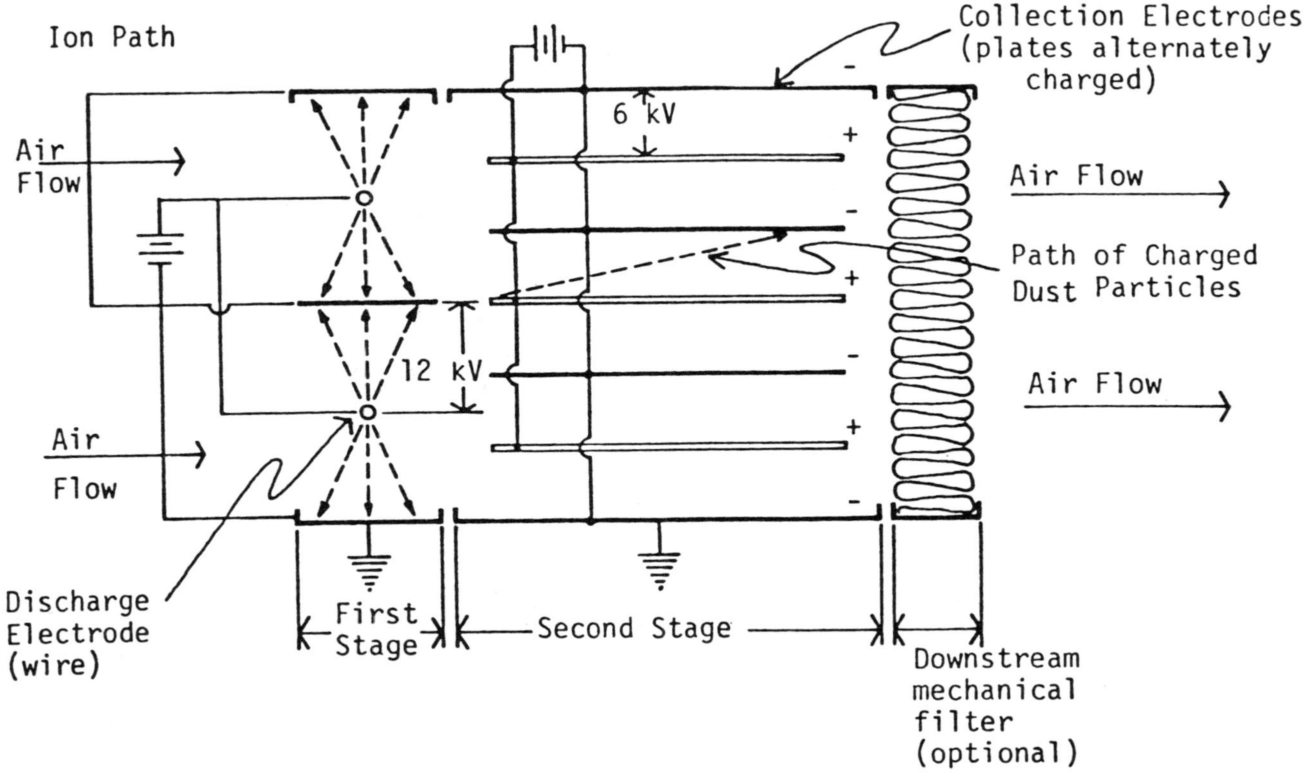

FIGURE IX-8 Schematic of a two-stage ionizing-plate electronic air-cleaner. After ASHRAE.[3]

• Charged-media, nonionizing type: A dielectric filtering medium (e.g., glass-fiber mat or cellulose mat) is supported on or in contact with an alternately charged and grounded gridwork (e.g., +12 kV d-c). Airborne particles are polarized in the resulting field and electromagnetically attracted to the filaments of the filter.

• Charged-media, ionizing type: Airborne particles are charged by positive ions from a discharge electrode. The charged particles are then collected on a charged filter mat downstream from the ionizer.

The performance of electrostatic precipitators is often evaluated in terms of the Deutsch equation:[47] $\eta = 1 - \exp[-Aw/Q]$, where η = particle removal efficiency, A = area of collecting surface (m^2), Q = gas volume flow rate (m^3/s), and w = migration velocity of particle (m/s). This equation can only approximate the actual removal efficiency. Manufacturers often publish performance data on removal efficiency, particle size, and an empirically derived migration velocity (called "precipitation rate parameter," w_p).[47(p. 211)] Unlike the migration velocity (w), the parameter w_p includes effects due to rapping losses (i.e., for industrial precipitators), gas flow distribution within the precipitator, particle size distribution, and dust resistivity.[47(p. 244)] A performance curve for a typical electrostatic precipitator is shown in Figure IX-9.

Conversely, the performance of electronic air-cleaners is not usually rated in terms of the Deutsch equation, but rather in terms of dust-spot efficiency[4] or the DOP method.[63] Their performance compares favorably with that of medium- to high-efficiency mechanical air filters, and their major advantage is the low resistance to air flow, compared with that of mechanical filters. However, this low resistance can be a disadvantage, if they are not installed so that there is uniform air velocity at the entrance of the cleaner. As shown in Table IX-8, compromises are often required with respect to removal efficiency, pressure drop, and space limitation when selecting electronic air-cleaners.

Electronic air-cleaners are often used for the same applications as described for medium-efficiency and HEPA filters and normally require the same type of prefiltering as medium- to high-efficiency mechanical filters to remove larger particles. Special care is needed in servicing, and any of three methods is acceptable:

• Removal of collection plates or charged media, washing with a detergent, and drying before reinstallation. This method is most common for residential units and small commercial applications.

• Washing of collecting plates in place with an integral washer. This method is commonly used in larger commercial installations.

• Collection of dry agglomerates dislodged from the collection plates. An automatic replaceable-medium filter is usually used for this purpose in large and small commercial installations (see Figure IX-8).

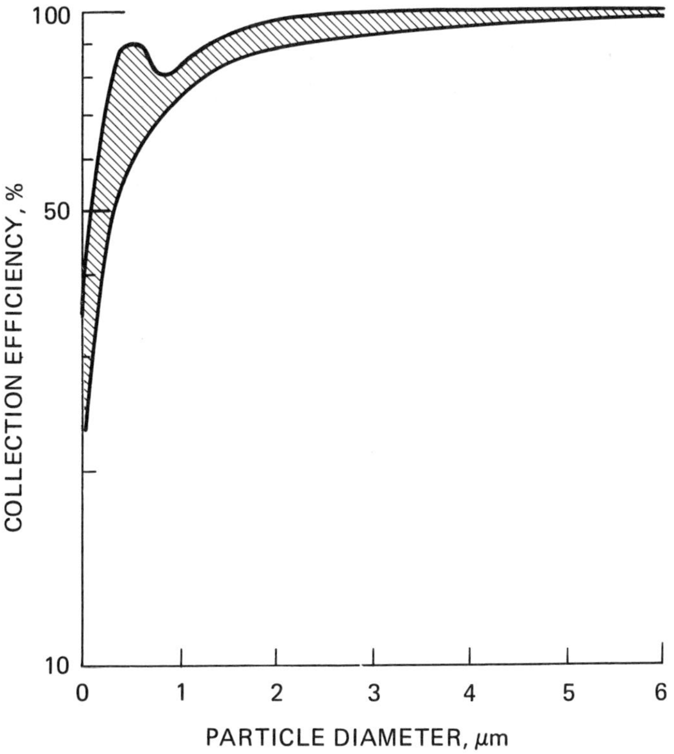

FIGURE IX-9 Typical removal efficiency of electronic air-cleaner. After ASHRAE.[3]

TABLE IX-8

Relative Performance of Electronic Air-Cleaners by Type[a]

Type of Cleaner	Removal Efficiency, %		Pressure Drop, Pa[a]	Relative Space Requirement
	Dust Spot Method	DOP Method		
Ionizing plate	85-90	65-80	25-40	Maximal
Charged media, nonionizing	25-35	10-20	40-75	Minimal
Charged media, ionizing	40-60	25-35	40-75	Moderate

[a]Values reported are approximate and were obtained by personal communication with C. W. Soltis, Filterlab, Inc., Houston, TX.

[b]At face velocities of 1.5-2.5 m/s. 1 pascal (Pa) = 1 newton/m^2.

There are some problems to be considered with electronic air-cleaners. There is a potential hazard associated with the high voltage at the ionizing electrode, collection plates and charged media;[3] ozone can be produced by the high voltage, even with the positive corona;[47(p. 196) 58] and soiling can be caused by charged dust particles that penetrate the air-cleaner.[48 57]

Disinfection with Germicidal Ultraviolet Radiation

Airborne contagion may be controlled with germicidal UV radiation produced by mercury-vapor discharge lamps. Modern germicidal lamps can be made of glass that blocks radiation in the ozone-producing range, but transmits the germicidal rays of 254-nm wavelength. This radiation is effective in disinfecting most pathogenic airborne bacteria and viruses, provided that the relative humidity does not exceed 70%; but it is less effective against fungi.

Direct exposure to germicidal UV radiation causes superficial skin and eye irritation, fading of fabric colors, and browning of plant leaves. The UV source must therefore be placed so as to prevent direct exposure. Usually, this is accomplished by irradiating the air above the heads of occupants. The UV lamp fixtures are relatively inexpensive to install and operate. Frequency of maintenance depends on dustiness of the environment. For example, in room installations with cold cathode tubes, yearly cleaning and biennial replacement are ordinarily adequate.

Upper-air irradiation is provided by lamp fixtures mounted on the wall or suspended from the ceiling at a height of about 7 ft.[51] Occupants of the room are protected from direct exposure to UV radiation by baffles. The effectiveness of disinfection depends on good mixing of the air in the upper and lower portions of the room; thus, stratification for energy conservation may be counterproductive for contamination control with UV irradiation. However, the concentration of airborne organisms in the breathing zone in a uniformly mixed space can be reduced to one-tenth to one-fifth that in the absence of UV radiation.

Upper-air irradiation is well suited to rooms with high ceilings if stratification is prevented, such as classrooms. Success in blocking the spread of measles has indicated that inactivation of measles virus with UV radiation is feasible. Improved UV lamps are being developed, but the modern trend toward low ceilings in homes will seriously limit the applicability of upper-air UV irradiation.

UV irradiation of air in supply-air ducts is technically easy, because intense radiation can be used there without hazard to people. The amount of radiation required depends on the size of the ducts and the supply-air flow rate.[18 78] When air is recirculated within the ventilation system, UV irradiation is very useful to reduce the concentrations of infectious organisms or droplet nuclei throughout the areas supplied by the forced-air system. UV irradiation in the ducts cannot stop the spread of infection in the room of a person with an infectious disease. When the source of infection is in the room, the concentration of infectious droplet nuclei is relatively high. Therefore, UV irradiation is required in the room itself, or other methods of contamination removal may be required.

Devices for Gas and Vapor Removal

Gas and vapor molecules cannot be effectively removed from airstreams by the mechanical and electrostatic principles so far described. Sorption is generally used for gas or vapor removal; there are three basic mechanisms, as described by Lieser:[35]

- Absorption: Penetration of molecules of the pollutant into the sorbent material (i.e., absorbing phase), which may be either solid or liquid. An example is absorption of nitrogen dioxide in a wet air-scrubber.
- Adsorption on external surfaces: Physical or chemical fixation on free surfaces. Examples of physical adsorption include fixation of noble gases or nitrogen on nonporous solids, such as aluminum oxide, graphite, ionic crystals, and metal foils. An example of chemical adsorption (chemisorption) is the uptake of carbon monoxide on transition metals, such as iron and nickel.
- Sorption on internal surfaces and in pores: Physical adsorption or chemisorption fixation on internal surfaces or within pores of porous solids. Capillary condensation within pores and occlusion of molecules or ions also occur. Examples of these sorbents are silica gel, aluminum hydroxide, activated carbon, clay minerals, and molecular sieves. An application of this mechanism would be radon sorption on activated carbon.

Although absorption is an important gas- or vapor-removal mechanism for industrial and military applications,[35] it is seldom used for environmental control in residential or commercial buildings, but is used to control odorous gases or vapors.

Adsorption is a dynamic process in which the net rate of adsorption is expressed as:[61] net rate of adsorption $\propto (ks - d)$, where s = rate of transfer of gas molecules to the adsorbent surface, k = fraction of molecules retained on the surface, and d = desorption rate. Because d increases with amount adsorbed, the performance of adsorption devices is not usually rated in terms of efficiency, but rather in terms of adsorption capacity and penetration time.[35,37,61]

Adsorption capacity is usually measured in terms of the amount (grams, moles, or cubic centimeters) adsorbed per gram of solid adsorbent as functions of the partial pressures of the adsorbates at constant temperatures (isotherms). Adsorption can be classified as physical, chemical, surface, or internal. Lieser[35] has reported on variations in adsorption efficiency as functions of relative partial pressure.

Because the adsorption process is dynamic, specification of the adsorption capacity is not sufficient to describe the effectiveness of the process. A measure of its desorption rate or its correlate, the penetration time, is also necessary. Penetration time has also been described as the "duration of adsorbent service before saturation"[61] and the "breakthrough time," or "time in which the maximum permissible concentration will not be exceeded."[76] Methods described by Turk[61] can provide a means to evaluate the dynamic performance of sorption devices that is consistent with those used to evaluate the performance of particle removal devices.

Standard test methods for evaluating the effectiveness of gas and vapor removal devices are not yet available in the heating, ventilating, and air-conditioning (HVAC) industry in the United States. However, some standard test procedures have been developed in western Europe.[76] The development of such a standard in the United States has recently been initiated by ASHRAE. The conventional method of selecting gas and vapor removal equipment has been to define the potential sources of contamination, describe the problem to equipment manufacturers, ask for equipment specifications, sometimes ask for a performance guarantee, and, less frequently, ask that the equipment be tested for compliance after it is installed (H. E. Burroughs, personal communication).

Performance of gas and vapor removal equipment depends on several factors:[61] concentration of the sorbate in the airstream, total surface area of the adsorbent, total volume of pores small enough to facilitate condensation of the adsorbed gases, presence of other gases or vapors (e.g., water vapor) that will compete with the adsorbate for a place on the adsorbent, physical and chemical characteristics of the adsorbate (weight, electric polarity, size, and shape), and electric polarity of the adsorbent surfaces.

Activated carbon consists mostly of neutral atoms of a single element and presents a surface with a relatively homogeneous electric charge. But oxygenated adsorbents (e.g., activated alumina, silica gel, and molecular sieves) contain nonhomogeneous distributions of electric charges and are polar. Oxygenated adsorbents have considerably greater selectivity than activated carbon and have much greater preferences for polar than for nonpolar molecules. Thus, oxygenated adsorbents are more useful for separation of pollutants, and activated carbon is generally more useful for overall decontamination. Because oxygenated adsorbents have a strong affinity for water vapor, which is highly polar, they are essentially ineffective for direct decontamination of moist air.[61]

To enhance adsorption, the adsorbent may be impregnated by other substances. Enhancement is achieved by chemical conversion of the pollutant by the impregnant to harmless or adsorbable products, the impregnant's functioning as a continuous catalyst for oxidation or decomposition, and the impregnant's functioning as an intermediate catalyst.[61]

Table IX-9 shows some examples of adsorbents and pollutants and the mechanisms of action. Except for formaldehyde, ammonia, and perhaps mercury, the pollutants listed may be found more commonly in the industrial environment. But this table illustrates the broad range of vapors and gases that might pollute the impregnated adsorbents. Note that activated carbon impregnated with sodium sulfite and activated

TABLE IX-9

Adsorbent Impregnations[a]

Adsorbent	Impregnant	Pollutant	Action
Activated carbon	Bromine	Ethylene; other alkenes	Conversion to dibromide, which remains on carbon
	Lead acetate	Hydrogen sulfide	Conversion to lead sulfide
	Phosphoric acid	Ammonia; amines	Neutralization
	Sodium silicate	Hydrogen fluoride	Conversion to fluorosilicates
	Iodine	Mercury	Conversion to mercuric iodide
	Sulfur	Mercury	Conversion to mercuric sulfide
	Sodium sulfite	Formaldehyde	Conversion to addition product
	Sodium carbonate or bicarbonate	Acidic vapors	Neutralization
	Oxides of copper, chromium, vanadium, etc.; noble metals (palladium, platinum)	Oxidizable gases, including reduced sulfur compounds, such as hydrogen sulfide, COS, and mercaptans	Catalysis of air oxidation
Activated alumina	Potassium permanganate	Easily oxidizable gases, especially formaldehyde	Oxidation
	Sodium carbonate or bicarbonate	Acidic gases	Neutralization

[a]Reprinted with permission from Turk.[61]

alumina impregnated with potassium permanganate are both effective in adsorbing formaldehyde. This point is important, because formaldehyde has been found indoors in concentrations above those considered acceptable by some European standards and ANSI/ASHRAE Standard 62-1981. Gas-cleaning devices with the required impregnated adsorbents are commercially available in the United States for use in residential and commercial buildings. For example, activated carbon does not adsorb carbon monoxide, but a combination of a desiccant (silica gel) and an oxidation catalyst, Hopcalite (a mixture of cupric oxide and manganese dioxide), has been reported to be effective,[75] although the Hopcalite must be kept scrupulously dry.

Preconditioning of the adsorbate can be used to enhance gas- or vapor-cleaning.[61] Some of the techniques are as follows:

- Use of a particulate prefilter to reduce loading of the adsorbent.
- Concentration of the adsorbate (e.g., by pressurizing the system).
- Removal of moisture from the airstream by dehumidification to a relative humidity below 50%.
- Cooling of the airstream to below 38°C (100°F).

Some of these techniques may be more energy-intensive than reducing pollutant concentration by dilution ventilation. Thus, limitations to the enhancement potential must be considered, including:

- Penetration time through the adsorbent is inversely proportional to the air flow rate and the concentration or vapor pressure of the adsorbate.[35, 61]
- Resistance to air flow (i.e., pressure drop) increases with the air flow rate or velocity, the mesh size of the adsorbent, and the thickness of bed.[61]
- Adsorption capacity decreases as the temperature or relative humidity of the airstream increases.[61, 75]

Servicing tehniques for gas and vapor removal equipment include replacement or regeneration of the sorbent and are similar to those for replacing mechanical filters. The sorbent may be contained in panel-like trays, canisters, or pleated retainers.[31] Discarding old sorbent and replacing with new is normally cost-effective, if the penetration time exceeds a month and the sorbate concentration is low (as in residential air-conditioning systems) or if the sorbent requires impregnation.[61] However, in some large HVAC systems, such as in airline terminals, where the adsorbate concentrations are relatively high (penetration times relatively short) and impregnated sorbents are not required, regeneration may be cost-effective. Regeneration is accomplished primarily by thermal methods, but techniques with ionizing radiation and chemical activation are also available.[14, 27, 31, 61] Regeneration may be accomplished on site, or the sorbent may be returned to the manufacturer for processing, as is usually done for air-conditioning applications.

SUMMARY

Air-cleaning devices for residential and commercial applications are commercially available. However, no single type of air-cleaning process is available that can control particles, gases, and vapors. Therefore, it is necessary to rely on multistage systems to obtain the desired control, if all three types of contaminants are to be removed. These systems consist of combinations of mechanical filters, electronic air-cleaners, and gas and vapor removal devices. The systems can be used as components within central air-conditioning systems or as unitary appliances (fan-filter modules).

Voluntary standards exist in the United States for methods of rating performance of particle-removal devices, and corresponding standards for gas- and vapor-removal devices are now being developed. However, no standards provide procedures for rating performance of fabricated or assembled air-cleaning subsystems, whether they are installed as components or used as unitary appliances.

Some effort toward predicting dynamic performance of gas-cleaning from liquids has been reported,[77] but dynamic modeling of indoor air-quality control systems has received little attention.

Many of the air-cleaning control devices impose substantial resistance to air flow (i.e., system pressure drop). This resistance requires additional energy consumption. Costs of installing and servicing these devices must also be considered. Thus, optimization techniques should be considered to decide between alternatives that will provide acceptable indoor air quality, energy consumption, and life-cycle costs.

STRATEGIES FOR CONTROL OF INDOOR POLLUTION

The demarcation between healthful and unhealthful air is not well defined. Although air-quality standards exist for outdoor air and for the industrial environment, few standards directly address the indoor, nonindustrial environment (see Appendix A). Occupants of indoor environments are expected to be exposed to long-term low concentrations and intermittent high concentrations of pollutants. Control strategies designed to limit indoor concentrations must consider the possible time dependence between exposure and effects, as well as the specifics of source configuration and contaminant characteristics (i.e., gas versus particles, reactivity versus nonreactivity, molecular weight, particle size).

Five general control strategies (see Table IX-10) have been identified and may be applied in indoor environments: ventilation, source removal, source modification, pollutant removal, and education. These strategies may not be mutually exclusive; combinations, such as source modification plus ventilation, may be preferred in some situations. This section briefly discusses these strategies and tabulates the relative effectiveness of some of them for various contaminants. More details on controls are provided in Chapter IV, with respect to individual contaminants or sources.

TABLE IX-10

Indoor-Pollution Control Strategies

I. Control by Ventilation

   A. General ventilation
   B. Spot (zone or localized) ventilation
   C. Infiltration

II. Control by Source Removal

   A. Material or product substitution
   B. Restrictions on source use, sales, and activities by type of indoor facilities

III. Control by Source Modification

   A. Change in combustion design
   B. Material substitution
   C. Reduction in emission rates by intervention of barriers

IV. Control by Air-Cleaning (Pollutant Removal)

   A. Particle filtering
   B. Gas and vapor removal
   C. Passive scavenging or absorption

V. Education

   A. Consumer information on products and materials
   B. Public information on health, soiling, productivity, and nuisance effects
   C. Resolution of legal rights and liabilities of consumer, tenant, manufacturer, etc., related to indoor quality

Ventilation is the principal means of controlling residential indoor pollution by dilution with outdoor fresh air. (See Figure IX-10 for schematic of HVAC system for nonresidential building.) Outdoor air is brought in and dilutes the indoor-generated pollution. This strategy is in conflict with energy conservation, in that the heat and humidity of the displaced indoor air are not conserved. Control by dilution requires a well-ventilated indoor space; energy conservation requires reduction in the amount of unconditioned air brought in from outdoors. Optimal air ventilation must be estimated, or energy must be conserved and indoor quality preserved without reducing ventilation rate. Heat-exchangers provide the technologic capabilities to conserve energy while not substantially reducing the ventilation rate.

Control of indoor air quality in many buildings depends on the HVAC system. Maintaining acceptable indoor air quality has not yet been a design feature of HVAC systems (see earlier sections of this chapter). The American Society of Heating, Refrigerating, and Air-Conditioning Engineers has recently included indoor air quality as additional design and operating criteria for HVAC in its new ventilation standard.[2] The use of HVAC systems as a means of controlling the quality of indoor air is promising.

HVAC systems can be divided into several categories, as listed in Table IX-11. Each category has advantages and disadvantages that broadly define the characteristics of the individual system.[3] Potentially, the mass concentration and thermodynamic requirements can be met by any of several HVAC systems. However, proper design will optimize equipment selection and component configuration and the functional and economic requirements of a building's HVAC system. Table IX-12 indicates typical applications for various HVAC systems. Local exhaust--e.g., forced ventilation near a known or well-defined source of indoor pollution--is widely used in industrial environments and to a lesser extent in residences. Control is achieved by exhausting the pollutant at the source to the outdoor environment. Examples of such residential use are bathroom vents and vents over cooking and heating facilities. It is important that exhaust air not be re-entrained into the building.

Some areas in a structure may have unique requirements for spot ventilation--for example, positive- and negative-air-pressure zones. Exhaust fans may be required to control moisture and odors in bathrooms without windows. Kitchens and kitchen stoves usually have some form of filtered, unvented forced draft for contaminant control, and unvented range hoods with charcoal filters have recently become popular. These may be effective for grease, odors, and other large molecules, but not for removing carbon monoxide and other small molecules. Furthermore, the filters are somewhat expensive and in normal practice are not changed often enough to be reliable. For these reasons, the unvented range hood cannot be considered a reliable pollution control device.

A further problem with the use of spot-ventilation exhaust fans in airtight houses is that too much powered exhaust reduces the natural draft in the furnace vent and can cause combustion products from the furnace to be drawn into the living space. Spot ventilation or exhaust is not provided for gas stoves, gas ovens, small gas water-heaters, and

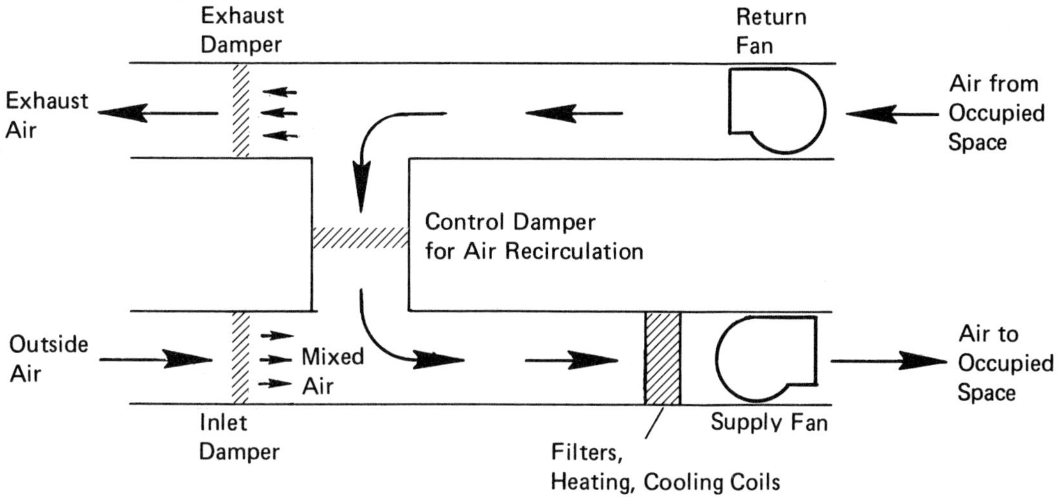

FIGURE IX-10 Schematic of conventional heating, ventilating, and air-conditioning (HVAC) air-supply system for nonresidential buildings.

TABLE IX-11

Air-Handling Systems

Category	Description	Advantages	Disadvantages
All air	Provides all required sensible and latent heat-exchange capacity in air supplied by the system	1. Centralized location of all major equipment 2. Removes major components from conditioned area 3. Greatest potential for use of free cooling 4. Flexibility under varied operating conditions 5. Easily adapted to heat recovery 6. Optimal distribution for air motion control 7. Suitable for large makeup-air requirements 8. Adaptable to automatic seasonal changeover 9. Adaptable to winter humidification	1. Additional duct clearance requirements 2. Additional fan energy required for perimeter load during unoccupied hours 3. Air balancing difficult 4. Accessibility to terminal devices required
Air-water	Both air and water distributed to space to provide required heating and cooling	1. Individual room temperature control 2. Flexibility under varied operating conditions 3. Low distribution space requirement 4. Reduced central equipment space requirement 5. Horsepower savings by using water instead of air 6. Reduction in fan power requirements during occupied period 7. Can eliminate cross-contamination 8. Long life of components	1. Low primary air quantities make design of two-pipe system critical for proper intermediate-season control 2. System changeover can be complicated 3. Usually limited to perimeter spaces 4. Controls tend to be more complex 5. Secondary air flows create high maintenance requirements 6. Primary air supply usually constant 7. Primary air provides all dehumidification, so low-dew-point air is provided 8. Not able to handle high-exhaust applications

Table IX-11 (contd)

Category	Description	Advantages	Disadvantages
All water	Provides space required for heating and cooling by distributing hot and chilled water to terminal units	1. Flexibility for adaptation to many building configurations 2. One of lowest-first-cost central-perimeter systems 3. Easy to retrofit to existing structures 4. Low system distribution requirements 5. Low cross-contamination potential 6. Individual room control with quick response to varying loads 7. No seasonal changeover need be required	1. Inadequate relative-humidity control 2. No positive ventilation for many types of designs 3. Through-the-wall units may be unsatisfactory in appearance on outside of building 4. Two-pipe systems require seasonal changeover 5. Maintenance and service work required in occupied space 6. Filters, coils, and condensate drain lines must be kept clean to limit bacterial growth
Unitary	Packaged system that can provide heating and cooling	1. Individual room control 2. Individual air distribution control 3. Heating and cooling independently controlled by zone 4. Individual ventilation air control 5. Usually space-saving 6. Usually low installation costs 7. Usually lower initial costs 8. Allows zone shutdown	1. Limited options available for size and control 2. Limited capability for exceptionally high or exceptionally low relative humidity 3. Acoustics must be carefully considered 4. Maintenance in occupied space required 5. Exterior building aesthetics may be affected 6. Higher operating costs

TABLE IX-12

HVAC System Typical Applications

Category	System	Typical Applications
All air	Single zone	Small department stores, individual shops, computer rooms, single-family residences, warehouses
	Variable air volume	Offices, institutional and public-assembly buildings
	Reheat	Offices, laboratories, hospitals
	Dual duct	Offices, institutional and public-assembly buildings
	Multizone	Small buildings, small offices
Air-water	Induction	Offices, hospitals
	Dual-duct induction	Offices
All water	Fan coil	Motels, hospitals, offices, apartments
	Valance units	Warehouses
Unitary	Window air conditioners	Residences, small office buildings
	Through-the-wall conditioners	Residences, small office buildings
	Rooftop	Department stores, malls, offices
	Water loop heat pumps	Large offices

space-heaters fueled with gas or oil; therefore, large amounts of carbon dioxide and water vapor are introduced indoors.

The indoor-pollution problems caused by the lack of spot ventilation or exhaust in single-family residences are only now being studied. The information is inadequate to assess the magnitude of the problems or to define the amount of ventilation air needed to abate the pollutants produced by these sources.

Source removal is the most effective means of controlling indoor pollution. Examples of source removal are no-smoking areas and prohibition of urea-formaldehyde foam insulation and kerosene heating units for indoor spaces occupied by people. These strategies are more effective when substitute products are available and less effective when they rely on enforcement to ensure compliance. Where source-removal strategies modify human behavior, conflict with consumer preference, or involve an economic penalty, they are less likely to be adopted by regulatory bodies. The adverse effects of indoor contaminant exposure must be well established in the public perception. Public debate centered on the restriction of smoking in public places illustrates the controversy that surrounds source-removal strategies for maintaining indoor air quality. However, when material or product substitution is not disruptive or expensive, source removal is clearly the strategy of choice. It is obvious that these decisions should be made early in the design stages of new facilities. If a material or product already in use is determined to be hazardous, removal may still be the strategy of choice. Source removal has been applied in the removal of lead from house paint both in the product and by paint removal. A current widespread effort to remove all asbestos from school buildings is another example of the source-removal control strategy. Cost consideration must be carefully compared with the likely benefits in reduced health risks and property damage and with other imputed benefits. Source removal may cause a displaced problem, such as occupational exposure during removal or a hazardous-waste disposal problem. These and other factors must be carefully considered before the institution of a program to remove an existing source.

Air-cleaning devices have been used in large indoor commercial, industrial, and institutional environments to eliminate or reduce indoor pollutants. This strategy has not been widely used in residences, because the devices are expensive to buy and operate and can be bulky and noisy. Small commercial electrostatic precipitators, ion generators, air filters, and gas absorbers (charcoal filters) are used to remove contaminants in some indoor environments. Many of these devices are advertised to provide particlefree and odorfree clean indoor environments. The efficiencies of these devices need to be evaluated by independent organizations.

Source modification is an alternative to source removal. The objective of source modification is to reduce the rate of pollutant emission into the indoor environment. Source modification includes maximizing the efficiency of gas cooking and heating facilities that reduce emission of some pollutants. Coating of lead-based paints and asbestos-containing building materials to seal the surface and prevent emission is effective and practical. Coating radon- and formaldehyde-emitting surfaces is promising and warrants further study. A source should not be modified when it can be assumed that the modification will cause emission of a different contaminant. The spraying of surfaces that are formaldehyde-emitters may itself constitute a source of indoor contamination.

Table IX-13 summarizes control strategies available for several types of pollutants. The table identifies strategies proved effective in controlling a pollutant, but interactive effects must be considered if several pollutants are to be controlled simultaneously. This requirement and the complexity of control strategies lead to the necessity of an overall systems designer.

Control of indoor contaminant concentrations by dilution with outdoor air will continue to be a major control strategy. Direct control of the ventilation system based on indoor contaminant concentration is the best means of achieving the optimal compromise between energy conservation and pollution control.

Some provision is needed to add or conserve moisture. Homes in cold climates need to conserve humidity in the indoor air in winter and reject as much water as possible to the outside in summer. Simple energy-conserving means for this kind of moisture control are not yet available, but the latent heat associated with moisture movement can represent substantial energy that is not conserved.

New ventilation control strategies are needed. Positive ventilation with heat recovery should be introduced in the building industry. Past practice fixed the temperature of the mixed air (outside air plus recirculated air). This simplified comfort control, but usually resulted in excessive energy loss. A floating mixed-air temperature based on outside-air temperature can provide closer control of the ventilation air and energy savings.

New sensors for optimal control of ventilation should be developed. Although laboratory instruments can measure the concentration of some indoor pollutants, often these instruments are too bulky, too expensive, too complex, and generally not suitable for extended, unattended use that might be required in measuring indoor environments.

Greater emphasis should be placed on controlling specific pollutants at their sources. Combustion-generated pollutants--including carbon dioxide, water vapor, carbon monoxide, and nitrogen oxides--can be removed at the source. New inexpensive, small, and uncomplicated pollutant control devices are also needed. New construction materials must be examined carefully for undesirable environmental effects.

The efficiency of each control strategy must be studied both in the laboratory and under "real-life" conditions. As indicated earlier, a systems approach may be required in large structures; however, less elaborate and inexpensive means of controlling contamination in indoor residential environments are conceptually possible, are needed, and can become practicable.

Some indoor pollution problems can be controlled through the marketplace choices of an educated consuming public. The general public must be informed of the sources of indoor contaminants and the

TABLE IX-13

Summary of Control Strategies for Sources of Indoor Air Pollutants[a]

Pollutants	Source Modification		Ventilation			Pollutant Removal		Other
	Material Substitution	Containment	Local Exhaust	General Ventilation	Spot Ventilation	Particle Removal	Gas and Vapor Removal	
Radon and progeny	R	R	NR	R	NR	R	NR	--
Formaldehyde and other organic substances	R	R	NR	R	NR	NR	R	--
Asbestiform minerals and fibrous glass	R	R	NR	NR	NR	R	NR	--
Combustion products	R[b]	NR	R[c]	R	R[c]	R[d]	R[d]	--
Consumer products	R	R	NR	R	NR	R[d]	R[d]	--
Tobacco smoke	R	NR	NR	R	NR	R[d]	R[d]	--
Odors	NR	R	R[c]	R	R[c]	R[d]	R[d]	Masking (R)
Biologic agents (particles)	R	R	R	R	R	R	NR	UV disinfection (R)
Carbon dioxide	NR	NR	R	R	R	NR	NR	--
Moisture	NR	NR	R	R	R	NR	R	--
Thermal extremes	R	R	R	R	R	NR	R	--

[a]R = recommended; NR = not recommended.

[b]Source of heating can be substituted (e.g., heat pump, electric range).

[c]Usually combined strategies for a given pollutant.

[d]Both cleaning methods usually required.

adverse consequences of acute and chronic exposures. It must be informed about the cost and effectiveness of various control options and the efficiencies of commercially available air-cleaning equipment. The public should be informed of its legal rights with respect to product liability. The obligation and rights under purchase and lease agreements pertaining to healthful indoor environments for residential, commercial, and public places must be defined. Education provides easy and inexpensive steps that help to improve indoor air quality. Such steps include reduction in indoor smoking, ban of potentially harmful indoor sprays, use of proper paint, changes in daily routines to avoid exposing all family members to pollutants, and the like. The efficiency of this control strategy cannot be estimated, but most would agree that only a properly educated public can require steps toward implementing one or more combinations of the other control strategies. Public-interest organizations, public utilities, professional societies, trade and manufacturing associations, and government agencies all have a responsibility to ensure that the public receives factual information related to indoor contaminants.

REFERENCES

1. American National Standards Institute. Constitution and Bylaws of the American National Standards Institute. New York: American National Standards Institute, 1978. 16 pp.
2. American National Standards Institute, and American Society of Heating, Refrigerating and Air-Conditioning Engineers. ANSI/ASHRAE Standard 62-1981. Ventilation for Acceptable Indoor Air Quality. New York: American Society of Heating, Refrigerating and Air Conditioning Engineers, Inc., 1981. 48 pp.
3. American Society of Heating, Refrigerating and Air-Conditioning Engineers. ASHRAE Handbook and Product Directory. 1979 Equipment, pp. 2.1-2.8. New York: American Society of Heating, Refrigerating and Air-Conditioning Engineers, Inc., 1979.
4. American Society of Heating, Refrigerating and Air-Conditioning Engineers. ASHRAE Standard 52-76. Method of Testing Air-Cleaning Devices Used in General Ventilation for Removing Particulate Matter. New York: American Society of Heating, Refrigerating and Air-Conditioning Engineers, Inc., 1976.
5. American Society of Heating, Refrigerating and Air-Conditioning Engineers. ASHRAE Standard 55-74. Thermal Environmental Conditions for Human Occupancy. New York: American Society of Heating, Refrigerating and Air-Conditioning Engineers, Inc., 1974. 12 pp.
6. American Society of Heating, Refrigerating and Air-Conditioning Engineers. ASHRAE Standard 62-73. Standards for Natural and Mechanical Ventilation. New York: American Society of Heating, Refrigerating and Air Conditioning Engineers, Inc., 1973. 17 pp.
7. American Society of Heating, Refrigerating and Air-Conditioning Engineers. ASHRAE Standard 90-75. Energy Conservation in New Building Design (Section 12). New York: American Society of Heating, Refrigerating and Air-Conditioning Engineers, Inc., 1977. 11 pp.

8. American Society of Heating, Refrigerating and Air-Conditioning Engineers. ASHRAE Handbook and Product Directory. 1978 Applications Volume, pp. 4.1-4.8. New York: American Society of Heating, Refrigerating and Air-Conditioning Engineers, Inc., 1978.
9. American Society of Heating, Refrigerating and Air-Conditioning Engineers, and American National Standards Institute. ANSI/ASHRAE Standard 62-1981. Ventilation for Acceptable Indoor Air Quality. New York: American Society of Heating, Refrigerating, and Air-Conditioning Engineers, Inc., 1981. 47 pp.
10. American Standards Association. American Standard Building Requirements for Light and Ventilation--A 53.1. New York: American Standards Association, 1946. 18 pp.
11. Arnold and O'Sheridan, Inc. Ventilation Practices and Energy Conservation. A Study of Ventilation Codes and Practices in the State of Wisconsin. Report to Wisconsin Department of Industry, Labor and Human Relations. Madison, Wisc.: Arnold and O'Sheridan, Inc., 1979. 50 pp.
12. Arthur D. Little, Inc. Impact Assessment of ASHRAE Standard 90-75, Energy Conservation and New Building Design. Federal Energy Administration Report. Washington, D.C.: U.S. Government Printing Office, 1976. Available from National Technical Information Service, Springfield, Va., as PB-252 639.
13. Banks, R. S., Ed. Hospital Ventilation Standards and Energy Conservation. Proceedings of the 1978 International Working Conference. Berkeley, Cal.: Lawrence Berkeley Laboratory, 1978. 310 pp.
14. Berg, R. Miscellaneous reactivation methods, pp. 337-374. In M. Bonnevie-Svendsen, Ed. Sorption and Filtration Methods for Gas and Water Purification. NATO Advanced Study Institutes Series E, Vol. 13. Leyden, Netherlands: Noordhoff International Publishing, a division of A. W. Sijthoff International Publishing Company, B. V., 1975.
15. Building Officials and Code Administrators International, Inc. BOCA Basic Building Code. Annual Supplement. 1976. Chicago: Building Officials and Code Administrators International, Inc., 1976.
16. Building Officials and Code Administrators International, Inc. The BOCA Basic Building Code 1975. 6th ed. Chicago: Building Officials and Code Administrators International, Inc., 1975. 497 pp.
17. Building Officials and Code Administrators International, Inc. (BOCA), International Conference of Building Officials (ICBO), National Conference of States on Building Codes and Standards, Inc. (NCSBCS), and Southern Building Code Congress International, Inc. (SBCCI). Model Code for Energy Conservation in New Building Construction. Washington, D.C.: U.S. Department of Energy, 1977. 76 pp. Available from National Technical Information Service, Springfield, Va., as SAN/1230-1.
18. Buttolph, L. J., and H. Haynes. Ultraviolet air sanitation. Report No. LD-11. Cleveland, Ohio: General Electric Lamp Division, 1950. 44 pp.
19. Carver, L. D. Particle size analysis. Ind. Res. 13(8):40-43, 1971.

20. Chang, P. K. Survey on Coanda flow. In Proceedings of the Fluid Amplification Symposium. Vol. 1. U.S. Department of the Army, Ordnance Corps, 1962. Distributed by U.S. Department of Commerce, Office of Technical Services, Washington, D.C.
21. Crawford, M. Air Pollution Control Theory. New York: McGraw-Hill Publishing Company, 1976. 624 pp.
22. Dorman, R. G. Dust Control and Air Cleaning. New York: Pergamon Press, Inc., 1974. 615 pp.
23. Fanger, P. O. Proposed Nordic standard for ventilation and thermal comfort. In F. E. de Oliveira, J. E. Woods, and A. Faist, Eds. Building Energy Management--Conventional and Solar Approaches. Proceedings of the International Congress, May 12-16, 1980, Povoa de Varzim, Portugal. New York: Pergamon Press, 1980.
24. Fanger, P. O. Thermal Comfort. Analysis and Applications in Environmental Engineering. Copenhagen: Danish Technical Press, 1970. 244 pp.
25. Gagge, A. P. Rational temperature indices of man's thermal environment and their use with a 2-node model of his temperature regulation. Fed. Proc. 32:1572-1582, 1973.
26. Geomet, Incorporated. The Status of Indoor Air Pollution Research 1976. U.S. Environmental Protection Agency (Environmental Monitoring and Support Laboratory) Report No. EPA-600/4-77-029. Washington, D.C.: U.S. Government Printing Office, 1977. 487 pp.
27. Halmo, T. Chemical regeneration, pp. 327-336. In M. Bonnevie-Svendsen, Ed. Sorption and Filtration Methods for Gas and Water Purification. NATO Advanced Study Institutes Series E, Vol. 13. Leyden, Netherlands: Noordhoff International Publishing, a division of A. W. Sijthoff International Publishing Company, B. V., 1975.
28. Iinoya, K., and C. Orr, Jr. Filtration, pp. 149-188. In A. C. Stern, Ed. Air Pollution. 3rd ed. Vol. IV. Engineering Control of Air Pollution. New York: Academic Press, Inc., 1977.
29. International Conference of Building Officials. Uniform Building Code Standards. 1979 Edition. Whittier, Cal.: International Conference of Building Officials, 1979. 1208 pp.
30. Janssen, J. E. Automatic Variable Ventilation Control System. Monthly Progress Report No. 12. Report to Lawrence Berkeley Laboratory. Contract No. 4500610. Minneapolis, Minn.: Honeywell Inc., April 1980.
31. Juhola, A. J. Thermal regeneration, pp. 303-326. In M. Bonnevie-Svendsen, Ed. Sorption and Filtration Methods for Gas and Water Purification. NATO Advanced Study Institutes Series E, Vol. 13. Leyden, Netherlands: Noordhoff International Publishing, a division of A. W. Sijthoff International Publishing Company, B. V., 1975.
32. Klauss, A. K., R. H. Tull, L. M. Roots, and J. R. Pfafflin. History of the changing concepts in ventilation requirements. ASHRAE J. 12(6):51-55, 1970.
33. Kusuda, T. Control of ventilation to conserve energy while maintaining acceptable indoor air quality. ASHRAE Trans. 82(Pt. 1):1169-1181, 1976.

34. Lapple, C. E. Particle-size analysis and analyzers. Chem. Eng. 75(11):149-156, 1968.
35. Lieser, K. H. Sorption mechanisms, pp. 91-145. In M. Bonnevie-Svendsen, Ed. Sorption and Filtration Methods for Gas and Water Purification. NATO Advanced Study Institutes Series E, Vol. 13. Leyden, Netherlands: Noordhoff International Publishing, a division of A. W. Sijthoff International Publishing Company, B. V., 1975.
36. Luciano, J. R. Air Contamination Control in Hospitals. New York: Plenum Press, 1977. 479 pp.
37. Maggs, F. A. P. The use of adsorptive filters in air purification, pp. 5-12. In M. Bonnevie-Svendsen, Ed. Sorption and Filtration Methods for Gas and Water Purification. NATO Advanced Study Institutes Series E, Vol. 13. Leyden, Netherlands: Noordhoff International Publishing, a division of A. W. Sijthoff International Publishing Company, B. V., 1975.
38. McIntyre, D. A. The effect of air movement on thermal comfort and sensation, pp. 541-560 (includes discussion). In P. O. Fanger and O. Valbjørn, Eds. Indoor Climate. Effects on Human Comfort, Performance, and Health in Residential, Commercial, and Light-Industry Buildings. Proceedings of the First International Indoor Climate Symposium, Copenhagen, August 30-September 1, 1978. Copenhagen: Danish Building Research Institute, 1979.
39. National Research Council, Committee on Revision of the Guide for Laboratory Animals Facilities and Care. Guide for the Care and Use of Laboratory Animals. DHEW Publication No. (NIH) 73-23. Washington, D.C.: U.S. Government Printing Office, 1972. 56 pp.
40. National Research Council, Institute of Laboratory Animal Resources. Standards for the Breeding, Care and Management of Laboratory Dogs. Washington, D.C.: National Academy of Sciences, 1964. 26 pp.
41. National Research Council, Institute of Laboratory Animal Resources, Committee on Standards. Standards and Guidelines for the Breeding, Care and Management of Laboratory Mice. Rev. ed. Washington, D.C.: National Academy of Sciences, 1967. 29 pp.
42. National Research Council, Institute of Laboratory Animal Resources, Committee on Standards. Standards for the Breeding, Care and Management of Laboratory Rabbits. Washington, D.C.: National Academy of Sciences, 1967. 12 pp.
43. Nevins, R. G. Air Diffusion Dynamics: Theory, Design and Application. Birmingham, Mich.: Business News Publishing Company, 1976. 136 pp.
44. Nevins, R. G., and P. L. Miller. Analysis, evaluation and comparison of room air distribution performance--A summary. ASHRAE Trans. 78(Pt. 2):235, 1972.
45. Nielsen, P. V. Flow in Air Conditioned Rooms. Technical University of Denmark, Dissertation, 1974.
46. Nielsen, P. V., A. Restivo, and J. H. Whitelaw. The velocity characteristics of ventilated rooms. J. Fluids Eng. 100:291-298, 1978.

47. Oglesby, S., Jr., and G. B. Nichols. Electrostatic precipitation, pp. 189-256. In A. C. Stern, Ed. Air Pollution. 3rd ed. Vol. IV. Engineering Control of Air Pollution. New York: Academic Press, Inc., 1977.
48. Penney, G. W., and N. G. Ziesse. Soiling of surfaces by fine particles. ASHRAE Trans. 74(Pt. 1):VI.3.1-VI.3.13, 1968.
49. Repace, J. L., and A. H. Lowrey. Indoor air pollution, tobacco smoke, and public health. Science 208:464-472, 1980.
50. Reynolds, G. L., and J. E. Woods. Building energy management programs in Iowa. In F. E. de Oliveira, J. E. Woods, and A. Faist, Eds. Building Energy Management--Conventional and Solar Approaches. Proceedings of the International Congress, May 12-16, 1980, Povoa de Varzim, Portugal. New York: Pergamon Press, 1980.
51. Riley, R. L., and S. Permutt. Room air disinfection by ultraviolet irradiation of upper air. Air mixing and germicidal effectiveness. Arch. Environ. Health 22:208-219, 1971.
52. Rohles, F. H., Jr., and R. G. Nevins. The nature of thermal comfort for sedentary man. ASHRAE Trans. 77(Pt. 1):239-246, 1971.
53. Rohles, F. H., Jr., J. E. Woods, and R. G. Nevins. The effects of air movement and temperature on the thermal sensations of sedentary man. ASHRAE Trans. 80(Pt. 1):101-118, 1974.
54. Sem, G. J., J. A. Borgos, K. T. Whitby, and B. Y. H. Liu. State-of-the-Art: 1971. Instrumentation for Measurement of Particulate Emissions from Combustion Sources. Vol. 3. Particle Size. U.S. Environmental Protection Agency Report No. APTD-1524. Washington, D.C.: U.S. Environmental Protection Agency, 1972. 84 pp. Available from National Technical Information Service, Springfield, Va., as PB-233 393.
55. Southern Building Code Congress International, Inc. Southern Standard Building Code. 1979 Edition. Birmingham, Ala.: Southern Building Code Congress International, Inc., 1979.
56. Spaite, P. W., and J. O. Burckle. Selection, evaluation, and application of control devices, pp. 43-95. In A. C. Stern, Ed. Air Pollution. 3rd ed. Vol. IV. Engineering Control of Air Pollution. New York: Academic Press, Inc., 1977.
57. Sutton, D. J., H. A. Cloud, P. E. McNall, Jr., K. Nodolf, and S. H. McIver. Performance and applications of electronic air cleaners in occupied spaces. ASHRAE J. 6(6):55-62, 1964.
58. Sutton, D. J., K. M. Nodolf, and K. K. Makino. Predicting ozone concentrations in residential structures. ASHRAE J. 18(9):21-26, 1976.
59. Syska and Hennesy, Engineers, and Tishman Research Corp. A Study of the Effects of Air Changes and Outdoor Air on Interior Environment, Energy Conservation, and Construction and Operating Costs: Phase I. Report to U.S. General Services Administration. New York: Syska and Hennesy, Engineers, 1973.
60. Thompson, G. P. Energy conservation and the law. ASHRAE J. 18(7):20, 1976.
61. Turk, A. Adsorption, pp. 329-363. In A. C. Stern, Ed. Air Pollution. 3rd ed. Vol. IV. Engineering Control of Air Pollution. New York: Academic Press, Inc., 1977.

62. U.S. Department of Commerce. Implementation of Federal voluntary standards policy; Proposed procedure for listing voluntary standards bodies eligible for Federal agency support and participation, and for a Department sponsored voluntary dispute resolution service for procedural complaints against listed voluntary standards bodies. Fed. Reg. 45:37374-37383, June 2, 1980.
63. U.S. Department of Defense, Office of the Assistant Secretary of Defense. Military Standard 282. Filter Units, Protective Clothing, Gas Mask Components and Related Products: Performance--Test Methods. Washington, D.C.: U.S. Government Printing Office, 1956. 72 pp.
64. U.S. Department of Energy. Emergency building temperature restrictions. Fed. Reg. 44:39354-39369, July 5, 1979.
65. U.S. Department of Energy, and U.S. Department of Housing and Urban Development. Energy performance standards for new buildings. Fed. Reg. 44:68218-68220, November 28, 1979.
66. U.S. Department of Health, Education, and Welfare, National Institutes of Health. Recombinant DNA research--Guidelines. Fed. Reg. 41:27902-27943, 1976.
67. U.S. Department of Health, Education, and Welfare, Health Resources Administration, Bureau of Health Facilities Financing, Compliance, and Conversion. Minimum Requirements of Construction and Equipment for Hospital and Medical Facilities. U.S. Department of Health, Education, and Welfare Publication No. (HRA) 79-14500. Washington, D.C.: U.S. Government Printing Office, 1979.
68. U.S. Department of Health, Education, and Welfare, Health Resources Administration. Minimum Requirements of Construction and Equipment for Hospital and Medical Facilities. DHEW Publication No. (HRA) 79-14500. Washington, D.C.: U.S. Government Printing Office, 1979.
69. U.S. Department of Housing and Urban Development. Minimun Property Standards for One- and Two-Family Dwellings. Vol. 1. Washington, D.C.: U.S. Department of Housing and Urban Development, 1973.
70. U.S. Department of Housing and Urban Development. Minimum Property Standards for Multi-Family Housing. Vol. 2. Washington, D.C.: U.S. Department of Housing and Urban Development, 1973.
71. U.S. Department of Housing and Urban Development. Minimum Property Standards for Care-Type Housing. Vol. 3. Washington, D.C.: U.S. Department of Housing and Urban Development, 1973.
72. U.S. Department of Labor, Occupational Safety and Health Administration. Occupational safety and health standards. Subpart Z--Toxic and Hazardous Substances. Code of Federal Regulations, Title 29, Part 1910.1001, July 1, 1980.
73. U.S. Office of Management and Budget. Federal participation in the development and use of voluntary standards. OMB Circular No. A-119. Fed. Reg. 45:4326-4329, January 21, 1980.
74. U.S. Public Law 94-279. Animal Welfare Act. United States Code, Title 7, Section 2131, et seq., 1976.
75. van Zelm, M., and P. C. Stamperius. Protection against toxic vapours, pp. 509-519. In M. Bonnevie-Svendsen, Ed. Sorption and Filtration Methods for Gas and Water Purification. NATO Advanced

Study Institutes Series E, Vol. 13. Leyden, Netherlands: Noordhoff International Publishing, a division of A. W. Sijthoff International Publishing Company, B. V., 1975.

76. van Zelm, M., R. H. van Dongen, and P. C. Stamperius. Tests for the performance of filters used in protection against toxic gases, pp. 253-269. In M. Bonnevie-Svendsen, Ed. Sorption and Filtration Methods for Gas and Water Purification. NATO Advanced Study Institutes Series E, Vol. 13. Leyden, Netherlands: Noordhoff International Publishing, a division of A. W. Sijthoff International Publishing Company, B. V., 1975.

77. Weber, W. J., Jr. Modeling, pilot tests, and control techniques: Numeric method for prediction and design, pp. 235-245. In M. Bonnevie-Svendsen, Ed. Sorption and Filtration Methods for Gas and Water Purification. NATO Advanced Study Institutes Series E, Vol. 13. Leyden, Netherlands: Noordhoff International Publishing, a division of A. W. Sijthoff International Publishing Company, B. V., 1975.

78. Westinghouse Lamp Division. Two new dimensions in forced air heating and air conditioning. ASC-170, Rev., Bloomfield, N.J.: Westinghouse Electric Corp. (undated)

79. Woods, J. E. Energy Efficient Indoor Air Quality Control in Hospitals. Paper presented at 1980 International Congress of Hospital Engineering, Washington, D.C., July 11, 1980.

80. Woods, J. E. Impact of ASHRAE Ventilation Standard 62-73 on energy use. ASHRAE Trans. 82(Pt. 1):1143-1153, 1976.

81. Woods, J. E. Influence of room air distribution on animal cage environments. ASHRAE Trans. 81(Pt. 2):559-571, 1975.

82. Woods, J. E. Interactions between primary (cage) and secondary (room) enclosures, pp. 65-83. In National Research Council, Institute of Laboratory Animal Resources. Laboratory Animal Housing. Proceedings of a Symposium Held at Hunt Valley, Maryland, September 22-23, 1976. Washington, D.C.: National Academy of Sciences, 1978.

83. Woods, J. E. Objective Criteria for Contamination Control. Paper presented at Engineering Foundation Conference on Ventilation vs. Energy Conservation in Buildings, Henniker, N.H., July, 1977.

84. Woods, J. E. The animal enclosure--A microenvironment. Lab. Animal Sci. 30:407-413, 1980.

85. World Health Organization. Health Aspects Related to Indoor Air Quality. Report on a WHO Working Group. EURO Reports and Studies No. 21. Copenhagen: World Health Organization, 1979. 34 pp.

86. Yaglou, C. P., E. C. Riley, and D. I. Coggins. Ventilation requirements. ASHVE Trans. 42:133-162, 1936.

87. Zegers, C. T. ASHRAE: Providing HVAC&R leadership. Consulting Eng. 54(1):78-81, 1980.

88. Zwemer, R. J., and J. Karibo. Use of laminar control device as adjunct to standard environmental control measures in symptomatic asthmatic children. Ann. Allergy 31:284-290, 1973.

APPENDIX A

AIR-QUALITY STANDARDS

The possibility of establishing standards for indoor air quality is under consideration, because its importance for protecting human health is recognized as a major national environmental issue. The ever-increasing cost of energy has heightened the need for considering such standards, inasmuch as a cost-effective method of reducing energy use in buildings is to reduce ventilation, an action that can increase indoor air pollution.

There is a regulatory indoor air standard for nonoccupational air in the United States only for ozone. There are voluntary standards for indoor air quality that may serve as guidelines to federal, state, or local government agencies on formaldehyde, carbon monoxide, chlorine, radon, carcinogenic aerosols, and other chemical substances. The ozone standard applies only to devices that produce ozone as a waste product. The radon standards and guidelines apply only to buildings that are contaminated as a result of uranium-processing (e.g., by the use of mill tailings as landfill) and buildings that are on phosphate land in Florida.

Tables A-1 through A-8 list a number of U.S. outdoor air-quality and occupational standards and some relevant foreign standards. They are presented not as an exhaustive list of air-quality standards, but rather to impart perspective to the many allusions to standards throughout this report.

TABLE A-1

National Primary Ambient-Air Quality Standards as Set by the U.S. Environmental Protection Agency

Contaminant	Long Term: Concentration, $\mu g/m^3$	Long Term: Averaging Time	Short Term: Concentration, $\mu g/m^3$	Short Term: Averaging Time	Reference
Sulfur oxides, measured as sulfur dioxide	80	1 yr	365[a]	24 h	15
Particulate matter	75[b]	1 yr	260[a]	24 h	14
Carbon monoxide	--	--	10,000[a]	8 h	16
			40,000[a]	1 h	16
Ozone	--	--	235[c]	1 h	20
Hydrocarbons	--	--	160	3 h[d]	17
Nitrogen dioxide	100	1 yr	--	--	19
Lead	1.5	3 mo[e]	--	--	18

[a]May be exceeded only once per year.

[b]Geometric mean.

[c]Standard is attained when expected number of days per calendar year with maximal hourly average concentrations above 0.12 ppm (235 $\mu g/m^3$) is equal to or less than 1, as determined by Appendix H to subchapter C, 40 CFR 50.

[d]3-h period is 6 a.m. to 9 a.m.

[e]3-mo period is a calendar quarter.

Table A-2

Additional Ambient Air Quality Guidelines[a]

Contaminant[b]	Long Term Concentration[c]	Long Term Time	Short Term Concentration[c]	Short Term Time[d]
Acetone - 0	7 mg/m^3	24 h	24 mg/m^3	30 min
Acrolein - 0	--	--	25 $\mu g/m^3$	C
Ammonia - 0	0.5 mg/m^3	Yr	7 mg/m^3	C
Beryllium	0.01 $\mu g/m^3$	30 d	--	--
Cadmium	2.0 $\mu g/m^3$	24 h	--	--
Calcium oxide (lime)	--	--	20-30 $\mu g/m^3$	C
Carbon disulfide - 0	0.15 mg/m^3	24 h	0.45 mg/m^3	30 min
Chlorine - 0	0.1 mg/m^3	24 h	0.3 mg/m^3	30 min
Chromium	1.5 $\mu g/m^3$	24 h	--	--
Cresol - 0	0.1 mg/m^3	24 h	--	--
Dichloroethane - 0	2.0 mg/m^3	24 h	6.0 mg/m^3	30 min
Ethyl acetate - 0	14 mg/m^3	24 h	42 mg/m^3	30 min
Formaldehyde - 0[e]	--	--	120 $\mu g/m^3$	C
Hydrochloric acid - 0	0.4 mg/m^3	24 h	3 mg/m^3	30 min
Hydrogen sulfide - 0	40-50 $\mu g/m^3$	24 h	42 $\mu g/m^3$	1 h
Mercaptans - 0	--	--	20 $\mu g/m^3$	1 h
Mercury	2 $\mu g/m^3$	24 h	--	--
Methyl alcohol - 0	1.5 mg/m^3	24 h	4.5 mg/m^3	30 min
Methylene	20 mg/m^3	Yr	150 mg/m^3	30 min
chloride - 0	50 mg/m^3	24 h		
Nickel	2 $\mu g/m^3$	24 h	--	--
Nitrogen monoxide	0.5 mg/m^3	24 h	1 mg/m^3	30 min
Phenol - 0	0.1 mg/m^3	24 h	--	--
Sulfates	4 $\mu g/m^3$	Yr	--	--
	12 $\mu g/m^3$	24 h		
Sulfuric acid - 0	50 $\mu g/m^3$	Yr	200 $\mu g/m^3$	30 min
	100 $\mu g/m^3$	24 h		
Trichloroethylene - 0	2 mg/m^3	Yr	16 mg/m^3	30 min
	5 mg/m^3	24 h		
Vanadium	2 $\mu g/m^3$	24 h	--	--
Zinc	50 $\mu g/m^3$	Yr	--	--
	100 $\mu g/m^3$	24 h		

[a]Reprinted with permission from ANSI/ASHRAE,[1] which states: "Outdoor air shall be considered unacceptable if it is known to contain any contaminant at a concentration above that listed in Table [A-2]. This table covers other common contaminants for which no EPA ambient air quality standards exist. These [concentrations] were selected from current practices in various states, provinces and other countries."

[b]Contaminants marked "0" have odors at concentrations sometimes found in outdoor air. The tabulated concentrations do not necessarily result in odorless conditions.

[c]Unless otherwise specified, all air quality measurements should be corrected to standard conditions of 25°C (77°F) temperature and 760 mm (29.92 in.) of mercury pressure (101.3 kPa).

[d]C, ceiling, or maximal allowable concentration.

[e]An industry organization has appealed the air quality limits of 120 $\mu g/m^3$ as shown in Tables 2 and 4 of Standard 62-1981. The appeal is under consideration. If any change in Standard 62-1981 results from the appeal, all original recipients will be informed by ASHRAE.

TABLE A-3

Selected Occupational-Safety and -Health Standards as Set by U.S. Occupational Safety and Health Administration[a]

Contaminant	Concentration,[b] ppm	mg/m^3
Carbon dioxide	5,000	9,000
Carbon monoxide	50	55
Formaldehyde	2	3
Nitric oxide	25	30
Nitrogen dioxide	5	9
Ozone	0.1	0.2
Sulfur dioxide	5	13
Inert or nuisance dust, respirable fraction	--	5
Asbestos	c	c

[a]Data from 29 CFR 1910.1000.[11]

[b]8-h time-weighted averages, except values for nitrogen dioxide, which are ceiling values.

[c]Fewer than two fibers longer than 5 μm per cubic centimenter.

Table A-6

Selected Guidelines for Air Contaminants of Indoor Origin[a]

Contaminant[b]	Concentration	Exposure Time	Comments
Acetone - 0	--	--	--
Ammonia - 0	--	--	--
Asbestos	--	--	Known human carcinogen; best available control technology
Benzene - 0	--	--	Known human carcinogen; best available control technology
Carbon dioxide	4.5 g/m^3	Continuous	--
Chlordane - 0	5 $\mu g/m^3$	Continuous	--
Chlorine	--	--	--
Cresol - 0	--	--	--
Dichloromethane - 0	--	--	--
Formaldehyde - 0[c]	120 $\mu g/m^3$	Continuous	W. German and Dutch guidelines
Hydrocarbons, aliphatic - 0	--	--	--
Hydrocarbons, aromatic - 0	--	--	--
Mercury	--	--	--
Ozone - 0	100 $\mu g/m^3$	Continuous	--
Phenol - 0	--	--	--
Radon	0.01 working level (WL)	Annual average	Background 0.002-0.004 WL
Tetrachloroethylene - 0	--	--	--
Trichloroethane - 0	--	--	--
Turpentine - 0	--	--	--
Vinyl chloride - 0	--	--	Known human carcinogen; best available control technology

[a]Reprinted with permission from ANSI/ASHRAE,[1] which states: "If the air is thought to contain any contaminant not listed [in various tables], guidance on acceptable exposure . . . should be obtained by reference to the standards of the Occupational Safety and Health Administration. For application to the general population the concentration of these contaminants should not exceed 1/10 of the limits which are used in industry. . . . In some cases, this procedure may result in unreasonable limits. Expert consultation may then be required." "These substances are ones for which indoor exposure standards are not yet available."

[b]Contaminants marked "0" have odors at concentrations sometimes found in indoor air. The tabulated concentrations do not necessarily result in odorless conditions.

[c]An industry organization has appealed the air quality limits of 120 $\mu g/m^3$ as shown in Tables 2 and 4 of Standard 62-1981. The appeal is under consideration. If any change in Standard 62-1981 results from the appeal, all original recipients will be informed by ASHRAE.

TABLE A-4

Radon Standards

Country	Average Annual Working Level	Action	Status	Reference
Indoor:				
United States:				
Sites contaminated by uranium-processing	0.015	Cost-benefit analysis required when level is only slightly above maximum	Interim and proposed clean-up standard for buildings contaminated by uranium-processing sites	13,21
Phosphate land, Florida:				
Existing housing	<0.02	Reduce to as low as reasonably achievable		
	>0.02	Action indicated	Recommendation to governor of Florida	12
New housing	Normal indoor background	--		
Canada:	>0.01	Investigate		
	>0.02	Primary action criterion	Policy statement by AECB	4
	>0.15	Prompt action		
Sweden:				
Max., existing buildings	200 Bq/m^3 [a]	--	Proposed standard	8
Max., new buildings	70 Bq/m^3 [a]	--		
Occupational:				
U.S. miners:				
Instantaneous maximum	1 WL	--	MSHA standard	22
Maximal cumulative dose	4 WLM/yr [b]			

[a] Assuming an equilibrium factor of 0.5, these values are 0.027 WL and 0.009 WL, respectively.
[b] Period is a calendar year. Dose for any month is defined as cumulative dose in WL-h divided by 173. Assuming 173 h worked per month (i.e., 2,076 h/yr), average annual working level is 1/3 WL.

TABLE A-5

Formaldehyde Standards

Country	Concentration, ppm[a]	Status	Reference
Indoor air:			
United States	b	b	--
Denmark	0.12 ppm maximum	Recommended	3
Netherlands	0.1 ppm maximum	Recommended by ministers of housing and health	5
Sweden	0.1 ppm maximum, new buildings 0.4 ppm minimum, old buildings[c] 0.7 ppm maximum, old buildings[c]	Proposed by National Board of Health and Welfare	23
Federal Republic of Germany	0.1 ppm maximum	Recommended by Ministry of Health	d
Occupational air:			
United States	3 ppm, 8-h time-weighted average	Promulgated by OSHA	11
	5 ppm, ceiling	Promulgated by OSHA	11
	1 ppm, 30-min maximum	Recommended by NIOSH	10

[a]0.1 ppm $\cong$ 120 $\mu g/m^3$.

[b]Several states have proposed indoor standards in the range of 0.2-0.5 ppm.

[c]0.4-0.7 ppm is a border range. Concentrations higher than 0.7 ppm do not meet the standard. Those lower than 0.4 ppm do meet the standard. Those within the range do not meet the standard if dwellers complain. In recently built houses, 0.7 ppm should be acceptable during first 6 mo.

[d]J. E. Woods (personal communication).

TABLE A-7

Other Indoor Air-Quality Standards

Contaminant	Concentration	Reference
United States:		
Ozone	0.05 ppm (100 $\mu g/m^3$)	9
Japan:		
Carbon dioxide	1,000 ppm (1,800 mg/m^3)	
Carbon monoxide	10 ppm (11 mg/m^3)	6
Particles	150 $\mu g/m^3$	

TABLE A-8

Ventilation Standards for Dwellings

U.S. STANDARDS

Area	ASHRAE Standard 62-73: Single-Unit Dwellings,[2] cfm/person		ANSI/ASHRAE Standard 62-1981: Single or Multiple Units,[1] cfm/room
	Minimum	Recommended	Minimum
General living areas	5	7-10	10
Bedrooms	5	7-10	10
Kitchens	20	30-50	100 (intermittent operation)
Toilets, bathrooms	20	30-50	50 (intermittent operation)
All other rooms	NA	NA	10
Basements, utility rooms	5	5	NA

PROPOSED NORTHERN EUROPEAN STANDARDS[7]

Area	Standard
General living areas	0.5 ach measured in spring and autumn, but not less than 4 L/s per bed[a]
Kitchens	10 L/s continuously,[a] plus: For an electric stove with more than two rings, an adjustable fan capable of removing at least 80% of the gaseous cooking products For other electric stoves, an exhaust fan of at least 30-L/s capacity For gas stoves, an exhaust fan of the size necessary to remove the combustion products
Toilets	10 L/s continuously,[a] plus an openable window or vent or an exhaust fan capable of 30 L/s

[a] 1 L/s equals approximately 2 cfm.

REFERENCES

1. American National Standards Institute, and American Society of Heating, Refrigerating and Air-Conditioning Engineers. ANSI/ASHRAE Standard 62-1981. Ventilation for Acceptable Indoor Air Quality. New York: American Society of Heating, Refrigerating and Air-Conditioning Engineers, Inc., 1981. 48 pp.
2. American Society of Heating, Refrigerating and Air-Conditioning Engineers. ASHRAE Standard 62-73. Standards for Natural and Mechanical Ventilation, p. 6. New York: American Society of Heating, Refrigerating and Air-Conditioning Engineers, Inc., 1973.
3. Andersen, I. Formaldehyde in the indoor environment--Health implications and the setting of standards, pp. 65-77. In P.O. Fanger, and O. Valbjørn, Eds. Indoor Climate. Effects on Human Comfort, Performance, and Health in Residential, Commercial, and Light-Industry Buildings. Proceedings of the First International Indoor Climate Symposium, Copenhagen, August 30-September 1, 1978. Copenhagen: Danish Building Research Institute, 1979.
4. Atomic Energy Control Board [Canada] (AECB). Criteria for Radioactive Clean-up in Canada. AECB Information Bulletin 77-2. Ottawa, Ont., Canada: Atomic Energy Control Board, 1977.
5. Baars, R. The formal aspects of the formaldehyde problem in the Netherlands, pp. 77-82. In P.O. Fanger, and O. Valbjørn, Eds. Indoor Climate. Effects on Human Comfort, Performance, and Health in Residential, Commercial, and Light-Industry Buildings. Proceedings of the First International Indoor Climate Symposium, Copenhagen, August 30-September 1, 1978. Copenhagen: Danish Building Research Institute, 1979.
6. National Technical Information Service. Building Control Law and Dust Collectors. (in Japanese; English abstract) 1974. APTIC No. 63252.
7. NKB. Forslag till Nordiska riktlinjer for byggnadsbestammelser rorande: Luftkvalitet. [Proposed Nordic Guidelines for Building Codes: Air Quality] Stockholm, Sweden: NKB, 1979.
8. Swedish Ministry of Agriculture. Preliminary Proposal for Measures to Minimize Radiation Risk in Buildings, Sections 3.2.2 and 3.2.4. Stockholm: Swedish Ministry of Agriculture, 1979.
9. U.S. Department of Health, Education, and Welfare, Food and Drug Administration. Standard for equipment producing ozone as a byproduct. Maximum acceptable level of ozone. Code of Federal Regulations, Title 21, Part 801.415, July 1, 1979.
10. U.S. Department of Health, Education, and Welfare, National Institute for Occupational Safety and Health. Criteria for a Recommended Standard....Occupational Exposure to Formaldehyde. DHEW (NIOSH) Publication No. 77-126. Washington, D.C.: U.S. Government Printing Office, 1977.
11. U.S. Department of Labor, Occupational Safety and Health Administration. Occupational safety and health standards. Air contaminants. Code of Federal Regulations, Title 29, Part 1910:1000, July 1, 1979.

12. U.S. Environmental Protection Agency. Indoor radiation exposure due to radium-226 in Florida phosphate lands: Radiation protection recommendations and request for comment. Fed. Reg. 44:38664-38670, July 2, 1979.
13. U.S. Environmental Protection Agency. Interim cleanup standards for inactive uranium processing sites. Fed. Reg. 45:27366-27368, April 22, 1980.
14. U.S. Environmental Protection Agency. National primary ambient air quality standards for particulate matter. Code of Federal Regulations, Title 40, Part 50.6, July 1, 1980.
15. U.S. Environmental Protection Agency. National primary ambient air quality standards for sulfur oxides (sulfur dioxide). Code of Federal Regulations, Title 40, Part 50.4, July 1, 1980.
16. U.S. Environmental Protection Agency. National primary and secondary ambient air quality standards for carbon monoxide. Code of Federal Regulations, Title 40, Part 50.8, July 1, 1980.
17. U.S. Environmental Protection Agency. National primary and secondary ambient air quality standard for hydrocarbons. Code of Federal Regulations, Title 40, Part 50.10, July 1, 1980.
18. U.S. Environmental Protection Agency. National primary and secondary ambient air quality standards for lead. Code of Federal Regulations, Title 40, Part 50.12, July 1, 1980.
19. U.S. Environmental Protection Agency. National primary and secondary ambient air quality standard for nitrogen dioxide. Code of Federal Regulations, Title 40, Part 50.11, July 1, 1980.
20. U.S. Environmental Protection Agency. National primary and secondary ambient air quality standards for ozone. Code of Federal Regulations, Title 40, Part 50.9, July 1, 1980.
21. U.S. Environmental Protection Agency. Proposed cleanup standards for inactive uranium processing sites. Fed. Reg. 45:27370-27375, April 22, 1980.
22. U.S. Mine Safety and Health Administration. Regulations and standards applicable to metal and nonmetal mining and milling operations. Code of Federal Regulations, Title 30, Part 57.5-38 and 57.5-39, July 1, 1979.
23. Wahren, H. Formaldehyde Indoor Air Standards in Sweden. Paper presented at the Consumer Product Safety Commission Technical Workshop on Formaldehyde, Washington, D.C., April 9-11, 1980.

APPENDIX B

ESTIMATING THE IMPACT OF RESIDENTIAL ENERGY-CONSERVATION MEASURES ON AIR QUALITY: A HYPOTHETICAL CASE

HYPOTHETICAL CASE STUDY

Two of the simplest and most cost-effective methods of reducing the energy consumption of a residence are to increase the insulation and to decrease air infiltration. However, infiltration is the primary source of ventilation for residences, and reducing it may adversely affect air quality. Therefore, although caulking and weatherstripping a home may reduce energy consumption, they may also adversely affect the health and reduce the comfort of the occupants, unless alternative methods of controlling air quality are applied.

Attempts to estimate the impact of residential energy-conservation measures on air quality in the home and, consequently, on the health and comfort of the residents are fraught with difficulty. Most troublesome is the issue of incommensurability: one cannot confidently compare the dollar costs of insulating a house and the associated reductions in fuel bills with the essentially nonquantifiable potential adverse effects on air quality, health, and comfort. Furthermore, numerous assumptions must be made. Some of the assumptions are relatively reliable; for example, demographic studies can provide evidence on average family size, lifestyle characteristics (such as smoking habits), and proportion of homes with a particular appliance (such as a gas oven). Other assumptions may be based on evidence and experience from the building trades--for example, the effectiveness of caulking the windows of a home. (Engineering analyses of related interactions have been performed.) Assumptions concerning the air quality in homes before and after the institution of energy-conservation measures can be based on evidence now being accumulated or on data already in hand.

The following case study is an unvalidated example of the type of analysis that might be considered to assist in making decisions concerning energy conservation versus indoor air quality. It is proposed not as a solution to the analytic problem, but as an approach subject to further study and refinement. As a discussion piece, it may assist in identifying the types of data needed for analysis, the most appropriate mathematical models, and, most important, the assumptions that may be validly applied. The reader must be aware that the models presented here have not been validated or tested in practical cases to determine their effectiveness in predicting results. This presentation is for the purpose of illustration and discussion of a possible approach.

EXISTING CONDITIONS

To evaluate the possible impact of energy-conservation measures on single-family residences, conditions in a hypothetical home in central Iowa are simulated. It is a 15-yr-old, split-level house with a basement and an attached two-car garage. The total heated floor space is 2,100 ft^2, of which 700 ft^2 is below grade. The house is of wood frame construction on a concrete-block foundation. It has insulation values of R7 in the walls and R11 in the ceilings, double-pane windows, and an infiltration rate of 0.8 air change per hour (ach) with windows and doors closed. The house is heated with a natural-gas, forced-air furnace and cooled with an electric central air-conditioning system. The house is occupied by a family of five: a father, who smokes cigarettes; his wife; her mother; and two children, 2 and 10 yr old. Appliances include a natural-gas stove, a gas clothes-dryer, an electric washing machine, an electric dishwasher, and a gas water-heater.

All this is assumed to be fairly typical of a middle-class family in central Iowa. These conditions were used as the basis of an energy and air-quality analysis of the home. The home was then reanalyzed for two mutually exclusive conservation measures, to determine the changes in energy consumption and air quality. The first measure was to reinsulate the walls to a value of R11 (1 additional inch of cellulose insulation) and the ceiling to R19 (2 additional inches of cellulose insulation); this measure was assumed to be accompanied by a reduction in the infiltration rate to 0.5 ach. The second measure was a higher insulation alternative in which the walls were increased to R11 and the ceiling to R30; the infiltration rate was assumed to decrease to 0.3 ach. Two other independent measures were analyzed for air-quality impact: the installation of an electronic air-cleaner and the cessation of cigarette-smoking. A summary of these alternatives is shown in Table B-1. The results of the energy-consumption and air-quality analyses for these alternatives were either directly or indirectly used in an economic model to determine the rate of return available to the homeowners for the various alteratives.

CASE ANALYSIS

Energy Consumption

The annual heat loss and heat gain for the building were calculated from a simple steady-state model, with an overall heat-transfer coefficient and annual degree-days for heating and cooling. The model was exercised for each of the three cases listed in Table B-2. Values

TABLE B-1

Summary of Scenario Analyzed in Hypothetical Example

Condition	Case	Insulation Wall	Insulation Ceiling	Infiltration (ach)	Air Cleaner	Smoker
Existing	P	R-7	R-11	0.8	No	Yes
Alternative	A-1	R-11	R-19	0.5	No	Yes
	A-2	R-11	R-19	0.5	Yes	Yes
	B-1	R-11	R-30	0.3	No	Yes
	B-2	R-11	R-30	0.3	Yes	Yes
	B-3	R-11	R-30	0.3	No	No

TABLE B-2

Insulation Alternatives

Case	Description	R Value ($h \cdot ft^2 \cdot {}^oF/Btu$) Wall	Ceiling	Infiltration
P	Existing condition	7	11	0.8
A	Low insulation	11	19	0.5
B	High insulation	11	30	0.3

for the overall heat-transfer coefficient (UA) were calculated in accordance with the method used by ASHRAE;[2] the results are listed in Table B-3.

The annual degree-days[11] are based on 30-yr averages for Des Moines, Iowa, and are based on 65°F. The values used for heating and cooling were 6,710 and 928 degree-days/yr, respectively. These values for the overall heat-transfer coefficient and degree-days were used in the following equation to calculate the annual heat loss and heat gain: $Q = 24(UA)(DD)$, where Q = annual heat loss or heat gain (Btu), UA = overall heat-transfer coefficient (Btu/h•°F), and DD = annual heating (cooling) degree-days. The results of these calculations are also listed in Table B-3.

To estimate more accurately the energy consumed for heating, a seasonal furnace efficiency had to be determined. This efficiency depends on the steady-state efficiency of the furnace and the amount by which it is oversized. As the heating load is reduced, owing to the conservation measures, the seasonal furnace efficiency is also reduced--by approximately 2% for each 10% oversize increment (John E. Janssen, personal communication). The seasonal furnace efficiencies used for each of the cases are shown in Table B-4. By dividing these efficiencies into the heating loads, the energy input to the house can be calculated; by applying the energy conversion factor for natural gas (100,000 Btu/ccf), the annual fuel consumption can be determined. These results are shown in Table B-4.

The annual electric consumption for cooling is calculated from the following equation: $Q_{elec} = 1.3Q/(COP)(3,412)$, where Q_{elec} = electric-energy consumption (kWh), Q = sensible heat gain (Btu), 1.3 = adjustment for latent load (assumed to be 30% of sensible load), COP = seasonal coefficient of performance (assumed to be 2.5), and 3,412 = conversion factor (Btu/kWh). The results of these calculations are listed in Table B-5.

Air Quality

The air quality in the conditioned space was evaluated for the three cases and for the two independent measures (installation of an air-cleaner and cessation of cigarette-smoking). The contaminants evaluated were carbon monoxide, nitrogen dioxide, formaldehyde, radon, and respirable suspended particles (RSP), which include dust and cigarette smoke. The models used in these evaluations are simple ones that have not been experimentally validated. There is a need to validate these findings not only experimentally, but also in practical test cases. The objective of these analyses was to determine the sensitivity of various parameters to the contaminant concentrations, and absolute values may only be assumed as approximate.

General Models. The general model used to calculate the contaminant concentration profiles (except that for radon) is shown

TABLE B-3

Overall Heat-Transfer Coefficients

Case	Description	UA (Btu/h • °F)	Heat Loss, 10^6 Btu/yr	Heat Gain, 10^6 Btu/yr
P	Existing condition	737	119	16
A	Low insulation	582	94	13
B	High insulation	502	81	11

TABLE B-4

Annual Natural-Gas Consumption for Heating

Case	Heat Loss, 10^6 Btu/yr	Seasonal Furnace Efficiency	Natural-Gas Consumption 10^6 Btu/yr	CCF/yr
P	119	0.60	198	1,980
A	94	0.56	168	1,680
B	81	0.53	153	1,530

TABLE B-5

Annual Electricity Consumption for Cooling

Case	Heat Gain, 10^6 Btu/yr	Electric Consumption, kWh
P	16	2,400
A	13	2,000
B	11	1,700

schematically in Figure B-1. The assumptions and nomenclature used are as follows:

- Equal infiltration and exfiltration rates ($\dot{V}_i$).
- Uniform contaminant concentration (C) in the occupied volume (V).
- Constant outdoor contaminant concentration (C_o).
- An electronic air-cleaner with an RSP-removal efficiency of ε, operating continuously with a constant-supply airflow rate ($\dot{V}_s$).
- A net contaminant generation rate (N)--decay rates are neglected. A mass balance equation that describes the air quality of the house is given as follows:

$$\dot{V}_i(C_o - C) + \dot{V}_s(C_s - C) + \dot{N} = V\frac{dC}{dt} \qquad (1)$$

and

$$C_s = (1 - \varepsilon)C.$$

This set of equations can be combined and rearranged to give the following differential equation for the indoor concentration (C):

$$V\frac{dC}{dt} + (\dot{V}_i + \varepsilon\dot{V}_s)C = \dot{N} + \dot{V}_i C_o \qquad (2)$$

rate of change in air quality	dilution and removal effects	generation and infiltration effects

The solution of this equation is:

$$C = \left[C_i - \left(\frac{\dot{N} + \dot{V}_i C_o}{\dot{V}_i + \varepsilon\dot{V}_s}\right)\right] \exp\left[-\left(\frac{\dot{V}_i + \varepsilon\dot{V}_s}{V}\right)t\right] + \left(\frac{\dot{N} + \dot{V}_i C_o}{\dot{V}_i + \varepsilon\dot{V}_s}\right), \qquad (3)$$

where C_i is the initial condition for the concentration. This equation is valid only for constant values of $\dot{N}$, $\dot{V}_i$, $\dot{V}_s$, ε, and C_o. For this analysis, the generation rate is assumed to vary by steps. Therefore, Equation 3 can be applied to each step separately, with the initial condition for a given step being the final concentration of the previous step.

A slightly different model is used for radon, because of the assumption of different concentrations above and below grade. The model is shown schematically in Figure B-2 and includes an air exchange ($\dot{V}_{ab}$) between the above- and below-grade spaces and no generation in

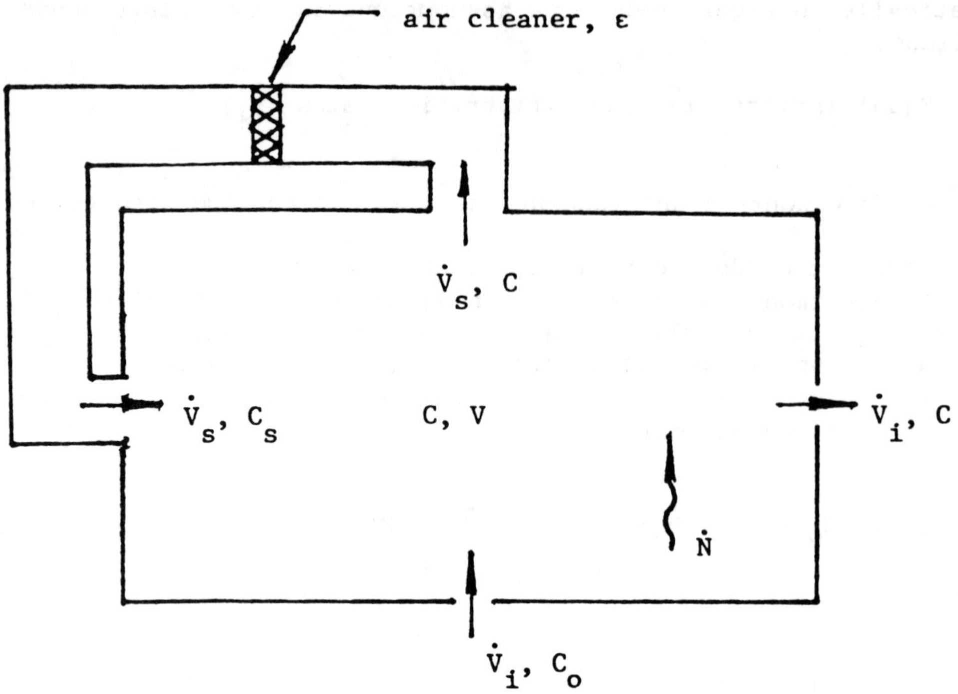

FIGURE B-1 General air-quality model for hypothetical single-family residence.

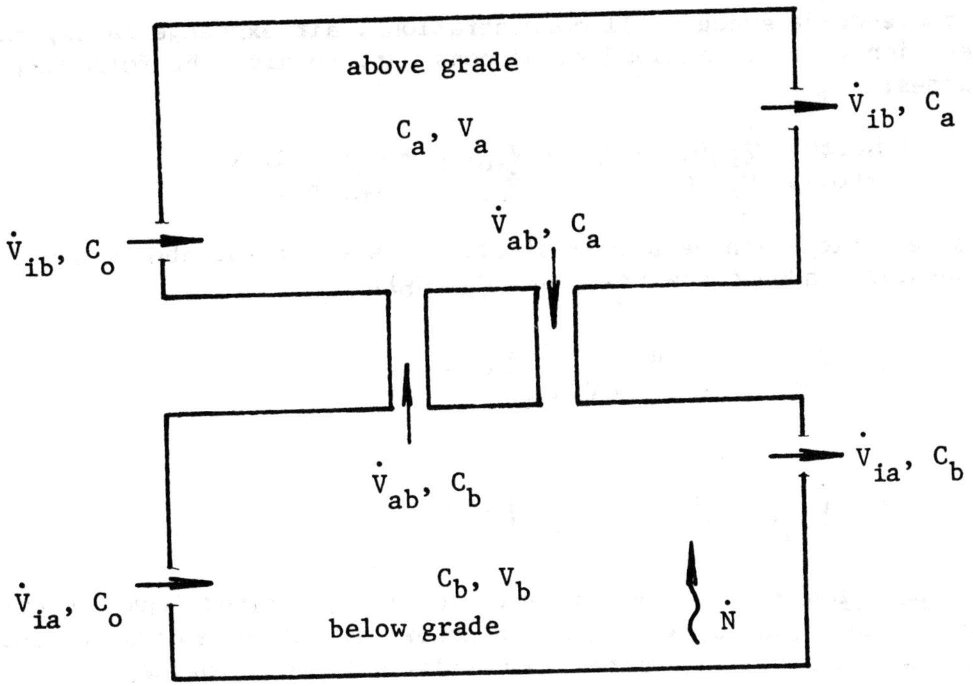

FIGURE B-2 Radon model.

the above-grade space. All concentrations, air-exchange rates, and the generation rate are assumed to be constant, to give the following mass balances:

$$\text{below: } \dot{V}_{ib}(C_b - C_o) + \dot{V}_{ab}(C_b - C_a) = \dot{N}, \text{ and} \qquad (4)$$
$$\text{above: } \dot{V}_{ia}(C_a - C_o) + \dot{V}_{ab}(C_a - C_b) = 0.$$

These equations can be solved simultaneously for the above- and below-grade radon concentrations (C_a, C_b):

$$C_a = \left(\frac{\dot{V}_{ab}}{\dot{V}_{ia}\dot{V}_{ib} + \dot{V}_{ia}\dot{V}_{ab} + \dot{V}_{ib}\dot{V}_{ab}}\right)\dot{N} + C_o \ , \qquad (5a)$$

$$C_b = \left(\frac{\dot{V}_{ia} + \dot{V}_{ab}}{\dot{V}_{ia}\dot{V}_{ib} + \dot{V}_{ia}\dot{V}_{ab} + \dot{V}_{ib}\dot{V}_{ab}}\right)\dot{N} + C_o \ . \qquad (5b)$$

Generation Rates. The contaminants are generated from several sources, including cooking (carbon monoxide, nitrogen dioxide, and formaldehyde), smoking (carbon monoxide, nitrogen dioxide, formaldehyde, and RSP), material outgassing (formaldehyde and radon), and indoor dust generation (RSP). The assumed daily generation profiles of these sources are shown in Figure B-3. The generation rate for cooking is assumed to be constant and occurs at 7 a.m., 12 noon, and 5 p.m. for 15, 30, and 60 min, respectively. Smoking occurs at 7 a.m. and 7:30 a.m. and every half-hour from 5:30 p.m. to 11:30 p.m., inclusive. The duration of each occurrence of smoking is 10 min.[9] Material outgassing is assumed to be constant throughout the day. Indoor dust is generated from 7 a.m. to 11 p.m., primarily owing to resuspension of particles from carpeting. The generation rates for all these sources are listed in Table B-6, with the outdoor concentrations.

Concentration Profiles. To determine the daily indoor-contaminant concentration profiles, a daily generation profile for each contaminant (carbon monoxide, nitrogen dioxide, formaldehyde, and RSP) was determined by summing the generation rates of the appropriate sources. This provides an overall generation profile consisting of step changes to which Equation 3 can be applied, as discussed previously. The solution is started by choosing an initial condition (usually $C_i = C_o$) at the beginning of a period (usually 7 a.m.) and applying Equation 3 to each interval of constant generation rate. The solution proceeds throughout the day and is repeated until no changes occur in the initial conditions from one day to the next.

The concentration profiles for carbon monoxide, nitrogen dioxide, and formaldehyde are shown in Figures B-4 through B-6 for cases P, A, and B (0.8, 0.5, and 0.3 ach). Figure B-7 shows the estimated concentration profiles for RSP for each of the infiltration rates and

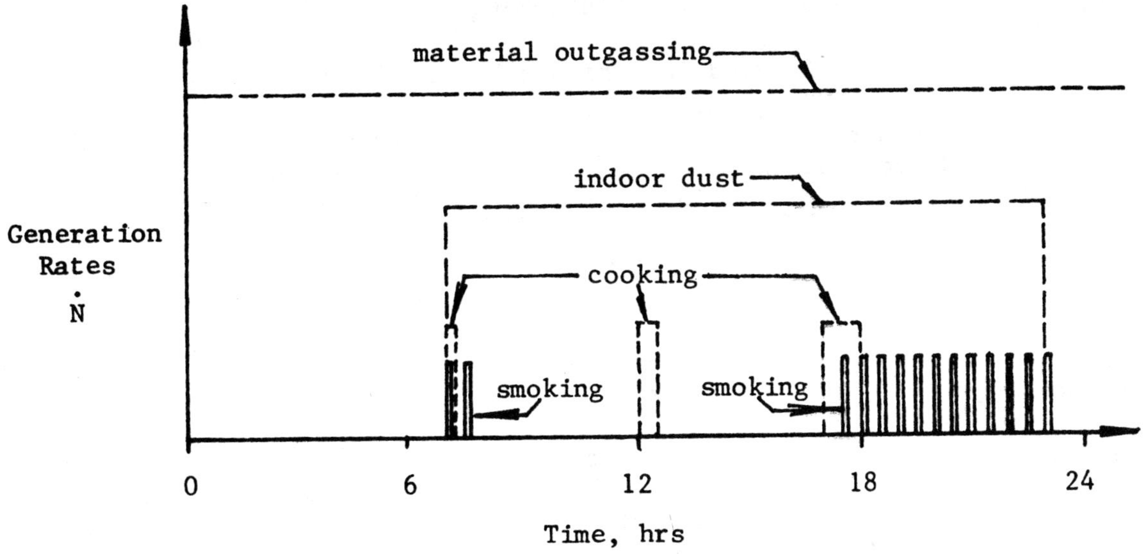

FIGURE B-3 Generation profiles of indoor pollutants for hypothetical single-family residence.

TABLE B-6

Assumed Contaminant Source Summary

Contaminant	Outdoor Concentration (C_o),[a] μg/m^3	Indoor Source	Indoor Generation Rate ($\dot{N}$), μg/h	Total Production, mg/d
Carbon monoxide	1,500	Cigarettes[b]	31,000	70
		Cooking[c]	1,000,000	1,750
Nitrogen dioxide	50	Cigarettes[b]	474	1
		Cooking[c]	57,000	100
Formaldehyde	5	Cigarettes[b]	684	2
		Cooking[c]	10,000	18
		Materials[d]	11,000	264
Respirable particles	20[e]	Cigarettes[b]	192,000	450
		Indoor activity[e]	8,600	140
Radon[e]	0	Above grade	0	0
		Below grade	256,000[f]	6,140[g]

[a]See Hollowell *et al.*[4]

[b]See Woods.[12]

[c]Calculated from data in Hollowell *et al.*;[6] oven at 350°F (177°C).

[d]Calculated from data in Hollowell *et al.*[5]

[e]Assumed values.

[f]pCi/h.

[g]nCi/d.

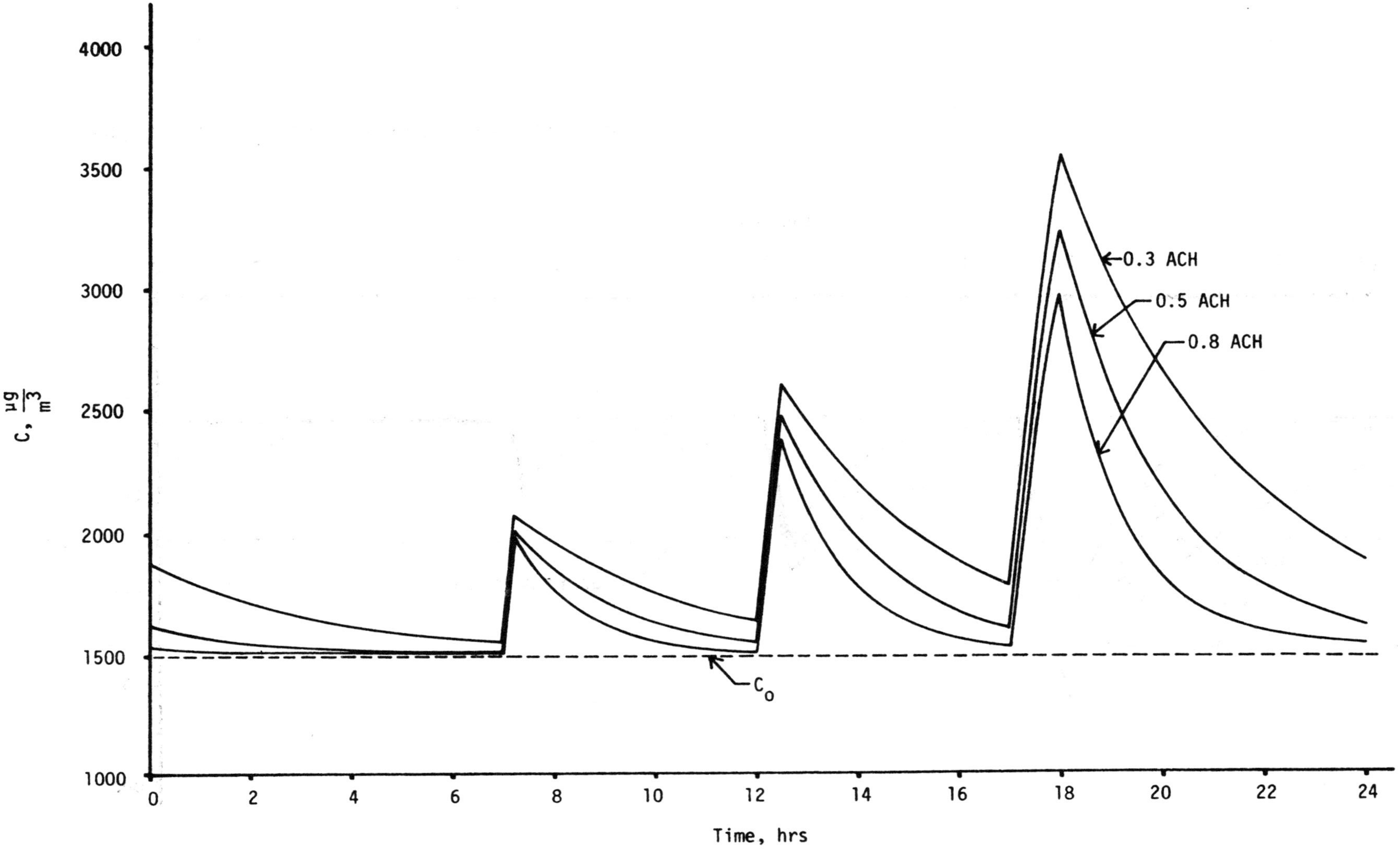

FIGURE B-4 Carbon monoxide concentrations.

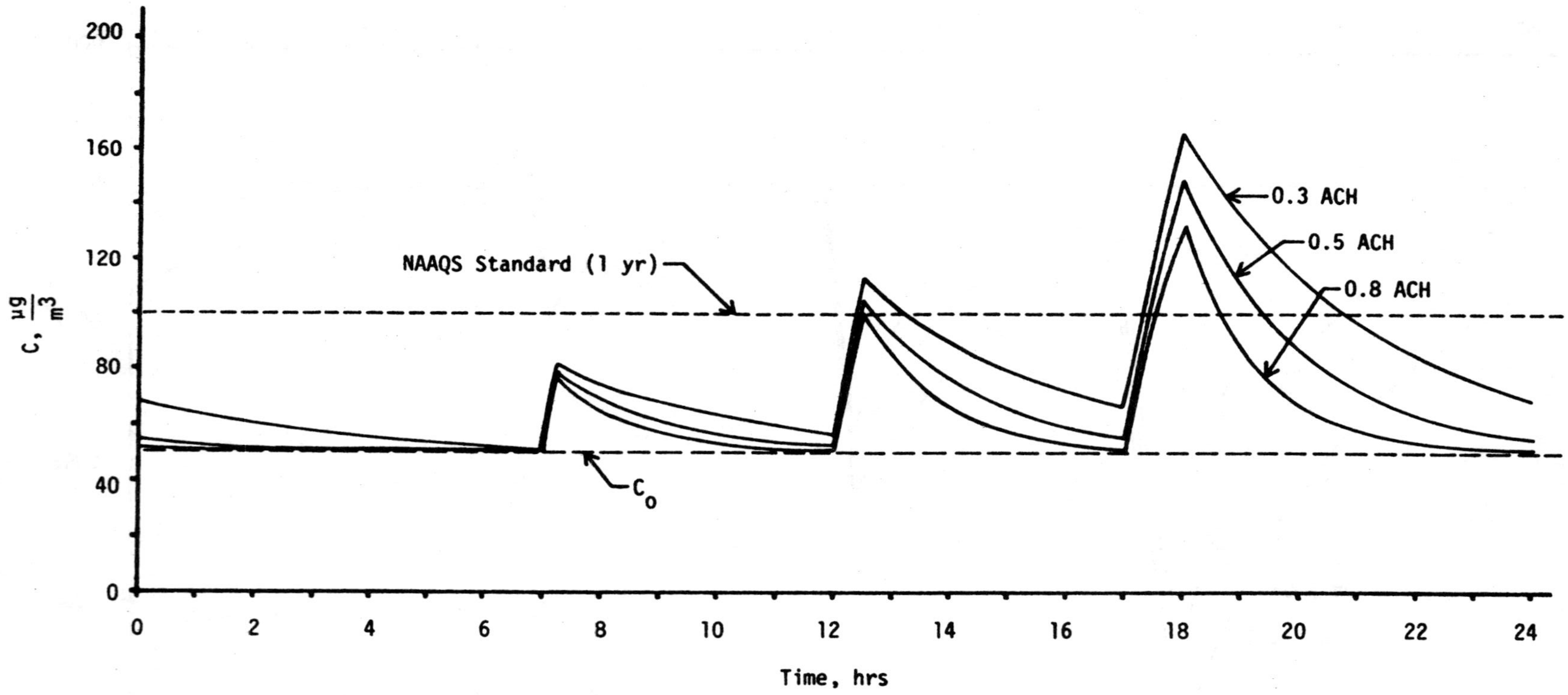

FIGURE B-5 Nitrogen dioxide concentrations.

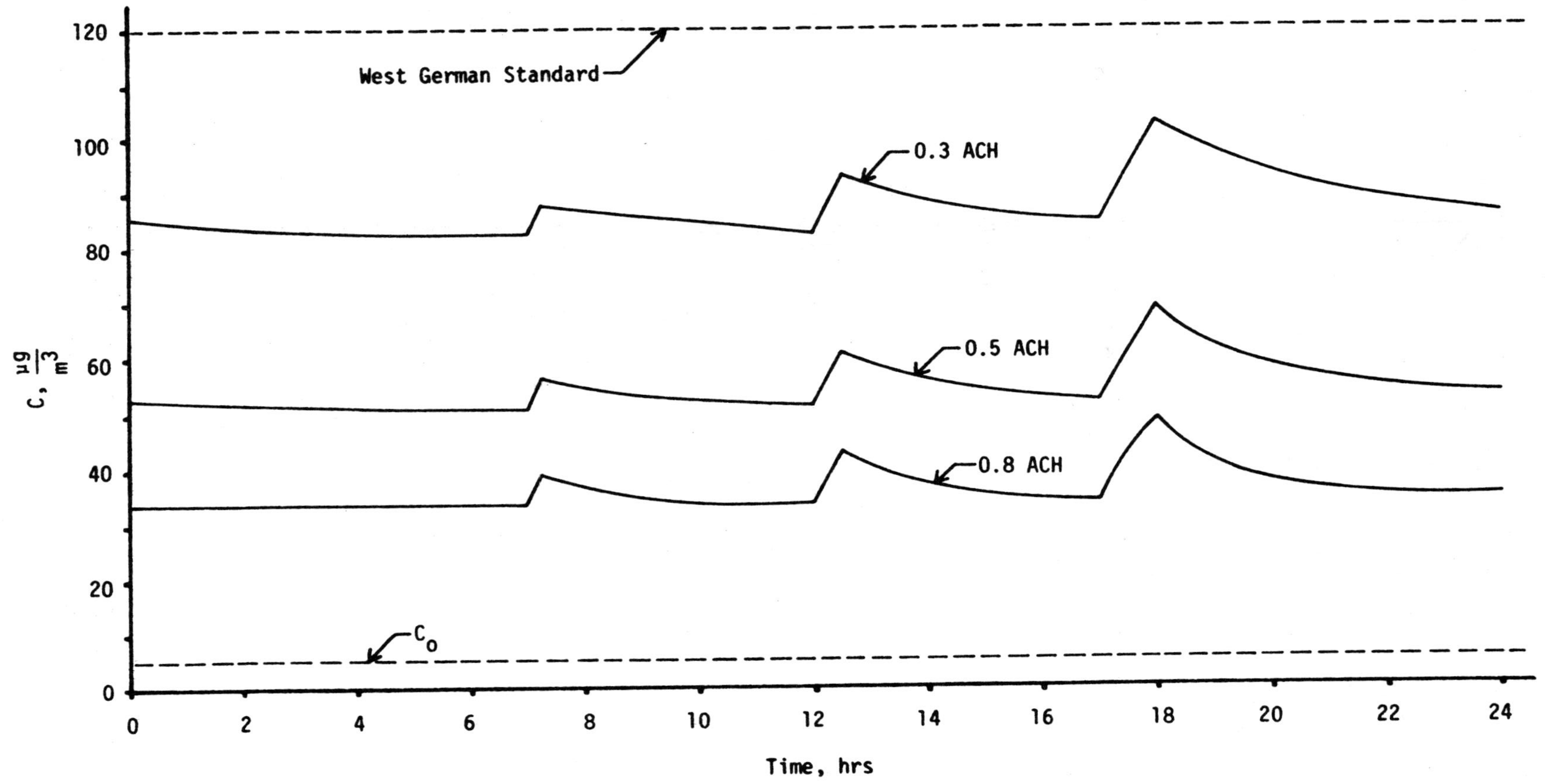

FIGURE B-6 Formaldehyde concentrations.

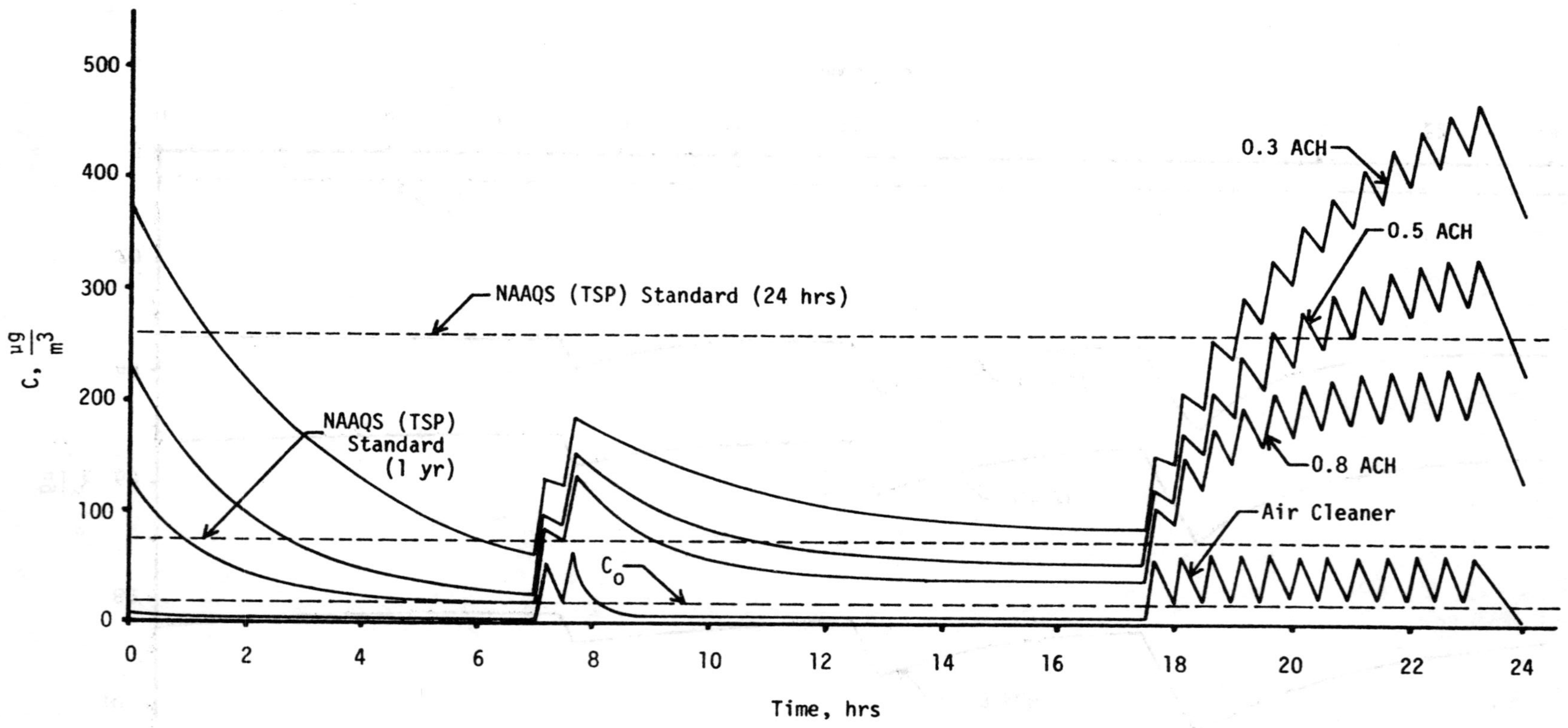

FIGURE B-7 Concentrations of respirable suspended particles.

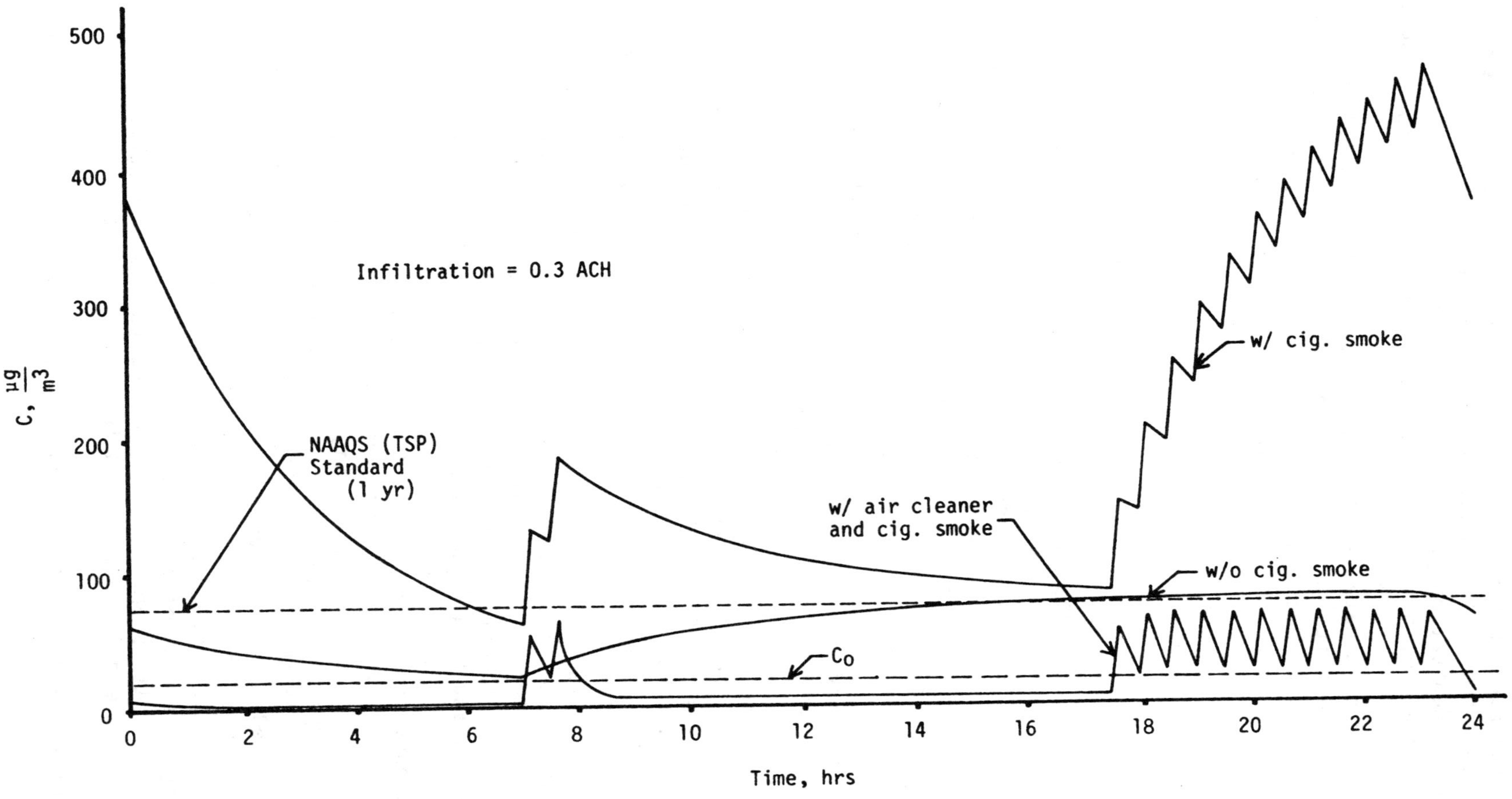

FIGURE B-8 Effects of cigarette smoke and air cleaner on concentrations of respirable suspended particles.

a fourth profile for the effect of the electronic air cleaner (ε = 0.93, $\dot{V}_s$ = 1,000 cfm).* The infiltration rate has a negligible effect on the concentration profile after inclusion of the air-cleaner; therefore, only one profile for all infiltration rates is shown.

The cessation of cigarette-smoking has no significant effect on the carbon monoxide, nitrogen dioxide, or formaldehyde concentration profiles. However, it does have a significant effect on RSP concentrations, as shown in Figure B-8 for an infiltration rate of 0.3 ach.

Radon, an inert gas, is generated from the decay of radium in the below-grade building materials (i.e., concrete).[7] The hazardous radiation effects of radon are due primarily to its progeny (RaA, RaB, and RaC). The combined radiation effect of these progeny is taken into account with the working level (WL) defined as:[7] WL = 0.00103 RaA + 0.00507 RaB + 0.00373 RaC, where RaA, RaB, and RaC are concentration in picocuries per liter. The decay rate of radon is 0.0075/h, which is negligible, compared with the assumed infiltration rates of 0.8, 0.5, and 0.3 ach. Therefore, infiltration was assumed to be the only method of radon removal in the model.

To calculate the radon concentration profiles, radiation of 0.5 and 1.0 pCi/L was assumed for the above- and below-grade spaces at the present condition (P). When these values are substituted for C_a and C_b in Equations 5 and it is assumed that the outdoor radon concentration is negligible, compared with the indoor concentration (C_o = 0), values of $\dot{V}_{ab}$ = 8,960 ft^3/h and $\dot{N}$ = 256,000 pCi/h are obtained. The values for the air-exchange rate between the above- and below-grade spaces and the radon generation rate are assumed to remain unchanged for cases A and B. If these values are substituted in Equations 5 with the appropriate infiltration rates, the above-grade radon concentrations can be calculated for cases A and B; they are listed in Table B-7.

Then, from the radon concentration, the corresponding working levels were calculated from WL = FRn/100, where Rn is the radon concentration in picocuries per liter and F is the "equilibrium" factor, which is a function of the progeny concentrations.[7] For test cases P, A, and B, a value of F = 0.84 was used to take into account plateout of the progeny to the walls, and F = 0.32 was used to take into account progeny removal with the electronic air-cleaner. These values for F are approximate comparisons with experimental data from Jonassen.[7] The resulting working levels are shown in Table B-7.

A major assumption is made in the remainder of this section: the predicted values are treated as real pollutant concentrations. The reader is reminded that the models used to estimate pollutant concentrations have not been validated against measured concentrations. All subsequent comparisons, therefore, are constrained by this lack of model validation.

Air-quality health standards for each of the contaminants considered in this section are listed in Table B-8. These minimal

---

*From manufacturer's data for a Lennox EAC 7-20 electronic air-cleaner.

TABLE B-7

Radon Concentrations and Working Levels

Case	Equilibrium Factor (F)	Above Grade		Below Grade	
		Radon Concentration, pCi/L	WL	Radon Concentration, pCi/L	WL
P	0.84	0.50	0.004	1.00	0.008
A	0.84	0.88	0.007	1.43	0.012
B	0.84	1.58	0.013	2.17	0.018
A filtered	0.32	0.88	0.003	1.43	0.005
B filtered	0.32	1.58	0.005	2.17	0.007

TABLE B-8

Air-Quality Standards[a]

Contaminant	Concentration	Time	Standard
Carbon monoxide	40 $\mu g/m^3$ 10 $\mu g/m^3$	1 h 8 h	NAAQS NAAQS
Nitrogen dioxide	100 $\mu g/m^3$	1 yr	NAAQS
Total suspended particles	75 $\mu g/m^3$ 260 $\mu g/m^3$	1 yr 24 h	NAAQS NAAQS
Formaldehyde	120 $\mu g/m^3$	Continuous	West German
Radon	0.01 WL	Continuous	37 FR 25918

[a]Derived from ANSI and ASHRAE.[1]

acceptable values are also plotted in Figures B-5, B-6, and B-7 for nitrogen dioxide, formaldehyde, and RSP, respectively, for comparison with the predicted indoor contaminant concentrations. The carbon monoxide health standard greatly exceeds the predicted values in Figure B-4 and thus are not shown there. Figure B-5 shows that the long-term standard (1 yr) for nitrogen dioxide concentration would be exceeded for approximately 1 h/d for the present condition (P), whereas cases A and B would exceed the standard for 2 and 4 h/d, respectively. The formaldehyde concentration in Figure B-6 would reach 41% of the short-term standard (continuous) for the present condition (P), 58% for case A, and 86% for case B. Figure B-7 shows that the RSP concentrations would exceed the short-term standard (24 h) for case A during 3.5 h/d and case B during 6 h/d, whereas the present condition never exceeds this standard. The long-term standard (1 yr) would be exceeded by cases P and A for 11 and 17 h/d, respectively, and case B would constantly exceed the short-term standard. The inclusion of the electronic air-cleaner would reduce the RSP concentration to a point below the long-term standard for all cases. The cessation of cigarette-smoking without the air-cleaner would also reduce RSP concentrations below this standard for cases P and A. However, case B would exceed the standard slightly for 13 h, owing to indoor dust generation, as shown in Figure B-8. Table B-7 shows that the short-term standard (continuous) of 0.01 WL for radon would be exceeded below grade for case A and above and below grade for case B, unless the electronic air-cleaner were used.

Economics

To perform the economic analysis, estimates for the installation costs and energy-cost savings were needed for each of the conservation measures considered.[3 8 10] Present annual energy costs for natural gas and electricity were calculated by multiplying the annual energy requirements (Tables B-3 and B-4) by the present fuel costs in Ames, Iowa (0.28 $/ccf for natural gas and 0.057 $/kWh for electricity) for each of the three cases (P, A, and B). Energy-cost savings for each of the insulation alternatives (A and B) over the present condition (P) were then calculated. The results are listed in Table B-9.

The assumed method of insulating for each of the alternatives was to add sufficient cellulose insulation to the walls and ceiling to obtain the desired R value. Installation and material cost estimates from a local insulation contractor in Ames, Iowa, were 0.20 $/ft^2 of ceiling area to upgrade from R11 to R19 (case A), 0.31 $/ft^2 of ceiling area to upgrade from R11 to R30 (case B), and 0.50 $/ft^2 of gross exterior wall area to upgrade from R7 to R11 (cases A and B). With an insulated ceiling area of 1,400 ft^2 and a gross exterior wall area of 1,576 ft^2, total installed insulation costs of $1,218 and $1,378 were obtained for cases A and B, respectively. These costs include $150 for caulking and weatherstripping. The only other first cost needed in the economic analysis was $726 for the installed cost of the electronic air-cleaner.

Salvage values at the end of the economic life for the various alternatives were also needed. The economic life used was the length of time that the present owner would continue to own the house. At the end of this life (assumed to be 7 yr), it was assumed that the salvage value of the insulation in terms of today's dollars would be the same as its first cost, owing to the increase in resale value of the house. The salvage value of the electronic air-cleaner was assumed to be $250.

Rates for electricity and natural gas were assumed to increase by 18 and 22%/yr, respectively, and the rate of general inflation was assumed to be 10%/yr.

The economic analysis was performed for three distinct situations, each containing two mutually exclusive alternatives, as shown in Table B-10.

The inflation-adjusted rates of return, shown in Table B-10, were calculated for each alternative over present condition, as well as the inflation-adjusted rate of return on the incremental costs for each pair of alternatives.

If the homeowner's marginally acceptable rate of return (MARR) were 10%, he should choose alternative B for situation I, alternative A for situation II, and alternative B for situation III.

SUMMARY

Care must be exercised when considering estimates based on models that have not been validated against measurements. In such cases, the magnitude of the estimated values may not be equivalent to that of the observed values. Model estimates can be used, however, for comparative studies to illustrate cause-effect relationships among various parameters. From this perspective, the scenarios described in this appendix show that energy-conservation measures may adversely affect the indoor air quality of single-family residences. The inclusion of the cost of air-quality control may reduce the economic attractiveness of some energy-conservation measures. Although these simulations have been based on several assumptions, they demonstrate the inter- relationships between energy conservation and indoor air quality. In addition, the simulations of this hypothetical residence focus attention on the factors that must be considered in the regulation of indoor environments.

RECOMMENDATIONS

Some parts of the models presented here have not been validated in practical cases that show their utility. Further research is needed to develop models and to test and validate their usefulness in assessing the relationships between air quality and energy conservation in residential and commercial buildings. A large program should be established to develop this research tool further and to demonstrate the usefulness of models in evaluating indoor

TABLE B-9

Present Annual Energy Costs

Case	Cost, $ Natural Gas	Electricity	Total
P	550	140	690
A	470	110	580
B	430	100	530
P - A	80	30	110
P - B	120	40	160

TABLE B-10

Inflation-Adjusted Rates of Return for Hypothetical Examples

Situation	Alternative	ROR over P cond, %	ROR on Incremental Investment over A, %
I	A-1	18.3	--
	B-1	23.0	55.1
II	A-1	18.3	--
	B-2	13.8	6.2
III	A-2	9.5	--
	B-2	13.8	55.1

environmental conditions. Models may be used in the design of future structures to ensure the health and comfort of the public and conservation of natural resources.

REFERENCES

1. American National Standards Institute, and American Society of Heating, Refrigerating and Air-Conditioning Engineers. ANSI/ASHRAE Standard 62-1981. Ventilation for Acceptable Indoor Air Quality. New York: American Society of Heating, Refrigerating and Air Conditioning Engineers, Inc., 1980. 48 pp.
2 American Society of Heating, Refrigerating, and Air Conditioning Engineers. ASHRAE Handbook and Product Directory. 1977 Fundamentals, Chapter 22. New York: American Society of Heating, Refrigerating, and Air Conditioning Engineers, Inc., 1977.
3. Engineering Research Institute. Manual of Procedures for Authorized Class A Energy Auditors in Iowa. Ames, Iowa: Iowa State University Press, 1979.
4. Hollowell, C. D., J. V. Berk, M. L. Boegel, R. R. Miksch, W. W. Nazaroff, and G. W. Traynor. Indoor air quality in residential buildings. In F. E. de Oliveira, J. E. Woods, and A. Faist, Eds. Building Energy Management--Conventional and Solar Approaches. Proceedings of the International Congress, May 12-16, 1980, Povoa de Varzim, Portugal. New York: Pergamon Press, 1980.
5. Hollowell, C. D., J. V. Berk, C. Lin, and I. Turiel. Indoor Air Quality in Energy Efficient Buildings. Lawrence Berkeley Laboratory Report LBL-8892. Berkeley, Cal.: Lawrence Berkeley Laboratory, 1979.
6. Hollowell, C. D., J. V. Berk, and G. W. Traynor. Impact of reduced infiltration and ventilation on indoor air quality. ASHRAE J. 21(7):49-53, 1979.
7. Jonassen, N. Indoor radon concentrations and building materials control of airborne radioactivity. In F. E. de Oliveira, J. E. Woods, and A. Faist, Eds. Building Energy Management--Conventional and Solar Approaches. Proceedings of the International Congress, May 12-16, 1980, Povoa de Varzim, Portugal. New York: Pergamon Press, 1980.
8. Montag, G. M. A commercial building ownership energy cost anaysis model. In F. E. de Oliveira, J. E. Woods, and A. Faist, Eds. Building Energy Management--Conventional and Solar Approaches. Proceedings of the International Congress, May 12-16, 1980, Povoa de Varzim, Portugal. New York: Pergamon Press, 1980.
9. Repace, J. L., and A. H. Lowrey. Indoor air pollution, tobacco smoke, and public health. Science 208:464-471, 1980
10. Smith. G. W. Engineering Economy. 3rd ed. Ames, Iowa: The Iowa State University Press, 1979.
11. U.S. Department of Commerce, National Climatic Center. Local Climatological Data. Asheville, North Carolina.
12. Woods, J. E., Ventilation, health and energy consumption: A status report. ASHRAE J. 21(7):23-27, 1979.